3075

22

PHILIP G. HILL
Department of Mechanical Engineering
Massachusetts Institute of Technology

CARL R. PETERSON
Research and Development, Ingersoll-Rand Company
Bedminster, New Jersey

MECHANICS AND THERMODYNAMICS OF PROPULSION

ADDISON-WESLEY PUBLISHING COMPANY
Reading, Massachusetts
Menlo Park, California · London · Amsterdam · Don Mills, Ontario · Sydney

This book is in the

ADDISON-WESLEY SERIES IN AEROSPACE SCIENCE

HOWARD W. EMMONS, S. S. PENNER, *Consulting Editors*

Third printing, November 1970

Library of Congress Catalog Card No. 65-10408

ISBN 0-201-02838-7
  MNOPQRST-MA-89876543210

Preface

This book is based on an introductory course in propulsion which has been given for several years to students in aeronautical and mechanical engineering at Massachusetts Institute of Technology. It deals mainly with the propulsion of aircraft and space vehicles.

The objective of the book is to show how basic principles may be used to predict the behavior of propulsion devices. There are, of course, many methods of propulsion, but the underlying physical principles are relatively few. Often there is a large gap between theoretical studies and their useful application to actual engineering problems. This book does not pretend to close such a gap, but it does stress the application of the student's theoretical knowledge to several real devices, pointing toward the engineering objectives of propulsion. Experience has shown that the application of fundamentals to these problems can reward the student not only with a valuable appreciation of their practical significance but also with a deeper understanding of the principles themselves. To this end, the book attempts to show how basic physical laws both describe and limit the performance of particular devices.

With the present trend toward increasingly basic engineering curricula there remains a definite need for at least one course in which a variety of fundamental ideas are brought together and made to interact so that their engineering implications can be clearly seen. Propulsion is an admirable subject for this purpose, since it significantly involves so many disciplines. This text may therefore serve a more general purpose than simply conveying knowledge of propulsion devices to the student.

In general, the most satisfactory performance of a given propulsion engine is determined not only by the laws of mechanics and thermodynamics, but also, to a large extent, by the behavior of materials. The main emphasis in this book is on the former, though material limitations are identified and discussed.

The mathematical portions of the text assume a knowledge of calculus. In addition, an introductory course in vector calculus is desirable, but not essential. An attempt has been made to minimize mathematical complication so that physical principles can be more readily discerned, and the problem sections at the ends of the chapters are designed to give the reader an idea of the practical applications of these principles.

Throughout the text much attention has been given to the "working fluid" of the propulsion engine. Generally it is considered to be a continuum, since the principal interest lies in its macroscopic behavior. Two exceptions to this rule occur. First, in the study of plasma, it is necessary to examine microscopic processes in order to understand and predict the macroscopic behavior. Second, in the study of ion or electron engines, it is often necessary to discuss the behavior of discrete particles.

The control volume method of analysis, whose usefulness for fluid flow has been demonstrated so thoroughly by Shapiro, has been used extensively. One-dimensional approximations are frequently employed, but certain two-dimensional phenomena like heat transfer and other boundary layer behavior are also discussed.

Basically, two classes of propulsion devices are considered: (1) air-breathing engines, and (2) rockets.

Air-breathing engines have been under very intensive development over the past 25 years. They are particularly instructive examples of engineering achievement, because of both the magnitude and the success of the development effort. It is most illuminating to see the power and the limitations of theoretical methods and to observe the mixture of art and science in their evolution. Although it has now approached some degree of completion, development continues toward better performance and new applications, so that air-breathing engines will remain a source of challenging engineering problems for some time to come.

Rocket engines, by comparison, are currently under much more active development. Much is now known about chemical rockets, but nuclear and electrical rockets are only entering the experimental stage. The latter are nonetheless discussed in this text, since current developments indicate that nuclear and electrical phenomena will be important in future propulsion engines. There is need for an introductory discussion of the application of these phenomena to propulsion.

Part 1 is concerned mainly with a review of those topics in thermodynamics, combustion, and fluid mechanics which are relevant to propulsion engines. One of the purposes of this review is to introduce necessary concepts and laws in the notation used in subsequent chapters. In Chapter 2 the thermodynamics of equilibrium combustion reactions are discussed in some detail because of their relevance to chemical rockets especially. Since propulsion systems generally deal with accelerating fluids, the gas dynamics of isentropic expansion are reviewed in Chapter 3, along with the modifying effects of friction at solid boundaries, and shocks. In many engines, e.g., the turbojet, the viscous boundary layer exercises a controlling effect on the performance of the engine; hence the question of boundary layer behavior is discussed in some detail in Chapter 4. Convective heat transfer is also briefly reviewed, chiefly because of its relevance to the cooling of rocket motors. Electrostatic and electromagnetic acceleration are reviewed in Chapter 5 because they are centrally important in the rising generation of propulsion engines for space flight. We also give an elementary description of the behavior of charged particle beams and conducting fluids, in order to aid understanding of relevant phenomena, though it is not entirely clear at present which phenomena will be of greatest significance in working electrical-thrust engines.

Part 2 is confined to air-breathing engines. Chapter 6 demonstrates that the performance of any air-breathing engine can be predicted from the laws of thermodynamics, given the maximum permissible temperature and the appropriate efficiencies of engine components, e.g., the compressor. Chapter 7 deals with the components of the ramjet (diffuser, combustor, and nozzle) and shows how their efficiencies depend principally on aerodynamic phenomena. The essential difference between the ramjet and the turbojet is simply the turbomachinery, which is the subject of Chapters 8 and 9. Fortunately the turbojet is so highly developed that several interesting examples of actual engine configurations and performance can be cited by way of illustration.

Part 3 is confined to rockets. Chapter 10 presents an elementary treatment of rocket vehicle performance in order to show the significance of specific impulse and other variables, and also to indicate the relative advantages of low- and high-thrust engines for various missions. The trajectories of space vehicles are briefly discussed in order to show what they require of the propulsion system. Chemical rockets are the subject of Chapters 11, 12, 13 and 14, which take up a series of pertinent topics. The combustion process, the gas dynamics of the nozzle, and the cooling of the wall are all examined in some detail, as well as auxiliary equipment such as the turbopump. Chapter 15 is concerned with the nuclear rocket, in which a nuclear reactor is used to heat a gas such as hydrogen to high temperature before it expands in the nozzle. The limitations of this engine and the phenomena that govern its performance are briefly discussed. Chapter 16 focuses on electrostatic and electromagnetic rockets, although these are in early stages of development. Considerable experimental work has been done on the components of the electrostatic rocket, and it appears relatively promising for future application, provided suitable electrical energy conversion machinery can be developed. The electrothermal and three kinds of possible electromagnetic rockets are briefly described.

This text may be used for either a one- or two-term course in propulsion. For a two-term sequence, the first term could be used for either air-breathing or rocket engines. A one-term course could be based on Chapters 6, 10, and 11, with selected portions of Parts 2 and 3 and appropriate review material.

The authors would like to record their indebtedness to Professor Edward S. Taylor, who has been our teacher, critic, and friend during our years of association in the M.I.T. Gas Turbine Laboratory. The text has been written largely as a result of his encouragement, and it may be that his influence will be recognizable in some of the more worthwhile portions. We are certainly grateful to him for writing the introductory chapter.

Throughout the text, there are a number of references to the work of Professors J. H. Keenan and Ascher H. Shapiro. Their expositions of thermodynamics and fluid mechanics were so enlightening to us as students that we could not help drawing heavily from them in presenting the subject of propulsion.

The preparation of earlier editions of classroom notes upon which the book is based was supported in part by a grant made to the Massachusetts Institute of Technology by the Ford Foundation for the purpose of aiding in the improvement of engineering education. This support is gratefully acknowledged.

The authors would also like to acknowledge the diligent and efficient work of Mrs. Madelyn Euvrard and Mrs. Rose Tedmon in piecing together many fragments and revisions of the text.

Cambridge, Massachusetts   P.G.H.
October 1964   C.R.P.

Contents

PART 1

Introduction and Review of Fundamental Sciences

Kitty Hawk, North Carolina, 1903: man's first plane flight. The Wright brothers' famous flying machine was the first plane to lift a man into the air. Orville Wright took the first airplane ride, lying face downward on the lower wing of the biplane. His ride, which was only a few feet above the ground, lasted 12 seconds and covered 120 feet before running into the ground. Photograph courtesy the Library of Congress and United Airlines.

1

Propulsion*

1-1 THEORY OF PROPULSION

The verb "propel" means to drive or push forward or onward. Clearly this definition implies the existence of a body, a force (drive or push), and a preferred direction (forward or onward). According to Newton's third law of motion, forces occur only in equal and opposite pairs, and the existence of a force on a body therefore entails a reaction. This reaction cannot be on the body itself, since a pair of equal and opposite forces on the same body will not "drive or push forward or onward." Therefore one or more additional bodies are essential to propulsion. For a self-propelled vehicle like an automobile, the second body is the earth itself, and the act of propulsion entails a transfer of momentum between the automobile and the earth. In cases of particular interest in this text, the second body is usually a fluid medium. In the instance of a propeller-driven airplane, this fluid is the surrounding air, while in the case of a rocket it is a fluid carried along with, and discharged from, the rocket itself. In a jet airplane, the major part of the fluid used in propulsion is the surrounding atmosphere, but the fuel, which may constitute 2% of the propulsive fluid, is carried along and discharged as in a rocket.

If we consider an airplane in straight level flight at uniform speed, it is clear that the horizontal component of the resultant force acting on the airplane must be zero, in order to satisfy Newton's second law of motion. We know that to propel the airplane it is necessary to overcome the drag (frictional and induced) of the fluid on the body. If this is done by means of a propeller, the force of the propeller on the airplane must be sufficient to cancel the drag forces. The propeller can produce this force only by transferring momentum to the stream of air which passes through it. The drag forces also transfer the momentum to the surrounding air; and indeed, the horizontal momentum change due to the propeller thrust and the horizontal momentum change due to drag forces must be equal and opposite,

---

* An introduction by Edward S. Taylor, Professor of Flight Propulsion, Department of Aeronautics and Astronautics, Massachusetts Institute of Technology.

whether we are speaking of the momentum of the airplane or of the fluid medium surrounding it. Thus, although the air is disturbed by the passage of such an airplane, there is no net change in its horizontal momentum before and after the passage.

If we consider an accelerating rocket (assumed, for the purpose of simplifying the argument, to be in gravity-free flight outside the atmosphere), the total momentum of the rocket and the material which it has discharged must remain constant. It is inconvenient to keep track of the momentum of the propellant after it leaves the rocket, and we shall see (Chapter 2) that in order to calculate acceleration of the rocket, it is necessary only to observe the momentum of the jet leaving the rocket at the nozzle discharge. Clearly the principles of propulsion are merely applications of Newton's laws of motion.

Long before Newton announced his laws of motion (1687) which now make the principles of propulsion comprehensible to us, devices for effecting propulsion were known and used. The Chinese had rockets as early as the twelfth century, and screw propellers are visible in some of Leonardo da Vinci's drawings. About 200 B.C., the Egyptian philosopher Hero invented and constructed a turbine which clearly shows the principle of jet propulsion. In propulsion, as in most other technological advances, the art preceded the science. However, it is highly improbable that man could ever have produced supersonic aircraft or rockets capable of putting vehicles in orbit around the earth without a very good theory of propulsion. By "theory" we mean a method of generalizing the results of experiments so that we can, with some confidence, predict the results of new experiments.

A good theory enables one to see the main points without getting confused by details, and makes it possible to perform limit analysis; e.g., to determine what could be done if the machinery could be made without losses. Furthermore, a theory facilitates communication between workers. Moreover, even poor theory may stimulate thought and experiment which will lead to new ways of accomplishing things. A good theory can prevent wasting effort on unworkable devices.

Many erroneous ideas of propulsion by jets were prevalent as late as the 1920's. It was well known at this time that a fireboat directing its streams of water at a fire would gradually move away from the fire. Eminent professors of engineering have been heard to scoff at this observation, saying that "you can't push on a rope." This statement, of course, reflects a profound ignorance of the problem, which has nothing to do with whether the streams from the fire nozzle ever reach any target.

Goddard, at about this time, found it necessary to demonstrate that his rockets had thrust even when fired in a vacuum. This points up the prevalence of the same fallacy as existed in the fireboat problem. It apparently was difficult for people to conceive of something propelling itself without some medium or body to push on. The idea of the device pushing on a medium which was being discarded was understood by only a few. Now it is a part of the knowledge of most if not all newly graduated mechanical and aeronautical engineers.

1-2 DEVELOPMENT OF PROPULSIVE DEVICES

The history of propulsion illustrates the interaction between economic or military need and technological advances. Although da Vinci's sketches of screw propellers date from the fifteenth (or possibly early sixteenth) century, the need for such a device did not arise until the development of the steam engine provided an appropriate driving mechanism. The first successful steamboats were driven by paddle wheels, but the paddle wheel must operate at the interface of two media (air and water) and is therefore clearly unsuitable for submarines, lighter-than-air ships, and airplanes. (It might also be said to be unsuitable for *any* ship when the sea is rough and the interface is shifting about.) The first successful submarine, the first airship, and the first airplane all had screw propellers. The airplane propeller still serves well when the speed of the aircraft is less than 350 miles per hour. When man began to want to fly at higher speeds (originally for military reasons), a lighter and smaller power plant was essential. The jet engine answered this need, while at the same time avoiding the difficulties encountered by propellers operating at supersonic tip velocities. It is interesting to note that there was no incentive for developing a jet engine until the art of airplane design had advanced enough to make high-speed flight practical. As a matter of fact, the jet is inappropriate to low-speed flight, and the propeller provided an essential first step without which we would almost certainly not have developed airplanes at all. An interesting history of the development of aircraft engines is to be found in Reference 4.

The jet engine is a good example of the solution of a technical problem—that of high-speed flight—which required that many conditions exist simultaneously. First the need had to exist; then there had to be sufficient mastery of the art of airplane design, including aerodynamics, structure, and control; and finally there had to be the knowledge of turbomachinery design and the availability of materials capable of withstanding high temperatures. These are only some of the major hurdles. Of course the solution of the problem also depended on a sophisticated manufacturing technique—forging, casting, cold-forming, welding, and machining—and on the availability of countless "minor" but essential items: bearings, seals, lubricants, pumps, control devices, etc. In any development as complex as the jet engine, special circumstances are bound to arise which require changes and advancement of the technology in these "minor" items. However, without a reasonable state of development in these as well as the major items, the development period would be likely to stretch out until it became clear that the project had better be postponed until the necessary supporting technology was able to catch up. John Barber patented a gas turbine in 1791. There is little doubt that Barber had insufficient knowledge of basic mechanics, and certainly insufficient knowledge of fluid mechanics to develop a successful gas turbine; but even if he had had all the fundamental knowledge we have today, his job would still have been impossible until the general technological level had risen enormously. Materials and manufacturing techniques were not adequate even to begin this task until well over a hundred years later. Furthermore, had he been able to build a gas turbine, there would have been no economically feasible use for it.

History is full of such examples of inventions made too soon. They emphasize the importance of the engineering task of determining when a need exists and when supporting technology, both in theory and practice, are sufficiently advanced to begin a new development. To determine these facts involves a knowledge or at least an estimate of what extensions of theory and what refinements of practice are essential to its success.

The first successful aircraft propulsion device was Henry Gifford's steam-engine, propeller-driven dirigible balloon, which went from Paris to Trappes in 1851 at an average speed of about seven miles per hour. However, for an airplane power plant, the necessary water supply (or condenser) and boiler would be so unreasonably heavy that even if the steam engine weighed nothing, only very poor airplane performance would be possible. The successful airplane, in a practical sense, had to wait for the development of the internal combustion engine, which eliminated the necessity for either a water supply (or condenser) or a boiler. At the time of the Wright brothers' experiments, available internal combustion engines were still too heavy, but they did exist, and the time was ripe for their development to a reasonable engine weight per horsepower. The Wright brothers had to take on this problem [1], along with the design and construction of the first powered airplanes to demonstrate controlled flight. Langley, who was facing the same problem at the same time, found an extraordinary engineer in Manley [2], who handled this part of the problem with outstanding success.

The possibility of propulsion by means of a jet had been well known [3] for many years before it was usefully applied by von Ohain (first flight August 27, 1939) and Whittle (first flight May 15, 1941). Before jet propulsion could become feasible, it was essential that airplanes be designed and built for high speed, and also it was necessary that the art of design and construction of high-temperature turbines and efficient compressors be advanced considerably to meet the stringent requirements (primarily light weight combined with high efficiency) of a useful aircraft engine.

The development of the sophisticated rocket is an illustration of a similar series of events. The Chinese seem to have used the rocket mostly for fireworks, although there is some indication of military use. In the early days of artillery, accuracy was so poor that even a crude unguided rocket could compete, and about the beginning of the nineteenth century the British officer Congreve was instrumental in developing rockets as a substitute for artillery. There is no doubt that these rockets were useful in frightening cavalry horses, and perhaps were not much less accurate than the artillery of the time. The "rocket's red glare" of the national anthem refers to Congreve rockets in the War of 1812.

Subsequently artillery was greatly improved and military rockets became obsolete. Rockets, in order to be effective, would have to be guided. The necessary understanding of control problems came about in the 1930's, and the requisite development of this technology came soon after.

While the German V-2 rocket, with a 2000-pound warhead and a range of about a hundred miles, had considerable value as a morale-builder for the Germans, the damage done by these rockets was minor and the accuracy was poor compared with the effort involved. However, with the advent of nuclear devices, the potential explosive power of a warhead of given size was multiplied by an enormous factor, and the effectiveness of a long-range missile with even poor accuracy became enormous. This, then, was the motivating force behind the development of rocket propulsion; it was inevitable that billions of dollars would be poured into the development of such devices. Except for the guidance devices (where enormous advantages were yet to be obtained by subsequent improvements), the art of rocketry was reasonably ready and the science had been ready for some time, with the exception of one factor: the reentry of a rocket at high speed. This problem was not sufficiently understood until much later, but practically speaking it proved to be less difficult than was feared. Thus in the forties the need was evident and the technology and science existed. With reasonable luck and not-unreasonable improvement in the technology, the job could be done.

1–3 FUTURE POSSIBILITIES

At the present writing (1964) we see successful chemical rockets, and the need, if we are to travel far in the solar system, for vastly improved means of propulsion. The possibilities now appear to be:

(1) Improvement in chemical rockets, which undoubtedly will take place, but may be insufficient to answer the needs for exploring the solar system.

(2) Nuclear rockets. These hinge on our ability to make nuclear reactors which will operate at very much higher temperatures than they do at present.

(3) Some form of electrical propulsion. This appears to require better understanding of the properties and behavior of high-temperature ionized gases, and considerable improvement in the means for producing electric power in space.

What route we shall take from here is unclear. It may be that improvement in conventional chemical rockets will offer the least expensive alternative. If we can judge from history, it may even be that we shall wait for new scientific discoveries, new materials, and new improvements in the technology of manufacturing before the task can be accomplished. In any case we must remember that a continuing strong desire to do the job will be essential to its accomplishment.

References

1. MEYER, R. B., JR., Annual Report of the Smithsonian Institution, 1961
2. "A Description of the Manley Engine," *Soc. of Automotive Engineers*, 1942
3. SCHLAIFER, R., and S. D. HERON, *Development of Aircraft Engines and Fuels*. Cambridge, Mass.: Graduate School of Business Administration, Harvard University, 1950
4. TAYLOR, C. F., "Aircraft Propulsion." Annual Report of the Smithsonian Institution, 1962, pages 245–298

2

Mechanics and
Thermodynamics of
Fluid Flow

2-1 INTRODUCTION

An understanding of fluid mechanics and thermodynamics is perhaps the most important prerequisite for the study of propulsion. In nearly all the propulsion methods discussed in this volume, thrust is developed by imparting momentum to fluid streams. Most of the methods involve thermal effects in one way or another. The purpose of this chapter is to set forth a concise statement of the principles of mass, momentum, energy, and entropy in a form which is suitable to the treatment of fluid streams.

We shall assume that the student is thoroughly familiar with the first and second laws of thermodynamics and their corollaries as applied to ordinary nonreacting continuum substances in equilibrium. Hence, after careful definition of the concepts of system and control volume, these laws will merely be stated in forms which the reader will find useful in the remainder of the text. The thermodynamics of equilibrium chemical reactions will then be treated in some detail, because of its relevance to combustors in propulsion engines. Considerable literature is available as background material for the subject matter of both this chapter and Chapter 3. This summary largely follows the material of Keenan [1] and Shapiro [2].

The concepts of *system* and *control volume*, as applied in the following pages and throughout the text, have very specific meanings. The purpose of these concepts is the specification either of a definite collection of material or of a region in space which is to be analyzed.

System. A system is a collection of matter of fixed identity. It may be considered enclosed by an invisible, massless, flexible surface through which *no matter* can pass. The boundary of the system may change position, size, and shape.

7

Control volume. A control volume is a region of constant shape and size which is fixed in space relative to the observer. One can imagine an invisible, massless, rigid envelope (the control surface) enclosing the control volume. The control surface offers no resistance to the passage of mass and is arbitrarily located. Analyses to be developed later will show that the most suitable location of a control surface in any particular problem is a matter of convenience.

Most problems may be treated by either system or control volume analyses. However, the proper choice often leads to substantial simplification. For example, most fluid-flow problems, and especially steady flows, are much more conveniently handled in terms of an appropriately specified control volume.

Continuum. In the following discussions the fluids under consideration will be treated as continua. That is, their atomic structure will be ignored and they will be considered as capable of being subdivided into infinitesimal pieces of identical structure. In this way it is legitimate to speak of the properties of a continuum— e.g., density, pressure, velocity—as point properties.

2–2 THE LAWS OF MECHANICS AND THERMODYNAMICS FOR FLUID FLOWS

Mass conservation and the continuity equation

The law of mass conservation applied to a system simply requires that the mass of the system remain constant regardless of its size or shape, the number of pieces into which it is divided, or the length of the time interval during which it is observed (for "nonrelativistic" system velocities).

Here, as in most cases, the physical law is stated as it applies to a system. This is inconvenient for the stationary observer of fluid flow. Such an observer is unable to continuously measure the behavior of a moving fluid particle (a small system) and must, therefore, content himself with measurements taken at specific points in the space through which the fluid moves. The basic law must then be expressed in terms of the required relationships between fluid properties at various stationary points; i.e., at points within a control volume.

Consider a general control volume defined within a region of fluid flow as shown in Fig. 2–1. The conservation of mass simply requires that the rate of change of mass stored within the control volume be equal to the net inflow of mass:

$$\frac{d}{dt} m_{cv} = \dot{m}_{in} - \dot{m}_{out},$$ (2–1)

where m_{cv} is the mass within the control volume and $\dot{m}$ is the indicated mass flow rate. These terms can be expressed in more convenient analytical forms. Consider the mass flow through a small surface element dA on the control surface. If the local velocity is $\mathbf{u}$ and a unit vector normal to dA is $\mathbf{n}$ (outward pointing) as in Fig. 2–1, the elemental mass flow rate is

$$d\dot{m} = \rho |\mathbf{u}| \cos \alpha \, dA,$$

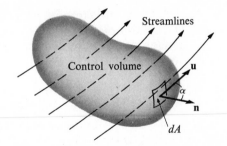

FIG. 2–1.   Fluid flow through a control volume.

where ρ is the local density and α is the angle between $\mathbf{u}$ and $\mathbf{n}$, the product $|\mathbf{u}|\cos\alpha$ being positive for outflow and negative for inflow. The mass within the control volume m_{cv} can be found by the integration

$$m_{cv} = \int_{cv} \rho \, d\upsilon,$$

where $d\upsilon$ is a volume element and the integral extends throughout the control volume. Finally, writing the product $|\mathbf{u}|\cos\alpha$ as the usual *dot product* $\mathbf{u}\cdot\mathbf{n}$, the *continuity equation* may be written

$$\frac{d}{dt}\int_{cv}\rho\,d\upsilon + \int_{cs}\rho\mathbf{u}\cdot\mathbf{n}\,dA = 0, \qquad (2\text{–}2)$$

where the second integral extends over the entire control surface.

Newton's second law and the momentum equation

For a system of mass m acted on by a force $\mathbf{F}$, Newton's law requires that

$$\mathbf{F} = K\frac{d}{dt}(m\mathbf{u}), \; = m\,\frac{d\bar{u}}{dt} + \bar{u}\,\frac{dm}{dt} \qquad (2\text{–}3)$$

where K is simply a proportionality constant whose numerical value depends entirely on the units selected for the three variables in the equation. It is possible to choose units of force, mass, length, and time such that K is unity, and this is done in all the following equations.

In a derivation very similar to the preceding (see, for example, Chapter 1 of Reference [2]), the consequences of this system equation can be written for a control volume in the following vector equation form, where the notation is that of Fig. 2–1:

$$\sum \mathbf{F} = \frac{d}{dt}\int_{cv}\rho\mathbf{u}\,d\upsilon + \int_{cs}\rho\mathbf{u}(\mathbf{u}\cdot\mathbf{n})\,dA. \qquad (2\text{–}4)$$

The term $\sum\mathbf{F}$ is the vector sum of *all* forces acting on the material within the control volume. Forces may be one of two kinds: surface forces (stresses) and body forces. The two types of stresses are pressure, which is a stress normal to the surface, and shear, which is a stress tangential to the surface. Body forces are those which are proportional to the mass or volume of the fluid and include gravitational, electrostatic, and electromagnetic forces.

The first law of thermodynamics and the energy equation

The first law of thermodynamics relates *heat* and *work* interactions between a system and its environment to changes in the state of the system. To avoid difficulty in application of this law it is necessary to give each of these terms a rather careful definition. The following definitions are not the only ones which may be used, but they are the basis of the well-established method of expressing the principle of the first law used in this volume.

Heat. When two systems at different temperatures are brought together, an interaction occurs. This *interaction* is called *heat*. The mathematical sign of the heat is conveniently taken as positive for the lower-temperature (cold) system and negative for the higher-temperature (hot) system. It is common to speak of heat *transfer*, saying that heat transfers from the hotter system to the colder. Thus heat transfer *to* a system is positive and *from* a system is negative. If this terminology is used, we must carefully avoid the implication that heat is a storable quantity. Heat exists only as a transfer process or interaction and *cannot be stored.* A process which does not involve a heat transfer is said to be adiabatic.

Work. A system is said to do work on its surroundings when the sole effect external to the system could be the rise of a weight in a gravitational field. Work done by a system on its surroundings is, by convention, positive; work done on a system by its surroundings is negative. Work, like heat, is an interaction between systems, and is not a storable quantity. In general, the magnitude of the work depends on the motion of the observer.

The first law of thermodynamics may be stated for a system in the form

$$Q = \Delta E_0 + W, \qquad (2\text{–}5)$$

in which ΔE_0 denotes the change in internal energy of a system subjected to heat and work interactions Q and W, respectively.*

The internal energy, as its name suggests, is a property of the system. The first law in a sense *defines* this property except for an arbitrary datum value. Internal energy *is* a storable quantity, in contrast to heat and work which may be considered energy transfers. The first law itself does not distinguish between heat and work;

---

* The notation throughout this text differs slightly from the more common fundamental texts in that it attempts to present a unified system for the several disciplines covered. E_0 (seldom used) refers to internal energy *including* kinetic and potential energies, while E refers to internal energy without these terms.

both contribute to changes in the internal energy. The second law, however, shows that they are fundamentally different.

It is customary to treat separately those portions of E_0 due to motion [kinetic energy $m(u^2/2)$], position in a gravitational field (potential energy mgz), charge in an electrical field (electrical potential energy qV), and other effects. For most ordinary fluid-mechanics problems it is sufficient to consider kinetic and potential energies only and write, for a system of mass m,

$$E_0 = E + m\frac{u^2}{2} + mgz, \qquad (2\text{–}6)$$

where z is the height above some arbitrary datum and g is the (magnitude of the) local acceleration due to gravity. Further, for continua, it is convenient to denote the values of E_0 and E *per unit mass* as e_0 and e, respectively. Using this and Eq. (2–6), Eq. (2–3) can be written

$$Q = \Delta \int_s \left(e + \frac{u^2}{2} + gz\right) \rho \, d\upsilon + W, \qquad (2\text{–}7)$$

where $\int_s$ denotes integration throughout the entire system. To express the first law of thermodynamics for a control volume, it is often convenient to introduce heat and work transfer rates. Equation 2–7 may be transformed (see again Chapter 1 of Reference [2]) in the same way as Eqs. (2–2) and (2–4), with the result that

$$\int_{cs} \dot{\mathcal{Q}} \, dA = \frac{d}{dt} \int_{cv} (e_0 \rho) \, d\upsilon + \int_{cs} \left(h + \frac{u^2}{2} + gz\right) \rho \mathbf{u} \cdot \mathbf{n} \, dA + \mathcal{P}_s - \int_{cv} \mathbf{X} \cdot \mathbf{u} \, d\upsilon, \qquad (2\text{–}8)$$

where $\dot{\mathcal{Q}}$ = local heat transfer rate (per unit time per unit area), positive *to* control volume,

$h = e + p/\rho$ = enthalpy of fluid,

$\mathcal{P}_s$ = net "shear power" (work transfer rate), positive *from* control volume, and

$\mathbf{X}$ = body force per unit volume on fluid within control volume.

The shear power $\mathcal{P}_s$ is that work done by the material within the control volume on the surroundings at control surface points where both shear stress and material motion (parallel to the control surface) occur. Usually this term reduces to *shaft power* transmitted by shear stresses in rotating shafts. However, there are situations where the fluid may move along a control surface accompanied by a shear stress (at points within a boundary layer, for example). The last integral, which we shall not often have to consider, represents work transfer via body forces. The dot product $\mathbf{X} \cdot \mathbf{u}$ is the rate (per unit volume) at which work is done *on* the fluid *by* the environment through body force $\mathbf{X}$ (hence the negative sign preceding it).

It is instructive at this point to dwell on certain essential similarities of the continuity, momentum, and energy equations for a control volume. The common form of these equations may be seen by arranging them as follows:

Mass:
$$\frac{d}{dt}\int_{cv}\rho\,d\mathcal{v} + \int_{cs}(\rho\mathbf{u}\cdot\mathbf{n})\,dA = 0.$$

Momentum:
$$\frac{d}{dt}\int_{cv}\rho\mathbf{u}\,d\mathcal{v} + \int_{cs}\mathbf{u}(\rho\mathbf{u}\cdot\mathbf{n})\,dA = \sum\mathbf{F}.$$

Energy:
$$\frac{d}{dt}\int_{cv}e_0\rho\,d\mathcal{v} + \int_{cs}e_0(\rho\mathbf{u}\cdot\mathbf{n})\,dA = \int_{cs}\dot{\mathcal{Q}}\,dA$$
$$- \mathcal{P}_s + \int_{cv}\mathbf{X}\cdot\mathbf{u}\,d\mathcal{v} - \int_{cs}p\mathbf{u}\cdot\mathbf{n}\,dA.$$

In each equation the first term represents the rate of change of a certain quantity contained within the control volume. The second term is its rate of convection out of the control volume. The right-hand side of each equation may be called a production or source term. The zero on the right-hand side of the first equation means, of course, that mass can be neither produced nor destroyed. On the other hand, momentum can be brought into being by the application of force. Correspondingly, energy can be changed by heat transfer or work.

These equations will generally be greatly simplified for application in the following chapters, usually by assuming steady flow and single inlet and outlet states, in which case the integrations become trivially easy.

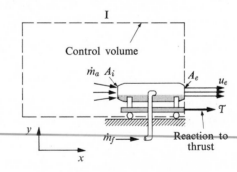

FIG. 2–2. Stationary jet engine.

To illustrate the use of these equations in a control-volume analysis, let us consider the generation of thrust by a stationary turbojet. Figure 2–2 shows the engine running steadily on a test stand. Air enters the inlet at a total mass flow rate $\dot{m}_a$ and fuel at a rate $\dot{m}_f$. Since the flow is steady, there cannot be a continuous accumulation of fluid within the engine, and the rate at which the fluid exhausts from the engine must be $(\dot{m}_a + \dot{m}_f)$. The exhaust velocity is u_e, and the force necessary to balance engine thrust is $\mathcal{T}$, as in Fig. 2–2.

In order to show how thrust depends on the properties of the jet, it is convenient to choose the control volume indicated in Fig. 2–2. Its surface intersects the exhaust jet in the plane of the nozzle, where the flow is nearly one-dimensional and the momentum flux can be measured more easily than at planes further downstream. The control surface extends far upstream of the engine and far from the floor on which it rests. Later we shall select another control volume, simply to illustrate the convenience of a suitable choice; but for the moment we proceed with the control volume of Fig. 2–2.

In applying Eq. (2–4) to this control volume, we confine our attention to those components of force and momentum parallel to the axis of the jet, so that the momentum equation may be written

$$\sum F_x = \int_{cs} \rho u_x (\mathbf{u} \cdot \mathbf{n})\, dA,$$

in which the subscript x denotes the direction indicated in Fig. 2–2. The time-derivative term in Eq. (2–4) has been omitted because the flow is steady.

No body forces act in the x-direction, so that the total x-force on the control volume is due to pressure distributed over its surface and the reaction to the thrust T. We assume the fuel hose to be flexible so that it does not transmit force to the control volume. The force T may be considered to be a normal force on the plane where the control surface intersects the structural member which holds the engine stationary. The pressure forces always act normal to the control surface, so that only the pressure forces on the two ends of the rectangular surface of Fig. 2–2 contribute to the x-summation of forces. Further, the pressure on these surfaces may be considered to have the constant value p_a, except possibly in the exhaust plane of the jet. There it may have a different value p_e, if the jet is sonic or supersonic. Thus the sum of the forces acting on the control volume in the x-direction can be written as

$$\sum F_x = T + A_e p_a - A_e p_e,$$

in which A_e is the nozzle exhaust area. The momentum flux integral may be evaluated simply by recognizing that the momentum flux is infinitesimal except in the exhaust plane of the jet. There the velocity is normal to the surface, so that

$$|\mathbf{u} \cdot \mathbf{n}| = u_x = u_e,$$

and the momentum flux integral is therefore

$$\int_{cs} \rho u_x (\mathbf{u} \cdot \mathbf{n})\, dA = \int_{A_e} \rho_e u_e^2\, dA,$$

in which ρ_e is the fluid density at the exhaust plane. Considering the flow in this plane to be uniform (i.e., one-dimensional), the integral may be simplified to $\rho_e u_e^2 A_e$. The mass flow rate at this plane is $\rho_e u_e A_e$, and this has already been shown to equal $(\dot{m}_a + \dot{m}_f)$. Thus the momentum flux from the engine is

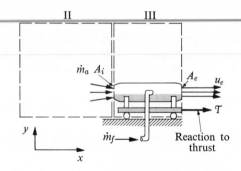

FIG. 2-3.   Stationary jet engine.

$(\dot{m}_a + \dot{m}_f)u_e$, and the thrust may be obtained from Eq. (2-4) as

$$T = (\dot{m}_a + \dot{m}_f)u_e + A_e(p_e - p_a).$$

Consider now the alternative control volume III indicated in Fig. 2-3. If we assume the flow in the engine inlet plane to be approximately one-dimensional, with pressure p_i and velocity u_i, then by reasoning very similar to the foregoing it may be established from control volume III that

$$T = (\dot{m}_a + \dot{m}_f)u_e - \dot{m}_a u_i + A_e(p_e - p_a) - A_i(p_i - p_a).$$

This thrust equation is of course not so convenient as the previous one, because it contains three additional terms: A_i, u_i, and p_i. Other choices of control volume could lead to even more inconvenient results.

The two thrust equations can easily be shown to be equivalent, however. Using control volume II and Eq. (2-4) again, it may be shown in a similar way that

$$0 = \dot{m}_a u_i - A_i(p_a - p_i).$$

Using this equation, we can easily transform the second thrust equation to the first one.

The second law of thermodynamics and the entropy equation

The second law of thermodynamics states that it is impossible for a system to describe a cyclic process (i.e., to go through a process and return to its original state) which produces work and exchanges heat with a single reservoir of uniform temperature. Although we shall not use this statement directly, we shall use several of its important corollaries. The absolute temperature scale which we shall use is in fact based on this law [1]. Further, the second law permits the definition of the property entropy. For a system,

$$dS \equiv \left(\frac{dQ}{T}\right)_{\text{rev}}, \qquad (2-9)$$

where dS is the change of entropy during a reversible heat exchange dQ at the temperature T. Keenan [1] and other thermodynamicists discuss the concept of thermodynamic reversibility extensively. For our purposes it is adequate to note that processes which involve any of the following are *not* reversible:

Not Reversible

(1) Friction,

(2) Heat transfer with finite temperature gradient,

(3) Mass transfer with finite concentration gradient,

(4) Unrestrained expansion.

In general, irreversibility lessens the engineering usefulness of any given process. Though it may be impossible to absolutely eliminate irreversibility from any real process, its effects may often be substantially reduced in practical devices, so that a reversible process is the limit of possible performance. For any process it can be shown that

$$dS \geq \frac{dQ}{T},\tag{2–10}$$

where the equality holds only for reversible processes. Thus if a process is reversible and adiabatic ($dQ = 0$), it must be *isentropic* ($dS = 0$), although the converse is not necessarily true. Finally, defining s as the entropy per unit mass of a continuum substance, the control-volume form of Eq. (2–10) can be written

$$\int_{cs} \frac{\dot{Q}dA}{T} \leq \frac{d}{dt}\int_{cv} \rho s \, dv + \int_{cs} s(\rho \mathbf{u} \cdot \mathbf{n}) \, dA,\tag{2–11}$$

where again the equality holds only for reversible processes.

2–3 THERMODYNAMICS OF GASES

A pure substance may be defined, in the thermodynamic sense, as a substance which has only two independent static properties (in the absence of electricity, magnetism, and capillarity). Most engineering fluids, including gases like oxygen, nitrogen, and vapors in equilibrium with their liquid phase are pure substances.

Consider a small system composed of a pure substance in the absence of gravity and motion (of system relative to observer). If the properties are uniform throughout the system, the first law for an incremental change of state is

$$dq = de + dw,$$

where q and w are the heat and work per unit mass, respectively. Suppose the system experiences a reversible process for which $dw = p \, dv$, in which v is the specific volume. Then, using Eq. (2–9), we write

$$T \, ds = de + p \, dv.\tag{2–12}$$

Although this equation has been derived for a reversible process it must also be true for irreversible ones, since there are only two independent static properties for a pure substance, and all the terms in this equation are properties. The entropy change ds depends only on the change in state of the system, which is fixed by the values of de and dv. The increment ds is therefore independent of the process required to produce the increments de and dv. Thus, with the specified restrictions, Eq. (2–12) must hold true for any process whose end states could possibly be connected by a process for which the work is $\int p\,dv$.

The solution of a thermodynamic problem involving pure substances must include sufficient information on the relationships between the properties of the substance. The existence of only two independent static properties greatly simplifies the presentation of this information. If it is given algebraically, the equations are known as the *equations of state* of the substance. Equations of state are merely statements of experimentally observed relationships. Often the empirical relationships cannot be conveniently reduced to equation form and the equations of state must be represented by graphs or tables. Equations, graphs, or tables are available for the properties of fluids of common interest. One of the simplest of these is the perfect gas.

Perfect gas. A perfect gas is here defined by the following equation of state:

$$pv = RT, \tag{2–13}$$

in which p is the absolute pressure, v the volume per unit mass, T the absolute temperature and R a constant, called the gas constant. The gas constant depends only on the identity of the gas. More generally, it is observed that

$$pv = \frac{\overline{R}}{\overline{M}}T, \tag{2–14}$$

in which $\overline{M}$ is the molecular weight and $\overline{R}$ is the universal gas constant which has the value

$$\overline{R} = 1545.43 \text{ ft·lb}_f/°\text{R·lb}_m\text{-mole}$$
$$= 1.9857 \text{ cal}/°\text{K·gm-mole.}$$

Most real gases have a large range of temperatures and pressures over which they obey Eq. (2–14) very closely. By use of the Maxwell equations it may be shown [1] that the internal energy of any gas whose state can be so described depends only on temperature:

$$e = e(T).$$

Since the enthalpy h is defined by

$$h = e + pv,$$

Eq. (2–13) indicates that the enthalpy also depends only on temperature

$$h = h(T).$$

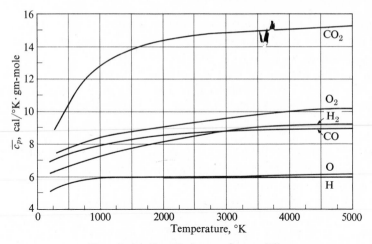

FIG. 2–4. Specific heats of gases [5].

In this case the relationship between internal energy, enthalpy, and temperature may be described by defining the "specific heats" at constant volume and temperature as follows (for a perfect gas):

$$c_v = \frac{de}{dT}, \qquad c_p = \frac{dh}{dT}. \tag{2–15}$$

These terms are actually misnomers, since heat is not a storable quantity. Also, the internal energy and the enthalpy can change in the complete absence of heat.

Using the definition of enthalpy and Eq. (2–13), we have

$$dh = de + R\,dT.$$

The relations (2–15) show that for a perfect gas the specific heats must be related by

$$c_p = c_v + R. \tag{2–16}$$

Typical variations with temperature of the specific heat of gases at constant pressure are shown in Fig. 2–4. Figure 2–5 illustrates the corresponding enthalpy-temperature dependence for a number of common gases where the zero has been arbitrarily set at 300°K.

Two types of perfect gas calculations may be demonstrated.

1. *Constant specific heats.* Figure 2–4 shows that for a sufficiently small range in temperature it may be acceptable to approximate the specific heats by constants. Using the perfect gas law, Eq. (2–12) may be written

$$ds = \frac{dh}{T} - \frac{dp}{\rho T}$$

or

$$ds = c_p \frac{dT}{T} - R \frac{dp}{p}. \tag{2–17}$$

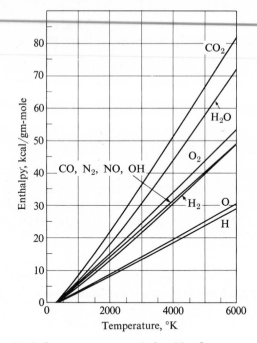

FIG. 2-5.   Enthalpy–temperature relationships for common gases [8].

For the particular case of constant specific heat and for an *isentropic* process, Eq. (2–17) may be integrated, with the results that

$$\frac{p}{p_1} = \left(\frac{T}{T_1}\right)^{c_p/R}.$$ (2–18)

Defining the specific heat ratio γ,

$$\gamma \equiv \frac{c_p}{c_v},$$

and using Eq. (2–16), Eq. (2–18) may be expressed in the form

$$\frac{p}{p_1} = \left(\frac{T}{T_1}\right)^{\gamma/(\gamma-1)}.$$ (2–19)

Using the perfect gas law, this can be transformed to

$$\frac{\rho}{\rho_1} = \left(\frac{T}{T_1}\right)^{1/(\gamma-1)}$$ (2–20)

or

$$\frac{p}{p_1} = \left(\frac{\rho}{\rho_1}\right)^{\gamma},$$ (2–21)

in which $\rho = 1/v$ is the fluid density.

2. *Variable specific heats.* The functions $h(T)$ and $e(T)$ are tabulated for a number of gases (see, for example, References [4] and [8]). By defining a function ϕ according to

$$\phi = \int \frac{c_p \, dT}{T},$$

Eq. (2–17) may be integrated in the form

$$s - s_1 = \phi - \phi_1 - R \ln \frac{p}{p_1}, \qquad (2\text{–}22)$$

in which the subscript 1 again denotes a reference value. The function ϕ depends only on temperature for a perfect gas, and hence may be tabulated with the enthalpy against the temperature argument for each gas. If the process is isentropic, Eq. (2–22) may be written

$$\frac{p}{p_1} = e^{(\phi - \phi_1)/R}.$$

For any given datum state 1 this *isentropic* pressure ratio p_r may be expressed by

$$p_r = \frac{p}{p_1} = e^{(\phi - \phi_1)/R},$$

which may also be tabulated as a function of temperature. Then if the temperature changes from any value T_A to another value T_B, the isentropic pressure change may be found from

$$\frac{p_B}{p_A} = \frac{p_r(T_B)}{p_r(T_A)},$$

in which $p_r(T_B)$ and $p_r(T_A)$ are obtained from tabulated functions.

The state of a gas having only two independent properties may be designated by a single point on a graph whose x- and y-axes correspond to any two properties. A process between states may be indicated by a line. The temperature-entropy diagram shown in Fig. 2–6 is an example of such a graph. The series of lines marked with various pressures correspond to possible constant-pressure processes. The shape of each line may be obtained from Eq. (2–17),

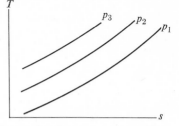

FIG. 2–6. Constant-pressure lines on a T-s diagram.

$$\left(\frac{\partial T}{\partial s}\right)_p = \frac{T}{c_p}.$$

From this it is clear that the lines have positive slopes, increasing with temperature. Further, if c_p is a function of T only, the slope of the curve is a function of T only. Hence the curves are of the same shape displaced horizontally.

From Eq. (2–17), it is also clear that

$$\left(\frac{\partial p}{\partial T}\right)_s = \frac{c_p}{R}\frac{p}{T}.$$

Since this derivative is always positive, the upper curves represent higher values of pressure.

2–4   EQUILIBRIUM COMBUSTION THERMODYNAMICS; CHEMICAL REACTIONS

Up to this point thermodynamic processes in which system components change their chemical identity have not been discussed.   Since combustion is of vital importance to many propulsion methods, it is necessary to show how chemical reactions are governed by thermodynamic laws.   Fortunately it is possible to simplify the discussion a great deal by adopting certain restrictions.   First, chemical reactions of interest in propulsion systems nearly always occur between gases which can be closely approximated in behavior by-a perfect gas.   Second, with the exception of a few very rapid expansion processes, the important chemical processes usually take place between states which are approximately in equilibrium.

The basic problem may be stated as follows: Given a mixture of gases capable of a chemical reaction and a set of conditions under which the reaction will occur, what will be the result of that chemical reaction? For example, one might be asked to determine the composition and temperature of the exhaust products of a constant-pressure adiabatic combustion process.   To formulate methods for answering this kind of question, it is convenient to consider three separate topics:

(1) Mixtures of gases,

(2) Chemical transformation,

(3) Composition of products.

Mixtures of gases

The reactants and products of a chemical process each generally exist as a gaseous mixture of chemical compounds. The properties of the mixture may be determined from the properties of the constituents by using the *Gibbs-Dalton Law*, which is expressed by Keenan [1] as follows:

1. The pressure of a mixture of gases is equal to the sum of the pressures of each constituent when each occupies alone the volume of the mixture at the temperature of the mixture.

2. The internal energy and the entropy of a mixture are equal, respectively, to the sums of the internal energies and the entropies of its constituents when each occupies alone the volume of the mixture at the temperature of the mixture.

Thus for a mixture of n constituents:

Temperature $T_m = T_1 = T_2 = \cdots = T_n,$ (2–23a)

Pressure $p_m = p_1 + p_2 + p_3 + \cdots + p_n,$ (2–23b)

Volume $\mathcal{V}_m = \mathfrak{M}_m v_m = \mathfrak{M}_1 v_1 = \mathfrak{M}_2 v_2 = \cdots = \mathfrak{M}_n v_n,$ (2–23c)

Energy $E_m = \mathfrak{M}_m e_m = \mathfrak{M}_1 e_1 + \mathfrak{M}_2 e_2 + \cdots + \mathfrak{M}_n e_n,$ (2–23d)

Entropy $S_m = \mathfrak{M}_m s_m = \mathfrak{M}_1 s_1 + \mathfrak{M}_2 s_2 + \cdots + \mathfrak{M}_n s_n,$ (2–23e)

Enthalpy $H_m = \mathfrak{M}_m h_m = \mathfrak{M}_1 h_1 + \mathfrak{M}_2 h_2 + \cdots + \mathfrak{M}_n h_n,$ (2–23f)

in which $\mathfrak{M}$ signifies mass, subscript m refers to the mixture and subscripts 1, 2, ..., n refer to a series of constituents.

Since $de = c_v\, dT$, then

$$c_{vm} = \frac{\mathfrak{M}_1 c_{v1} + \mathfrak{M}_2 c_{v2} + \cdots + \mathfrak{M}_n c_{vn}}{\mathfrak{M}_m}. \qquad (2\text{–}24)$$

Similarly,

$$c_{pm} = \frac{\mathfrak{M}_1 c_{p1} + \mathfrak{M}_2 c_{p2} + \cdots + \mathfrak{M}_n c_{pn}}{\mathfrak{M}_m}.$$

The pressures p_1, p_2, p_3, p_n that would be exerted by each constituent alone in the volume of the mixture at the temperature of the mixture are called the *partial pressures* of each constituent. The partial pressure is related to the number of moles of a constituent present. For any constituent i,

$$p_i v_i = R_i T_m$$

or, in terms of the universal gas constant $\bar{R}$,

$$p_i v_i = \frac{\bar{R}}{\bar{M}_i} T_m,$$

where $\bar{M}_i$ is the molecular weight of constituent i. Since

$$v_i = \frac{\mathcal{V}_m}{\mathfrak{M}_i},$$

$$p_i = \frac{\mathfrak{M}_i \bar{R} T_m}{\bar{M}_i \mathcal{V}_m} = n_i \frac{\bar{R} T_m}{\mathcal{V}_m},$$

in which n_i is the *number of moles* of constituent i present in the mixture. Since, according to the Gibbs-Dalton Law, the mixture pressure is the sum of the partial pressures,

$$\frac{p_i}{p_m} = \frac{n_i}{n_1 + n_2 + \cdots + n_n}. \qquad (2\text{–}25)$$

Thus the ratio of the partial pressure to the mixture pressure is just the ratio of the number of moles of the constituent to the total number of moles. This ratio is called the *mole fraction*, χ:

$$\chi_i = \frac{n_i}{n_1 + n_2 + \cdots + n_n}. \tag{2–26}$$

The entropy change of the mixture is the sum of the entropy changes of all the constituents, each behaving as though the other constituents were not there. It is interesting to note that during a process which is isentropic for the mixture, the constituents do not necessarily undergo isentropic changes. There is generally a redistribution of entropy among the constituents.

In determining partial pressures, it is convenient to use moles rather than unit masses. For this purpose one can define molal specific heats by the equations

$$\bar{c}_p = \overline{M} c_p, \qquad \bar{c}_v = \overline{M} c_v,$$

To avoid confusion the values per mole of a property are usually denoted with a bar, as

$$d\bar{h} = \overline{M} dh = \bar{c}_p \, dT.$$

Chemical transformation

A general chemical reaction, in which α moles of reactant A and β moles of reactant B, etc., combine to form μ moles of product M plus ν moles of product N, etc., may be represented by

$$\alpha A + \beta B + \cdots \rightarrow \mu M + \nu N + \cdots. \tag{2–27}$$

The coefficients α, β, etc., are known from the mixture ratio of the reactants, and it may be assumed for the present that the product composition, that is, μ, ν, etc., is also known.

Accompanying a chemical reaction there is generally an exchange of energy with the surroundings. Those reactions which tend to give off heat are called *exothermic* and once started they will, under the proper conditions, proceed unaided. The other kind, called *endothermic*, require an energy input to proceed. Combustion processes, for instance, are exothermic.

The magnitude of these energy interactions can be determined from the composition and the heats of formation of the reactants and products. The constant-pressure heat of formation of a substance is the heat interaction which occurs when one mole of the substance is formed at constant pressure and temperature from its elements as they occur in nature. For example, the heat of formation of carbon dioxide (-94.052 kcal/gm-mole) is the heat interaction accompanying the following reaction from reactants to products at 300°K:

$$C_{\text{solid}} + O_{2\,\text{gas}} \rightarrow CO_{2\,\text{gas}}.$$

In this example, as in all exothermic reactions, the heat of formation is negative, since the system must give off heat to maintain (or return to) the reference temperature.

As another example, the formation of monatomic oxygen from naturally occurring oxygen is $\frac{1}{2}O_2 \rightarrow 0$. In this endothermic reaction the heat of formation (+59.162 kcal/gm-mole at 300°K) is positive, since the heat transfer to the system must be positive.

These heat quantities may be determined experimentally with a simple constant-pressure container. Practically, it may be difficult to produce a reaction at constant temperature and pressure. It is not necessary, in any event, since from the standpoint of thermodynamics only the end states of the reaction are of interest. The intermediate conditions do not matter (although they may be important to the preservation of laboratory equipment).

A *steady-flow process* is another experimental method for determining the heat of reaction Q_f. Equation (2–8) states that, for a steady-flow process with no shaft work, the heat transfer of the fluid flowing through a control volume is equal to the change in enthalpy (neglecting changes in velocity and elevation). Thus

$$Q = \Delta H$$

or, per *mole* of product formed,

$$Q = Q_f = H_{\text{product}} - H_{\text{reactants}}. \tag{2–28}$$

For example, the formation of carbon dioxide in steady flow is represented by Fig. 2–7.

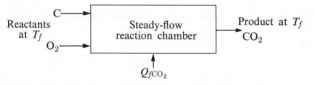

FIG. 2–7. Schematic of steady-flow process for formation of CO_2, with T_f being the reference temperature.

Table 2–1 gives the heats of formation for some common products of combustion, T_f being the reference temperature at which Q_f is measured. The heats of formation of naturally occurring elements are, by definition, zero.

In contrast, most common reactions are complex in the sense that the reactants are not naturally occurring elements and the products include more than one substance. Also, they do not generally occur at constant temperature. Complex reactions can be treated as a combination of simpler reactions by the following method. Consider the steady-flow reaction indicated in Fig. 2–8, where Q is the heat interaction necessary to maintain constant temperature. It is convenient to consider this reaction to be the sum of distinct decomposition and formation reactions.

Table 2–1   Heats of Formation

Compound	Formula	Heat of formation, kcal/gm-mole		T_f, °K	Reference number
Aluminum oxide	AL_2O_3	−399.09	(g)	298	5
Ammonium nitrate	NH_4NO_3	− 87.27	(s)	298	5
Ammonium perchlorate	NH_4CLO_4	− 69.42	(s)	298	5
Aniline	$C_6H_5NH_2$	+ 6.11	(l)	300	9
Boron	B_2	+124.5	(g)	298	5
Boron atom	B	+ 97.2	(g)	298	5
Bromine atom	Br	+ 23.050	(g)	300	6
Carbon	C	+171.698	(g)	298	5
Carbon (graphite)	C	0	(s)	300	
Carbon dioxide	CO_2	− 94.052	(g)	300	7
Carbon monoxide	CO	− 26.413	(g)	300	7
Dimethylhydrazine (unsymmetrical) (UDMH)	$(CH_3)_2NNH_2$	+ 12.734	(l)	298	10
Fluorine	F_2	− 3.0	(l)		
		− 0	(g)	300	
Fluorine atom	F	+ 18.906	(g)	300	6
Hydrazine	N_2H_4	+ 12.000	(l)	298	10
Hydrogen	H_2	− 1.92	(l)	300	9
Hydrogen atom	H	+ 52.092	(g)	300	7
Hydrogen bromide	HBr	− 8.66	(g)	298	5
Hydrogen fluoride	HF	− 64.20	(g)	298	5
Hydrogen peroxide	H_2O_2	− 33.74	(g)	300	8
		− 44.84	(l)		
		− 47.36	(s)		
Hydrazine hydrate	$N_2H_4H_2O$	− 10.3	(l)	300	12
Hydroxyl	OH	+ 10.063	(g)	300	7
Isobutane (2-methylpropane)	C_4H_{10}	− 31.489	(g)	300	7
		− 36.169	(l)		
JP–3	$C_{7.366}H_{11.442}$	− 26.66	(g)	300	13
JP–4	H/C 1.93	− 0.423*	(l)	300	8
Methyl alcohol	CH_3OH	− 48.08	(g)	298	5
		− 57.04	(l)		14
Monomethylhydrazine	CH_3NHNH_2	+ 13.109	(l)	298	10
Nitric acid	HNO_3	− 41.35	(l)	300	8
		− 31.92	(g)		

* kcal/gm in this case.

Table 2–1 Heats of Formation (*Continued*)

Compound	Formula	Heat of formation, kcal/gm-mole		T_f, °K	Reference number
Nitric acid, red fuming	14% NO_2 1% H_2O	− 45.3	(l)	300	8
Nitric oxide	NO	21.6	(g)	298	5
Nitrogen	N_2	0	(g)	300	
Nitrogen atom	N	+112.5	(g)	300	15
Nitrogen dioxide	NO_2	+ 8.091	(g)	298	5
Nitrogen pentoxide	N_2O_5	+ 3.6	(g)	298	5
Nitrogen tetroxide	N_2O_4	+ 2.30	(g)	300	8
Nitrogen trioxide	N_2O_3	+ 20.0	(g)	298	5
Nitrous oxide	N_2O	+ 19.55	(g)	298	5
N-octane	C_8H_{18}	− 49.88	(g)	300	7
		− 59.795	(l)		
Oxygen	O_2	0	(g)	300	
		+ 3.08	(l)		
Oxygen atom	O	+ 59.162	(g)	300	7
Ozone	O_3	+ 34.00	(g)	300	8
		+ 30.3	(l)		
Sulfur dioxide	SO_2	− 70.96	(g)	298	5
Water	H_2O	− 57.802	(g)	300	7

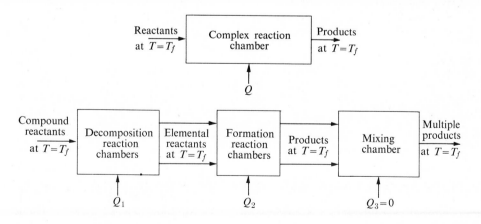

Fɪɢ. 2–8. Schematic of complex reaction at constant temperature, showing equivalent elementary reactions.

The decomposition reaction is just the reverse of the formation reaction; hence it is a simple one. However hypothetical the reverse process may be, the first law tells us that the decomposition heat interaction will be equal and opposite to the formation heat interaction. Thus, summing over all the decomposition reactions,

$$Q_1 = - \sum_i (n_i Q_{fi})_{\text{reactants}}, \tag{2–29}$$

where n_i is the number of moles of reactant i and Q_{fi} is its heat of formation per mole.

Once the reactants are in elemental form, the remainder of the reaction may be treated as the sum of separate formations, so that

$$Q_2 = \sum_j (n_j Q_{fj})_{\text{products}}, \tag{2–30}$$

where n_j is the number of moles of product j. Since the mixing process in the third chamber is adiabatic, the net heat interaction for the complex reaction at constant temperature T_f is

$$Q_R = Q_1 + Q_2 = \sum_j (n_j Q_{fj})_{\text{products}} - \sum_i (n_i Q_{fi})_{\text{reactants}}. \tag{2–31}$$

Also, using Eq. (2–28),

$$Q_R = H_{\text{products}} - H_{\text{reactants}} = \sum_j (n_j Q_{fj})_{\text{products}} - \sum_i (n_i Q_{fi})_{\text{reactants}}.$$

This difference in enthalpy is given the symbol $H_{\text{RP}f}$ and may be interpreted as the change in enthalpy from products to reactants at the constant reference temperature T_f. Hence

$$H_{\text{RP}f} = H_{\text{P}f} - H_{\text{R}f} = \sum_j (n_j Q_{fj})_{\text{products}} - \sum_i (n_i Q_{fi})_{\text{reactants}}. \tag{2–32}$$

The value of $H_{\text{RP}f}$ is tabulated for many common reactions, but if such a tabulation is not available for a particular reaction it can be calculated from Eq. (2–32) and a table such as Table 2–1. Note that this calculation requires knowledge of the composition of the product.

If the products or reactants are not at the reference temperature T_f, the foregoing procedure must be modified somewhat. Hypothetically, it is possible to bring the reactants to the reference temperature by heat transfer, allow the reaction to occur at constant temperature, and then bring the products to the final temperature by another heat transfer. The heat interactions required are indicated in Fig. 2–9.

The net heat interaction from reactants at T_1 to products at T_2 may be written as the sum of the three heat-transfer rates indicated in Fig. 2–9. Thus

$$Q = (H_{\text{P}2} - H_{\text{P}f}) - (H_{\text{R}1} - H_{\text{R}f}) + H_{\text{RP}f}. \tag{2–33}$$

Each of the parenthesized terms is a change in enthalpy from the actual state to

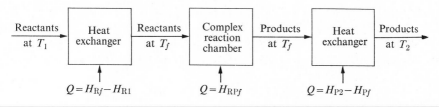

$$Q = H_{Rf} - H_{R1} \qquad Q = H_{RPf} \qquad Q = H_{P2} - H_{Pf}$$

FIG. 2–9. Reaction at variable temperatures.

the reference state (T_f), and may be calculated from tables of the mixture properties or by the methods of Section 2–3. The reference temperature T_f is simply that temperature for which the heat of reaction, H_{RPf}, is available.

Let us take a typical problem: we wish to determine the adiabatic flame temperature for a given reactant mixture. Once we know the product composition and determine the reference temperature, we can calculate H_{R1}, H_{Rf}, and H_{RPf}. Then, since $Q = 0$, we can determine the product enthalpy H_{P2}, and from this the product temperature.

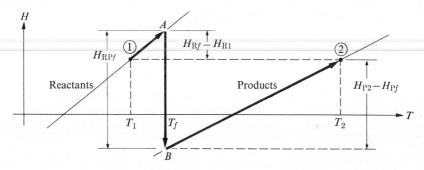

FIG. 2–10. Enthalpy-temperature diagram for adiabatic combustion.

This can all be presented graphically on an enthalpy-temperature diagram which shows the relationship between enthalpy and temperature for both product and reactant mixtures. Figure 2–10 describes a hypothetical path of the process. The reactants at state ① are brought to T_f at state A, after which conversion to products occurs at constant temperature along line AB. Then the products are heated to the outlet state at ②, and T_2 is the outlet temperature. Since there is no net heat transfer in this particular example, H_{R1} must therefore equal H_{P2} [as may be seen by inserting the definition of H_{RPf} in Eq. (2–33) and setting Q equal to zero]. From this diagram it can be seen that if the specific heats* of reactant and product mixtures are unequal, then the heat of reaction H_{RPf} must be a function of temperature.

---

* The specific heat of the mixture is given by the slope of its enthalpy-temperature curve in Fig. 2–9.

Composition of products

In all the aforementioned relationships, our calculations have depended on knowledge of the composition of the products of reaction. Once this composition is known, calculating the state of the product is a rather simple matter. However, determining the product composition is not always easy. The product composition depends on the temperature, which in turn has just been shown to depend on the composition. This suggests that a trial-and-error procedure may be required for the solution.

For certain reactions, it is relatively simple to determine product composition. For example, at sufficiently low temperature, the combustion of a particular non-stoichiometric mixture of octane and oxygen yields the following products:

$$C_8H_{18} + 15O_2 \rightarrow 8CO_2 + 9H_2O + 2.5O_2.$$

In this case, the composition is easily determined by requiring that the number of atoms of each element taking part in the reaction be constant.

Actually, at high temperatures this expression may be a considerable over-simplification, because the compounds CO_2, H_2O, and O_2 dissociate to some extent into free atoms and radicals. A more general statement is

$$C_8H_{18} + 15O_2 \rightarrow aCO_2 + bH_2O + cO_2 + dH_2 + eC + fCO$$

$$+ gH + iO + jOH + kC_8H_{18} + \text{other hydrocarbons}, \qquad (2\text{–}34)$$

where a, b, c, d, etc., are the unknown numbers of moles of each substance. A reasonable first approximation may be made by recognizing that for all practical purposes C_8H_{18} and the "other hydrocarbon" molecules are present in negligible quantities. However, there are still unknowns—a, b, c, d, e, f, g, i, j—for which we can obtain only three equations in terms of numbers of atoms. Atom balances yield:

For carbon, $a + e + f = 8$;

For hydrogen, $2b + 2d + g + j = 18$;

For oxygen, $2a + b + 2c + f + j + i = 30.$

Six more independent equations are needed in order to determine the product composition. They may be obtained from the requirements for chemical equilibrium.

Chemical equilibrium. Consider a reaction in which a stoichiometric combination of A and B reacts to form products M and N with no A and B left:

$$\alpha A + \beta B \rightarrow \mu M + \nu N. \qquad (2\text{–}35)$$

At the same time this reaction is taking place, there is generally some tendency for the reverse reaction,

$$\mu M + \nu N \rightarrow \alpha A + \beta B$$

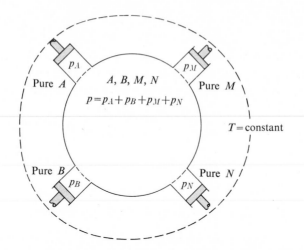

Fɪɢ. 2–11.   Reaction chamber for derivation of equilibrium constants.

to occur, especially at high temperatures. Equilibrium is that state at which both forward and reverse reactions are occurring at equal rates. When equilibrium obtains, the concentrations of A, B, M, and N are constant. Using the second law of thermodynamics, we can express the conditions for equilibrium in algebraic form.

Suppose that an equilibrium mixture resulting from the combination of A and B is contained in a chamber which communicates through semipermeable membranes with each individual constituent substance in a pure form, as in Fig. 2–11. The entire apparatus is surrounded by a constant-temperature bath.

A semipermeable membrane is a hypothetical surface through which one substance may pass freely, though all others are prevented from passing. Such membranes are only approximated by certain real substances, but they are nevertheless useful analytical concepts. For example, in Fig. 2–11, substance A may pass freely into or out of the mixture at cylinder A, and the pressure of the pure A will be equal to the partial pressure of A in the mixture, whereas B, M, and N cannot pass through this membrane. Similarly, the membranes of cylinders B, M, and N admit only substances B, M, and N respectively.

The amount of any particular substance within the mixture chamber can be controlled by simply pushing or pulling on the piston in the cylinder containing the pure substance. This control process is reversible if done slowly.

Starting with an equilibrium mixture resulting from the combination of A and B in any proportion, consider the introduction of additional small amounts of A and B through the semipermeable membranes into the mixture chamber, in the stoichiometric ratio $k\alpha A + k\beta B$, where k is some small number. Once in contact within the main chamber, the A and B tend to react, forming M and N. There are two ways to restore the contents of the main chamber to their original condition. One is to withdraw from the chamber $k\alpha A + k\beta B$. The other is to withdraw $k\mu M + k\nu N$. In both cases the total number of atoms of each element in the chamber will have returned to its original value. After the mixture has again reached equilibrium, its composition will be restored to its former value. As a result of the reaction

$$k\alpha A + k\beta B \rightarrow k\mu M + k\nu N,$$

there will be a heat interaction of magnitude kH_{RP} with the constant-temperature bath surrounding the apparatus, since the reaction occurs at the constant temperature T of the bath. In the limit this heat interaction will be reversible. Then, for a system composed of all the gases in the mixture and in the cylinders,

$$dS = \left(\frac{dQ}{T}\right)_{\text{rev}} = \frac{kH_{RP}}{T}.$$

For perfect gases it may be shown that H_{RP} is a function of temperature only. Thus for a given value of k the change in entropy of the system is a function of temperature only: $dS = kf(T)$.

Since the temperature, volume, mass, and composition of the mixture do not change, there will be no change of state of the system in the mixture chamber. Thus

$$dS_{\substack{\text{mixture} \\ \text{chamber}}} = 0.$$

In addition, the partial pressure of each constituent does not change and hence the pressure in each cylinder does not change. Since the temperature is also constant in each cylinder, the entropy *per mole* of each pure substance must be constant and the change in entropy within the cylinders may be written

$$dS_A = -k\alpha \bar{S}_A, \qquad dS_M = +k\mu \bar{S}_M,$$
$$dS_B = -k\beta \bar{S}_B, \qquad dS_N = +k\nu \bar{S}_N,$$

where $\bar{S}$ is the entropy per mole of pure substance.

The total entropy change is equal to the sum of the changes in the mixture chamber and in the cylinders:

$$dS = -k\alpha \bar{S}_A - k\beta \bar{S}_B + k\mu \bar{S}_M + k\nu \bar{S}_N = k\frac{H_{RP}}{T}$$

or

$$\alpha \bar{S}_A + \beta \bar{S}_B - \mu \bar{S}_M - \nu \bar{S}_N = -\frac{H_{RP}}{T} = f(T). \tag{2–36}$$

The entropy of a perfect gas *per mole* can be obtained by integrating Eq. (2–17) with the result:

$$\bar{S} = -\bar{R} \ln p + \int_{T1}^{T} \bar{c}_p \frac{dT}{T} + \bar{R} \ln p_1,$$

where the subscript 1 refers to an arbitrary datum state. Thus, for a given datum state, Eq. (2–36) can be written

$$-\alpha\overline{R}\ln p_A - \beta\overline{R}\ln p_B + \mu\overline{R}\ln p_M + \nu\overline{R}\ln p_N = g(T),$$

where $g(T)$ is another function of temperature only. Adding the logarithmic terms and writing $g(T) = \overline{R}\ln K_p(T)$ leads to the following relationship between the partial pressures:

$$\frac{p_M^\mu p_N^\nu}{p_A^\alpha p_B^\beta} = K_p(T). \qquad (2\text{--}37)$$

The term K_p is called the *equilibrium constant* for this mixture in terms of partial pressures.

Figure 2–12 gives typical values of the equilibrium constant K_p. Several general statements should be made regarding equilibrium constants before their use is demonstrated in a detailed example.

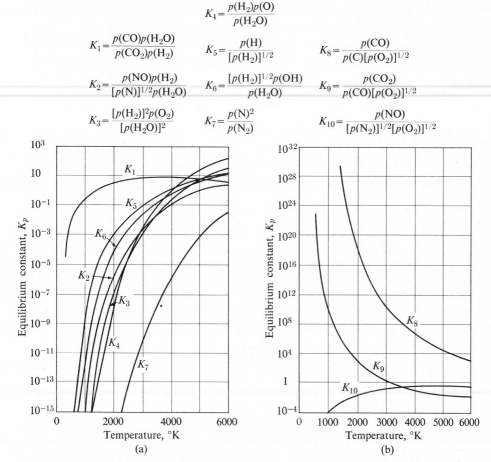

$$K_4 = \frac{p(H_2)p(O)}{p(H_2O)}$$

$$K_1 = \frac{p(CO)p(H_2O)}{p(CO_2)p(H_2)} \qquad K_5 = \frac{p(H)}{[p(H_2)]^{1/2}} \qquad K_8 = \frac{p(CO)}{p(C)[p(O_2)]^{1/2}}$$

$$K_2 = \frac{p(NO)p(H_2)}{[p(N)]^{1/2}p(H_2O)} \qquad K_6 = \frac{[p(H_2)]^{1/2}p(OH)}{p(H_2O)} \qquad K_9 = \frac{p(CO_2)}{p(CO)[p(O_2)]^{1/2}}$$

$$K_3 = \frac{[p(H_2)]^2 p(O_2)}{[p(H_2O)]^2} \qquad K_7 = \frac{p(N)^2}{p(N_2)} \qquad K_{10} = \frac{p(NO)}{[p(N_2)]^{1/2}[p(O_2)]^{1/2}}$$

FIG. 2–12. Equilibrium constants [8] in terms of partial pressures. Pressure in atmospheres.

First, it may be noted that the partial pressures of the product appear in the numerator and those of the reactants in the denominator, each with an exponent taken from Eq. (2–35). Qualitatively, then, one can expect a large K_p for sub-reactions which approach completion and small K_p for those that do not. The number of constituents considered in this derivation is four, but the result can easily be extended to include any number of reactants and products.

Second, the exponents are the *stoichiometric* coefficients. For example, consider the relationship between carbon dioxide, carbon monoxide, and oxygen which would be of interest in our octane-combustion example:

$$CO_2 \rightarrow CO + \tfrac{1}{2}O_2.$$

In this case the equilibrium equation is

$$\frac{p(CO)[p(O_2)]^{1/2}}{p(CO_2)} = K_p.$$

It may be noted that these exponents are not directly related to a, c, or f of Eq. (2–34) for the overall reaction. This equilibrium equation provides one of the six relationships necessary to the determination of product composition after the reaction (2–34). In a similar way we can state five other equilibrium requirements to govern the relative partial pressures of the products.

Third, since the object of this calculation is to determine composition, it is convenient to define an equilibrium constant K_n, based on *mole fractions* rather than partial pressures:

$$K_n = \frac{\chi_M^\mu \chi_N^\nu}{\chi_A^\alpha \chi_B^\beta}. \qquad (2\text{–}38)$$

Since, according to the Gibbs-Dalton Law, $\chi_i = p_i/p_m$, it may be shown that

$$K_n = p_m{}^{\alpha+\beta-\mu-\nu} K_p, \qquad (2\text{–}39)$$

where K_p is the equilibrium constant based on partial pressures. Note that K_n, while more convenient in calculating mole fractions, is a function of two variables, pressure and temperature, and is thus inconvenient to tabulate. In Eq. (2–39), p_m must be in the units for which K_p is tabulated.

The actual use of these concepts can best be demonstrated by an example which contains a considerably simplified description of an actual chemical reaction.

Example 1

One mole of H_2O is raised in temperature to 6000°F, and it may be assumed that it dissociates to an equilibrium mixture of H_2O, H_2, and O_2 at a total pressure of 1 atm according to $H_2O \rightarrow H_2 + \tfrac{1}{2}O_2$. The equilibrium constant for this reaction at 6000°F is $K_p = 0.23$ atm$^{1/2}$. The concentration of the mixture may be determined as follows. During the dissociation, x moles of H_2O break down to form x moles of H and $x/2$ moles of O_2, according to the reaction

$$xH_2O \rightarrow xH_2 + \frac{x}{2} O_2.$$

The relative concentrations of the constituents of the mixture may be tabulated as shown below.

Constituents	Moles present at equilibrium	Mole fraction X_i
H_2O	$1 - x$	$\dfrac{1 - x}{1 + x/2}$
H_2	x	$\dfrac{x}{1 + x/2}$
O_2	$\dfrac{x}{2}$	$\dfrac{x}{2(1 + x/2)}$
Total	$1 + x/2$	

The relative concentrations may be determined as follows:

$$K_n = \frac{X_{H_2}(X_{O_2})^{1/2}}{X_{H_2O}} = K_p p_m^{-1/2}.$$

Substituting the equilibrium concentrations, we obtain

$$\frac{\left(\dfrac{x}{1 + x/2}\right)\left(\dfrac{x/2}{1 + x/2}\right)^{1/2}}{\dfrac{1 - x}{1 + x/2}} = 0.23.$$

The quantity x may then be determined by trial and error and the mole fractions of H_2O, O_2, and H_2 at equilibrium obtained from the above tabulation.

The foregoing calculation oversimplifies the dissociation. Strictly speaking, the following three dissociation reactions are also of importance at 6000°F:

$$H_2O \rightarrow HO + \tfrac{1}{2}H_2, \qquad O_2 \rightarrow 2O, \qquad H_2 \rightarrow 2H.$$

Determination of the equilibrium concentrations of H_2O, H_2, O_2, O, and H then requires the simultaneous solution of three equilibrium conditions with known values of the three equilibrium constants for the above reactions.

Example 2 (from Reference [3])

Consider a complete set of six products for a hydrogen-oxygen reaction:

$$AH_2 + BO_2 \rightarrow n_{H_2O}H_2O + n_{O_2}O_2 + n_{H_2}H_2 + n_OO + n_HH + n_{OH}OH.$$

$$(2\text{–}40)$$

Since two equations can be obtained from the conservation of hydrogen and oxygen atoms, four equilibrium relationships are required.

The following method is due to Penner [3], who has shown that it is convenient to define yet another equilibrium constant $\Re_n$, based on *number of moles* rather than mole fractions. The use of this equilibrium constant results in simplified calculations, and it is introduced here because the technique has value beyond this example.

Using again the typical reaction (2–35),

$$\Re_n \equiv \frac{n_M^\mu n_N^\nu}{n_A^\alpha n_B^\beta}, \tag{2–41}$$

and using Eq. (2–25),

$$\Re_n = \left(\frac{n}{p_m}\right)^{\mu+\nu-\alpha-\beta} K_p. \tag{2–42}$$

Strictly speaking, the use of this equilibrium constant introduces another unknown, n, along with an additional equation which simply states that n is the total number of moles in the actual product mixture (not the subreaction). However, the simplification attained by this approach lies in the fact that n can be rather easily approximated, and in the fact that the results of the calculation are not strongly dependent on n. Thus, as illustrated in this example, n is at first assumed known and later checked. An iterative procedure can be used if greater accuracy is desired.

Returning to the example in Eq. (2–40), the two equations resulting from conservation of atoms are:

For hydrogen, $N_H = 2n_{H_2O} + 2n_{H_2} + n_H + n_{OH} = 2A,$ (2–43)

For oxygen, $N_O = n_{H_2O} + 2n_{O_2} + n_O + n_{OH} = 2B,$ (2–44)

where N_H and N_O are the *known* numbers of hydrogen and oxygen atoms, respectively.

The equilibrium information available is of the following form:

$$\tfrac{1}{2}O_2 \rightarrow O, \qquad K_{p1} = \frac{p_O}{p_{O_2}^{1/2}},$$

$$\tfrac{1}{2}H_2 \rightarrow H, \qquad K_{p2} = \frac{p_H}{p_{H_2}^{1/2}},$$

$$\tfrac{1}{2}H_2 + \tfrac{1}{2}O_2 \rightarrow OH, \qquad K_{p3} = \frac{p_{OH}}{p_{O_2}^{1/2}p_{H_2}^{1/2}},$$

$$H_2 + \tfrac{1}{2}O_2 \rightarrow H_2O, \qquad K_{p4} = \frac{p_{H_2O}}{p_{H_2}p_{O_2}^{1/2}},$$

and the corresponding relationships in terms of moles may be written from Eq. (2–41):

$$\mathfrak{K}_{n1} = \frac{n_O}{n_{O_2}^{1/2}}, \qquad \mathfrak{K}_{n2} = \frac{n_H}{n_{H_2}^{1/2}}, \qquad \mathfrak{K}_{n3} = \frac{n_{OH}}{n_{O_2}^{1/2} n_{H_2}^{1/2}}, \qquad \mathfrak{K}_{n4} = \frac{n_{H_2O}}{n_{H_2} n_{O_2}^{1/2}}.$$

A procedure for obtaining a solution for the six unknowns is now outlined.

STEP 1: From the equilibrium equations write the number of moles of each *new* species appearing in the product in terms of the reactants and the proper equilibrium constant. In terms of $\mathfrak{K}_n$:

$$n_O = \mathfrak{K}_{n1} n_{O_2}^{1/2}, \qquad\qquad n_H = \mathfrak{K}_{n2} n_{H_2}^{1/2},$$
$$n_{OH} = \mathfrak{K}_{n3} n_{O_2}^{1/2} n_{H_2}^{1/2}, \qquad n_{H_2O} = \mathfrak{K}_{n4} n_{H_2} n_{O_2}^{1/2}. \tag{2–45}$$

STEP 2: Use the equilibrium relationships (2–45) to eliminate the number of moles of new species in one of the atom-conservation equations. Thus, substituting Eq. (2–45) in (2–43), we have

$$2\mathfrak{K}_{n4} n_{H_2} n_{O_2}^{1/2} + 2n_{H_2} + \mathfrak{K}_{n2} n_{H_2}^{1/2} + \mathfrak{K}_{n3} n_{O_2}^{1/2} n_{H_2}^{1/2} = N_H.$$

Then, solving for n_{O_2} in terms of n_{H_2}:

$$n_{O_2} = \left(\frac{N_H - 2n_{H_2} - \mathfrak{K}_{n2} n_{H_2}^{1/2}}{2\mathfrak{K}_{n4} n_{H_2} + \mathfrak{K}_{n3} n_{H_2}^{1/2}} \right)^2. \tag{2–46}$$

STEP 3. Use this expression to eliminate n_{O_2} from Eqs. (2–45).

$$n_O = \mathfrak{K}_{n1} \left(\frac{N_H - 2n_{H_2} - \mathfrak{K}_{n2} n_{H_2}^{1/2}}{2\mathfrak{K}_{n4} n_{H_2} + \mathfrak{K}_{n3} n_{H_2}^{1/2}} \right),$$

$$n_H = \mathfrak{K}_{n2} n_{H_2}^{1/2},$$

$$n_{OH} = \mathfrak{K}_{n3} n_{H_2}^{1/2} \left(\frac{N_H - 2n_{H_2} - \mathfrak{K}_{n2} n_{H_2}^{1/2}}{2\mathfrak{K}_{n4} n_{H_2} + \mathfrak{K}_{n3} n_{H_2}^{1/2}} \right), \tag{2–47}$$

$$n_{H_2O} = \mathfrak{K}_{n4} n_{H_2} \left(\frac{N_H - 2n_{H_2} - \mathfrak{K}_{n2} n_{H_2}^{1/2}}{2\mathfrak{K}_{n4} n_{H_2} + \mathfrak{K}_{n3} n_{H_2}^{1/2}} \right).$$

STEP 4: Use these expressions along with Eq. (2–46) to eliminate all but n_{H_2} from the remaining atom conservation equation (2–44).

$$(\mathfrak{K}_{n4} n_{H_2} + \mathfrak{K}_{n1} + \mathfrak{K}_{n3} n_{H_2}^{1/2}) \left(\frac{N_H - 2n_{H_2} - \mathfrak{K}_{n2} n_{H_2}^{1/2}}{2\mathfrak{K}_{n4} n_{H_2} + \mathfrak{K}_{n3} n_{H_2}^{1/2}} \right)$$

$$+ 2 \left(\frac{N_H - 2n_{H_2} - \mathfrak{K}_{n2} n_{H_2}^{1/2}}{2\mathfrak{K}_{n4} n_{H_2} + \mathfrak{K}_{n3} n_{H_2}^{1/2}} \right)^2 = N_O. \tag{2–48}$$

Equation (2–48) can now be solved for n_{H_2} if the various $\mathfrak{K}_n$ are known. The $\mathfrak{K}_n$'s may be evaluated from Eq. (2–42) for known values of K_p, p_m, and n. The total number of moles in the mixture is actually unknown, but it can be rather easily approximated, and the results are not very sensitive to n anyway. The actual product composition is not far from that resulting from "complete" reaction (unless the temperature is very high) so that n is approximately equal to the number of moles resulting from "complete" reaction. For this example, it is

$$AH_2 + BO_2 \rightarrow AH_2O + (2B - A)O_2$$

if there is excess O_2, or

$$2BH_2O + (A - 2B)H_2$$

if there is excess H_2. Hence n can be readily approximated in terms of the known reactant quantities.

For a numerical example, consider the equilibrium composition of $2H_2 + O_2$ at $3500°K$ (centigrade absolute) and 300 psia. Then $N_H = 4$ and $N_O = 2$. For complete reaction, the number of moles of products would be 2, and we write

$$2H_2 + O_2 \rightarrow 2H_2O.$$

However, since the temperature is rather high, n is assumed to be 2.2. Assumptions of this nature come from experience in this type of calculation; but as we shall see, iterations are easily made and the assumption is not critical. From the known K_p (pressure in atmospheres in this case) and pressure (20.42 atm), and the assumed n, the new equilibrium terms can be calculated. Using Eq. (2–42),

$$\mathfrak{K}_{n1} = K_{p1}\left(\frac{p_m}{n}\right)^{-1/2} = 0.17003, \qquad \mathfrak{K}_{n2} = K_{p2}\left(\frac{p_m}{n}\right)^{-1/2} = 0.19362,$$

$$\mathfrak{K}_{n3} = K_{p3} = 1.3046, \qquad \mathfrak{K}_{n4} = K_{p4}\left(\frac{p_m}{n}\right)^{1/2} = 15.020.$$

These values are then used with various assumed values of n_{H_2} to calculate the left-hand side of Eq. (2–48). The correct value is of course $N_O = 2$. Trial results are tabulated below.

Assumed n_{H_2}	Calculated N_O from Eq. (2–48)
0.2500	2.2561
0.3400	2.9549
0.3000	2.0224
0.3200	1.9917

By linear interpolation between the last two figures, n_{H_2} was found to be $n_{H_2} = 0.3170$. Then Eqs. (2–46) and (2–47) give

$$n_{H_2} = 0.3170, \qquad n_{O_2} = 0.1008, \qquad n_O = 0.05399,$$
$$n_H = 0.1090, \qquad n_{OH} = 0.2333, \qquad n_{H_2O} = 1.5119.$$

Table 2-2   Enthalpies and Heats of Formation for a Given Product Composition

Reactants	n	$\Delta\bar{h}$, kcal/gm-mole	$n\,\Delta\bar{h}$	Q_f, kcal/gm-mole	nQ_f
H_2	2	0	0	0	0
O_2	1	0	0	0	0
			$\overline{0}$		$\overline{0}$
Products					
H_2	0.317	25.693	8.145	0	0
O_2	0.1008	28.263	2.849	0	0
O	0.05399	16.028	0.865	59.162	3.194
H	0.1090	15.897	1.733	52.092	5.678
OH	0.2333	25.907	6.044	10.063	2.348
H_2O	1.5119	36.920	55.819	−57.802	−75.118
			$\overline{75.455}$		$\overline{-63.898}$

Adding these gives us the total number of moles (which was assumed to be 2.2):

$$n = 2.326.$$

The next step is to recalculate n_{H_2}, etc., assuming $n = 2.326$. When we do this, we find no appreciable change in composition. Hence even an inexperienced person could probably complete the calculation in two, or at most three, iterations.

This solution method is not limited to reactions of only two reactants. The generalization of the method for additional reactants is discussed in Reference [3].

In order to take 2 moles of H_2 and 1 mole of O_2 from 300°K and transform them at constant pressure to the above composition at 3500°K, it would be necessary to accompany the reaction by a heat transfer Q which can be determined by Eq. (2–33). If we use Table 2–1 to compute the heat of reaction, we must let $T_f = 300$°K. Since the inlet temperature is also 300°K in this example, $H_{R1} - H_{Rf} = 0$. Table 2–2 shows how the terms $(H_{P2} - H_{Pf})$ and H_{RPf} are computed from enthalpy-temperature information (e.g. Fig. 2–5), heats of formation (Table 2–2), and the given compositions. In the table, n is the number of moles and $\Delta\bar{h}$ the enthalpy per mole of each constituent above the datum temperature T_f. As before, Q_f is the heat of formation of each constituent, per mole. From Table 2–2, the heat of reaction H_{RPf} is −63.898 kcal/gm-mole of oxygen reactant. Also the enthalpy difference $(H_{P2} - H_{Pf})$ is 75.455 kcal/gm-mole of oxygen reactant. Therefore, from Eq. (2–33), the heat transfer necessary to take two moles of H_2 and one mole of O_2 at 300°K and transform them to the product composition of Table 2–2 at 3500°K is:

$$Q = (H_{P2} - H_{f2}) - (H_{R1} - H_{Rf}) + H_{RPf}$$
$$= 75.455 - 0 - 63.898 = 11.6 \text{ kcal/gm-mole·}O_2.$$

References

1. KEENAN, J. H., *Thermodynamics*. New York: John Wiley & Sons, 1941

2. SHAPIRO, ASCHER H., *The Dynamics and Thermodynamics of Compressible Fluid Flow*, Volume I. New York: The Ronald Press Company, 1953; Chapters 1–8 and 16

3. PENNER, S. S., *Chemistry Problems in Jet Propulsion*. New York: Pergamon Press, 1951

4. KEENAN, J. H. and J. KAYE, *Gas Tables*. New York: John Wiley & Sons, 1957

5. ROSSINI, F. D., *et al.*, "Selected Values of Properties of Hydrocarbons." Circular No. 500, National Bureau of Standards, Washington, D. C., Feb. 1, 1952

6. "Selected Values of Chemical Thermodynamic Properties." Series III, National Bureau of Standards, Washington, D. C., April 1, 1954

7. ROSSINI, F. D., *et al.*, "Selected Values of Properties of Hydrocarbons." Circular No. 461, National Bureau of Standards, Washington, D. C., Nov. 1947

8. MARTINEZ, J. S. and G. W. ELVERUM, JR., Memorandum No. 20–121, "A Method of Calculating the Performance of Liquid-Propellant Systems Containing the Species C, H, O, N, F, and One Other Halogen, with Tables Required; Thermochemical Properties to 6000°K." Jet Propulsion Laboratory, California Institute of Technology, Pasadena, California, Dec. 6, 1955

9. RICE, H. E., "Performance Calculations of RFNA-Aniline and Three-Component Systems Using Hydrogen." Memorandum No. 9–2, Jet Propulsion Laboratory, Pasadena, California, June 30, 1947

10. "Heat of Formation of Hydrazine, Unsymmetrical Dimethylhydrazine, and Monomethylhydrazine." Inorganic Research and Development Dept., FMC Corporation, Princeton, N. J., May 14, 1959

11. JOHNSTON, S. A., "A Short-Cut Method for Calculating the Performance of Fuels Containing C, H, O, and N with HNO_3, O_2, or NH_4NO_3." Progress Report No. 20–202. Jet Propulsion Laboratory, Pasadena, California, November 9, 1953

12. CLARK, C. E., "Hydrazine," first edition. Technical Bulletin No. 123.1, Olin Mathieson Chemical Corporation, Baltimore, Maryland, 1953

13. CARTER, J. M., and M. PROTTEAU, "Thermochemical Calculations on the White Fuming Nitric Acid and JP-3 Propellant Combination." Research Technical Memorandum No. 58. Aerojet-General Engineering Corporation, Azusa, California, April 5, 1950

14. PERRY, J. H., *Chemical Engineer's Handbook*, third edition. New York: McGraw-Hill, 1950

15. GAYDON, A. G., *Dissociation Energies and Spectra of Diatomic Molecules*, second edition. London: Chapman and Hall, 1953

Problems

1. A perfect gas of molecular weight $\overline{M} = 20$ and specific heat ratio $\gamma = 1.2$ expands adiabatically from a pressure of 1000 psia, temperature 5000°R, and velocity 1000 fps to a pressure of 14.7 psia.

   (a) If the final temperature is 2600°R and the final velocity negligible, how much work has been done by the gas during expansion?

   (b) If no work is done by or on this gas during expansion to 14.7 psia, what is the maximum possible velocity at the end of the process?

   (c) If any adiabatic expansion process to the same final pressure is permissible, is it possible for the exit temperature to be 300°R? State your reasoning.

2. An inventor suggests a propulsion device shown schematically in the figure. Would such a scheme be feasible for steady operation over long periods of time? Discuss rigorously and briefly.

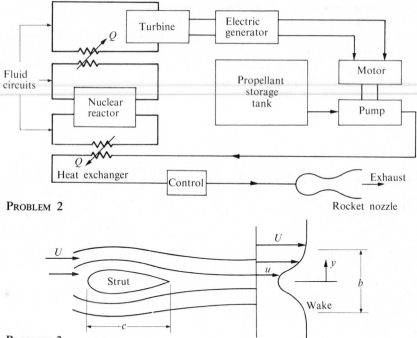

PROBLEM 2

PROBLEM 3

3. A strut of chord c is placed in an incompressible flow which is everywhere uniform except for the illustrated effect of the strut. Velocity measurements at a certain plane far downstream of the strut indicate the presence of a wake, as shown. The velocity distribution across the wake is approximately a cosine function; that is,

$$\frac{u}{U} = 1 - a\left(1 + \cos\frac{2\pi y}{b}\right) \quad \text{for} \quad -\frac{b}{2} < y < \frac{b}{2},$$

where U is the velocity of the uniform flow field far from the strut and its wake. If the measured values of a and b/c are 0.10 and 2.0, respectively, determine the drag coefficient $C_{\mathfrak{D}}$ of the strut. The drag coefficient is defined by $C_{\mathfrak{D}} = \mathfrak{D}/\frac{1}{2}\rho U^2 c$, where $\mathfrak{D}$ is the total drag per unit length on the strut due to pressure and viscous forces, and ρ is the fluid density.

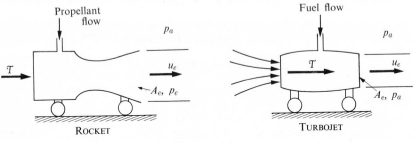

PROBLEM 4

4. Calculate the static thrusts of the rocket and turbojet engines described in the figure. The thrust T is the force necessary to prevent horizontal movement of the engine. Both engines exhaust a mass flow of 120 lb/sec. The ratio of air and fuel mass flowing into the turbojet is 50:1 and in its exhaust plane the velocity is 1500 ft/sec and the pressure is the same as the ambient pressure. The rocket contains all its own propellant and exhausts it at a velocity of 9000 ft/sec through an area of 2 ft². The pressure in the exhaust plane of the rocket is 22.5 psia and ambient pressure is 14.7 psia.

5. An idealized supersonic ramjet diffuser consists of an axisymmetric center body located in a cylindrical duct as shown in the figure. The flow at stations ①and②may be assumed uniform at the values indicated and the stream tube which enters the inlet has a diameter D far upstream.

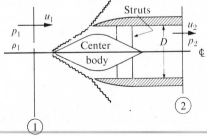

   Show that the aerodynamic drag on the center body and struts is

$$\mathfrak{D} = -\frac{\pi}{4} D^2 [\rho_1 u_1 (u_2 - u_1) + p_2 - p_1].$$ PROBLEM 5

6. A mixture of gases containing 10 lb of nitrogen, 10 lb of hydrogen, and 15 lb of helium is contained at a pressure of 100 psia and temperature 500°R. If the constituents are taken to be perfect gases and the Gibbs-Dalton law holds, what is the molecular weight and specific heat ratio of the mixture?

Gas	Mol wt, $\bar{M}$	Specific heat ratio, γ
Nitrogen	28	1.4
Hydrogen	2	1.4
Helium	4	1.67

7. Calculate the combustion temperature of an oxygen-hydrogen mixture whose mass ratio is 3:1 ($O_2:H_2$). Since this means a considerable excess of hydrogen over the stoichiometric proportion, assume that the product temperature will be low enough to prevent dissociation. The temperature of the reactants entering the chamber is 20°C, and the heat transfer to the coolant from the chamber walls is 2 kcal per mole of product. As a first approximation, the mean specific heats may be taken as in the table below, and the heat of reaction as −57.802 kcal/gm-mole at 300°K (see Table 2–1).

	Average specific heat, cal/gm·mole·°C		
	H_2	O_2	H_2O
Reactants	6.5	7.5	
Products	7.4	8.3	11.0

8. A mixture of air and a hydrocarbon fuel whose average composition is indicated by $CH_{5.4}$ undergoes complete combustion. Twice as much air is present as needed to burn all the fuel. How high is the final temperature? The heats of formation are given in the table below. It may be assumed that air is 79% nitrogen and 21% oxygen by volume. The reactants have a temperature of 77°F before combustion. The adiabatic flame temperature may be found either by using average values of the specific heats, or enthalpy tables for each constituent. A very approximate estimate may be made by simply assuming that the product mixture has the same specific heat as air at the same temperature.

Constituent	Q_f, Btu/lb-mole at 14.7 psia and 77°F
$CH_{5.4}$	−130,000
CO_2 (g)	−169,300
O_2 (g)	0
N_2 (g)	0
H_2O	−104,000

9. Find the flame temperature of the products of combustion of hydrogen and oxygen whose composition and average specific heats are given in the table below. The reactants enter the adiabatic combustion chamber at 0°F. In this table, $\overline{C}_{pp}$ is the average molar specific heat of the component between 77°F and the flame temperature, $\overline{C}_{pr}$ is the average molar specific heat of the component between 0°F and 77°F, and Q_f is the heat of formation at 77°F.

Component	Moles	$\overline{C}_{pp}$, Btu/lb-mole·°R	$\overline{C}_{pr}$	Q_f, Btu/lb-mole
O_2	0.1008	8.8	7.0	0
H_2	0.3170	8.0	7.0	0
O	0.054	5.0		+105,400
H	0.109	5.0		+ 92,910
OH	0.233	8.0		+ 18,000
H_2O	1.512	11.2		−104,000

10. One mole of CO_2 is heated at atmospheric pressure to 5000°F. What is the equilibrium mixture composition if only CO, O_2, and CO_2 are present?

11. An inventor claims that a secret device shown figuratively below can take airstreams ① and ② and convert them into output streams ③ and ④ without energy exchange with its environment. Inside the device are mechanisms for mass, momentum, and energy exchange between the streams. Is it possible that such a device can be made to operate in the steady state? It may be assumed that velocities at all stations are negligible.

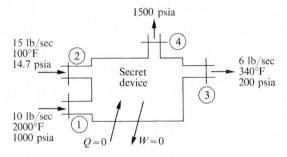

PROBLEM 11

12. Show that if any constant pressure line p_1 is given on a temperature-entropy plane, a line corresponding to any other constant pressure p can be obtained by horizontal shift of magnitude (for a perfect gas):

$$\Delta S = -R \ln \frac{p}{p_1}.$$

13. Two streams of air mix in a constant-area mixing tube. The primary stream enters the mixing tube at station ① with a velocity of 1000 ft/sec and a temperature of 1500°R. The secondary stream enters with velocity 100 ft/sec and temperature 500°R. The flow at stations ① and ② may be assumed one-dimensional. The pressure at station ① is 15 psia and the ratio of primary to secondary flow areas at station ① is 1:3.

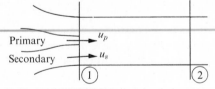

PROBLEM 13

(a) Using continuity, momentum, and energy equations, along with the perfect-gas law, show how the flow at ② may be determined from conditions at ①.
(b) Determine the velocity, temperature, and pressure at station ②.

3

Steady
One-Dimensional Flow
of a Perfect Gas

3-1 INTRODUCTION

In most of the propulsion methods discussed in this volume, the working fluid can be quite satisfactorily approximated by a perfect gas. In this chapter the basic laws stated in the first chapter are applied to perfect gases flowing in channels. For algebraic convenience the specific heat of the gas is usually assumed constant herein.

Useful analyses of actual channel flows are relatively easy if it is assumed that the fluid conditions vary in the streamline direction only. In this case, the flow is said to be one-dimensional. The flow in the immediate vicinity of a wall can never be one-dimensional, since the velocity at the wall surface is zero and there are significant property variations across the streamlines. However, with the exception of a thin layer of fluid next to the wall, fluid conditions are often fairly uniform over a large part of a flow. Even if the bulk of the flow is not uniform on any surface normal to the streamlines, the one-dimensional approximation may still lead to useful solutions for the streamwise variation of average fluid properties. One-dimensional solutions exhibit the major characteristics of many important flows, including those in nozzles, diffusers, and combustors.

3-2 GENERAL ONE-DIMENSIONAL FLOW
OF A PERFECT GAS

Fluid flows are governed by the equations of continuity, momentum, and energy, along with an appropriate equation of state. In this section these equations are applied to a one-dimensional differential control volume within a duct, as indicated in Fig. 3–1. The rates of heat and work transfer are $d\dot{Q}$ and $d\mathcal{P}_s$, respectively. A body force of magnitude X per unit volume of fluid may act in the stream direction.

For steady flow the continuity requirement [from Eq. (2–2)] is

$$\int_{cs} \rho \mathbf{u} \cdot \mathbf{n}\, dA = 0.$$

Since the flow of Fig. 3–1 is assumed one-dimensional, this expression may be reduced to $(d/dx)(\rho u A) = 0$, or

$$\frac{d\rho}{\rho} + \frac{dA}{A} + \frac{du}{u} = 0 \tag{3–1}$$

for the control volume of infinitesimal length indicated in Fig. 3–1.

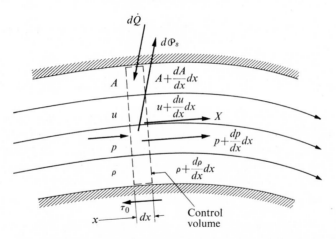

FIG. 3–1.   One-dimensional flow through a control volume.

The momentum equation for steady flow [from Eq. (2–4)] is

$$\sum \mathbf{F} = \int_{cs} \rho \mathbf{u}(\mathbf{u} \cdot \mathbf{n})\, dA.$$

Writing $\sum \mathbf{F}$ as the sum of pressure, wall shear, and body forces, and again considering the flow of Fig. 3–1, this equation can be written

$$-\left(A\, dp + \tau_0 c\, dx \right) + XA\, dx = \rho u A\, du$$

or

$$-\left(\frac{dp}{dx} + \frac{\tau_0 c}{A} \right) + X = \rho u \frac{du}{dx}, \tag{3–2}$$

where τ_0 is the wall shear stress and c is the duct circumference.

For steady flow the energy requirement [from Eq. (2–8)] is

$$d\dot{Q} = \int_{cs} \dot{\mathcal{Q}}\, dA_{cs} = \int_{cs} \left(h + \frac{u^2}{2}\right)\rho\mathbf{u}\cdot\mathbf{n}\,dA - \int_{cv} \mathbf{X}\cdot\mathbf{u}\,d\mho + d\mathcal{P}_s,$$

where A_{cs} indicates the entire control surface area (as opposed to the area A of the duct cross section). The gravitational potential energy term has been neglected, as is usually permissible. Further, except for special cases to be considered in Chapters 5 and 16, the body force $\mathbf{X}$ will be zero. In this case the energy equation reduces to

$$\dot{m}(dh + u\,du) = d\dot{Q} - d\mathcal{P}_s \tag{3–3a}$$

or

$$dh + u\,du = dq - dw, \tag{3–3b}$$

where $\dot{m}$ is the mass flow rate $\rho u A$ (a constant) and q and w are the heat and work transfer per unit mass.

If the fluid can be assumed a perfect gas (as it will be in this chapter), the equation of state is

$$p = \rho R T. \tag{3–4}$$

If the flow area, heat transfer, work, shear stress, and body force are known, changes in the state of the fluid stream may be obtained from these four equations. In general, solutions cannot be expressed in simple algebraic form. There are a number of important special cases, however, which can be integrated in closed form. Before we proceed to these special cases, let us introduce two concepts useful for all cases: the stagnation state and the Mach number.

Stagnation state

The *stagnation state* is defined as that state which would be reached by a fluid if it were brought to rest reversibly, adiabatically, and without work. The energy equation for such a deceleration is obtained from Eq. (3–3b),

$$dh + u\,du = 0,$$

which may be integrated to

$$h_0 \equiv h + \frac{u^2}{2}. \tag{3–5}$$

The constant of integration, h_0, is called the stagnation enthalpy.

For any adiabatic flow viewed from a reference frame in which no work occurs, the stagnation enthalpy is constant. Note particularly that this conclusion holds for both reversible and irreversible flows. Unlike enthalpy, the pressure reached when there is adiabatic zero-work deceleration of a fluid is decreased by irreversible processes. This is illustrated in Fig. 3–2, an enthalpy-entropy diagram for a typical gas. The stagnation state corresponding to state ① is ⓪① and the stagnation pressure is p_{01}, which is the pressure attained by the fluid after it has

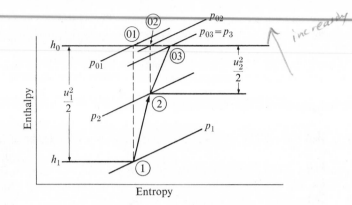

FIG. 3–2. Definition of stagnation state.

been stopped isentropically. If the flow is brought to rest irreversibly, as at ③ (the increased entropy is a measure of the irreversibility), the stagnation enthalpy is the same as before but the pressure attained is less than p_{01}. For any process, such as ① to ②, the stagnation state generally undergoes a change like the one from ⑩ to ⑫. The decrease in pressure from p_{01} to p_{02} is, in the absence of work, an indication of the irreversibility of the process. If the fluid is a perfect gas, the stagnation temperature is related to the stagnation enthalpy through the relation

$$h_0 - h = \int_T^{T_0} c_p\, dT.$$

The stagnation temperature T_0 of the fluid is that temperature which would be reached upon adiabatic, zero-work deceleration. For a perfect gas with constant specific heats the stagnation temperature and pressure are related [using Eq. (2–19)] by

$$\frac{p_0}{p} = \left(\frac{T_0}{T}\right)^{\gamma/(\gamma-1)},\tag{3–6}$$

since the deceleration process is, by definition, isentropic. Other properties of the fluid at the stagnation state may be determined from p_0, T_0, and the equation of state. The stagnation conditions may be regarded as local fluid properties. Aside from analytical convenience, the definition of the stagnation state is useful experimentally since T_0 and p_0 are relatively easily measured. It is usually much more convenient to measure the stagnation temperature T_0 than the temperature T.

Mach number

A very convenient variable in compressible-flow problems is the Mach number M, defined as

$$M = \frac{u}{a},\tag{3–7}$$

where a is the local speed of sound in the fluid. The speed of sound is the speed of propagation of very small pressure disturbances. For a perfect gas it is given by [1]

$$a = \sqrt{\gamma RT}. \tag{3-8}$$

3-3 ISENTROPIC FLOW

Many actual processes, such as flows in nozzles and diffusers, are ideally isentropic. It is, therefore, worthwhile to study the isentropic flow of a perfect gas in the absence of work and body forces. The simple results obtained with the constant specific heat approximation are useful even for large temperature changes, so long as appropriate average values of c_p and γ are used.

For constant specific heat, Eq. (3-5) may be written

$$\frac{T_0}{T} = 1 + \frac{u^2}{2c_p T} \tag{3-9}$$

which, with the definition of Mach number, becomes

$$\frac{T_0}{T} = 1 + \frac{\gamma - 1}{2} M^2. \tag{3-10}$$

Equation (3-6) may then be written

$$\frac{p_0}{p} = \left(1 + \frac{\gamma - 1}{2} M^2\right)^{\gamma/(\gamma-1)}. \tag{3-11}$$

From the perfect gas law and Eqs. (3-10) and (3-11),

$$\frac{\rho_0}{\rho} = \left(1 + \frac{\gamma - 1}{2} M^2\right)^{1/(\gamma-1)}. \tag{3-12}$$

For *any* flow of a perfect gas these can be used to relate the local conditions (T, p, ρ) to the *local* stagnation conditions T_0, p_0, ρ_0, since this relationship is by *definition* isentropic whatever the actual flow considered may be.

The mass flow per unit area is

$$\frac{\dot{m}}{A} = \rho u.$$

Using Eqs. (3-7), (3-8), and (3-10), we may express the velocity as

$$u = M \sqrt{\frac{\gamma R T_0}{1 + \frac{\gamma - 1}{2} M^2}}.$$

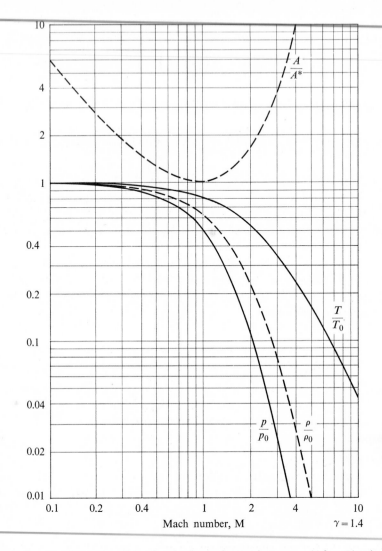

FIG. 3–3. One-dimensional isentropic flow of a perfect gas. (After Shapiro [1].)

Then, using this expression with Eq. (3–12), we may write the mass flow rate as

$$\frac{\dot{m}}{A} = \frac{p_0\sqrt{\gamma}}{\sqrt{RT_0}} M \left(\frac{1}{1 + \dfrac{\gamma - 1}{2} M^2} \right)^{(\gamma+1)/2(\gamma-1)}. \tag{3–13}$$

For a given fluid (γ, R) and inlet state (p_0, T_0), it may readily be shown that the mass flow per unit area is maximum at $M = 1$. If we indicate those properties of the flow at $M = 1$ with an asterisk, the maximum flow per unit area is, from Eq. (3–13),

$$\frac{\dot{m}}{A^*} = \frac{p_0}{\sqrt{RT_0}} \sqrt{\gamma} \left(\frac{2}{\gamma + 1}\right)^{(\gamma+1)/2(\gamma-1)}. \tag{3-14}$$

Combining Eqs. (3–13) and (3–14), we have

$$\frac{A}{A^*} = \frac{1}{M} \left[\frac{2}{\gamma + 1}\left(1 + \frac{\gamma - 1}{2}M^2\right)\right]^{(\gamma+1)/2(\gamma-1)}. \tag{3-15}$$

For a given isentropic flow (given γ, R, p_0, T_0, $\dot{m}$), it is clear that A^* is a constant so that it may be used as in Eq. (3–15) to normalize the actual flow area A. Equations (3–10), (3–11), and (3–12) relate the fluid properties to the Mach number M and Eq. (3–15) shows how the Mach number depends on the flow area. Expressing flow variables in terms of Mach number is a matter of great convenience, since the various flow functions can be plotted or tabulated as a function of M for the appropriate γ. Figure 3–3 gives the one-dimensional isentropic functions for a perfect gas of $\gamma = 1.4$. Tabulations for other γ are given by Shapiro [1], Keenan and Kaye [2], and others.

3–4 FRICTIONLESS CONSTANT-AREA FLOW WITH STAGNATION TEMPERATURE CHANGE

Another special case for which the basic equations may be solved in closed form is frictionless flow in a constant-area duct in which a stagnation enthalpy change occurs. Four processes of interest in which the stagnation temperature of a moving stream changes are:

1. Combustion,
2. Evaporation or condensation of liquid drops traveling with stream,
3. Flux of electric current through a fluid of finite conductivity (Joule heating), and
4. Wall heat transfer.

In any real flow, frictional effects are always present, especially near solid boundaries. As we shall discuss in Chapter 4, wall heat transfer and wall friction are in fact so closely related that it is not realistic to discuss the former without at the same time talking about the latter. Nevertheless, the study of frictionless flow in a constant-area duct in which a T_0 change occurs illustrates some important features of the real flows in which the first three effects above are present. Combustion and evaporation or condensation processes may result in variations of molecular weight and gas constants, but for the sake of simplicity these may be considered negligible. The stagnation-enthalpy change in the processes can be determined from

$$\Delta h_0 = q - w,$$

where q and w are net heat and work transfers per unit mass of fluid. For this case the continuity, momentum, and energy equations are from Eqs. (3–1), (3–2), and (3–3), respectively:

$$\text{Continuity:} \quad \frac{d\rho}{\rho} + \frac{du}{u} = 0,$$

$$\text{Momentum:} \quad dp = -\rho u \, du,$$

$$\text{Energy:} \quad dh_0 = dh + u \, du.$$

The first two of these can be readily integrated to

$$\rho u = \rho_1 u_1,$$

$$p - p_1 = -(\rho u)(u - u_1) = -(\rho_1 u_1)(u - u_1),$$

in which the subscript 1 signifies an initial condition. The second of these may also be expressed as

$$p - p_1 = \rho_1 u_1^2 - \rho u^2.$$

By introducing the Mach number, we can transform this to

$$\frac{p}{p_1} = \frac{1 + \gamma M_1^2}{1 + \gamma M^2}.$$

Further, by use of Eq. (3–11), we can derive the stagnation pressure ratio from this equation, with the result that

$$\frac{p_0}{p_{01}} = \left(\frac{1 + \gamma M_1^2}{1 + \gamma M^2}\right) \left(\frac{1 + \dfrac{\gamma - 1}{2} M^2}{1 + \dfrac{\gamma - 1}{2} M_1^2}\right)^{\gamma/(\gamma - 1)}. \tag{3–16}$$

Thus the change in stagnation pressure is directly related to the Mach number.

The dependence of Mach number on stagnation enthalpy (or stagnation temperature) can in turn be obtained as follows: Using the perfect-gas law,

$$\frac{T}{T_1} = \frac{p}{p_1} \frac{\rho_1}{\rho},$$

or invoking the continuity condition, $\rho u = \rho_1 u_1$, it may be seen that

$$\frac{T}{T_1} = \frac{p}{p_1} \frac{u}{u_1}.$$

Now if the Mach number relation $M = u/\sqrt{\gamma R T}$ is introduced, it can easily be shown that

$$\frac{T}{T_1} = \left(\frac{p}{p_1} \frac{M}{M_1}\right)^2$$

and, using the above relation between static pressure ratio and Mach number,

$$\frac{T}{T_1} = \left[\frac{1 + \gamma M_1^2}{1 + \gamma M^2}\left(\frac{M}{M_1}\right)\right]^2.$$

From Eq. (3–10) and this expression, we can show that the stagnation temperature ratio is governed by

$$\frac{T_0}{T_{01}} = \left[\frac{1 + \gamma M_1^2}{1 + \gamma M^2}\left(\frac{M}{M_1}\right)\right]^2\left(\frac{1 + \dfrac{\gamma - 1}{2}M^2}{1 + \dfrac{\gamma - 1}{2}M_1^2}\right). \tag{3–17}$$

Thus the Mach number depends uniquely on the ratio of stagnation temperatures and the initial Mach number.

It is desirable in this case, as in the isentropic case, to simplify these relationships by the choice of a convenient reference state. Since stagnation conditions are *not* constant, the stagnation state is not suitable for this purpose. However, the state corresponding to unity Mach number is suitable since, as Eqs. (3–16) and (3–17) show, conditions there *are* constant for a given flow (i.e., for a given p_{01}, T_{01}, M_1). Again using an asterisk to signify properties at $M = 1$, it can be shown from the equations above that

$$\frac{T}{T^*} = \left(\frac{1 + \gamma}{1 + \gamma M^2}\right)^2 M^2, \tag{3–20}$$

$$\frac{T_0}{T_0^*} = \frac{2(\gamma + 1)M^2\left(1 + \dfrac{\gamma - 1}{2}M^2\right)}{(1 + \gamma M^2)^2}, \tag{3–21}$$

$$\frac{p}{p^*} = \frac{1 + \gamma}{1 + \gamma M^2}, \tag{3–22}$$

$$\frac{p_0}{p_0^*} = \left(\frac{2}{\gamma + 1}\right)^{\gamma/(\gamma-1)}\left(\frac{1 + \gamma}{1 + \gamma M^2}\right)\left(1 + \frac{\gamma - 1}{2}M^2\right)^{\gamma/(\gamma-1)}. \tag{3–23}$$

In this way the fluid properties may be presented as a function of a single argument, the local Mach number. Equations (3–20) through (3–23) are shown graphically in Fig. 3–4 for the case $\gamma = 1.4$. Given particular entrance conditions, T_{01}, p_{01}, M_1, the exit conditions after a given change in stagnation temperature may be obtained as follows: The value M_1 fixes the value of T_{01}/T_0^* and thus the value T_0^*, since T_{01} is known. The exit state is then fixed by T_{02}/T_0^*, determined by

$$\frac{T_{02}}{T_0^*} = \frac{T_{01}}{T_0^*} + \frac{\Delta T_0}{T_0^*} \quad \text{or} \quad \frac{T_{02}}{T_0^*} = \frac{T_{01}}{T_0^*} + \frac{q - w}{c_p T_0^*}.$$

Then $M_2, p_2/p^*, p_{02}/p_0^*$ are all fixed by the value of T_{02}/T_0^* (for a given value of γ).

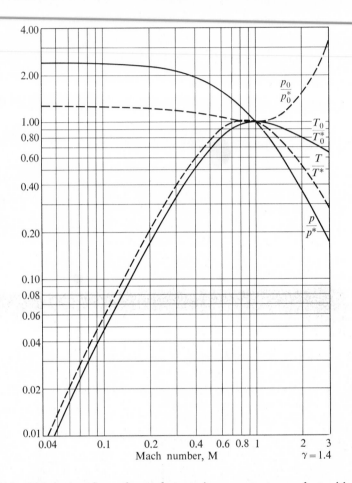

$$\frac{p_0}{p_0^*}$$

$$\frac{T_0}{T_0^*}$$

$$\frac{T}{T^*}$$

$$\frac{p}{p^*}$$

Mach number, M

$\gamma = 1.4$

FIG. 3–4. Frictionless flow of a perfect gas in a constant-area duct with stagnation temperature change. (After Shapiro [1].)

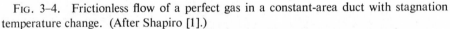

It may be seen from Fig. 3–4 that increasing the stagnation temperature drives the Mach number toward unity whether the flow is supersonic or subsonic. After the Mach number has approached unity in a given duct, further increase in stagnation enthalpy is possible only if the initial conditions change. The flow may be said to be thermally choked.

It is interesting to note the variations in stagnation pressure as the stream is subjected to energy transfer. The stagnation pressure always drops when energy is added to the stream and rises when energy is transferred from the stream. Thus, whether the flow is subsonic or supersonic, there may be significant loss of stagnation pressure due to combustion in a moving stream. Conversely, cooling tends to increase the stagnation pressure.

3–5 CONSTANT-AREA FLOW WITH FRICTION

Another solution of Eqs. (3–1) through (3–4) can be obtained for constant-area adiabatic flow with friction but no body force or work. In this case, from Eqs. (3–1), (3–2), and (3–3) the continuity, momentum, and energy relations are, respectively,

$$\text{Continuity:} \qquad \frac{d\rho}{\rho} + \frac{du}{u} = 0,$$

$$\text{Momentum:} \qquad \frac{dp}{p} = -\frac{\tau_0 c\, dx}{pA} - \frac{\rho u\, du}{p},$$

$$\text{Energy:} \qquad 0 = dh + u\, du.$$

For flow in long pipes it will be shown (Fig. 4–14) that the wall shear stresses τ_0 can be correlated as follows:

$$\frac{\tau_0}{\rho u^2/2} = f\left(\frac{\rho u D}{\mu}, \frac{\Re}{D}\right)$$

or

$$C_f = f\left(\text{Re}, \frac{\Re}{D}\right),$$

in which C_f is the skin friction coefficient $\tau_0/(\rho u^2/2)$, Re is the Reynolds number $\rho u D/\mu$, $\Re$ is the average height of surface roughness elements, and D is the duct diameter. For Mach numbers greater than unity, the Mach number is also a significant variable. The effects of roughness (or at least of variations in roughness) are usually fairly unimportant for surfaces which have a reasonably smooth finish. This point is discussed in more detail in Chapter 4.

The way in which friction affects the Mach number may be shown as follows. First the conservation equations are transformed (by introducing the perfect gas law and the definition of Mach number) to read:

$$\text{Continuity:} \qquad \frac{dp}{p} - \frac{dT}{T} + \frac{1}{2}\frac{du^2}{u^2} = 0,$$

$$\text{Momentum:} \qquad \frac{dp}{p} = \frac{-\gamma M^2}{2}\left(4\frac{C_f\, dx}{D}\right) - \frac{u\, du}{RT},$$

$$\text{Energy:} \qquad 0 = \frac{dT}{T} + \frac{\gamma-1}{\gamma}\frac{u\, du}{RT}.$$

Since

$$\text{M} = \frac{u}{\sqrt{\gamma RT}},$$

then

$$\frac{d\text{M}^2}{\text{M}^2} = \frac{du^2}{u^2} - \frac{dT}{T}.$$

And now, if we use the continuity equation to eliminate du^2/u^2, it follows that

$$\frac{d\mathrm{M}^2}{\mathrm{M}^2} = -2\frac{dp}{p} + \frac{dT}{T}.$$

Also, combining the momentum and energy equations, we can easily show that

$$\frac{dp}{p} = \frac{-\gamma \mathrm{M}^2}{2}\left(4\frac{C_f\,dx}{D}\right) + \frac{\gamma}{\gamma - 1}\frac{dT}{T}.$$

Combining these last two expressions yields

$$\frac{d\mathrm{M}^2}{\mathrm{M}^2} = \gamma \mathrm{M}^2\left(4C_f\frac{dx}{D}\right) - \frac{\gamma + 1}{\gamma - 1}\frac{dT}{T}$$

which, for constant stagnation temperature [using Eq. (3–10)], becomes†

$$\frac{d\mathrm{M}^2}{\mathrm{M}^2} = \frac{\gamma \mathrm{M}^2\left(1 + \dfrac{\gamma - 1}{2}\mathrm{M}^2\right)}{1 - \mathrm{M}^2}\left(\frac{4C_f\,dx}{D}\right). \tag{3–24}$$

Equation (3–24) shows that $d\mathrm{M}^2 > 0$ for $\mathrm{M} < 1$ and $d\mathrm{M}^2 < 0$ for $\mathrm{M} > 1$. That is, friction *always* changes the Mach number toward unity. For constant C_f, Eq. (3–24) can be integrated between the limits $\mathrm{M} = \mathrm{M}$ and $\mathrm{M} = 1$ to yield

$$\frac{4C_fL^*}{D} = \frac{1 - \mathrm{M}^2}{\gamma \mathrm{M}^2} + \frac{\gamma + 1}{2\gamma}\ln\frac{(\gamma + 1)\mathrm{M}^2}{2\left(1 + \dfrac{\gamma - 1}{2}\mathrm{M}^2\right)}, \tag{3–25}$$

in which L^* is the length of duct necessary to change the Mach number of the flow from M to unity.

Consider a duct of given cross-sectional area and variable length. If the inlet state, mass flow rate, and average skin friction coefficient are fixed, there is a maximum length of duct which can transmit the flow. Since the Mach number is unity at the exhaust plane in that case, the length is designated L^* and the flow may be said to be friction-choked. From Eq. (3–25), we can see that at any point 1 in the duct, the variable C_fL^*/D depends only on M_1 and γ. Since D is constant and C_f is assumed constant, then at some other point a distance x ($x < L^*$) downstream from point 1,

$$\left(\frac{4C_fL^*}{D}\right)_x = \left(\frac{4C_fL^*}{D}\right)_1 - \frac{4C_fx}{D}.$$

From this we can determine M_x. Other relationships which may be derived [1]

---

† For noncircular ducts, the hydraulic diameter $D_{II} = 4A/c$ can be used as an approximation in place of the diameter D.

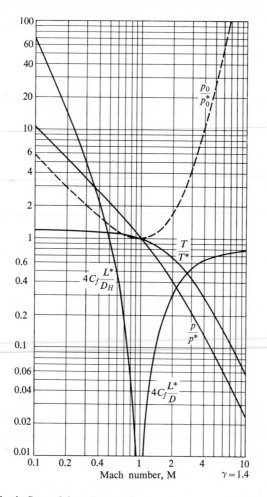

FIG. 3–5. Adiabatic flow of a perfect gas in a constant-area duct with friction. (After Shapiro [1].)

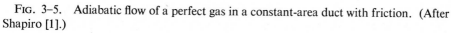

by using the definition of Mach number and the continuity and energy relations are:

$$\frac{T}{T^*} = \frac{\gamma + 1}{2\left(1 + \dfrac{\gamma - 1}{2}M^2\right)}, \tag{3-26}$$

$$\frac{p}{p^*} = \frac{1}{M}\left[\frac{\gamma + 1}{2\left(1 + \dfrac{\gamma - 1}{2}M^2\right)}\right]^{1/2}, \tag{3-27}$$

$$\frac{p_0}{p_0^*} = \frac{1}{M}\left[\frac{2}{\gamma + 1}\left(1 + \frac{\gamma - 1}{2}M^2\right)\right]^{(\gamma+1)/2(\gamma-1)}. \tag{3-28}$$

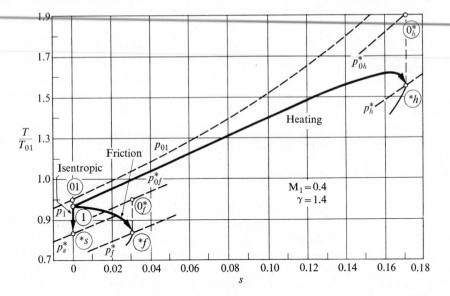

FIG. 3–6. Typical temperature-entropy diagram for choking process (subsonic).

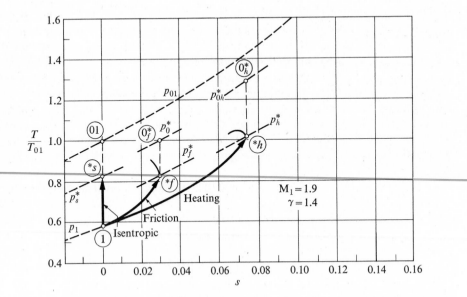

FIG. 3–7. Typical temperature-entropy diagram for choking process (supersonic).

Equations (3–26), (3–27), and (3–28) are plotted in Fig. 3–5 for $\gamma = 1.4$. Again the asterisk denotes properties at that point where $M = 1$.

In Sections 3–3, 3–4, and 3–5, three special problems have been treated to show the separate effects of area change, energy transfer, and friction. In each case, reference was made to a state where $M = 1$ and properties at that point were distinguished by an asterisk. Note that properties at that point are used as convenient normalizing quantities only, and the state need not actually exist in the real flow studied.

For a given inlet condition, the $M = 1$ conditions represent *three distinctly different states*. This is clearly indicated in Figs. 3–6 and 3–7. For two initial states, one subsonic and the other supersonic, three processes to reach a Mach number of one are shown on a typical T-s diagram. The energy-transfer path is called a Rayleigh line and the friction path is called a Fanno line. On both these lines the * condition occurs on the nose of the curve. The notation *s, *f, *h indicates the $M = 1$ states reached by isentropic, friction, and heating processes, respectively. Subsonic states are above the nose of the appropriate curve, while supersonic states are below it. Note also that travel in either direction is possible along the isentropic and energy-transfer curves, while only one direction is possible on the friction curve.

If it is important to study combined effects of these processes, then simple solutions are not available. However, the complete equations may be integrated numerically according to procedures described by Shapiro [1].

3–6 SHOCKS

A shock is a discontinuity in a (partly) supersonic flow fluid. Fluid crossing a stationary shock front rises suddenly and irreversibly in pressure and decreases in velocity. It also changes its direction except when passing through a shock which is perpendicular (normal) to the approaching flow direction. Such plane normal shocks are the easiest to analyze.

Normal shocks

For flow through a normal shock, with no direction change, area change, or work done, the continuity, momentum, and energy equations are

$$\text{Continuity:} \qquad \rho_1 u_1 = \rho_2 u_2,$$

$$\text{Momentum:} \qquad p_1 - p_2 = \rho_1 u_1 (u_2 - u_1),$$

$$\text{Energy:} \qquad T_{01} = T_{02},$$

where the subscripts 1 and 2 indicate initial and final states, respectively. Two solutions to these equations are possible, one of which states simply that no change occurs. In terms of Mach number, the interesting solution may be stated [1] that

$$M_2^2 = \frac{M_1^2 + \dfrac{2}{\gamma - 1}}{\dfrac{2\gamma}{\gamma - 1} M_1^2 - 1},$$

$$\frac{p_2}{p_1} = \frac{2\gamma}{\gamma + 1} M_1^2 - \frac{\gamma - 1}{\gamma + 1},$$

$$\frac{p_{02}}{p_{01}} = \frac{\left[\left(\dfrac{\gamma + 1}{2} M_1^2\right)\Big/\left(1 + \dfrac{\gamma - 1}{2} M_1^2\right)\right]^{\gamma/(\gamma-1)}}{\left(\dfrac{2\gamma}{\gamma + 1} M_1^2 - \dfrac{\gamma - 1}{\gamma + 1}\right)^{1/(\gamma-1)}},$$

$$\frac{T_2}{T_1} = \frac{\left(1 + \dfrac{\gamma - 1}{2} M_1^2\right)\left(\dfrac{2\gamma}{\gamma - 1} M_1^2 - 1\right)}{\dfrac{(\gamma + 1)^2}{2(\gamma - 1)} M_1^2}.$$

In Fig. 3–8, M_2 and p_{02}/p_{01} are plotted for the case $\gamma = 1.4$. Note especially that rather large losses of stagnation pressure occur across Mach numbers greater than about 1.5. Tables of all the functions for various γ are provided by Keenan and Kaye [2].

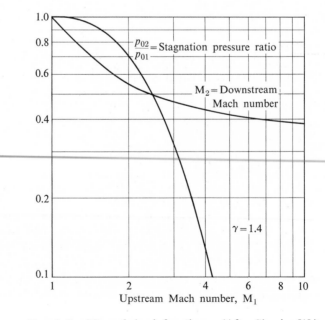

FIG. 3–8. Normal shock functions. (After Shapiro [1].)

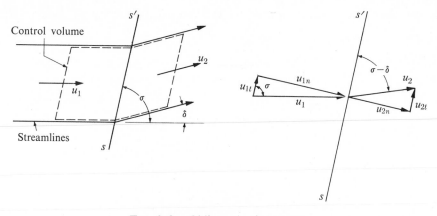

FIG. 3–9. Oblique shock geometry.

Oblique shocks

If a plane shock is inclined at an angle to the flow, the fluid passing through suffers not only a sudden rise in pressure and decrease in speed but also a sudden change of direction. The situation is illustrated in Fig. 3–9 for an oblique shock s-s' in one-dimensional flow. In passing through the shock the fluid is deflected through an angle δ. The basic equations applied to the indicated control volume are:

$$\text{Continuity:} \quad \rho_1 u_{1n} = \rho_2 u_{2n},$$
$$\text{Energy:} \quad T_{01} = T_{02}.$$

Two momentum equations may be written for this flow, one for changes in momentum perpendicular to the shock, the other for changes parallel to it:

(a) Momentum normal to the shock, $p_1 - p_2 = \rho_2 u_{2n}^2 - \rho_1 u_{1n}^2,$

(b) Momentum parallel to the shock, $0 = \rho_1 u_{1n}(u_{2t} - u_{1t}),$

in which the subscripts n and t indicate directions normal to and parallel to the shock, respectively. From (b),

$$u_{2t} = u_{1t}.$$

Since the velocity component parallel to the shock is the same on both sides of it, it may be seen that an oblique shock becomes a normal shock relative to a coordinate system moving with velocity $u_{1t} = u_{2t}$. This fact may permit us to use the normal shock equations to calculate oblique shocks. For example, given the upstream (or initial) Mach number M_1 and the overall stagnation pressure ratio, the components M_{1n} and M_{2n} could be determined from Fig. 3–8. The pressure and temperature ratios p_2/p_1 and T_2/T_1 could then be obtained by replacing M_1 with M_{1n} in the appropriate normal shock equations. The shock angle σ could

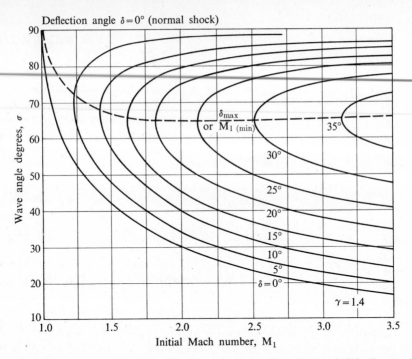

FIG. 3–10. Shock angle versus inlet Mach number and turning angle [1]. Curves above dashed line hold for $M_2 < 1$ and curves below hold for $M_2 > 1$. (After Shapiro[1].)

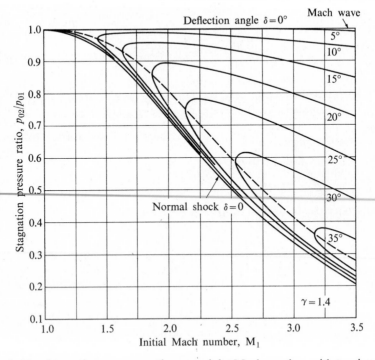

FIG. 3–11. Stagnation pressure ratio versus inlet Mach number, with turning angle as parameter [1]. Curves above dashed line hold for $M_2 > 1$ and curves below hold for $M_2 < 1$. (After Shapiro [1].)

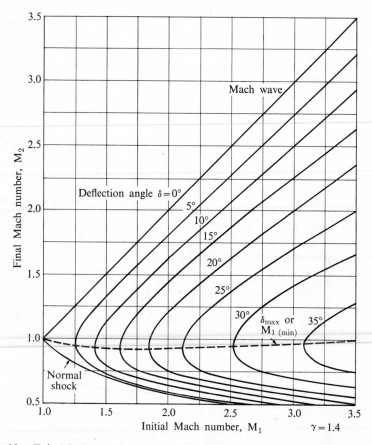

FIG. 3–12. Exit Mach number versus inlet Mach number, with turning angle as parameter [1]. Curves above dashed line correspond to small σ and curves below correspond to large σ. (After Shapiro [1].)

be determined from M_{1n} and M_1 and the downstream Mach number from M_{2n}, T_2/T_1 and the condition $u_{1t} = u_{2t}$.

Alternatively, the above four equations, along with the equation of state, can be reduced to a set of four independent equations relating the variables M_1, M_2, p_2/p_1, ρ_2/ρ_1, σ, and δ [1]. If any two variables are given, for example M_1 and the turning angle δ, then all others are determined and the downstream (or final) conditions can be found in terms of the upstream conditions.

The solution of these equations from Shapiro [1] is given in Figs. 3–10, 3–11 and 3–12 for the case $\gamma = 1.4$ and for various values of δ.

Axisymmetric oblique shocks are discussed by Shapiro [3], who presents a complete solution for the case of the conical shock as well as methods of treatment of the general axisymmetric shock problem.

References

1. SHAPIRO, ASCHER H., *The Dynamics and Thermodynamics of Compressible Fluid Flow*, Volume I. New York: The Ronald Press Company, 1953; Chapters 1–8, 16

2. KEENAN, J. H., and J. KAYE, *Gas Tables*. New York: John Wiley & Sons, 1957

3. SHAPIRO, ASCHER H., *op. cit.*, Volume II, Chapter 17

Problems

1. Show that for a pure substance the stagnation pressure cannot increase for an adiabatic zero-work process. During what physical processes might it increase for a zero-work process?

2. Show that, for a perfect gas, the Mach number at the nose of the Fanno and Rayleigh lines is indeed unity. The nose is defined by $ds/dT = 0$.

3. A perfect fluid expands in a frictionless nozzle from stagnation conditions $p_0 = 600$ psia, $T_0 = 5000°R$, to ambient pressure of 15 psia. If the expansion is isentropic, determine the following conditions at the final pressure: (1) velocity, (2) Mach number, (3) temperature, and (4) area per unit mass flow. How does the final flow area compare with the throat area for a given mass flow? The specific heat ratio γ is 1.4 and the molecular weight $\overline{M}$ is 30.

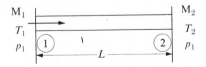

PROBLEM 4

4. Air flows through a cylindrical combustion chamber of diameter 1 ft and length 10 ft. Stations ① and ② are the inlet and outlet, respectively. The inlet stagnation temperature is 500°R and the inlet stagnation pressure is 200 psia.

   (a) If the skin friction coefficient is $C_f = 0.0040$ (approximately uniform) and no combustion takes place, what will the inlet Mach number be if it is subsonic and the exit Mach number is unity? What are the values of static and stagnation pressures at the exit?

   (b) If combustion takes place and the heat of reaction is 300 Btu per pound of mixture, neglecting frictional effects, and if the inlet Mach number is 0.25, what is the exit Mach number?

   (c) If, in a similar problem, the effects of combustion and skin friction on the flow are of the same order of magnitude, how might their simultaneous effects be estimated? The fluid may be considered a perfect gas with $\gamma = 1.4$ and $\overline{M} = 29$.

5. A Mach 2 flow passes through an oblique shock as shown, and deflects 10°. A second oblique shock reflects from the solid wall. What is the pressure ratio across the two-shock system? It may be assumed that there is no boundary layer near the wall; i.e., the flow is uniform in each of the regions bounded by the shocks.

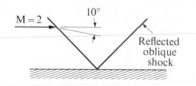

PROBLEM 5

6. Air enters a constant-area duct at Mach 3 and stagnation conditions 1300°R and 185 psia. In the duct it undergoes a frictionless energy-transfer process such that the exit Mach number is unity. Consider two cases: (a) normal shock at inlet to the duct, and (b) shockfree supersonic heating. Determine the stagnation temperature and pressure at the exit in each case. Is there any reason why the total energy transfer should differ (or be equal) in the two cases?

7. A uniform mixture of very small solid particles and a perfect gas expands adiabatically from a given chamber temperature through a given pressure ratio in a nozzle. The particles are so small that they may be considered to travel with the local gas speed. Since the density of the solid material is much higher than the gas phase, the total volume of the particles is negligible. The ratio of solid to gas flow rates is μ, where $0 < \mu < 1$. Consider two cases:

(a) no heat transfer between solid and gaseous phases, and (b) the solid particles have the local gas temperature at all points in the flow. Qualitatively, how does the presence of the solid particles affect the velocity at the end of the expansion? Show that for both cases the mixture is equivalent to the adiabatic expansion of a homogeneous gas of different molecular weight $\bar{M}$ and specific heat ratio γ.

8. A 30-ft^3 rocket-propellant chamber is filled with combustion gases at 6000°R and 1000 psia at the instant combustion ceases. If the throat area of the nozzle is 1 ft^2, estimate the time it takes for the chamber pressure to drop to 100 psia. Assume that at the instant combustion ceases the chamber propellant supply stops completely, and also that $\gamma = 1.2$ and $\bar{M} = 20$.

9.   (a) Compare the work of compression per unit mass of air for both reversible adiabatic and reversible isothermal compression through a pressure ratio of 10 with initial conditions 500°R and 14.7 psia. Is the importance of specific heat variation large? Why should cooling during compression reduce the work?

(b) In an effort to provide a continuous cooling process during compression, water is sprayed into an airstream entering the axial compressor of a gas turbine engine. The compression may be taken to be reversible and the evaporation rate is approximately $dw/dT = k$, where w is the water-air mass ratio, T is the mixture temperature, and k is a constant. If $w \ll 1$, the process may be simplified by considering the sole effect of the evaporation to be energy transferred from the gas phase. Show how the final temperature and the work of compression depend on k.

4

Boundary Layer
Mechanics and
Heat Transfer

4-1 THE VISCOUS BOUNDARY LAYER

Historically the development of boundary layer theory followed the development of mathematical solutions for flows assumed to be nonviscous. In the latter half of the nineteenth century—after Euler had formulated the basic equations of motion—Helmholtz, Kelvin, Lamb, Rayleigh, and others applied them to a number of physical situations. By neglecting the effects of viscosity they obtained elegant mathematical descriptions of various flow fields. It was argued that a fluid like air, for example, which is so rarefied as to be invisible, must have negligible viscosity. However, with some exceptions, the so-called perfect-fluid theory disagreed strongly with the results of experiment.

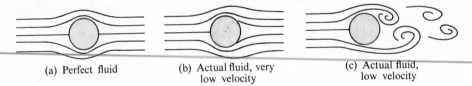

(a) Perfect fluid   (b) Actual fluid, very   (c) Actual fluid,
                               low velocity   low velocity

FIG. 4-1.   Flow patterns about a cylinder.

One particular example, which came to be known as *d'Alembert's paradox*, attracted a great deal of attention. As an illustration, the flow of a fluid over a cylinder as predicted by nonviscous theory is shown in Fig. 4-1(a). The theory predicts that the velocity and pressure fields are both symmetrical about a plane normal to the upstream velocity vectors. Thus the fluid can exert no net forces (lift or drag) on the cylinder. Moreover, this conclusion is independent of the shape of the body; if the fluid is truly inviscid, it can be shown mathematically

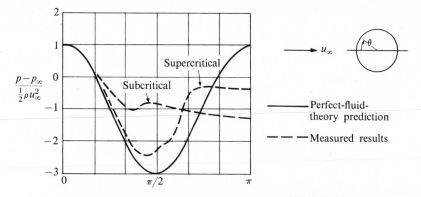

FIG. 4–2. Pressure distribution on a cylinder. (After Flachsbart [11].)

that for any body shape the net integrated pressure force upon the body will be zero. Of course, this conclusion directly contradicts experience. Everyone knows that it takes a finite force to hold a body immersed in a moving stream or to propel it through a stationary mass of fluid; hence the paradox.

Observations of the actual streamline pattern around the cylinder reveal something like Fig. 4–1(b), which suggests that the streamline symmetry is grossly disturbed. On the rear side of the cylinder the fluid seems almost stagnant when the stream velocity is quite low. For higher speeds, fairly violent oscillations take place, which affect the entire streamline pattern near the cylinder. Under certain conditions it is possible to observe a periodic shedding of vortices alternately from the top and bottom sides, as suggested by Fig. 4–1(c). This downstream flow pattern has been called a *vortex street*.

The drag of the actual fluid on the cylinder is largely due to the asymmetry of the pressure distribution on its surface. Figure 4–2 shows measured pressure distributions on a cylinder and, for comparison, the prediction of perfect-fluid theory. The symbol ρ stands for fluid density and the pressure and velocity far upstream are denoted by p_∞ and u_∞. It can be seen that two experimental pressure distributions may be measured (the difference between the two will be discussed later) and that both differ greatly from the theoretical result. The fact that the average pressure on the cylinder is much higher on the upstream hemi-surface explains most of the drag. In addition to this, the viscous force of the fluid on the cylinder imposes a drag force which, in this particular case, is considerably smaller than the pressure drag. For "streamlined" bodies, which have much less pressure drag, the viscous force is relatively more important.

For quite a period the discrepancy between existing theory and physical results was unexplained, and a marked rift developed between the theoreticians, who continued to expound the perfect-fluid theory, and the engineering-oriented workers like Bernoulli and Hagen, who actively developed the empirical body of knowledge generally called hydraulics. The equations of motion which do take viscosity into account—the Navier-Stokes equations—were formulated but were so very complicated that they could not, in general, be solved. It appeared exceedingly difficult to improve mathematically on the perfect-fluid theory except in a few very special cases.

Then Prandtl (in 1904) made a major contribution to the solution of the problem by introducing the concept of the viscous boundary layer. He showed that even for fluids of vanishingly small viscosity there is a thin region near the wall in which viscous effects cannot be neglected. Since the velocity of the fluid particles on the wall is zero, there may be, under certain conditions, a region of large velocity gradient next to it. Then, even though the viscosity μ may be small, the shear stress on a fluid layer, $\tau = \mu(\partial u/\partial y)$, may be large in that region since $\partial u/\partial y$ is large. This region is called the viscous boundary layer. A boundary layer of thickness δ is indicated in Fig. 4–3. Outside the layer, the velocity gradient $\partial u/\partial y$ will be so small that viscous shear may be quite unimportant, so that the fluid behaves as though it actually had zero viscosity.

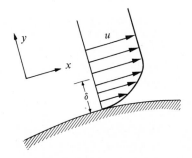

Fig. 4–3. Boundary layer velocity distribution.

Thus Prandtl showed that the flow field may be broken into two parts; a thin viscous zone near the wall, and an outer zone where the nonviscous theory is adequate. This suggested the major simplification of the Navier-Stokes equations which is the basis of boundary layer theory. For fairly high-speed flow, in which the boundary layer is so thin that the pressure gradient normal to the wall is negligible, Prandtl and others showed how to treat the boundary layer mathematically for special cases, many of which have been solved in the past sixty years.

Given that a boundary layer exists on a body, it is easy to see qualitatively why the case of vanishing viscosity is fundamentally different from the case of zero viscosity. Figure 4–2 shows that for zero viscosity a particle on the surface streamline of the cylinder rises in pressure to the stagnation value when it impinges upon the cylinder ($\theta = 0$). Then the pressure falls to a minimum (and the velocity rises to a maximum) at the position $\theta = \pi/2$. From that point it travels to the rearward stagnation point ($\theta = \pi$), having just enough momentum to climb the "pressure hill" and arrive at the stagnation point.

However small the viscosity, the viscous case is fundamentally different. The fluid near the wall has been slowed by viscous action. When a particle in the boundary layer reaches $\theta = \pi/2$ it will have reduced momentum so that only a small pressure rise will stop its forward motion, or actually send it moving back-

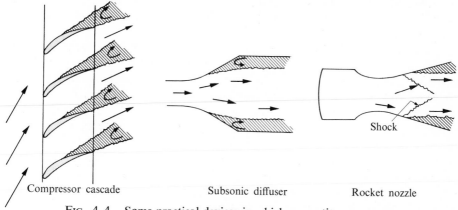

Compressor cascade   Subsonic diffuser   Rocket nozzle

Fig. 4–4. Some practical devices in which separation may occur.

ward. For this reason the particle can no longer follow the contour of the wall. The accumulation of stagnant fluid on the back part of the cylinder deflects the outer streamlines, as suggested in Fig. 4–1(b). This in turn greatly modifies the wall pressure distribution, as shown in Fig. 4–2. When the streamlines in the vicinity of the wall cease to follow it, they are said to *separate*.

This accumulation of stagnant or near-stagnant fluid and the resultant gross distortion of the streamlines means that the boundary layer behavior can have an important influence on overall flow behavior even though the quantity of fluid directly affected by viscosity is a small fraction of the total flow. Three separated flows of practical significance are shown in Fig. 4–4. Both the compressor cascade and the subsonic diffuser are designed so that the mainstream velocity decreases in the flow direction; hence the flow is against an adverse pressure gradient. If the pressure rise is too great the boundary layer will separate, creating regions of near-stagnant fluid as shown. Within rocket nozzles the pressure gradient is usually favorable, so that separation need not be a problem. However, under certain off-design conditions shocks can occur within rocket nozzles and the sudden pressure rise across the shock can cause separation.

Separation and its consequences in these devices will be discussed in the following chapters where appropriate. Our purpose in this chapter is to establish the nature and cause of separation and to indicate how one might predict its occurrence (although this is seldom possible with purely theoretical methods). Separation is, of course, only one phase of boundary layer behavior. We shall be interested also in the behavior of non-separated boundary layers, since they are of importance to heat transfer and other phenomena.

The classic picture of separation is illustrated in Fig. 4–5, which shows a series of velocity profiles in a boundary layer as it approaches and passes the separation point. As the pressure rises the free-stream velocity U falls, as the Bernoulli equation would predict. However, the effect of a given change in pressure is greater on the slow-moving fluid near the wall than on the free-stream fluid.

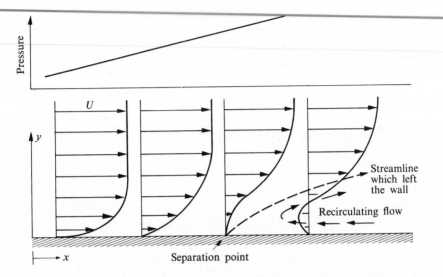

FIG. 4–5.   A simplified picture of the development of separation.

The fluid near the wall can be slowed to a stagnant state and then be made to flow backward if the adverse pressure gradient continues.

The separation point is generally defined as that point where a streamline very near the wall leaves the wall. In two-dimensional flow this corresponds to the point at which the velocity profile has zero slope at the wall. That is,

$$\left(\frac{\partial u}{\partial y}\right)_{y=0} = 0 \quad \text{(two-dimensional separation)}.$$

Figure 4–5 suggests that, past the separation point, there is a region of recirculating flow. In practice this region is highly unstable. It may be quite difficult to determine analytically, since once separation occurs there is a strong interaction between the body of stagnant fluid and the free stream. Often there results an adjustment of the free-stream geometry such that the adverse pressure gradient is reduced to practically zero, in which case there is no force to drive the recirculating fluid.

The effects of *pressure gradient* on the shape of the boundary layer, which are shown in Fig. 4–5, may be explained in a simplified manner as follows. Let the boundary layer be replaced by a hypothetical discontinuous flow shown in Fig. 4–6, in which the continuous variation in velocity normal to the wall is replaced by two step changes. Further, in order to focus on the effect of pressure gradient, let us for the moment neglect shear stresses. (The effects of shear stresses will be dis-

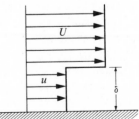

FIG. 4–6.   Discontinuous boundary layer.

cussed subsequently.) With these assumptions, the momentum equation requires that

$$\frac{dp}{\rho} = -U\,dU \text{ (free stream)}, \quad \frac{dp}{\rho} = -u\,du \text{ (boundary layer)},$$

where, in keeping with boundary layer assumptions, the free stream and the boundary layer experience the *same* pressure change. Equating the two expressions for dp and defining the ratio $\alpha = u/U$, there results

$$d\alpha = -\frac{dp}{\rho U^2}\left(\frac{1 - \alpha^2}{\alpha}\right).$$

Hence if the pressure rises $(dp > 0)$, α decreases and the boundary layer fluid is decelerated more than the free stream. Noting that $-dp/\rho U^2$ is simply dU/U, this expression can be integrated (for constant density), yielding

$$\frac{1 - \alpha^2}{1 - \alpha_1^2} = \frac{U_1^2}{U^2},$$

where the subscript 1 refers to an initial condition. The thickness δ of the boundary layer can be found from the equation for continuity which requires that

$$\rho u \delta = \rho u_1 \delta_1$$

or, again for constant density,

$$\frac{\delta}{\delta_1} = \frac{u_1}{u} = \frac{\alpha_1 U_1}{\alpha U}.$$

The two quantities α and δ are plotted against U/U_1 in Fig. 4–7 for the particular case $\alpha_1 = \frac{1}{2}$. It may be seen that this boundary layer separates $(\alpha \to 0, \ \delta \to \infty)$ when the free-stream velocity has decreased by less than 14%.

With this picture of the relatively large effect of a given pressure rise on the more slowly moving fluid, it would appear that *any* pressure rise applied to a real boundary layer would cause separation since, if the velocity goes continuously to zero at the wall, there will always be *some* fluid with insufficient momentum to negotiate the pressure rise. However, this tendency toward separation is resisted, up to a point, by viscous stress. That is, the slowly moving fluid can be "dragged" through a pressure rise by the faster fluid farther from the wall. An important method of reducing the separation tendency is to enhance this "drag" force by increasing the effective shear stress through turbulent motion.

The mention of turbulent motion introduces another important aspect of boundary layer behavior. In 1883 Reynolds showed that under certain conditions dye particles injected into a pipe flow moved smoothly along streamlines without lateral fluctuations. In the same pipe at higher velocities, this smooth laminar motion broke down into turbulent motion consisting of rapid random fluctua-

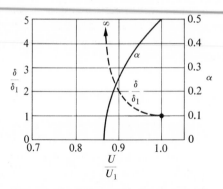
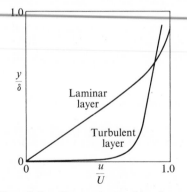

FIG. 4–7. "Growth" of step boundary layer near separation for $\alpha_1 = \frac{1}{2}$.

FIG. 4–8. Comparison of laminar and turbulent velocity profiles.

tions superposed on the mean motion. After the flow had become turbulent the dye lines rapidly mixed with the surrounding fluid, the macroscopic turbulent mixing process being at least an order of magnitude faster than molecular diffusion (the only agency for mixing in laminar flow).

The same transition behavior takes place in the boundary layer, resulting in a significant reduction in its tendency to separate. The reason for this is that the turbulent fluctuations constitute a mechanism for transporting high momentum from the outer part of the layer to the region near the wall. The effect is as if there were an increased shear stress or, as will be shown, an increased coefficient of viscosity. This tends to raise the wall shear stress, but it also means that the turbulent boundary fluid can climb higher up the "pressure hill" than the fluid in the laminar layer. A comparison of typical turbulent and laminar boundary layer velocity profiles is indicated in Fig. 4–8.

The fact that the turbulent layer tends to have much higher momentum (and shear stress) near the wall gives it an increased resistance to separation. As a method of specifying the resistance of a boundary layer to separation, it is useful to define a pressure coefficient C_p as

$$C_p = \frac{\Delta p}{\frac{1}{2}\rho U^2},$$

where U is the free-stream velocity at the point where the pressure begins to rise and Δp is the total increase in pressure up to the point in question.

For an approximate comparison of cases in which the adverse pressure gradient is not abrupt (i.e., where the shear stress has time to act) it may be said that the laminar boundary layer will support a pressure rise given by $0.15 < C_p < 0.2$, whereas the turbulent boundary layer can tolerate without separation a pressure rise corresponding to $0.4 < C_p < 0.8$, depending on the Reynolds number and the upstream pressure distribution of the flow.

The onset of turbulence, or transition as it is called, is dependent in a given device on fluid properties and velocity, the general or upstream turbulence level, and the geometry of the particular device. In the case of pipe flow, for example, geometry is completely specified (far from the entrance) by pipe diameter D and wall roughness. The pertinent fluid properties are density ρ, viscosity μ, and the average velocity $\bar{U}$. These variables can be arranged in a dimensionless form known as the Reynolds number, Re, so that

$$\mathrm{Re} \equiv \frac{\rho \bar{U} D}{\mu}.$$

Reynolds found that for "smooth" pipes (commercial pipes and tubing are usually "smooth") and very quiet inlet conditions, the transition process could in fact be correlated with a critical Reynolds number below which the flow was laminar and above which it was turbulent.

More generally it is found that transition in any series of *geometrically similar* devices can be correlated with the Reynolds number, where D is replaced by some characteristic length L and $\bar{U}$ by some characteristic velocity U. Thus

$$\mathrm{Re} = \frac{\rho U L}{\mu}.$$

Obviously, a critical value for Re can have meaning only within the geometry for which it was defined. With these facts in mind, the explanation of the two pressure distributions observed on the cylinder of Fig. 4–2 becomes straightforward. For reasonably low velocities (below a "critical" Reynolds number) the boundary layer remains laminar and separates readily on the back of the cylinder before much pressure rise has taken place. At the critical Reynolds number, the boundary layer becomes turbulent and is able to flow considerably further around the back of the cylinder before separation. As it does so it rises in pressure, thus reducing the discrepancy between inviscid and actual pressure distributions. Since the drag on a blunt body such as a cylinder is primarily a pressure force (as opposed to skin friction), the onset of turbulent flow in this case is actually accompanied by decreased drag.

In most practical fluid machines to be discussed in this book the boundary layers are turbulent, and so long as this is true the influence of Reynolds-number variations on overall performance is usually minor. However, under extreme conditions, as for example at very high altitudes, the Reynolds number can become low enough that portions of the flow revert to laminar flow. There usually follows a rather substantial reduction in efficiency of performance, due to the increased tendency toward separation.

4–2 THE BOUNDARY LAYER EQUATIONS

The general equations for the flow of Newtonian viscous fluids are called the Navier-Stokes equations [1, 2]. Because of their mathematical complexity, it is possible to obtain exact solutions of them for only a few physical situations. In many cases, however, it is legitimate to make simplifying approximations and thereby obtain solutions which have a useful range of validity. The boundary

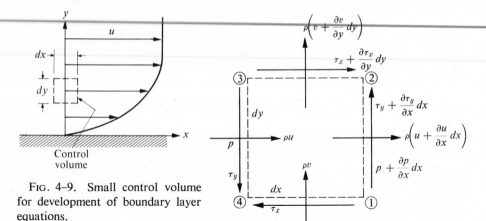

FIG. 4–9. Small control volume for development of boundary layer equations.

layer equations are much more tractable than the Navier-Stokes equations. They may be derived by neglecting certain terms in the Navier-Stokes equations which can be shown to be quite unimportant in boundary layer flows. The method is developed by Schlichting [1] in careful detail.

In the following discussion, equations will be developed less rigorously, but more directly, for the special case of the steady, two-dimensional, incompressible boundary layer. Shapiro [3] uses this approach to develop equations for compressible flow, and includes a good discussion of the range of validity of the simplifications. The object in view is to lay a foundation for a quantitative description of both laminar and turbulent boundary layers, with particular emphasis on skin friction, separation, and heat transfer. Although the solutions of the boundary layer equations which are available are seldom directly applicable to the complex flows in real machines, they at least illustrate the forces at work and the behavior typical of the boundary layer.

Consider a very small control volume within a boundary layer, as shown in Fig. 4–9. Velocities in the x- and y-directions are u and v, respectively, while the shear stresses in these directions are τ_x and τ_y. To apply the continuity and momentum equations to this control volume, we shall need to know the mass and momentum flux across each surface and the force (shear and pressure) acting on each surface.

The outward fluxes of mass and x-momentum are shown in Table 4–1. Since the flow is taken to be two-dimensional, no fluid crosses the surfaces parallel to the xy-plane.

The continuity equation for steady, two-dimensional, incompressible flow requires that

$$\rho\left(u + \frac{\partial u}{\partial x}dx\right)dy - \rho u\,dy + \rho\left(v + \frac{\partial v}{\partial y}dy\right)dx - \rho v\,dx = 0$$

or

$$\frac{\partial u}{\partial x} + \frac{\partial v}{\partial y} = 0. \tag{4-1}$$

Table 4–1

Surface	Mass flux	Flux of x-momentum
①②	$\rho\left(u + \dfrac{\partial u}{\partial x}\,dx\right)dy$	$\rho\left(u + \dfrac{\partial u}{\partial x}\,dx\right)^2 dy$
③④	$-\rho u\,dy$	$-\rho u^2\,dy$
②③	$\rho\left(v + \dfrac{\partial v}{\partial y}\,dy\right)dx$	$\rho\left(v + \dfrac{\partial v}{\partial y}\,dy\right)\left(u + \dfrac{\partial u}{\partial y}\,dy\right)dx$
④①	$-\rho v\,dx$	$-\rho vu\,dx$

The momentum equation in the x-direction, using Table 4–1 to obtain the net outward momentum flux, is (applying Eq. (2–4))

$$\sum F_x = \rho\left(u + \frac{\partial u}{\partial x}\,dx\right)^2 dy - \rho u^2\,dy + \rho\left(v + \frac{\partial v}{\partial y}\,dy\right)\left(u + \frac{\partial u}{\partial y}\,dy\right)dx - \rho uv\,dx.$$

Expanding and disregarding second-order terms, we have

$$\sum F_x = \rho\left(2u\frac{\partial u}{\partial x} + u\frac{\partial v}{\partial y} + v\frac{\partial u}{\partial y}\right)dx\,dy.$$

Using the continuity equation, we may simplify this to

$$\sum F_x = \rho\left(u\frac{\partial u}{\partial x} + v\frac{\partial u}{\partial y}\right)dx\,dy.$$

Since only pressure and shear forces act on the control surface, the force summation may be written, in accordance with Fig. 4–9, as

$$\sum F_x = p\,dy - \left(p + \frac{\partial p}{\partial x}\,dx\right)dy + \left(\tau_x + \frac{\partial \tau_x}{\partial y}\,dy\right)dx - \tau_x\,dx$$

$$= \left(-\frac{\partial p}{\partial x} + \frac{\partial \tau_x}{\partial y}\right)dx\,dy.$$

Thus the momentum equation may be written

$$\rho\left(u\frac{\partial u}{\partial x} + v\frac{\partial u}{\partial y}\right) = -\frac{\partial p}{\partial x} + \frac{\partial \tau_x}{\partial y}.$$

The shear stress τ developed in a Newtonian fluid is proportional to the rate of shearing strain. For the particular case of the boundary layer, in which there is a large velocity gradient close to the wall, the shear stress is very nearly equal to

$$\tau_x = \mu \frac{\partial u}{\partial y},$$

where μ is the coefficient of viscosity.

Using this definition, the momentum equation for the x-direction can be written

$$\rho \left(u \frac{\partial u}{\partial x} + v \frac{\partial u}{\partial y} \right) = -\frac{\partial p}{\partial x} + \mu \frac{\partial^2 u}{\partial y^2}.$$

As a consequence of the thinness of the boundary layer, pressure changes normal to the wall may be neglected inside the layer, so that it may be assumed that p is a function of x only. Then the momentum equation may be written

$$u \frac{\partial u}{\partial x} + v \frac{\partial u}{\partial y} = -\frac{1}{\rho} \frac{dp}{dx} + v \frac{\partial^2 u}{\partial y^2}, \qquad (4\text{–}2)$$

where $v \equiv \mu/\rho$ is the kinematic viscosity.

The momentum equation as derived here assumes a fluid of constant viscosity and constant density. Although these conditions may be satisfied within many low-speed laminar boundary layers, we shall see that the "effective" viscosity in turbulent flow may vary strongly through the boundary layer. Further, ρ and μ could vary as the result of temperature variation within the boundary layer, due either to high heat-transfer rates or, in compressible flow, to the variation of static temperature in response to velocity variation. In such cases a complete set of equations must include at least an energy equation and an equation of state. However, even in cases not requiring the simultaneous solution of an energy equation, Eqs. (4–1) and (4–2) can present considerable difficulties in the solution for u and v (where p is assumed a known function of x). It is desirable, indeed necessary in most cases, to utilize approximate methods for solving the boundary layer equations.

An extremely useful approximation, the *momentum integral method*, has been developed and widely applied. The object of the method is to find a solution which will satisfy the integral of the momentum equation across the boundary layer even though it may fail to satisfy the equation at particular points. The integration of the momentum equation (4–2) may proceed as follows:

$$\int_0^h \left(u \frac{\partial u}{\partial x} + v \frac{\partial u}{\partial y} \right) dy = \int_0^h -\frac{1}{\rho} \frac{dp}{dx} dy + \int_0^h v \frac{\partial^2 u}{\partial y^2} dy,$$

where h is an undefined distance from the wall outside the boundary layer. In the free stream,

$$-\frac{1}{\rho} \frac{dp}{dx} = U \frac{dU}{dx}.$$

Thus this equation may be written

$$\int_0^h \left(u\frac{\partial u}{\partial x} + v\frac{\partial u}{\partial y} - U\frac{dU}{dx} \right) dy = \int_0^h v\frac{\partial^2 u}{\partial y^2}\, dy.$$

Integrating the right-hand term and noting that the wall shear stress is given by

$$\tau_0 = \mu \left(\frac{\partial u}{\partial y} \right)_{y=0},$$

while $\partial u/\partial y$ is zero outside the boundary layer, we can write (for constant v)

$$\int_0^h \left(u\frac{\partial u}{\partial x} + v\frac{\partial u}{\partial y} - U\frac{dU}{dx} \right) dy = -\frac{\tau_0}{\rho}.$$

Integrating the second term in the integral by parts, we have

$$\int_0^h v\frac{\partial u}{\partial y}\, dy = vu\Big|_0^h - \int_0^h u\frac{\partial v}{\partial y}\, dy.$$

From the continuity equation, v at $y = h$ is given by

$$v = \int_0^h \frac{\partial v}{\partial y}\, dy = -\int_0^h \frac{\partial u}{\partial x}\, dy.$$

Thus

$$\int_0^h v\frac{\partial u}{\partial y}\, dy = -U\int_0^h \frac{\partial u}{\partial x}\, dy + \int_0^h u\frac{\partial u}{\partial x}\, dy.$$

The momentum equation may then be written

$$\int_0^h \left(2u\frac{\partial u}{\partial x} - U\frac{\partial u}{\partial x} - U\frac{dU}{dx} \right) dy = -\frac{\tau_0}{\rho}$$

or, rearranging,

$$\int_0^h \frac{\partial}{\partial x}[u(U - u)]\, dy + \frac{dU}{dx}\int_0^h (U - u)\, dy = +\frac{\tau_0}{\rho}. \qquad (4\text{–}3)$$

Both the integrands are zero outside the boundary layer, so h can be indefinitely large.

Equation (4–3) is arranged in this particular form because of the physical significance (and the common usage) of the individual integrals. In the second integral, $(U - u)$ is the mass flux *defect* (actually $1/\rho$ times the defect) that occurs as the result of the deceleration of the boundary layer fluid. In the case of flow within a duct, for example, if there is a mass flux defect near the walls there must be an *increased* mass flux within the free stream. This effect on the free stream would be the same if there were no boundary layer but the duct walls were displaced inward an amount δ^* so that

$$U\delta^* = \int_0^\infty (U - u)\, dy.$$

With respect to mass flux then, the boundary layer has the effect of making the thickness δ^* unavailable for free-stream flow; δ^* is called the *displacement thickness* of the boundary layer.

The first integral can be treated in a similar manner. Note first that since the limits of integration are not functions of x, the partial differentiation $\partial/\partial x$ can be carried outside the integral where it becomes d/dx. In this case we interpret u as $1/\rho$ times the actual mass flux, while $(U - u)$ is the *momentum defect* per unit mass of the boundary layer fluid. Hence the integral is the momentum defect in the boundary layer which, as above, could be accounted for by the loss of free-stream thickness θ such that

$$U^2\theta = \int_0^\infty u(U - u)\, dy.$$

Here θ is called the *momentum thickness* of the boundary layer. The two terms are then defined as follows (for incompressible flow):

$$\delta^* = \int_0^\infty \left(1 - \frac{u}{U}\right) dy, \tag{4–4a}$$

$$\theta = \int_0^\infty \frac{u}{U}\left(1 - \frac{u}{U}\right) dy. \tag{4–4b}$$

With these definitions, the momentum integral equation, Eq. (4–3), may be written

$$\frac{d}{dx}(U^2\theta) + \frac{dU}{dx}(U\delta^*) = \frac{\tau_0}{\rho}$$

or

$$\frac{d\theta}{dx} + \left(2 + \frac{\delta^*}{\theta}\right)\frac{\theta}{U}\frac{dU}{dx} = \frac{\tau_0}{\rho U^2}. \tag{4–5}$$

As it stands this equation applies to both laminar and turbulent boundary layers. The solution, say for a given $U(x)$, requires knowledge of the relationship between δ^* and θ, which may be found if an approximate velocity profile shape is assumed (δ^*/θ is often called the *shape factor H*), and the skin friction τ_0 is known. Note that shear stress is evaluated only at the wall where, even in turbulent flow, it is given by $\mu(\partial u/\partial y)$, since turbulent fluctuations must go to zero at the wall.

4–3 LAMINAR BOUNDARY LAYER SOLUTIONS

A variety of exact† solutions of Eqs. (4–1) and (4–2) are available, two of which will be mentioned here.

---

† Exact in the sense that they do not neglect any of the terms in the boundary layer equations. Most solutions are, however, expressed in terms of series expansions.

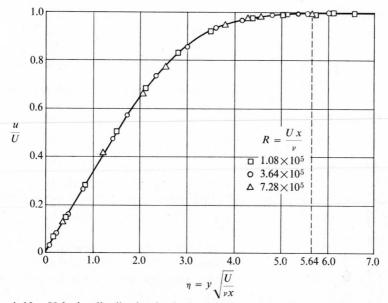

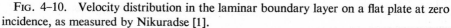

FIG. 4–10. Velocity distribution in the laminar boundary layer on a flat plate at zero incidence, as measured by Nikuradse [1].

The simplest flow is that over a flat plate where dp/dx is zero. In this case, Eqs. (4–1) and (4–2) reduce to

$$u\frac{\partial u}{\partial x} + v\frac{\partial u}{\partial y} = \nu\frac{\partial^2 u}{\partial y^2}, \qquad \frac{\partial u}{\partial x} + \frac{\partial v}{\partial y} = 0,$$

with the boundary conditions

$$y = 0: \qquad u = v = 0,$$

$$y = \infty: \qquad u = U = \text{constant}.$$

A stream function ψ may be defined such that

$$u = \frac{\partial \psi}{\partial y}, \qquad v = -\frac{\partial \psi}{\partial x}.$$

The convenience of this form is that it automatically satisfies the continuity equation so that the above two equations reduce to

$$\frac{\partial \psi}{\partial y}\frac{\partial^2 \psi}{\partial x\,\partial y} - \frac{\partial \psi}{\partial x}\frac{\partial^2 \psi}{\partial y^2} = \nu\frac{\partial^3 \psi}{\partial y^3}. \tag{4–6}$$

This equation may be considerably simplified by substituting

$$\eta = y\sqrt{U/\nu x} \qquad \text{and} \qquad \psi = \sqrt{\nu x U}\,f(\eta),$$

with the result that

$$ff'' + 2f''' = 0,$$

where the primes signify differentiation of the function f with respect to η. The transformed boundary conditions are

$$\eta = 0: \qquad f = f' = 0,$$
$$\eta = \infty: \qquad f' = 1.$$

This equation is an ordinary differential equation, but the fact that it is nonlinear makes it hard to solve. The method of solution employed by Blasius, a series expansion, is described by Schlichting [1]. Only the results of his calculation will be presented here.

The velocity profile u/U obtained from the Blasius solution is shown in Fig. 4–10. Very careful experiments have verified this result almost exactly. As the figure shows, a single curve can be used to represent in dimensionless form the velocity profiles corresponding to all locations on the flat plate. The velocity profiles are said to be similar.

The solution gives the growth of the boundary layer with distance x from the leading edge as

$$\delta \simeq 5.0\sqrt{\nu x/U}, \qquad\qquad (4\text{–}7a)$$

$$\delta^* = 1.72\sqrt{\nu x/U}, \qquad\qquad (4\text{–}7b)$$

$$\theta = 0.664\sqrt{\nu x/U}, \qquad\qquad (4\text{–}7c)$$

where δ is defined as that point where $u = 0.99\,U$. The skin friction is

$$\frac{\tau_0}{\rho U^2} = 0.332\sqrt{\nu/Ux}. \qquad\qquad (4\text{–}7d)$$

The drag on a flat plate may be computed by integrating the skin friction over the entire area. It may be seen that the local wall shear stress decreases with distance downstream but approaches zero only at infinity. Thus separation cannot occur in flow over a flat plate of finite length unless there is an adverse pressure gradient.

A flow in which separation can occur was examined by Howarth [1], who assumed a free-stream velocity variation

$$U = U_1 - \alpha x.$$

Such a velocity would occur in a two-dimensional duct of diverging ($\alpha > 0$) straight walls in which the initial free-stream velocity is U_1 at $x = 0$. Defining the parameters

$$\eta = \frac{y}{2}\sqrt{\frac{U_1}{\nu x}}, \qquad \xi = \frac{\alpha x}{U_1} = -\frac{x}{U_1}\frac{dU}{dx},$$

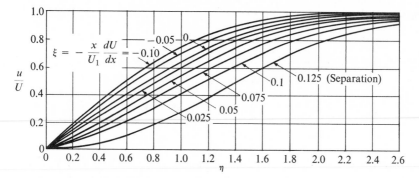

FIG. 4–11. Velocity distribution in the laminar boundary layer for the potential flow given by $U = U_1 - \alpha x$. (After Howarth [1].)

Howarth calculated a series expansion solution from which the velocity profiles shown in Fig. 4–11 are derived. Separation occurs when ξ reaches the value 0.125. (More accurate solutions give a slightly different value [1].)

 In contrast to the foregoing cases, the velocity profiles at various locations on the wall for a given pressure distribution cannot be made to lie on one nondimensional curve, so they are not similar in this sense. Again the Blasius solution appears, this time for the case $\xi = 0$. Separation of the flow occurs with $\xi = 0.12$, i.e., a 12% reduction of the initial free-stream velocity.

 Other "exact" solutions are described in Reference [1].

 The Karman-Pohlhausen method. The momentum integral equation may also be used to calculate the development of the laminar boundary layer. Again defining $\eta = y/\delta$, one could assume a velocity profile of the form

$$\frac{u}{U} = \alpha_0 + \alpha_1 \eta + \alpha_2 \eta^2 + \alpha_3 \eta^3 + \alpha_4 \eta^4,$$

where $\alpha_0, \ldots, \alpha_4$ may be functions of x. Whatever the form of the profile expression, it must satisfy certain boundary conditions. To obtain five boundary conditions for the above series, one could specify that the velocity go to zero at the wall, that it go continuously to U at $y = \delta$, and that the shear force and its derivative go continuously to zero at $y = \delta$. Then,

$$\text{at } y = 0: u = 0;$$

$$\text{at } y = \delta: u = U, \qquad \frac{\partial u}{\partial y} = 0, \qquad \frac{\partial^2 u}{\partial y^2} = 0.$$

In addition, very near the wall where u and v are small, the pressure and viscous forces must balance so that, from Eq. (4–2),

$$\text{at } y = 0, \qquad \nu \frac{\partial^2 u}{\partial y^2} = \frac{1}{\rho} \frac{dp}{dx} = -U \frac{dU}{dx}.$$

Imposing these conditions on the above series, the α's must be such that

$$\frac{u}{U} = (2\eta - 2\eta^3 + \eta^4) + \left(\frac{\delta^2}{\nu} \frac{dU}{dx}\right) \eta \frac{(1 - \eta)^3}{6}.$$

With the definitions of δ^*, θ, and τ_0, it may be shown from Eqs. (4–4) and the definition of wall shear stress that this expression leads, for given values of U, dU/dx, and ν, to

$$\delta^* = f_1(\delta), \qquad \theta = f_2(\delta), \qquad \frac{\tau_0}{\rho U^2} = f_3(\delta),$$

where f_1, f_2, and f_3 are known functions. With these relationships, Eq. (4–5) can be regarded as an ordinary differential equation in δ. Once the equation has been solved for δ, the velocity profile, δ^*, θ, and wall shear stress at any x are determined.

The Karman-Pohlhausen method, in a similar procedure, employs the thickness θ instead of δ, the former being more precisely defined and obviously more convenient for use in Eq. (4–5). Again, δ^*/θ and $\tau_0/\rho U^2$ can be expressed as known functions of θ (and dU/dx) so that Eq. (4–5) becomes an ordinary differential equation in θ. Using the definition of momentum thickness θ (Eq. 4–4) and the above velocity profile, it may easily be shown that

$$\frac{\theta}{\delta} = f\left(\frac{\delta^2}{\nu} \frac{dU}{dx}\right),$$

where f is a known function. Thus

$$\frac{\theta^2}{\nu} \frac{dU}{dx} = \frac{\delta^2}{\nu} \frac{dU}{dx} \left(\frac{\theta}{\delta}\right)^2 = g\left(\frac{\delta^2}{\nu} \frac{dU}{dx}\right)$$

or

$$\frac{\delta^2}{\nu} \frac{dU}{dx} = h\left(\frac{\theta^2}{\nu} \frac{dU}{dx}\right),$$

and the velocity profile can be expressed as

$$\frac{u}{U} = (2\eta - 2\eta^3 + \eta^4) + h\left(\frac{\theta^2}{\nu} \frac{dU}{dx}\right) \frac{\eta(1 - \eta)^3}{6},$$

where h is a known function.

Again using Eqs. (4–4) and the definition of wall shear stress, it may be shown that

$$\frac{\delta^*}{\delta} = g_1\left(\frac{\theta^2}{\nu} \frac{dU}{dx}\right), \qquad \frac{\theta}{\delta} = g_2\left(\frac{\theta^2}{\nu} \frac{dU}{dx}\right).$$

Thus

$$\frac{\delta^*}{\theta} = g_3\left(\frac{\theta^2}{\nu} \frac{dU}{dx}\right).$$

Table 4–2 **Auxiliary Functions for the Approximate Calculation of Laminar Boundary Layers**   **(After Holstein, Bohlen [14])**

$\dfrac{\theta^2}{\nu}\dfrac{dU}{dx}$	$\dfrac{\delta^2}{\nu}\dfrac{dU}{dx}$	$\dfrac{\delta^*}{\theta}$	$\dfrac{\tau_0\theta}{\mu U}$	$\dfrac{\theta^2}{\nu}\dfrac{dU}{dx}$	$\dfrac{\delta^2}{\nu}\dfrac{dU}{dx}$	$\dfrac{\delta^*}{\theta}$	$\dfrac{\tau_0\theta}{\mu U}$
0.0884	15	2.279	0.346	0.0497	4	2.392	0.297
0.0928	14	2.262	0.351	0.0385	3	2.427	0.283
0.0941	13	2.253	0.354	0.0264	2	2.466	0.268
0.0948	12	2.250	0.356	0.0135	1	2.508	0.252
0.0941	11	2.253	0.355	0	0	2.554	0.235
0.0919	10	2.260	0.351	−0.0140	−1	2.604	0.217
0.0882	9	2.273	0.347	−0.0284	−2	2.647	0.199
0.0831	8	2.289	0.340	−0.0429	−3	2.716	0.179
0.0819	7.8	2.293	0.338	−0.0575	−4	2.779	0.160
0.0807	7.6	2.297	0.337	−0.0720	−5	2.847	0.140
0.0794	7.4	2.301	0.335	−0.0862	−6	2.921	0.120
0.0781	7.2	2.305	0.333	−0.0999	−7	2.999	0.100
0.0770	7.052	2.308	0.332	−0.1130	−8	3.085	0.079
0.0767	7	2.309	0.331	−0.1254	−9	3.176	0.059
0.0752	6.8	2.314	0.330	−0.1369	−10	3.276	0.039
0.0737	6.6	2.318	0.328	−0.1474	−11	3.383	0.019
0.0721	6.4	2.323	0.326	−0.1567	−12	3.500	0
0.0706	6.2	2.328	0.324	−0.1648	−13	3.627	−0.019
0.0689	6	2.333	0.321	−0.1715	−14	3.765	−0.037
0.0599	5	2.361	0.310	−0.1767	−15	3.916	−0.054

Likewise

$$\frac{\tau_0\theta}{\mu U} = g_4\left(\frac{\theta^2}{\nu}\frac{dU}{dx}\right),$$

where $g_1, \ldots, g_4$ are known functions. Thus Eq. (4–5) may also be regarded as an ordinary differential equation in

$$\left(\frac{\theta^2}{\nu}\frac{dU}{dx}\right),$$

or simply θ, since $U(x)$ and ν are known.

Walz [2] showed that the solution of Eq. (4–5) is approximately

$$\frac{U\theta^2}{\nu} = \frac{0.470}{U^5}\int_0^x U^5\,dx.$$

Thus if the pressure distribution on any surface is known, the potential flow velocity may be calculated from

$$U\frac{dU}{dx} = -\frac{1}{\rho}\frac{dp}{dx},$$

and the momentum thickness from the previous expression. The relationship
between θ, τ_0, δ^* and δ is given in Table 4–2.

The convenience and accuracy of this method have caused it to be used widely.
It is more accurate in negative than in positive pressure gradients, and in fact
may become somewhat misleading near the point of separation. Schubauer [4]
measured the pressure distribution in two-dimensional flow past an ellipse. On
the basis of this pressure distribution, the Karman-Pohlhausen method did not
predict the separation actually observed by means of smoke visualization and
hot-wire velocity measurements. Thwaites [5] has developed another method of
solving the integral momentum equation which should predict laminar separation
more accurately. For high pressure gradients the Karman-Pohlhausen velocity
profile assumes an unrealistic shape, and the calculation method developed by
Launder [16] is superior. In many applications the integral momentum method
has proved very useful indeed in calculation of the overall behavior of the boundary
layer.

4–4 THE TURBULENT BOUNDARY LAYER

In most engineering applications where boundary layers are important, the
Reynolds numbers are high enough that most of the flow is turbulent. A very
advantageous consequence of this fact is the superior pressure recovery which
can be obtained with a turbulent boundary layer as compared with a laminar one.
This means that airfoils, compressor blades, and diffusers will perform very much
better than they would if the flow remained laminar. On the other hand, there is
the fact that the skin friction of a turbulent layer is generally much greater.

Analytically, the onset of turbulence renders important features of the flow,
such as skin friction and separation, essentially indeterminable, except by semi-
empirical methods. At the present time it cannot be said that there exists even the
framework of a satisfactory purely theoretical treatment of turbulence. Nor is
there a very bright prospect of one coming into being. It is unfortunately true
also that the empirical methods are very limited in their usefulness. It may be
said, for example, that fully developed turbulent pipe flow is well known, since
general correlations of the measurements of many investigators are available.
But developing flows, especially with large pressure gradients, are not well understood at the present time. In particular, though several methods of predicting the
separation point exist, none of them seem to be very reliable.

For flows in which the geometry and free-stream velocity can each be described
by a single parameter, i.e., within geometrically similar devices, the technique of
dimensional analysis has been of inestimable usefulness in correlating measurements of friction, drag, and heat transfer. Examples of general correlations of
this kind will follow shortly. For more complex situations in which the wall pressure or free-stream velocity varies in an arbitrary manner, existing analytical
methods usually employ the momentum integral equation (4–5), or an energy
integral equation along with some suitable assumption about the turbulent diffusion

of momentum. Except for the fact that empirical information is introduced in the determination of the factors δ^*/θ and $\tau_0/\rho v^2$ in Eq. (4–5), existing methods are very similar to methods employed for laminar flow; e.g., the Karman-Pohlhausen method.

The discussion of turbulent flow is greatly facilitated by the assumption that all velocity components at each point may be considered the sum of two parts, one steady and one fluctuating, according to

$$u(x, y, t) = \bar{u}(x, y) + u'(x, y, t),$$
$$v(x, y, t) = \bar{v}(x, y) + v'(x, y, t),$$
$$w(x, y, t) = \bar{w}(x, y) + w'(x, y, t).$$

In each case the first term on the right-hand side is the time-mean value defined, for the u velocity component, by

$$\bar{u} = \frac{1}{T} \int_0^T u \, dt,$$

where T is a time period that is very long compared with the period of the fluctuation. The second term is the instantaneous fluctuating component (which can be measured by a velocity probe with a very rapid response, such as a hot-wire anemometer). Even in a flow which is two-dimensional on the average, that is, $\bar{w} = 0$, it is necessary to recognize that three-dimensional disturbances do, in general, exist.

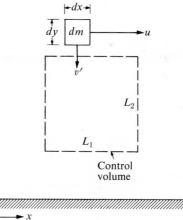

FIG. 4–12. The turbulent transfer of momentum.

Generally speaking, the turbulent fluctuations are greater in the boundary layer than in the free stream, as we shall see. Measurements have shown that the outer third or half of the boundary layer may be only intermittently turbulent. Spasms of turbulence may be separated by periods of negligible fluctuation intensity.

The effect of turbulence in changing the shape of the velocity profile can be ascribed to the operation of the so-called turbulent shear stress. Consider an element of fluid dm indicated in Fig. 4–12, which through turbulent motion is about to enter a control volume situated within a boundary layer and bounded by sides L_1 and L_2. As the element dm crosses the control surface, it adds x momentum to the control volume in the amount dmu. This momentum is added in a time $dt \approx dy/-v'$, so that the momentum addition *rate* is

$$\frac{dmu}{dt} = \frac{\rho(dx\,dy\,dz)}{dy/-v'}u = -\rho uv'\,dx\,dy,$$

which is equivalent to a shear force acting on the area $dx\,dy$. The equivalent or turbulent shear stress τ_t is then, averaging over a time which is long compared with the period of the individual fluctuations,

$$\tau_t = -\rho\overline{uv'}.$$

Writing u as $\bar{u} + u'$, and noting that $\overline{\bar{u}v'}$ is zero (since $\bar{u}$ is a constant and $\overline{v'} = 0$), we obtain

$$\tau_t = -\rho\overline{u'v'},$$

where $\overline{u'v'}$ is called the *correlation* of the velocity fluctuations u' and v'. Assuming that the turbulent motion does not seriously alter the molecular (laminar) transport of momentum, it is permissible to treat a turbulent boundary layer in much the same manner mathematically as a laminar layer, if the shear stress is written as

$$\tau = \mu\frac{\partial u}{\partial y} - \rho\overline{u'v'}.$$

The second term may be two orders of magnitude larger than the first in the outer part of the boundary layer, though it must be negligible in an exceedingly thin region near the wall where the fluctuations approach zero. The great difficulty with the term $\rho\overline{u'v'}$ is that no general method has been found for predicting its magnitude. Measurements with two hot wires set up to measure fluctuation components v' and u' simultaneously can provide accurate knowledge of the correlation in a particular flow. A typical variation of the average and fluctuating velocities and their correlation is shown in Fig. 4–13. Note that the fluctuations are much smaller in the free stream than in the boundary layer.

The existence of this turbulent-shear-stress concept has led to several simplifying hypotheses, all of which have been shown to lack universal validity, though they can be of considerable usefulness in certain situations. In each, the effect of the turbulence on the boundary layer is expressed by a single variable, or at most by two variables.

For example, Boussinesq postulated that, by analogy with the viscous shear stress, the turbulent shear stress τ_t could be written

$$\tau_t = -\rho\overline{u'v'} = \rho\epsilon\frac{d\bar{u}}{dy},$$

in which ϵ is the so-called eddy viscosity. If ϵ can be assumed constant, then any

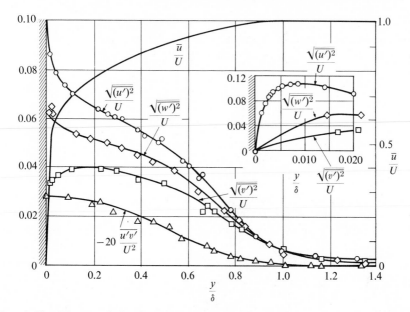

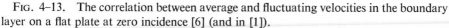

FIG. 4–13. The correlation between average and fluctuating velocities in the boundary layer on a flat plate at zero incidence [6] (and in [1]).

boundary layer problem which has a solution for laminar flow can be solved for turbulent flow, if the appropriate value of ϵ is used. Provided great accuracy is not required, the constant-eddy-viscosity solution may be quite useful at some distance from the wall. Very near the wall, the fluctuations and the eddy viscosity must approach zero. The utility of this concept is greatly restricted by virtue of the fact that the eddy viscosity is not a property of the fluid.

Prandtl proposed, in what is called the mixing length theory, that the eddy viscosity ϵ should be proportional to the mean velocity gradient $d\bar{u}/dy$. His reasoning was based on the idea that as a small "package" of fluid moves suddenly from one point in a velocity gradient to another, it moves a finite distance l before mixing out with adjacent fluid. The idea is analogous to some extent with the kinetic-theory explanation of viscosity, in which the "mixing length" is the mean free path. Prandtl proposed that

$$\tau_t = -\rho\overline{u'v'} = \rho l^2 \left|\frac{d\bar{u}}{dy}\right|\frac{d\bar{u}}{dy}.$$

The mixing length, which must be experimentally deduced, is not itself constant. In fully developed pipe flow it is found to be proportional to distance from the wall in the region near the wall.

These hypotheses and other similar ones do not reveal any fundamental information on the general problem of turbulence. They are, however, useful in certain instances in correlating experimental results for application to engineering problems.

Boundary layer calculations based on a direct solution of the momentum and continuity equations have not been as successful for turbulent flow as for laminar flow. The difficulty, of course, lies in properly evaluating the turbulent shear stress. To be sure one can, for an assumed variation of turbulent shear stress, obtain a solution of these equations, usually by numerical techniques on high-speed computers. However, even if the assumptions prove correct in one situation there is little likelihood that they will be correct for other situations.

For many applications it is sufficient to determine the overall development of the boundary layer as described by the momentum integral equation, and this approach has found wide use. The Karman-Pohlhausen solution to the momentum integral equation, discussed previously, is not mathematically restricted to laminar flow, since the turbulent boundary layer velocity profile must satisfy the same boundary conditions. Generally, approximate empirical velocity profiles are used to determine the ratio δ^*/θ for turbulent boundary layers. In addition, since the approximate profiles may not have the correct velocity gradient at the wall, empirical information may be necessary to evaluate wall shear stress, τ_0.

A great deal of the empirical information in use comes from investigations of fully developed turbulent pipe flow, and it happens to be useful in a surprisingly wide variety of flows. Pipe flow is said to be fully developed when, far from the pipe entrance, the velocity profile does not change with axial position. Under these conditions shear stress, which is ordinarily difficult to measure, is easily determined.

Consider a control volume of length L which fills a pipe of radius R. For fully developed flow there is no net momentum flux out of the control volume, so that the momentum equation gives the wall shear stress as

$$\sum F_x = \pi R^2 \, \Delta p - 2\pi R L \tau_0 = 0$$

or

$$\tau_0 = \frac{\Delta p R}{2L},$$

where Δp is the pressure drop in length L. Similarly, for a smaller control volume of radius $r < R$ within the pipe, the local shear stress τ is given by

$$\tau = \frac{\Delta p r}{2L},$$

so that the variation of τ across the pipe is

$$\frac{\tau}{\tau_0} = \frac{r}{R} = 1 - \frac{y}{R},$$

where $y = R - r$ is the distance from the wall. Hence with a simple pressure measurement one can determine shear stress, and from a determination of the velocity profile one can deduce the variations in eddy viscosity, mixing length, and so on.

The velocity profile within the pipe can be broken into two or three regions. Very near the wall the flow must be laminar, since the presence of the wall prohibits turbulent velocities. Here we can write (and from now on we omit the bar for the temporal mean velocity u),

$$\tau \approx \tau_0 = \mu \frac{du}{dy} \approx \mu \frac{u}{y}$$

or, for convenient comparison with subsequent results,

$$\frac{u}{\sqrt{\tau_0/\rho}} = \frac{y\sqrt{\tau_0/\rho}}{\nu}. \tag{4-8}$$

This is the so-called *laminar sublayer*. The group $\sqrt{\tau_0/\rho}$, which has the dimensions of velocity, is often called the *friction velocity*.

Farther from the wall, where turbulent shear is dominant, the velocity is given by the empirical relation [1]

$$\frac{u}{\sqrt{\tau_0/\rho}} = 2.5 \ln \frac{y\sqrt{\tau_0/\rho}}{\nu} + 5.5. \tag{4-9}$$

This expression has been called the *law of the wall*, since it shows the velocity u to depend only on conditions near the wall.

Near the center of the pipe the velocity is given by

$$\frac{U_c - u}{\sqrt{\tau_0/\rho}} = -2.5 \ln \frac{y}{R}, \tag{4-10}$$

where U_c is the center-line velocity. The form of these equations may be derived by various semi-theoretical methods, as described by Schlichting [1], but the numerical constants are obtained experimentally.

The velocity profiles of Eqs. (4-9) and (4-10) do not depend explicitly on Reynolds number and surface roughness variations. However, both these affect the wall shear stress in general and thus do have an effect on the velocity profiles.

Over limited Reynolds-number ranges the velocity profile may be approximated (except close to the wall) by the very much simpler form

$$\frac{u}{U_c} = \left(\frac{y}{R}\right)^{1/n}, \tag{4-11}$$

where n is a constant within a reasonable range of Reynolds number. For a pipe Reynolds number of about 1×10^5, n should be about 7 [1], and this value is often used to represent turbulent boundary layers.

It can be shown (see Problem 9) that the requirement that Eqs. (4-9) and (4-10) both apply in an intermediate range of y is sufficient to specify τ_0 (actually the skin friction coefficient $\tau_0/\frac{1}{2}\rho U_c^2$) as a function of the Reynolds number $U_c R/\nu$. For other profile assumptions, such as the above power law, it is necessary to have additional information on τ_0 to solve the momentum integral equation.

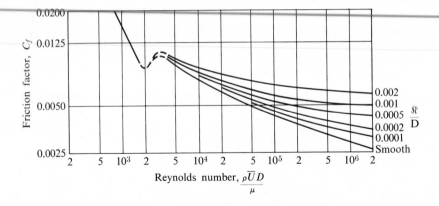

FIG. 4–14.   Curves showing relation between friction factor and Reynolds number for pipes having various values of relative roughness. (Adapted from Moody [15].)

Physically, τ_0 must depend on the following variables for fully developed pipe flow:

$$\tau_0 = f(\rho, \mu, \bar{U}, D, \text{roughness}),$$

where $\bar{U}$ is the bulk mean velocity $(\dot{m}/\pi R^2 \rho)$. Defining the *skin friction* coefficient C_f, dimensional analysis yields

$$C_f = \frac{2\tau_0}{\rho \bar{U}^2} = f\left(\frac{\rho \bar{U} D}{\mu}, \text{roughness}\right) \quad \text{or} \quad C_f = f\left(\text{Re}, \frac{\Re}{D}\right),$$

where $\Re/D$ is the ratio of average wall roughness height to pipe diameter. This function is shown in Fig. 4–14, which includes both laminar and turbulent flows. The laminar flows have Reynolds numbers less than about 2000, and a friction factor which may be obtained theoretically as

$$C_f = 16\left(\frac{\bar{U} D}{\nu}\right)^{-1} \quad \text{(laminar).} \tag{4–12}$$

The pipe roughness is characterized by a ratio of average roughness height $\Re$ to diameter D. It has been shown that roughness elements have no effect on the flow if they are smaller in size than the laminar sublayer thickness, which is approximately

$$y = \frac{5\nu}{\sqrt{\tau_0/\rho}}.$$

For smooth pipes the friction factor for turbulent flow can be approximated over a fairly wide range of Reynolds numbers [7] ($10^4 < \text{Re} < 10^6$) by

$$C_f = \frac{2\tau_0}{\rho \bar{U}^2} = 0.046\left(\frac{\bar{U} D}{\nu}\right)^{-1/5} \quad \text{(turbulent).} \tag{4–13}$$

With these equations one could, for any Reynolds number, determine C_f and hence τ_0/ρ so that pipe velocity profiles could be constructed from Eqs. (4–8) through (4–10).

The experimental data available for turbulent flow over plates is much less than that for flow in pipes. However, a method for utilizing pipe measurements to predict the behavior of the two-dimensional turbulent flow on plates has been suggested by Prandtl and Von Karman. The boundary layer of thickness δ over a plate is taken to be identical to the flow in a pipe of radius R equal to δ when the free-stream velocity of the former is the same as the center-line velocity of the latter. The two cases are not exactly equivalent because the pipe will always have some pressure gradient. However, the difference can be negligible if the Reynolds number is reasonably large.

The equation for the skin friction coefficient for fully developed pipe flow, Eq. (4–13), can be written

$$C_f = \frac{2\tau_0}{\rho \overline{U}^2} = 0.040 \left(\frac{\overline{U}R}{\nu}\right)^{-1/5},$$

where R is the pipe radius. The ratio of bulk mean velocity to center-line velocity, $\overline{U}/U_c$, depends on the exact shape of the velocity profile. For the present purposes, however, it is legitimate to use the approximation $\overline{U}/U_c = 0.8$ for all Reynolds numbers. In this way the pipe friction may be written

$$C_f = \frac{2\tau_0}{\rho U_c^2} = 0.029 \left(\frac{U_c R}{\nu}\right)^{1/5},$$

and in accordance with the Prandtl-Karman hypothesis the local friction coefficient for the flat plate (identifying U_c with U_∞ and R with δ) is then

$$C_f = \frac{2\tau_0}{\rho U_\infty^2} = 0.029 \left(\frac{U_\infty \delta}{\nu}\right)^{-1/5}, \tag{4–14}$$

where U_∞ is the free-stream velocity and δ is the boundary layer thickness. With the approximation

$$\frac{u}{U_\infty} = \left(\frac{y}{R}\right)^{1/7},$$

this can be expressed in terms of momentum thickness. Using Eq. (4–4), we have $\theta/\delta = 7/72$. Substituting in Eq. (4–14), we get

$$C_f = 0.018 \left(\frac{U_\infty \theta}{\nu}\right)^{-1/5}. \tag{4–15}$$

The momentum integral equation becomes, in the absence of pressure gradient,

$$\frac{d\theta}{dx} = \frac{\tau_0}{\rho U_\infty^2}. \tag{4–16}$$

Equations (4-15) and (4-16) may be solved for the growth of the two-dimensional boundary layer of zero pressure gradient, yielding

$$\theta = 0.023 \left(\frac{Ux}{\nu}\right)^{-1/6} x, \tag{4-17}$$

$$\delta = 0.24 \left(\frac{Ux}{\nu}\right)^{-1/6} x, \tag{4-18}$$

if the turbulent boundary layer begins to grow at $x = 0$.

Similarly, the velocity profile for turbulent flow over a flat plate can be assumed, in the same way as that within a pipe. That is, Eqs. (4-8) through (4-10) or (4-11) may be used if R is simply replaced by δ and U_c by U_∞.

For surfaces on which the pressure is *not* constant it may be assumed, in the absence of better information, that the skin friction may be calculated from an expression like Eq. (4-15). An alternative proposed by Prandtl [2] is

$$C_f = 0.0256 \left(\frac{U\theta}{\nu}\right)^{-1/4}, \tag{4-19}$$

where U is the local free-stream velocity. Another, due to Ludwieg and Tillman [2], is

$$C_f = 0.246 \times 10^{-0.678H} \left(\frac{U_\infty\theta}{\nu}\right)^{-0.268}, \tag{4-20}$$

where H is the shape factor δ^*/θ.

If the momentum integral equation is to be used to calculate the growth of momentum thickness along the surface, it is necessary to have relationships for both C_f and H. Since the results of a number of quite different arbitrary methods are much the same, only the simplest one will be described here.

The value of H for a flat-plate turbulent boundary layer is about 1.4 for zero pressure gradient. Substituting this value and the skin friction equation (4-19) into the momentum integral equation, we obtain

$$\frac{d\theta}{dx} + 3.4 \frac{\theta}{U} \frac{dU}{dx} = 0.0128 \left(\frac{U\theta}{\nu}\right)^{-1/4},$$

which may be integrated in the form

$$\theta \left(\frac{U\theta}{\nu}\right)^{1/4} = U^{-4} \left(C_1 + 0.016 \int_{x=x_t}^{x} U^4 \, dx\right),$$

where the constant C_1 depends on the momentum thickness at the transition point $x = x_t$. This result agrees well with measured data except very near the separation point.

Unfortunately, procedures like the foregoing which give approximate solutions have not yielded a satisfactory separation criterion, and more complicated methods which consider variable H have not been successful beyond predicting that separation generally occurs for $1.8 < H < 2.4$.

Thus at the present time reliable general methods of predicting turbulent boundary layer separation do not exist, and the designer of fluid machines generally relies on a statement of experience (such as that given in Section 4–1) that separation is associated with a certain range of pressure coefficients. For a fairly extensive review of turbulent boundary layers, see Reference [10].

4–5 BOUNDARY LAYER HEAT TRANSFER

When a temperature difference exists between a wall and the fluid flowing past it, heat transfer occurs between the two by conduction and radiation. Conduction accelerated by motion of the fluid is called *convection*, and it is this process which will concern us in this section. Further, we shall limit the discussion to so-called forced convection, in which the fluid motion is generated by an external device such as a fan, pump, or propulsion engine. For more detailed discussions of the material reviewed here and for treatment of radiative heat transfer, the reader is referred to complete texts on the subject, such as References [7, 8, 9].

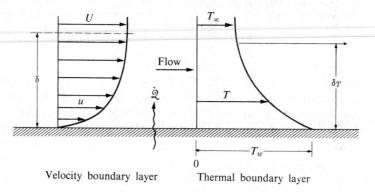

Velocity boundary layer   Thermal boundary layer

FIG. 4–15.   Velocity and thermal boundary layers.

Whenever there is heat transfer between a fluid and a wall, there is generally a smooth variation in fluid temperature in the region near the wall, as suggested by Fig. 4–15. It will be shown that in a large number of interesting cases the thickness of the layer in which the temperature variation is large, the so-called thermal boundary layer thickness, δ_T, has the same order of magnitude as the velocity boundary layer of thickness δ.

Assuming the wall heat transfer rate per unit area $\dot{Q}_0$ to be proportional to the overall temperature difference $T_w - T_\infty$, a *heat transfer film coefficient* h_f may be defined by

$$\dot{Q}_0 = h_f(T_w - T_\infty), \qquad (4\text{–}22)$$

where h_f is the coefficient of heat transfer across the liquid or gas film (i.e., boundary layer) near the wall, and T_w and T_∞ are the wall and free-stream temperatures, respectively. The shape of the temperature profile and the determination of h_f require the use of the energy equation within the boundary layer.

In general, viscosity and density depend on temperature, so that temperature variations in the fluid affect the flow field to some extent. If these properties are essentially constant, the velocity distribution may be determined from Eqs. (4–1) and (4–2), which are then independent of temperature.

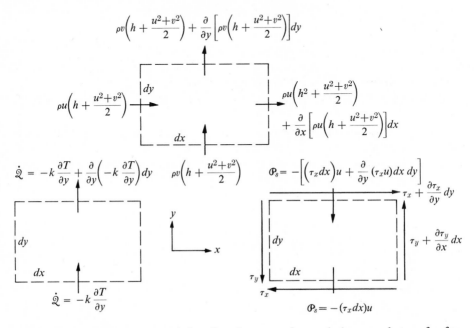

FIG. 4–16. Approximate enthalpy flux, heat transfer, and shear work transfer for elemental control volume.

The appropriate energy equation for two-dimensional steady flow may be derived by considering a control volume within the boundary layer, as shown in Fig. 4–16, where enthalpy flux, heat transfer, and shear work are shown separately. In Fig. 4–16 the velocity components parallel and normal to the wall are u and v, respectively. The symbol h designates the local enthalpy per unit mass while k is the thermal conductivity. As before, τ is the local shear stress. If the fluid is a gas (the case of principal interest in the following chapters), the heat transfer is almost entirely normal to the wall. Therefore heat transfer parallel to the wall has been neglected in Fig. 4–16. Cases where this is not permissible (for instance, heat transfer to liquid metals of high conductivity) are treated in heat-transfer texts. (See, for example, Reference 7.) Also the shear work is neglected on the surfaces normal to the wall since, as we have seen, both v and τ_y are much less than u and τ_x.

The net enthalpy flux from the system is

$$\frac{\partial}{\partial y}\left[\rho v\left(h + \frac{u^2 + v^2}{2}\right)\right] dy\, dx + \frac{\partial}{\partial x}\left[\rho u\left(h + \frac{u^2 + v^2}{2}\right)\right] dx\, dy,$$

which may be expanded to

$$\left(h + \frac{u^2 + v^2}{2}\right)\left[\frac{\partial}{\partial x}(\rho u) + \frac{\partial}{\partial y}(\rho v)\right] dx\, dy + \rho u \frac{\partial}{\partial x}\left(h + \frac{u^2 + v^2}{2}\right) dx\, dy$$

$$+ \rho v \frac{\partial}{\partial y}\left(h + \frac{u^2 + v^2}{2}\right) dx\, dy.$$

From the continuity equation, the term in square brackets is zero.

The net heat transfer *to* the control volume is simply $+(\partial/\partial y)[k(\partial T/\partial y)]\, dy\, dx$, where k is the thermal conductivity of the fluid.

Shear work is done on or by the fluid in the control volume through the shear stresses

$$\tau_x \qquad \text{and} \qquad \tau_x + \frac{\partial \tau_x}{\partial y}\, dy,$$

since the fluid along these boundaries is *moving* in the shear direction. The net work transfer rate *from* the control volume is $-(\partial/\partial y)(\tau_x u)\, dx\, dy$.

The energy equation for steady flow requires that the net enthalpy flux from the control volume be equal to the net heat flux minus the net shear work. Thus, combining these expressions, we have

$$\rho u \frac{\partial}{\partial x}\left(h + \frac{u^2 + v^2}{2}\right) dx\, dy + \rho v \frac{\partial}{\partial y}\left(h + \frac{u^2 + v^2}{2}\right) dx\, dy$$

$$= \frac{\partial}{\partial y}\left(k\frac{\partial T}{\partial y}\right) dy\, dx + \frac{\partial}{\partial y}(\tau_x u)\, dx\, dy.$$

This equation can be expressed in neater form by using the momentum equation to eliminate the velocity derivatives. Multiplying Eq. (4–2) by ρu, we obtain

$$\rho u \left(u\frac{\partial u}{\partial x} + v\frac{\partial u}{\partial y}\right) = \rho u \frac{\partial}{\partial x}\frac{u^2}{2} + \rho v \frac{\partial}{\partial x}\frac{u^2}{2}$$

$$= -u\frac{dp}{dx} + \frac{\partial}{\partial y}\left(\mu u \frac{\partial u}{\partial y}\right) - \mu\left(\frac{\partial u}{\partial y}\right)^2.$$

Then, assuming laminar flow so that $\tau_x = \mu(\partial u/\partial y)$ in the energy equation, neglecting the small terms

$$\rho u \frac{\partial}{\partial x}\frac{v^2}{2} \qquad \text{and} \qquad \rho v \frac{\partial}{\partial y}\frac{v^2}{2},$$

and assuming a perfect gas, the energy equation becomes

$$\rho u c_p \frac{\partial T}{\partial x} + \rho v c_p \frac{\partial T}{\partial y} = u \frac{dp}{dx} + \frac{\partial}{\partial y}\left(k\frac{\partial T}{\partial y}\right) + \mu\left(\frac{\partial u}{\partial y}\right)^2. \qquad (4\text{-}23)$$

If we chose to consider an incompressible fluid rather than a perfect gas, then

$$dh \approx c\,dT + (1/\rho)\,dp,$$

in which c is the specific heat of the fluid. Using this expression in Eq. (4–23) yields

$$\rho u c \frac{\partial T}{\partial x} + \rho v c \frac{\partial T}{\partial y} = \frac{\partial}{\partial y}\left(k\frac{\partial T}{\partial y}\right) + \mu\left(\frac{\partial u}{\partial y}\right)^2. \qquad (4\text{-}24)$$

In either case the last term, $\mu(\partial u/\partial y)^2$, represents viscous dissipation, or work done on the fluid within the control volume. In many cases it is negligible compared with the other terms.

An interesting and useful analogy can be drawn between wall heat transfer $\dot{\mathcal{Q}}_0$ and wall shear stress τ_0 if the momentum and energy equations are compared for the special case of no axial pressure gradient, negligible viscous dissipation, and constant fluid properties. The equations are then,

$$\text{Momentum:} \qquad u\frac{\partial u}{\partial x} + v\frac{\partial u}{\partial y} = \nu\frac{\partial^2 u}{\partial y^2},$$

$$\text{Energy:} \qquad u\frac{\partial T}{\partial x} + v\frac{\partial T}{\partial y} = \frac{k}{\rho c_p}\frac{\partial^2 T}{\partial y^2}.$$

It is evident that the equations are quite similar if

$$\nu = \frac{k}{\rho c_p} \qquad \text{or} \qquad \frac{\mu c_p}{k} = 1.$$

Since $\mu c_p/k$, the Prandtl number, is in fact of the order of unity for most gases, this similarity is worth examining. It means that a solution for u from the momentum equation will be a solution for T from the energy equation if velocity and temperature are expressed in terms of variables which have the *same* boundary conditions. Suitable variables are u/U and $(T_w - T)/(T_w - T_\infty)$, since both go from zero at the wall to unity in the free stream.* In this case the variations of u/U and $(T_w - T)/(T_w - T_\infty)$ are the same, and the thermal and velocity boundary layers are of the same thickness. Thus, to relate $\dot{\mathcal{Q}}_0$ to τ_0 for the case of constant wall temperature T_w,

$$\frac{\tau_0}{\dot{\mathcal{Q}}_0} = \frac{\mu(\partial u/\partial y)_{y=0}}{-k(\partial T/\partial y)_{y=0}} = \frac{U\mu}{(T_w - T_\infty)k}\frac{[(\partial/\partial y)(u/U)]_{y=0}}{\left[\dfrac{\partial}{\partial y}\left(\dfrac{T_w - T}{T_w - T_\infty}\right)\right]_{y=0}} = \frac{\mu}{k}\frac{U}{(T_w - T_\infty)}.$$

---

* Since both U and T_∞ are constant, then the case examined is for T_w constant.

Rearranging, this can be written

$$\frac{\tau_0}{\rho U^2} = \left(\frac{\mu}{\rho U x}\right)\left[\frac{\dot{\mathcal{Q}}_0 x}{(T_w - T_\infty)k}\right],$$

where x, the distance from where the boundary layer began to grow, is introduced because of its known importance in the development of the velocity boundary layer. With Eq. (4–22),

$$\frac{\tau_0}{\rho U^2} = \left(\frac{\mu}{\rho U x}\right)\left(\frac{h_f x}{k}\right).$$

Defining the Nusselt number Nu_x (based on length x) as $h_f x/k$, and with the Reynolds number Re_x (also based on x) and skin friction coefficient C_f, this can be written

$$\mathrm{Nu}_x = \tfrac{1}{2}\,\mathrm{Re}_x\,C_f.$$

Thus, for laminar flow of a fluid of Prandtl number unity, one can determine heat-transfer characteristics from skin friction and Reynolds number. In fact, with Eq. (4–7d),

$$\mathrm{Nu}_x = 0.332\sqrt{\mathrm{Re}_x}. \qquad (4\text{--}25)$$

For laminar flow over a flat plate with constant fluid properties there are "exact" solutions available for any Prandtl number, as discussed in heat transfer texts. For the case of constant wall temperature the Pohlhausen solution, together with the Blasius solution for the velocity boundary layer, gives [7]

$$\mathrm{Nu}_x = 0.332\sqrt{\mathrm{Re}_x}\,\sqrt[3]{\mathrm{Pr}}, \qquad (4\text{--}26)$$

where Pr is the Prandtl number. This result is valid for $\mathrm{Pr} > 0.5$; that is, for most gases. It can be seen that Eqs. (4–25) and (4–26) are in agreement for $\mathrm{Pr} = 1$. For the case of uniform $\dot{\mathcal{Q}}_0$ rather than constant T_w, the constant changes from 0.332 to 0.458.

Turbulence affects the thermal boundary layer much as it does the velocity boundary layer. This would seem reasonable since it can be expected that turbulent motion would enhance the transverse transfer of enthalpy very much as it affects the transfer of momentum. By analogy then, one can define an energy transfer diffusivity [7] ϵ_h to account for the turbulent conductivity in the form

$$\frac{\dot{\mathcal{Q}}}{\rho c_p} = -\left(\frac{k}{\rho c_p} + \epsilon_h\right)\frac{dT}{dy} = -(\alpha + \epsilon_h)\frac{dT}{dy}, \qquad (4\text{--}27)$$

where $\alpha = k/\rho c_p$ is called the thermal diffusivity of the fluid. Lacking functional relationships for ϵ_h (just as for ϵ), we must use empirical knowledge in the treatment of turbulent heat transfer.

Since momentum and enthalpy transfer are affected similarly by turbulence, one might expect a heat transfer–skin friction analogy as in laminar flow. The analogy can in fact be demonstrated in fully developed turbulent flow in pipes. For such flow we have seen that $\tau/\tau_0 = 1 - y/R$. Similarly, for a fully developed temperature profile, the variation of heat transfer with radius can be approximated [7] by $\dot{Q}/\dot{Q}_0 = 1 - y/R$, while the velocity and temperature profiles are also similar, so that

$$\frac{\tau}{\dot{Q}} \approx \frac{\tau_0}{\dot{Q}_0} \tag{4-28}$$

and

$$\frac{d}{dy}\left(\frac{u}{\overline{U}}\right) = \frac{d}{dy}\left(\frac{T_w - T}{T_w - \overline{T}}\right), \tag{4-29}$$

where $\overline{U}$ and $\overline{T}$ are the bulk mean velocity and temperature, respectively. The ratio of shear stress to heat transfer rate is

$$\frac{\tau}{\dot{Q}} = \frac{\rho(\nu + \epsilon)(du/dy)}{-\rho c_p(\alpha + \epsilon_h)(dT/dy)}$$

$$= \frac{\rho(\nu + \epsilon)\overline{U}(d/dy)(u/\overline{U})}{\rho c_p(\alpha + \epsilon_h)(T_w - \overline{T})(d/dy)[(T_w - T)/(T_w - \overline{T})]}.$$

Assuming turbulent flow where $\epsilon \gg \nu$ and $\epsilon_h \gg \alpha$, and using Eqs. (4–28) and (4–29), we have

$$\frac{\tau_0}{\dot{Q}_0} = \frac{\epsilon \overline{U}}{c_p \epsilon_h(T_w - \overline{T})}$$

which can be written

$$\frac{\dot{Q}_0}{T_w - \overline{T}}\frac{1}{\rho \overline{U} c_p} = \frac{\epsilon}{\epsilon_h}\frac{\tau_0}{\rho \overline{U}^2}$$

or

$$\frac{h}{\rho c_p \overline{U}} = \frac{(\epsilon/\epsilon_h)C_f}{2}. \tag{4-30}$$

It has been common in the past to assume that $\epsilon/\epsilon_h \approx 1$, but measurements have revealed values in the range [7] of $0.9 < \epsilon/\epsilon_h < 1.7$.

Equation (4–30) relates measured friction losses and heat transfer quite well. If the variation in the ratio of $(\nu + \epsilon)/[(k/\rho c_p) + \epsilon_h]$ through the zones of laminar and turbulent flow is taken into account, an improved relationship can be obtained. This has been done in different ways by Taylor, Von Karman, and Martinelli, as discussed in Reference [7].

Experimentally it has been found that heat transfer data for long smooth tubes correlate very well, with

$$\frac{h_f D}{k_b} = 0.023 \left(\frac{\rho \overline{U} D}{\mu}\right)_b^{0.8} \left(\frac{\mu c_p}{k}\right)_b^{0.33}, \tag{4-31}$$

where the subscript b denotes evaluation at the bulk mean fluid temperature.

This may also be written

$$\frac{h_f D}{k} = 0.023 \left(\frac{\rho \bar{U} D}{\mu}\right)^{0.8} \left(\frac{\mu c_p}{k}\right)^{0.33}$$

or

$$\frac{h_f}{\rho c_p \bar{U}} = 0.023 \left(\frac{\bar{U} D}{\nu}\right)^{-0.2} \left(\frac{\mu c_p}{k}\right)^{-0.67}. \tag{4-32}$$

Comparing Eq. (4–32) with Eq. (4–13), we may see that

$$\left(\frac{h_f}{\rho c_p \bar{U}}\right) \mathrm{Pr}^{0.67} = \frac{C_f}{2}. \tag{4-33}$$

Written in this way, the turbulent heat transfer–skin friction analogy takes the form due to Colburn,

$$\mathrm{St}\,\mathrm{Pr}^{0.67} = \frac{C_f}{2},$$

where $\mathrm{St} = h_f / \rho c_p \bar{U}$ is the Stanton number.

Colburn's form of the analogy is used to calculate the heat transfer to the turbulent boundary layer of a flat plate as follows [7]:

$$\frac{h_f}{\rho c_p U_\infty} \mathrm{Pr}^{2/3} = \frac{C_f}{2}.$$

To evaluate the skin friction coefficient C_f in this case, we may use Eqs. (4–15), (4–19), or (4–20). Experimental data has shown that this analogy is useful in a fairly wide range of pressure gradient.

References

1. SCHLICHTING, H., *Boundary Layer Theory*. New York: McGraw-Hill, 1960

2. PRANDTL and TIETJENS, *Fundamentals of Hydro- and Aero-Mechanics*. New York: McGraw-Hill, 1934

3. SHAPIRO, ASCHER H., *The Dynamics and Thermodynamics of Compressible Fluid Flow*, Volume II. New York: The Ronald Press Company, 1954; Chapter 26

4. SCHUBAUER, G. B., "Airflow in a Separating Laminar Boundary Layer," *NACA Report No.* 529, 1935

5. THWAITES, B., "Approximation of the Laminar Boundary Layer," *Aero Quarterly*, I, 245–280, 1949

6. KLEBANOFF, P. S., "Characteristics of Turbulence in a Boundary Layer with Zero Pressure Gradient," *NACA Report No.* 1247, 1955

7. ROHSENOW, W. M., and H. Y. CHOI, *Heat Mass and Momentum Transfer*. New York: Prentice-Hall, 1961

8. ECKERT, E. R. G., and R. M. DRAKE, *Heat and Mass Transfer*. New York: Mc-Graw-Hill, 1959

9. MCADAMS, W. H., *Heat Transmission*. New York: McGraw-Hill, 1954

10. CLAUSER, F. H., "The Turbulent Boundary Layer," *Advances in Applied Mechanics,* IV. New York: Pergamon Press, 1956

11. FLACHSBART, O., "Winddruck auf Gasbehälter," Reports of the *Aerodynamische Versuchsanstalt in Göttingen,* fourth series, pages 134–138, 1932

12. NIKURADSE, J., *Laminare Reibungsschichten an der längsangeströmten Platte.* Monograph Zentrale für Wissenschaftliches Berichtswesen, Berlin, 1942

13. HOWARTH, L., "On the Solution of the Laminar Boundary Layer Equations," Proc. Royal Society, London, *A*164, 547, 1938

14. HOLSTEIN, H. and T. BOHLEN, "Ein einfaches Verfahren zur Berechnung laminarer Reibungsschichten, die dem Näherungsansatz von K. Pohlhausen genügen," *Lilienthal Bericht, S*10, 5–16, 1940

15. MOODY, L. F., "Friction Factors for Pipe Flow," *ASME*, Trans. No. 66, pages 671–784, 1944

16. LAUNDER, B. E., "An Improved Pohlhausen-Type Method of Calculating the Two-Dimensional Laminar Boundary Layer in a Pressure Gradient." Trans. *ASME, J. Heat Transfer*, **86,** Series C, No. 3, August 1964

Problems

1. Show that for fully developed laminar flow in a round tube, the velocity profile is given by

$$\frac{u}{U} = 2\left[1 - \left(\frac{r}{R}\right)^2\right]$$

and the pressure gradient is given by

$$\frac{dp}{dx} = -\frac{8\mu U}{R^2},$$

in which U is the mean velocity, $\dot{m}/A\rho$, R is the tube radius and u is the velocity at any radius r.

Notes: (i) For axisymmetric laminar flow, the momentum equation which replaces Eq. 4–2 (which was derived for two-dimensional flow) is

$$u\frac{\partial u}{\partial x} + v\frac{\partial u}{\partial r} = -\frac{1}{\rho}\frac{\partial p}{\partial x} + \frac{v}{r}\frac{\partial}{\partial r}\left(r\frac{\partial u}{\partial r}\right),$$

and the continuity equation which replaces Eq. 4–1 is

$$\frac{\partial u}{\partial x} + \frac{1}{r}\frac{\partial(rv)}{\partial r} = 0.$$

(ii) If the flow is fully developed the axial velocity component u will (by definition) be a function of r only.

2. Fuel flows in a 10-ft pipe from a fuel tank to the pump of a jet engine. The flow rate is 40 gal/min. At high altitudes the fuel-tank pressure is low and there may be considerable danger of the fuel "boiling" or cavitating as it enters the pump. In such a case the pump would be vapor-locked and would cease to deliver fuel to the engine. Show qualitatively how the pressure drop in the line varies with its diameter (over a wide range). What would the minimum pipe diameter be if a pipe pressure loss of only 0.5 psi were tolerable? The pipe has no sharp bends or elbows. The kinematic viscosity of the fluid is 2.6×10^{-5} ft^2/sec and its specific gravity is 0.8.

3. Show that the definitions of displacement thickness δ^* and momentum thickness θ which will have the same physical significance for compressible flow as Eqs. (4–4) do for incompressible flow are

$$\delta^* = \int_0^\delta \left(1 - \frac{\rho u}{\rho' U}\right) dy,$$

$$\theta = \int_0^\delta \frac{\rho u}{\rho' U}\left(1 - \frac{u}{U}\right) dy,$$

where ρ' is the free-stream density and U is the free-stream velocity. Calculations of the boundary layer on the wall of a regeneratively cooled rocket motor may indicate a negative δ^*. Is this physically reasonable? Would a negative value of θ be physically reasonable?

4. A long duct of fairly complex geometry is to transmit a given mass flow of hot gas at a constant average velocity. The pressure drop for the duct is to be determined experimentally by passing cold gas at an equal mass flow rate through it and measuring the pressure drop, the mean pressure level being the same in the two cases. The temperatures of the hot and cold gas streams are 500°R and 2000°R, respectively. How should the pressure drop in the two cases compare? The Mach number is small in both cases, and the Reynolds number is high enough so that changing it by a factor of three or less would have an insignificant effect on the flow pattern in the duct.

5. Estimate the pressure drop due to friction in a tube twenty diameters long in which air flows at 600 ft/sec. The inlet air temperature and pressure are 1000°R and 100 psia, respectively, and the tube diameter is 1 ft. If the wall temperature were maintained at 2000°R, how much increase in fluid temperature could be expected during flow through the tube? As an approximation the velocity profile at entrance may be assumed fully developed.

6. An airstream of speed 500 ft/sec and temperature 5000°R travels on the inside of a 1-ft I.D. steel tube whose wall thickness is 0.100 in. On the outside of the tube, water coolant flows coaxially in an annular space 0.250 in. thick. The coolant velocity is 30 ft/sec and it has a local temperature of 40°F. Both flows are approximately fully developed. The pressure of the airstream is around 20 psia. Estimate the maximum wall temperature of the tube.

7. It is empirically observed that the best divergence angle for a two-dimensional subsonic diffuser of a given area ratio is about 7°. Such a diffuser is sketched in solid lines in the figure at the top of page 100. Two other diffusers for the same area ratio are shown, having divergence angles of 15° and 3°. It is observed that the flow in the 15° cone is separated and that there is a large stagnation pressure loss. The flow

PROBLEM 7

in the 3° cone, on the other hand, is not separated, but the stagnation pressure loss is again large relative to the loss in the 7° case. Why would you expect the flow in the 15° cone to be more likely to separate than that in the 7° cone? How would you explain the larger losses of the 3° cone?

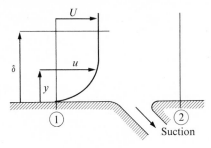

PROBLEM 8

8. An incompressible boundary layer is subjected to a pressure rise ΔP between stations ① and ② which are very close together. The shape of the boundary layer at station ① is satisfactorily described by $u/U = (y/\delta)^{1/7}$, in which U and δ are known quantities. Neglecting the action of viscous forces between stations ① and ②, how much boundary layer fluid would need to be removed by suction to prevent separation at station ②? How would viscosity tend to affect this result?

9. Show that, if the two expressions for velocity of Eqs. (4–9) and (4–10) are to match in some intermediate y-region, then the skin-friction coefficient is in fact determined by this velocity profile, and that (taking $\bar{U}/U_c \simeq 0.8$) it can be expressed as a function of the Reynolds number as follows:

$$\frac{1}{\sqrt{C_f}} = 3.3 \log_{10}\left[\mathrm{Re}\sqrt{C_f}\right] + 1.7,$$

in which

$$\mathrm{Re} = \frac{\bar{U}D}{\nu} \quad \text{and} \quad C_f = \frac{2\tau_0}{\rho \bar{U}^2}.$$

For comparison, the experimental result given by Nikuradze is

$$\frac{1}{\sqrt{C_f}} = 4.0 \log_{10}\left(\mathrm{Re}\sqrt{C_f}\right) - 0.4.$$

10. Show how to estimate the boundary layer thickness in a nozzle in which the free-stream velocity is proportional to the distance from the point where the boundary

layer begins to grow. Neglect compressibility, and consider two cases: (1) completely laminar boundary layer, and (2) completely turbulent boundary layer.   Indicate how the skin friction may be determined. Estimate very roughly the boundary layer displacement thickness at the throat of a nozzle the length of whose converging section is 1 ft. The diameter of the nozzle is also 1 ft and the throat velocity is 1000 ft/sec. The average density and viscosity of the fluid are 0.1 lb_m/ft^3 and 5×10^{-4} ft^2/sec, respectively.

5

Electrostatics, Magnetostatics, and Plasmas

5-1 INTRODUCTION

Electrostatic and electromagnetic forces may be used to accelerate a variety of propellants in devices which are described in Chapter 16. The propellant may consist of separate beams of ions and electrons, each capable of acceleration in an electrostatic field; or, alternatively, it may be a mixture of ions, electrons, and neutral particles flowing in a single stream. Such a stream, though electrically neutral, may conduct an electrical current so that the fluid, or plasma as it is called, is subject to an electromagnetic driving force much the same as a current-carrying conductor in an electric motor.

The basic equations of electrostatics and magnetostatics are stated briefly in this chapter, it being assumed that these relationships are familiar to the student. The effects of motion relative to the electric and magnetic fields will be treated in more detail, however. This chapter also includes a simplified discussion of the internal mechanisms of current conduction and momentum exchange in plasmas; it is assumed that the student is familiar with elementary kinetic theory concepts. The intent is to present, in simple form, an introductory analysis of the important phenomena. A full quantitative treatment, including many important but complicated effects, would require a much lengthier discussion and, at the same time, be out of place in a text of this scope, since the application of plasmas to propulsion has yet to be proven feasible. References are provided to literature on the subject of plasma behavior.

5-2 BASIC EQUATIONS OF ELECTROSTATICS AND MAGNETOSTATICS

Derivations and applications of the following relationships are contained in many texts on the subject, such as References [1], [2], and [3] (named in the approximate order of their increasing level of treatment). Reference [2] is especially good for an elementary vector treatment of the concepts.

The force on two charges is given by Coulomb's law, which states that

$$F \propto \frac{q_1 q_2}{r^2},$$

where q_1 and q_2 are the charges and r is the distance between them. Noting that forces are vectorially additive, the force on a charge q due to n other charges is given by

$$\mathbf{F}_q = \frac{q}{4\pi\epsilon} \sum_{i=1}^{n} \frac{q_i}{r_i^2} \mathbf{r}_i,$$

where $\mathbf{r}_i$ is a unit vector pointing from q_i to q, and $1/4\pi\epsilon$ is a proportionality constant (depending on the units of F, q, r, and the substance separating the charges). The constant ϵ is called the permittivity of the material. For a vacuum, ϵ has the particular value ϵ_0, called the permittivity of free space. In the convenient rationalized MKS system (whose units have been implied by the factor $1/4\pi$ above), in which F is in newtons, r in meters, and q in coulombs, ϵ_0 has the value 8.85×10^{-12} (coulomb)^2 per (newton)·(meter)^2. The permittivity of gases is very close to ϵ_0.

If the number of charges n is large (as is almost always the case), it is convenient to consider the charge as being continuously distributed through space, or perhaps over surfaces. Then, defining a volume charge density* ρ_q and a surface charge density σ_q, this summation becomes the integral equation (for gas or vacuum),

$$\mathbf{F}_q = \frac{q}{4\pi\epsilon_0} \left(\int_{\mathcal{V}} \frac{\rho_q \mathbf{r}}{r^2} \, d\mathcal{V} + \int_A \frac{\sigma_q \mathbf{r}}{r^2} \, dA \right), \tag{5–1}$$

in which $\mathcal{V}$ and A are the entire volume and surface of interest.

It is convenient to define the *electric field strength* $\mathbf{E}$ at a point as the force per unit (positive) charge at that point due to whatever other charges may be present. That is,

$$\mathbf{E} \equiv \frac{\mathbf{F}_q}{q} = \frac{1}{4\pi\epsilon_0} \left(\int_{\mathcal{V}} \frac{\rho_q \mathbf{r}}{r^2} \, d\mathcal{V} + \int_A \frac{\sigma_q \mathbf{r}}{r^2} \, dA \right). \tag{5–2}$$

For cartesian coordinates with unit vectors $\mathbf{x}$, $\mathbf{y}$, $\mathbf{z}$, we have

$$\mathbf{E} = E_x \mathbf{x} + E_y \mathbf{y} + E_z \mathbf{z}.$$

If Eq. (5–2) is applied to an infinitesimal volume of size $dx\,dy\,dz$ with an internal charge density ρ_q (and no surface charge distribution), it may be shown that

$$\frac{\partial E_x}{\partial x} + \frac{\partial E_y}{\partial y} + \frac{\partial E_z}{\partial z} = \frac{\rho_q}{\epsilon_0}. \tag{5–3}$$

---

* Like any continuum concept, charge density has meaning so long as the spacing between charges is much less than the smallest significant dimension in the problem considered.

By using the vector operator $\mathbf{\nabla}$, Eq. (5–3) may be written in the more general form

$$\mathbf{\nabla} \cdot \mathbf{E} = \frac{\rho_q}{\epsilon_0}. \tag{5–4}$$

This statement is known as Gauss's Law.

The work necessary to move a charge q in any electrostatic field is dependent only on the endpoints of the displacement. Hence the field may be described by a potential V such that

$$dV = -\mathbf{E} \cdot d\mathbf{s}, \tag{5–5}$$

where $d\mathbf{s}$ is the (vector) increment of displacement. In the MKS system the units of V, work per unit charge, are newton-meters per coulomb, commonly called *volts*. If an isolated charged particle moves through an electrostatic field, the sum of its electrical potential and kinetic energies remains constant (neglecting changes of gravitational potential energy). Thus for a particle of mass m and charge q,

$$\frac{mu^2}{2} + qV = \text{const.} \tag{5–6}$$

Equation (5–5) may be written, for cartesian coordinates,

$$dV = -(E_x\,dx + E_y\,dy + E_z\,dz),$$

so that

$$E_x = -\frac{\partial V}{\partial x}, \qquad E_y = -\frac{\partial V}{\partial y}, \qquad E_z = -\frac{\partial V}{\partial z}.$$

Again using the vector operator $\mathbf{\nabla}$, we may relate the field strength and the potential gradient by

$$\mathbf{E} = -\mathbf{\nabla}V. \tag{5–7}$$

Combining Eqs. (5–4) and (5–7), we obtain Poisson's equation relating potential and charge density,

$$\nabla^2 V = -\frac{\rho_q}{\epsilon_0}, \tag{5–8}$$

where $\nabla^2 = \mathbf{\nabla} \cdot \mathbf{\nabla}$ is the Laplacian operator.
In cartesian coordinates this is

$$\frac{\partial^2 V}{\partial x^2} + \frac{\partial^2 V}{\partial y^2} + \frac{\partial^2 V}{\partial z^2} = -\frac{\rho_q}{\epsilon_0}.$$

The Laplace equation is simply the special case of this expression in which ρ_q is zero.

Moving charges sometimes experience forces other than those which they would experience at the same position without motion. These are *magnetic forces*, and the regions in which they occur are called *magnetic fields*. Magnetic forces depend on the magnitude of the charge, the magnitude and direction of the charge velocity, and the strength of the magnetic field. In general the magnetic force on a charge q at velocity $\mathbf{u}$ is given by the vector product

$$\mathbf{F} = q\mathbf{u} \times \mathbf{B}. \qquad (5\text{–}9)$$

Equation (5–9) may be considered a definition of the magnetic field intensity vector $\mathbf{B}$, which is also called the *magnetic induction* or *magnetic flux density*. Its units, in the MKS system, are

$$\frac{\text{newtons/coulomb}}{\text{meters/second}} = \frac{\text{volt-seconds}}{(\text{meter})^2} = \frac{\text{webers}}{(\text{meter})^2}.$$

The gauss is another common unit for magnetic flux density. It is related to the rationalized MKS unit by

$$1 \text{ weber/m}^2 = 10^4 \text{ gauss.}$$

The earth's very weak magnetic field is only about 0.5 gauss, while "strong" permanent magnets may produce 5 to 10 thousand gauss at a pole face, and very strong electromagnets may produce 10^6 gauss.

Since an electrical current consists of charges in motion, the force on a current-carrying element can be determined from the forces on the individual charges. The current density vector $\mathbf{j}$ is defined as the rate per unit area at which charge travels through the current-carrying element. It is measured in coulombs per second per square meter or amperes per square meter. The current density is related to the volume charge density ρ_q of the moving charges and their velocity $\mathbf{u}$ by

$$\mathbf{j} = \rho_q\mathbf{u}. \qquad (5\text{–}10)$$

The electromagnetic force $d\mathbf{F}$ on the charge $\rho_q \, d\upsilon$ contained within any increment of volume $d\upsilon$ is

$$d\mathbf{F} = (\rho_q \, d\upsilon)\mathbf{u} \times \mathbf{B} = (d\upsilon)\mathbf{j} \times \mathbf{B}.$$

Therefore the force per unit volume is

$$\mathbf{F} = \mathbf{j} \times \mathbf{B}. \qquad (5\text{–}11)$$

Magnetic fields are induced by electrical currents, according to the law of Biot and Savart. The contribution $d\mathbf{B}$ to the magnetic field at a point in space is related to the current I in an increment of conductor $d\mathbf{L}$ at another point by

$$d\mathbf{B} = \frac{\mu I}{4\pi r^2} \, d\mathbf{L} \times \mathbf{r}. \qquad (5\text{–}12)$$

Here r is the distance between field point and conductor, while $\mathbf{r}$ is a unit vector pointing from the current element to the field point. The proportionality constant μ is called the *permeability*. Its magnitude depends on the units employed and the medium separating the two points. In the rationalized MKS system, the permeability of free space μ_0 has the value

$$\mu_0 = 4\pi \times 10^{-7} \frac{\text{newton}}{(\text{ampere})^2}.$$

The permeability of ferromagnetic materials (such as nickel and iron) is much larger than μ_0, but the permeability of most other materials is extremely close to μ_0.

The magnetizing field vector **H** is often used in place of **B**. In vacuum it is defined simply as

$$\mathbf{H} \equiv \frac{\mathbf{B}}{\mu_0} \quad \text{(vacuum)},$$

with units in amperes per meter. The vector **H** depends only on the distribution of applied currents. If a material medium is present, the magnetic field **B** is related to **H** by

$$\mathbf{B} = \mu_0\mathbf{H} + \mathbf{M},$$

where **M** is called the magnetization of the material. For many materials **M** is proportional to **H** (hence the name "magnetizing field"), so that **B** may be expressed in the form

$$\mathbf{B} = \mu_0\mathbf{H}(1 + x_m) = \mu\mathbf{H},$$

in which the constant x_m is called the magnetic susceptibility and μ is the permeability of the material in question. For some materials (permanent magnets, for example), **M** is not a function of **H** so that this last expression is not always applicable. In any event, the local **B** determines the magnetic force on a moving charge.

5–3 MOTION WITHIN ELECTRIC AND MAGNETIC FIELDS

In the following we shall have occasion to treat electric and magnetic effects as observed from a reference frame which is in motion with respect to the devices which create the fields. Hence we should investigate the effect of motion between the "field" and the observer. Consider the motion of a charge q through a uniform electric field **E** and magnetic field **B** as seen by two observers, one at rest with respect to the magnetic field and the other at rest with respect to the charge. Both see the same charge and both observe the same forces on the charge (so long as the velocity is small compared to the velocity of light). For the observer at rest, Eq. (5–9) and the definition of the electrostatic field intensity allow the total force on the particle to be written

$$\mathbf{F} = q(\mathbf{E} + \mathbf{u} \times \mathbf{B}).$$

To the moving observer the charge appears to have no velocity. He therefore must suppose that all this force is due to an "apparent" electrostatic field **E'**.

Equating the two forces, $\mathbf{F} = q(\mathbf{E} + \mathbf{u} \times \mathbf{B}) = q\mathbf{E}'$, or

$$\mathbf{E}' = \mathbf{E} + \mathbf{u} \times \mathbf{B}. \tag{5–13}$$

The foregoing analysis intentionally avoids the effect of motion on the observation of the magnetic field by specifying that the charge has no motion relative to the moving observer. Consider next the force on a charge q moving with a velocity $\mathbf{u}_q$ relative to the $\mathbf{B}$-field as seen by an observer moving with a different velocity $\mathbf{u}_0$ relative to the $\mathbf{B}$-field. Since the force does not depend on the motion of the observer,

$$\mathbf{F} = q[\mathbf{E}_0 + (\mathbf{u}_q - \mathbf{u}_0) \times \mathbf{B}_0] = q(\mathbf{E} + \mathbf{u}_q \times \mathbf{B}),$$

where $\mathbf{E}_0$ and $\mathbf{B}_0$ are the fields seen by the moving observer and $(\mathbf{u}_q - \mathbf{u}_0)$ is the charge velocity relative to him. *By definition*, $\mathbf{E}_0$ is the force per unit charge on a charge stationary *with respect to the observer*. Thus, from Eq. (5–13), we have

$$\mathbf{E}_0 = \mathbf{E} + \mathbf{u}_0 \times \mathbf{B}.$$

Therefore

$$q[(\mathbf{E} + \mathbf{u}_0 \times \mathbf{B}) + (\mathbf{u}_q - \mathbf{u}_0) \times \mathbf{B}_0] = q(\mathbf{E} + \mathbf{u}_q \times \mathbf{B})$$

or

$$(\mathbf{u}_q - \mathbf{u}_0) \times \mathbf{B}_0 = (\mathbf{u}_q \times \mathbf{B}) - (\mathbf{u}_0 \times \mathbf{B}) = (\mathbf{u}_q - \mathbf{u}_0) \times \mathbf{B},$$

and it follows that

$$\mathbf{B}_0 = \mathbf{B}.$$

Thus the motion of an observer in a magnetic field can be accounted for entirely by a change in the electrostatic field (so long as his velocity is small compared with the speed of light).

5–4 PLASMAS

A *plasma* is an electrically conducting gas. It consists of a collection of neutral atoms, molecules, ions, and electrons. The number of ions and the number of electrons are equal so that, on the whole, electrical neutrality prevails. Because of its ability to conduct electrons, the plasma can be subjected to electromagnetic -forces in much the same way as solid conductors of electric motors. Given the practicability of electromagnetic acceleration or deceleration, plasmas may be used as the working fluid in propulsion or electrical power generation.

In this brief treatment it will be assumed that the plasma can be considered a fluid continuum. Nevertheless, in order to understand the conduction of electrons through such a fluid, it is necessary to consider the behavior of both the charged and the neutral constituent particles. This is especially so in the presence of magnetic fields.

A gas in general is composed of particles which are very widely spaced relative to the particle dimensions. These particles move about in a random fashion, suffering collisions with one another and with the walls of their container.

The relationships between macroscopic (i.e., continuum) properties and the microscopic particle motions are given by kinetic theory. We shall simply summarize here the results obtained in many texts (as, for example, Reference [4]). The pressure exerted by a gas containing n particles per unit volume, each particle being of mass m, is given by

$$p = \tfrac{1}{3} n m \overline{u^2}, \tag{5–14}$$

where $\overline{u^2}$ is the average squared velocity given by

$$\overline{u^2} = \frac{\int_0^n u^2 \, dn}{n},$$

in which dn is the number of particles having velocities from u to $u + du$ and n is the total number of particles. The average squared velocity and temperature are related by

$$\frac{1}{3} m \overline{u^2} = \frac{\overline{R}}{N_0} T = kT,$$

where N_0 = Avogadro's number, 6.0251×10^{26} molecules/kg-mole,

$\overline{R}$ = universal gas constant = 8.3144×10^3 joules/kg-mole·°K,

k = Boltzmann's constant = 1.3803×10^{-23} joule/molecule·°K.

This is usually expressed in terms of the average particle kinetic energy,

$$\tfrac{1}{2} m \overline{u^2} = \tfrac{3}{2} kT. \tag{5–15}$$

In these expressions, u is measured relative to the bulk motion of the gas so that p and T are, as the notation indicates, the *static* properties of the gas. In addition to the root-mean-square velocity, which is sufficient for these expressions, we shall be interested in two additional average particle velocities. For a gas in equilibrium (i.e., one which has been at the same state for some time), the velocity distribution function, defined by $dn = nf(u) \, du$, is given by [4]

$$f(u) = \frac{4}{\sqrt{\pi}} \left(\frac{m}{2kT} \right)^{3/2} u^2 e^{-mu^2/2kT}, \tag{5–16}$$

which is shown graphically in Fig. 5–1.

It may be verified from Eq. (5–16) that the root-mean-square velocity, consistent with Eq. (5–15), is

$$u_{\text{rms}} \equiv \sqrt{\overline{u^2}} = \sqrt{3kT/m}.$$

The average or arithmetic velocity is

$$u_{\text{ave}} = \frac{1}{n} \int_0^n u \, dn = \sqrt{(8/\pi)(kT/m)}, \tag{5–17}$$

and the most probable velocity (that with the highest dn/du) is

$$u_{\mathrm{mp}} = \sqrt{2(kT/m)}. \tag{5–18}$$

Another important variable in gas or plasma behavior is the collision frequency ν. Here, a collision is defined as that which occurs when one particle enters the force field of another and therefore undergoes a change in its motion.

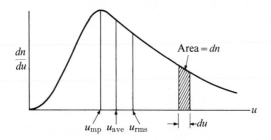

FIG. 5–1. Maxwell velocity distribution, showing most probable, arithmetic mean, and root-mean-square velocities.

Consider the motion of a single particle through a "sea" of surrounding particles. The frequency with which it collides with other particles is proportional to its average speed u_{ave} and the number of particles per unit volume:

$$\nu = Qnu_{\mathrm{ave}}.$$

The proportionality constant Q is measured in units of area and is called the *collision cross section* for a particular type of collision. Collisions may be classified according to the result of the interaction. For example, one may be interested only in the frequency of collisions which result in ionization, or in the loss of a certain increment of momentum, etc. This information can be collected by defining separate proportionality constants, called cross sections, for the separate processes such as "ionization cross section," etc. If there are several "target" species, the total collision frequency of that type is just the sum of the separate terms,

$$\nu = u_{\mathrm{ave}} \sum_{j} Q_j n_j. \tag{5–19}$$

Figure 5–2 shows typical experimentally determined cross sections of gas and alkali metal. Note that Q is quite energy-dependent.

It is possible (though in some instances misleading) to interpret the cross section as the geometric area of the target. Consider the simple case of collisions between rigid spheres. As a "projectile" of radius r_1 travels through a sea of "targets" of radius r_2, it is clear that any targets whose centers are within a cylinder of radius $r_1 + r_2$ centered on the path of the projectile will be hit. As the projectile moves, it sweeps out a collision volume at the rate $\pi(r_1 + r_2)^2 u_{\mathrm{ave}}$, and collisions occur

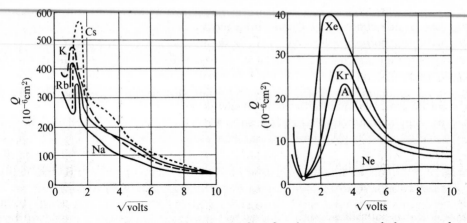

FIG. 5–2. Typical total-collision cross sections for electron-atom and electron-molecule collisions, from Delcroix [5].

at the rate

$$\pi(r_1 + r_2)^2 \, u_{\text{ave}} n.$$

Hence, for this case,

$$Q = \pi(r_1 + r_2)^2. \tag{5-20}$$

This simple picture does not account for the motion of the targets or for the distribution of projectile velocities. Computations which allow for these effects yield a correction factor for Eq. (5–19) which is of order of magnitude unity. For the particular case of electron projectiles, it is quite reasonable to assume that the targets are at rest since almost all the relative motion is due to the much faster electrons. However, if the treatment is to be accurate, it is necessary to take into account the distribution of electron velocities.

Associated with the collision frequency is the *mean free path* λ, or average distance between collisions; it can be obtained by dividing the total distance traveled by the number of collisions during the travel time t. Thus

$$\lambda = \frac{u_{\text{ave}} t}{\nu t}$$

or

$$\lambda = \frac{1}{\sum_j Q_j n_j}. \tag{5-21}$$

The collision cross section may be subdivided according to the nature of the interaction. The simplest interaction is an *elastic collision* in which the total kinetic energy of the particles is conserved. In *inelastic collisions*, energy is also conserved, of course, but part of the energy is absorbed by the particles, causing, for example, ionization or radiation.

As we shall see, the conductivity of a gas is primarily due to the motion of electrons through it. To be a conductor, it is necessary that the gas be at least partially ionized. The degree of ionization α of a plasma is defined by

$$\alpha = \frac{n}{N - n},$$

in which n is the number of electrons, or ions, per unit volume and N is the number of neutral particles per unit volume. In general, plasmas can be roughly separated into two groups [5]:

Weakly ionized gases: $\alpha < 10^{-4}$,

Strongly ionized gases: $\alpha > 10^{-4}$.

The basis for this classification is as follows. In the weakly ionized gas, collisions between electrons and neutral particles are very much more frequent than collisions between electrons and ions, so that it is possible to ignore the latter kind of collision altogether. In calculations of electrical conductivity, for example, this is a considerable simplification. Electron–ion or "coulomb" collisions are more complex because the significant extent of the force field between the two is not well defined. Most laboratory plasmas are weakly ionized. In pure argon the coulomb collisions become significant only at temperatures of the order of 7000°K [6]. Alkali metals dissociate at much lower temperatures, and are often introduced in small quantities into a gas stream in order to raise the ionization. In this case also, coulomb collisions become important at an ionization of about 10^{-4} [5]. The treatment of plasmas in this chapter and in Chapter 16 is restricted almost entirely to weakly ionized gases.

A plasma consists of equal numbers of ions and electrons mixed with neutral particles. An estimate of the order of magnitude of the minimum dimension over which the mixture can be considered electrically neutral is given as follows. Suppose that all the electrons should happen to leave a certain spherical control volume $\Delta \mathcal{V}$, so that it contains a collection of ions at the average ion or electron density n_e of the plasma. The potential variation within the volume can be found from Eq. (5–7) by

$$\nabla^2 V = -\frac{en_e}{\epsilon_0},$$

where en_e is the charge density. For spherical symmetry, this equation may be written

$$\nabla^2 V = \frac{1}{r^2}\frac{d}{dr}\left(r^2\frac{dV}{dr}\right) = -\frac{en_e}{\epsilon_0}.$$

The solution of this equation is

$$V = \frac{en_e}{6\epsilon_0}(r_0^2 - r^2),$$

where r_0 is the outside radius of the assumed sphere and $V(r_0) = 0$. An electron cannot leave this region (on the average) unless its thermal energy, $3/2kT$, is greater than the energy difference eV. Thus a rough estimate of the dimensions over which thermal energy may disturb electrical neutrality can be given by setting these two energies equal:

$$eV = \frac{e^2 n_e r_0^2}{6\epsilon_0} = \frac{3}{2} kT.$$

This yields $r_0 = \sqrt{9\epsilon_0 kT/e^2 n_e}$, which can be thought of as the order of magnitude of the minimum distance over which the plasma may be considered neutral. Except for the numerical factor, this radius is equal to the so-called Debye length, h:
$h = \sqrt{\epsilon_0 kT/e^2 n_e}$.

In the following treatment it will be assumed that the Debye length is much smaller than the significant flow dimension; e.g., the channel width. If this condition is not satisfied, the fluid cannot be considered a plasma.

5–5 CONDUCTION IN PLASMAS IN THE ABSENCE OF MAGNETIC FIELDS

Because magnetic effects are complex, it is helpful to consider initially the mechanism of plasma conduction in the absence of magnetic fields. Since high electron densities correspond to unrealistically high plasma temperatures (for propulsion or power generation applications) it will be assumed that the electron density is very low. Further, to reduce the problem to bare essentials, it will be assumed that the plasma has uniform properties in a uniform electric field.

To be a conductor, a plasma must contain free electrons. Such electrons are made available by an ionization process in which a neutral atom A_n decomposes into a positive ion I^+ and a free electron, e^-:

$$A_n \rightarrow I^+ + e^-.$$

It requires an energy input, called the *ionization potential*, to bring about this process. If the free electrons are to persist within the plasma they must, in general, be in equilibrium with the remainder of its constituents. The energy input must then come about through a heating of the entire plasma. The possibility of ionization at high temperature was mentioned in Chapter 2 in connection with equilibrium composition calculations. By the methods of that chapter, one would expect the equilibrium electron content to be described in terms of the electron partial pressure by an equation of the form

$$\frac{p_e p_I}{p_n} = k_p(T),$$

where　　p_e = electron partial pressure,

　　　　p_I = ion partial pressure, and

　　　　p_n = neutral atom partial pressure.

Proceeding as in Chapter 2 and identifying the energy exchange with the ionization potential, one can deduce the proper form of the relationship between the equilibrium electron concentration and the state of the mixture. However, continuum methods are insufficient to evaluate certain constants which appear in the relationship. Saha [7] has shown that the proper expression is

$$n_e^2 = \left(\frac{2\pi m_e}{h^2}\right)^{3/2} (kT)^{3/2} e^{-\epsilon_i/kT} n_n \qquad (5\text{-}22)$$

in which n_e = free electron density, electrons/m^3,

   m_e = electron mass (9.11×10^{-31} kg),

   h = Planck's constant (6.62377×10^{-34} joule-sec),

   ϵ_i = ionization potential of the material considered,

   n_n = equilibrium neutral atom density.

The strong influence of ionization energy ϵ_i in determining the temperature necessary for significant ionization can be seen from this equation. Inert materials such as argon are attractive as working fluids, except that they generally possess high ionization potentials (15.68 eV for argon). On the other hand, the low-ionization-potential materials such as sodium, potassium, and cesium (ϵ_i = 5.12, 4.318, and 3.87 eV respectively) are not attractive as working fluids in themselves. For propulsion applications a suitable compromise is reached by "seeding" a relatively inert material with a small percentage of easily ionized material.

As free or conduction electrons move through a plasma in response to an applied electric field, their motion is retarded by a type of "drag" force. On the macroscopic scale this is manifest as electrical resistance. On the microscopic scale it is the result of collisions between electrons and other plasma constituents. In the absence of magnetic fields, Ohm's law for a plasma may be written as

$$j = \sigma_0 E,$$

in which j is the current density, E is the electric field strength, and σ_0 is the conductivity, the subscript 0 referring to the absence of magnetic fields. The conductivity is the reciprocal of the resistivity.

As a very good approximation, the electron-atom collision may be considered an encounter between a high-speed particle of negligible dimensions and a massive stationary particle; at low electron energies (< 1 eV [8]), both particles can be considered hard spheres. This being the case, we may assume that the average collision destroys all of the electron's forward momentum. Such a collision is said to be *isotropic*. If the electrons have energies greater than 1 eV, they may be observed to retain, on the average, a fraction of their pre-collision momentum [8]. However, even at temperatures of several thousands of °K, the electron thermal energy will be less than one eV.

Under the influence of an electric field, the free electrons in a plasma experience a drift motion which, of course, accounts for the conduction of a current. The drift velocity u_d is superposed on the thermal motion, and is of such magnitude that on the average the electrostatic and drag forces on an electron just balance. The drag force may be estimated by multiplying the electron collision frequency by the average momentum loss per collision. As a first approximation, consider only the drift motion of the electrons. Since electron drift velocity increases linearly with time between collisions and goes to zero at each collision, the average collision velocity is $2u_d$ and the momentum loss per collision is $2m_e u_d$. Thus, equating electrostatic and drag forces, we have

$$-eE = \nu_e(2m_e u_d),\qquad\qquad(5\text{–}24)$$

where $e =$ electronic charge magnitude (1.602×10^{-19} coul), and $\nu_e =$ electron collision frequency. There is no force arising from a variation of electron partial pressure, since the plasma is assumed uniform. The current density is given in terms of drift velocity by

$$j = -eu_d n_e.\qquad\qquad(5\text{–}25)$$

Thus, combining Eqs. (5–23), (5–24), and (5–25), we have

$$\sigma_0 = \frac{j}{E} = \frac{e^2 n_e}{2\nu_e m_e}.\qquad\qquad(5\text{–}26)$$

With Eqs. (5–17) and (5–19), for ν_e (the collision frequency in the real case being primarily a function of the *thermal* motion and hence not appreciably altered by the drift motion), we have, using T_e to denote the electron temperature,

$$\sigma_0 = \frac{e^2 n_e}{2\sqrt{(8/\pi)(kT_e/m_e)}\,m_e \sum_j n_j Q_j}$$

or

$$\sigma_0 = 0.313\,\frac{e^2}{(m_e kT_e)^{1/2}}\,\frac{n_e}{\sum n_j Q_j},\qquad\qquad(5\text{–}27)$$

where the summation on j extends over all neutral species. More rigorous calculations, taking the electron thermal motion into account, give results differing by a constant from this result. For a Maxwell electron velocity distribution, Lin, et al. [6] give

$$\sigma_0 = 0.532\,\frac{e^2}{(m_e kT_e)^{1/2}}\,\frac{n_e}{\sum n_j Q_j}\qquad\qquad(5\text{–}28)$$

for the simplifying assumption that Q_j does not vary appreciably over the range of electron velocities encountered.

At this point one can compare the relative contributions to conductivity of ion and electron motions. A similar analysis would yield the same expression for the ion conductivity σ_{0i} except that m_i and T_i would replace m_e and T_e, and the collision cross section would change. For the hard-sphere collision model, Eq. (5–20) gives the collision cross section. For electron-atom collisions, since $r_e \ll r_i$,

$$Q_j = \pi(r_e + r_i)^2 \approx \pi r_i^2.$$

For ion-atom collisions, where $r_i \approx r_j$,

$$Q_j \approx \pi(2r_i)^2 = 4\pi r_i^2.$$

Combining these expressions, the ratio of ion to electron conductivity becomes

$$\frac{\sigma_{0i}}{\sigma_0} = \frac{1}{4}\left(\frac{m_e T_e}{m_i T_i}\right)^{1/2}.$$

Thus, unless $T_e \gg T_i$ (see subsequent comments on nonequilibrium effects), the ion contribution to conductivity is negligible since

$$\frac{m_e}{m_i} = \frac{1}{1836\,A},$$

where A is the atomic weight of the ion species.

The amount of seeding material necessary to provide suitable conductivity is very small. In fact, it can be shown to have an optimum value [6]. Equation (5–28), for an essentially inert gas g, seeded with an easily ionized material s, becomes

$$\sigma_0 = 0.532\,\frac{e^2}{\sqrt{(m_e k T_e)}}\,\frac{n_e}{n_s Q_s + n_g Q_g},$$

where n_s is the density of the un-ionized seed material. Equation (5–22) indicates that the electron density is proportional to the square root of n_n or, in this case, since the seed material is the only ionizable species, of n_s. Thus

$$\sigma_0 \propto \frac{n_s^{1/2}}{n_s Q_s + n_g Q_g}.$$

This expression is maximized when $n_s/n_g = Q_g/Q_s$. As it happens that the cross section of the seed material is usually much greater than that of the inert gases, this corresponds to very low seed concentration. For example [9], for potassium, $Q_s \approx 3 \times 10^{-14}$ cm^2, while for argon, $Q_g \approx 6 \times 10^{-17}$ cm^2. Thus the maximum conductivity of this combination corresponds to about 0.2% neutral potassium.

Figure 5–3 indicates the calculated conductivity, as a function of temperature and density, of argon seeded with 1% potassium. Standard temperature and pressure refer to $p_0 = 1$ atm, $T_0 = 20°C$. The parameter $\omega_e \tau_e$ will be discussed in Section 5–6.

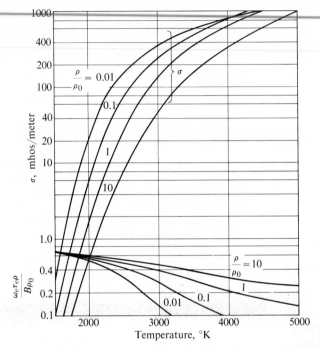

FIG. 5–3. Conductivity for argon seeded with 1% potassium for zero electric field, from Rosa [9]; B in webers/m^2.

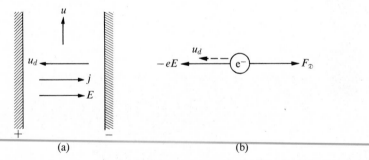

FIG. 5–4. Conduction at right angles to a moving plasma.

Consider conduction through a uniform plasma as it flows with velocity u in a rectangular channel, as in Fig. 5–4. To an observer moving with the fluid, the electrons have drift velocity u_d. Figure 5–4 illustrates the force balance on an electron in terms of the electrostatic force $-eE$ and an average (steady) drag force $F_{\mathcal{D}}$ as seen by an observer moving with the fluid. The fluid motion is introduced here simply to emphasize that it is electron motion relative to the plasma which is of importance. In this sense the flowing plasma may be considered a moving conductor.

5-6 CONDUCTION AND MOMENTUM EXCHANGE IN PLASMAS WITH MAGNETIC FIELDS

Electromagnetic interactions similar to those experienced in conventional motors and generators are possible when a plasma conducts a current in the presence of a magnetic field.   The electromagnetic force acts primarily on the free electrons which carry the current.   It is therefore important to examine the mechanism of momentum exchange from electrons to neutral particles, as well as the conductivity of the plasma.   Before introducing the complications due to collisions, it is helpful to consider the path of a collisionless electron.   Figure 5–5 illustrates the crossed electric and magnetic field configuration contemplated for many plasma devices. Suppose an electron with a velocity u_{e0} in the xy-plane is injected into such a region.   Consider first the simple case in which E equals zero.   Since the magnetic force is always normal to the electron velocity, its effect is to change the direction but not the magnitude of the velocity.   As a result the electron moves at constant speed in a circular path whose radius may be found by equating centrifugal and magnetic forces:

$$|e\mathbf{u}_{e0} \times \mathbf{B}| = eu_0 B = m_e \frac{u_{e0}^2}{r}$$

or

$$r = \frac{m_e u_e}{eB}. \tag{5-29}$$

The radian frequency ω_e of this motion, called the cyclotron frequency of the electron, is

$$\omega_e = \frac{u_e}{r} = \frac{eB}{m_e}. \tag{5-30}$$

Note that this is independent of the electron speed.

The presence of the electrostatic field tends to pull the electron in the $-E$-direction.   Conservation of energy [Eq. (5–6)] requires that the speed increase according

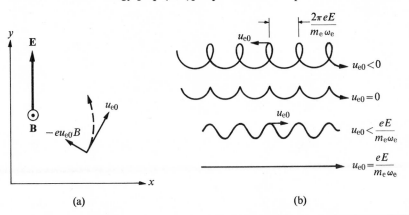

| (a) | (b) |

FIG. 5–5.   Electron trajectories in crossed uniform E and B fields.

to the local electrostatic potential. From Eq. (5–29), it can be seen that the instantaneous radius of curvature of the path increases as the speed increases, hence the path is not circular. In fact, the variable turning due to the magnetic field results in a net drift motion perpendicular to both **E** and **B** with no net motion in the **E**-direction. To treat this trajectory, it is convenient to consider the orthogonal components u_{ex} and u_{ey} of $\mathbf{u}_e$. Summing forces in the x-direction, for the fields of Fig. 5–5, we have

$$F_x = -e\dot{y}B,$$

where $\dot{y} = u_{ey}$. The resultant acceleration $du_{ex}/dt = \ddot{x}$ is

$$\ddot{x} = -\frac{e}{m_e}\dot{y}B = -\omega_e\dot{y}. \tag{5–31}$$

Similarly, in the y-direction, $F_y = -eE + e\dot{x}B$, and

$$\ddot{y} = \omega_e\dot{x} - \frac{eE}{m_e}. \tag{5–32}$$

The solution to these equations [10] is

$$x = -r\sin\omega_e t + \frac{eE}{\omega_e m_e}t, \qquad y = r\cos\omega_e t, \tag{5–33}$$

where

$$r = \frac{eE}{\omega_e^2 m_e} - \frac{u_{e0}}{\omega_e} \quad \text{and, at } t = 0, \quad x = 0 \text{ and } y = y_{\max}.$$

Typical solutions for various values of u_{e0} are shown in Fig. 5–5(b). For greater values of u_{e0} such that $u_{e0} > [(e/m_eE)/\omega_e]$, another set of curves, the mirror image of those shown, results with $y = y_{\min}$ at $t = 0$. The trajectory of an electron between collisions will be a portion of a curve of this type.

Consider again conduction through a plasma flowing in a rectangular duct as in Fig. 5–6(a). In this case a magnetic field **B** is imposed at right angles to both

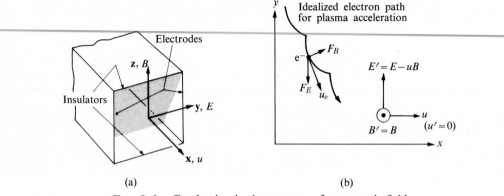

(a)   (b)

Fig. 5–6.   Conduction in the presence of a magnetic field.

u and **E**, as shown. Figure 5–6(b) illustrates an xy-plane within the channel as seen by an observer moving with the plasma. Note that for positive **u**, **E**, and **B** associated with the positive x-, y-, and z-directions, respectively,

$$E_y = |\mathbf{E}'| = |\mathbf{E} + \mathbf{u} \times \mathbf{B}| = E - uB.$$

Neglecting the thermal motion of the electrons, the path (relative to the gas) of an electron is a series of curved lines between collisions. If the average electron-atom collision is again assumed to result in total loss of electron momentum, then the average electron starts moving in the (minus) **E**′-direction after each collision, as shown in Fig. 5–6(b) [$u_{e0} = 0$, Fig. 5–5(b)]. As the electron gains velocity (energy) from the E' field, the magnetic force $-e\mathbf{u}_e \times \mathbf{B}$ bends the electron path. In this case the bending is in the positive x-direction. At each collision then, the electron-atom momentum exchange is such as to "push" the gas in the u-direction; hence a forward force acts on the plasma. If B is increased (or E decreased) sufficiently, E', and hence the electron motion, will reverse. The current then flows against the external field and the electrons tend to retard the gas, the result being extraction of electrical power, as in a generator.

It can be seen that the resultant current or electron drift is not parallel to the electric field. The component of current in the x-direction is called the *Hall current*. At this point it might seem that a Hall current would be necessary for force transmission, but we shall see that this is not so.

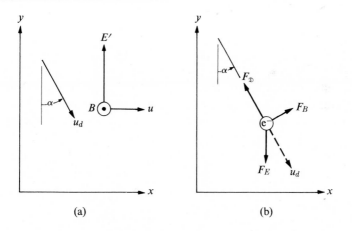

(a)   (b)

FIG. 5–7. Simplified electron drift and force balance for conduction in magnetic field.

For analytical simplicity consider the somewhat complex motion of Fig. 5–6(b) (which is already a simplified picture of reality) to be smoothed into a straight drift motion of velocity $\mathbf{u}_d$ at some angle α to the electric field, as shown in Fig. 5–7(a). For an electron moving at this (constant) velocity, the net force will be zero. Thus

$$-e(\mathbf{E}' + \mathbf{u}_d \times \mathbf{B}) = \mathbf{F}_\mathfrak{D}.$$

Making use of Eq. (5–24), with uniform acceleration between collisions, the drag force is simply $2\nu_e m_e \mathbf{u}_d$, and therefore

$$\mathbf{E}' + \mathbf{u}_d \times \mathbf{B} = -\frac{2\nu_e m_e \mathbf{u}_d}{e}.$$

Using Eq. (5–25) for $\mathbf{j}$, this becomes

$$\mathbf{E}' - \frac{\mathbf{j} \times \mathbf{B}}{en_e} = \frac{2\nu_e m_e}{e^2 n_e}\mathbf{j}$$

and, with Eq. (5–26) for σ_0, we obtain

$$\mathbf{E}' - \frac{\mathbf{j} \times \mathbf{B}}{en_e} = \frac{\mathbf{j}}{\sigma_0}. \qquad (5\text{–}34)$$

Taking the cross product of Eq. (5–34) with $\mathbf{B}$, we get

$$\mathbf{E}' \times \mathbf{B} - \frac{(\mathbf{j} \times \mathbf{B}) \times \mathbf{B}}{en_e} = \frac{\mathbf{j} \times \mathbf{B}}{\sigma_0}.$$

Then, using Eq. (5–34) to eliminate the term on the right-hand side, we have

$$\mathbf{E}' \times \mathbf{B} - \frac{(\mathbf{j} \times \mathbf{B}) \times \mathbf{B}}{en_e} = \frac{en_e}{\sigma_0}\left(\mathbf{E}' - \frac{\mathbf{j}}{\sigma_0}\right).$$

If $\mathbf{B}$ and $\mathbf{j}$ are perpendicular, the second term becomes $B^2\mathbf{j}/en_e$, so this relation may be solved for $\mathbf{j}$, with the following result:

$$\mathbf{j} = \frac{\sigma_0[(\sigma_0/n_e e)\mathbf{B} \times \mathbf{E}' + \mathbf{E}']}{1 + [(\sigma_0 B/n_e e)]^2} = \frac{\sigma_0[(\sigma_0 B/n_e e)(\mathbf{B}/B) \times \mathbf{E}' + \mathbf{E}']}{1 + [(\sigma_0 B/n_e e)]^2}.$$

The group $\sigma_0 B/n_e e$ is customarily expressed in terms of the cyclotron frequency of the electron ω_e, given by Eq. (5–30). With Eq. (5–26) for σ_0 again,

$$\frac{\sigma_0 B}{n_e e} = \frac{e^2 n_e}{2\nu_e m_e}\frac{B}{n_e e} = \frac{\omega_e}{2\nu_e} = \frac{1}{2}\omega_e \tau_e,$$

where $\tau_e = 1/\nu_e$ is the average time between electron collisions. Thus

$$\mathbf{j} = \frac{\sigma_0[\mathbf{E}' + \frac{1}{2}\omega_e\tau_e(\mathbf{B}/B) \times \mathbf{E}']}{1 + \frac{1}{4}\omega_e^2\tau_e^2}. \qquad (5\text{–}35)$$

With the coordinate system of Fig. 5–6, $\mathbf{B}/B = \mathbf{z}$, and $\mathbf{E}' = E_x'\mathbf{x} + E_y'\mathbf{y}$ (thus far $E_x' = 0$, but we shall need this term later). The $\mathbf{x}$- and $\mathbf{y}$-components of $\mathbf{j}$ are

$$j_x = \frac{\sigma_0}{1 + \frac{1}{4}\omega_e^2\tau_e^2}\left(E_x' - \frac{1}{2}\omega_e\tau_e E_y'\right), \qquad j_y = \frac{\sigma_0}{1 + \frac{1}{4}\omega_e^2\tau_e^2}\left(E_y' + \frac{1}{2}\omega_e\tau_e E_x'\right),$$

$$(5\text{–}36)$$

and the magnitude of the current is related to the magnitude of $\mathbf{E}'$ by

$$|\mathbf{j}| = \sqrt{j_x^2 + j_y^2} = \frac{\sigma_0}{(1 + \frac{1}{4}\omega_e^2\tau_e^2)^{1/2}} |\mathbf{E}'|. \tag{5–37}$$

For the present case, $E_x' = 0$, and the angle between $\mathbf{E}'$ and $\mathbf{j}$ (see Fig. 5–7) is given by

$$\tan \alpha = \frac{j_x}{j_y} = \frac{1}{2} \omega_e \tau_e$$

or

$$\alpha = \tan^{-1} \left(\tfrac{1}{2}\omega_e\tau_e\right). \tag{5–38}$$

From Eq. (5–37) for this case,

$$j = \frac{\sigma_0}{(1 + \frac{1}{4}\omega_e^2\tau_e^2)^{1/2}} E'.$$

The effect of the Hall current, apart from contributing a desired momentum exchange, is thus a reduction in apparent conductivity* by the factor $1/(1 + \frac{1}{4}\omega_e^2\tau_e^2)^{1/2}$. Considering the plasma a simple resistance, we can see that the effect of the magnetic field is to force the electrons to follow a longer path, in fact just $(1 + \frac{1}{4}\omega_e^2\tau_e^2)^{1/2}$ times as long. Thus the path resistance increases by this factor, or the conductivity decreases by its reciprocal.

In addition to adversely affecting conductivity, the Hall current affects the direction of the net force on the gas. Macroscopically this can be seen from Eq. (5–11), where the net force per unit volume of conductor (plasma, metal, or any other) was found to be $\mathbf{F} = \mathbf{j} \times \mathbf{B}$. Since $\mathbf{j}$ is not perpendicular to $\mathbf{u}$, it follows then that the net force on the fluid is *not* parallel to $\mathbf{u}$ as would be desirable. Microscopically this follows from the fact that the ions do not reach sufficient drift velocities to be affected by the magnetic field. Thus, summing the two drag forces (ion and electron) *on the gas*, it can be seen that the net force is simply F_B of Fig. 5–7(b).

In Chapter 16 we shall see that Hall current does in fact adversely affect the performance of many magnetohydrodynamic devices, as has been intimated here. For the present let us accept this possibility and inquire into the methods to reduce or eliminate Hall current.

Conventional electric motors utilize bundles of insulated wires extending in the y-direction, effectively reducing the conductivity in the x-direction to zero. This is clearly impossible in a flowing plasma. However, since the current is carried by the electrons, it can be seen by referring to Fig. 5–7(b) that the electrons might be forced to move only in the y-direction by the application of a suitable axial x-force. Such a force can be applied through an axial field E_x', and Eqs. (5–36) indicate

---

* Most writers use $\omega_e\tau_e$ instead of the $\frac{1}{2}\omega_e\tau_e$ which appears in the above expressions. However, this is not a matter of great consequence, since the collision frequency is hardly likely to be known within a factor of 2 anyway.

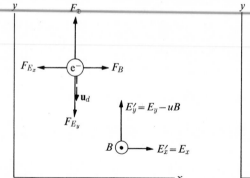

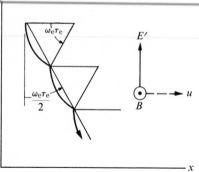

FIG. 5–8. Electron force balance for zero Hall current.

FIG. 5–9. Constant-speed model of electron trajectory.

just what this field must be. That is, to reduce the Hall current j_x to zero,

$$E'_x = \tfrac{1}{2}\omega_e \tau_e E'_y.$$

For this situation the force diagram for an electron is shown in Fig. 5–8. Note that the electron-gas momentum exchange is entirely in the y-direction. One might then inquire into the origin of the axial force, since Eq. (5–11) states that the force $\mathbf{j} \times \mathbf{B}$ exists regardless of the method of momentum exchange. The answer lies in the electrostatic force on the ions, since its axial component is not counterbalanced by a magnetic force as is the case for the electrons. The axial force on the ions is equal and opposite to the axial electrostatic force on the electrons and this, in turn, is equal and opposite to the magnetic force, $\mathbf{j} \times \mathbf{B}$, on the electrons. Thus, by the application of a suitable "Hall voltage," the Hall current can be eliminated without destroying the axial-force interaction. The practical attainment of a suitable axial field may be difficult, though segmented electrodes of varying potential distribution may produce an adequate potential distribution.

The physical significance of the so-called "Hall parameter," $\omega_e \tau_e$, can be seen in terms of a second simplified model. Suppose that the electron travels at the *constant* speed $|\mathbf{u}_d|$ between collisions. Its trajectory will then be circular and of radian frequency ω_e. As shown in Fig. 5–9, the average electron will sweep out an angle $\omega_e \tau_e$ radians between collisions. Thus, with the electron starting each arc in the (minus) E'-direction (the case for $E'_x = 0$ is shown), the drift velocity makes an angle $\omega_e \tau_e/2$ with E'. This is in good agreement with Eq. (5–38) for $\omega_e \tau_e < 1$.

The foregoing analyses contain many simplifications which render the results quantitatively incorrect, although they serve to illustrate the nature of electron current conduction and momentum exchange. The most obvious discrepancy between analyses and facts is the neglect of electron thermal motion when, in fact, this motion is much larger than the drift motion. Inclusion of the thermal motion introduces rather complicated averaging calculations which change the numerical results but do not change the qualitative results [compare Eqs. (5–27) and (5–28), for example]. In addition, we have neglected electron-ion or "coulomb" collisions, even though at high electron (and ion) concentrations these become dominant. Hence the conductivity expressions are qualitatively correct only at low temperatures; say the portions of the curves in Fig. 5–3 below 3000°K.

Our expressions have dealt with an "equilibrium" plasma in which the electron and neutral-particle temperatures were the same. Actually, appreciable non-equilibrium effects may be present because the energy input to a plasma goes almost entirely to the electrons as they move through the electrostatic field, and they in turn must transfer this energy to the gas as a whole. Owing to their very low mass relative to that of atoms, electrons lose very little energy per collision. For collisions with monatomic molecules, the relative energy loss per collision is [8] of the order of $2m_e/m_a$, in which m_e and m_a are the masses of the electron and molecule, respectively. This may be verified by considering a collision between elastic spheres of greatly differing mass. Collisions with polyatomic molecules can result in the excitation of rotational and vibrational motion within the molecule, and substantially more energy may be transferred per collision. If the electrons are to remain in equilibrium with the gas *and* the electric field, their energy (or "temperature") must rise relative to that of the gas until their energy loss per collision is just equal to that gained from the field between collisions. For monatomic gases this energy difference can be appreciable, and if it is assumed that the ionization phenomena are primarily influenced by the *electron* energy, the significant temperature to be employed in Eq. (5–22) is the electron "temperature." Reference [15] is a detailed theoretical and experimental study of the effect of elevated electron temperature on plasma conductivity.

We have considered that conduction is due only to the motion of electrons, after showing this to be a reasonable approximation in the absence of magnetic effects. However, the effect of the magnetic field on electrons is to cause them to drift at right angles to the electric field. This reduces the net electron motion in the E-direction (to zero if no collisions occur). Thus, for high values of either ω_e (high **B**) or τ_e (few collisions), the electron conductivity can be reduced to the point where ion conduction becomes appreciable. (The ions, moving at lower speeds and suffering more collisions, are not as adversely affected by the magnetic field.) Cowling [12] states that ion motion becomes appreciable for $\omega_e \tau_e$ of order of magnitude greater than ten.

5–7 ONE-DIMENSIONAL STEADY FLOW OF A PLASMA

The foregoing analyses present approximate expressions which are inadequate for accurate calculation of plasma properties. Suppose for the moment, however, that by better calculations or perhaps experimental measurement such properties of a plasma are known.

Considering the plasma a continuum, the one-dimensional steady-flow equations of motion, Eqs. (3–1), (3–2), (3–3), and (3–4), are applicable. A differential

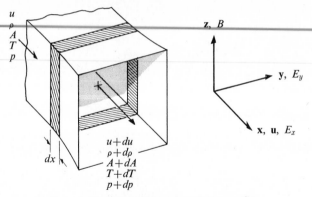

Fig. 5–10.   One-dimensional plasma flow.

control volume for one-dimensional flow is illustrated in Fig. 5–10, along with the **u**, **E**, and **B** convention of this chapter. The equations of continuity and of state (assuming the plasma behaves thermodynamically as a perfect gas) are unchanged. Thus

$$p = \rho RT, \tag{5–39}$$

$$\frac{d\rho}{\rho} + \frac{dA}{A} + \frac{du}{u} = 0. \tag{5–40}$$

For the momentum equation, the general body force **X** per unit volume in the **x**-direction may be written $\mathbf{X} = (\mathbf{j} \times \mathbf{B}) \cdot \mathbf{x}$ and, neglecting wall shear stress, the **x**-momentum equation becomes

$$\frac{dp}{dx} + \rho u \frac{du}{dx} = (\mathbf{j} \times \mathbf{B}) \cdot \mathbf{x}. \tag{5–41}$$

In a one-dimensional analysis the y- and z-components of $\mathbf{j} \times \mathbf{B}$ are ignored, but these terms will intentionally be minimized or eliminated in many devices anyway. The **z**-component is zero so long as $\mathbf{B} = B\mathbf{z}$. The energy equation becomes

$$\dot{m}(dh + u\,du) = d\dot{Q} - d\mathcal{O} = \dot{\mathcal{Q}}c\,dx + \mathcal{O}A\,dx,$$

where $\dot{\mathcal{Q}}$ is the rate of heat transfer per unit of surface of the control volume, c is the duct circumference, and $\mathcal{O}$ is the rate of electrical energy *addition* (per unit volume) to the control volume. The electrostatic and electromagnetic forces on the plasma are both summations of forces on individually charged particles. Since the electromagnetic force on a charged particle is always normal to that particle's velocity, it cannot increase the energy of the particle. Thus all the energy transfer to the gas is directly due to electrostatic forces acting on charged particles. The energy of the particle is increased by the action of the force component in the direction of its motion. Considering the plasma as a whole, the rate per unit volume of addition of energy is then $\mathcal{O} = \mathbf{j} \cdot \mathbf{E}$. Thus, since the mass flow rate

$\dot{m}$ equals $\rho u A$, the steady-state energy equation may, in the absence of heat transfer $\dot{Q}$, be written

$$\rho u c_p \frac{dT}{dx} + \rho u^2 \frac{du}{dx} = \mathbf{j} \cdot \mathbf{E}, \tag{5–42}$$

where c_p is the specific heat at constant pressure. Of this total electrical power input, only a fraction is transformed directly to fluid kinetic energy. The remainder appears as enthalpy. The part that is converted into kinetic energy is, per unit volume, equal to the accelerating force per unit volume times the fluid velocity. Its magnitude is

$$\mathbf{F} \cdot \mathbf{u} = (\mathbf{j} \times \mathbf{B}) \cdot \mathbf{u}.$$

The remainder of the electrical power input is consumed by the flow of current through the electrical resistance of the plasma, and thus may be termed *ohmic* or *joule heating*. To an observer moving with the fluid this apparent heat is the total power input to the fluid, since he is unable to observe changes in kinetic energy. To him the heating rate is $\mathbf{j} \cdot \mathbf{E}'$, since $\mathbf{E}'$ is the local electrostatic field intensity in his coordinate system. The sum of these two terms is $\mathbf{j} \cdot \mathbf{E}' + (\mathbf{j} \times \mathbf{B}) \cdot \mathbf{u}$. However, since

$$\mathbf{E}' = \mathbf{E} + \mathbf{u} \times \mathbf{B} \quad\text{and}\quad \mathbf{j} \cdot (\mathbf{u} \times \mathbf{B}) = -\mathbf{u} \cdot (\mathbf{j} \times \mathbf{B}),$$

it may be seen that the sum of these two terms is simply $\mathbf{j} \cdot \mathbf{E}$, which is the same as the right-hand side of Eq. (5–42).

In addition to these conventional equations we need Ohm's law, Eq. (5–36), to describe the current-voltage-conductivity relationship.

For the particular coordinate system employed here,

$$\mathbf{u} = u\mathbf{x}, \quad \mathbf{E} = E_x\mathbf{x} + E_y\mathbf{y}, \quad \mathbf{j} = j_x\mathbf{x} + j_y\mathbf{y}, \quad \mathbf{B} = B\mathbf{z},$$

and these equations reduce to the following:

State:
$$p = \rho RT, \tag{5–43}$$

Continuity:
$$\frac{d\rho}{\rho} + \frac{dA}{A} + \frac{du}{u} = 0, \tag{5–44}$$

Energy:
$$\rho u c_p \frac{dT}{dx} + \rho u^2 \frac{du}{dx} = j_x E_x + j_y E_y, \tag{5–45}$$

Momentum:
$$\frac{dp}{dx} + \rho u \frac{du}{dx} = j_y B, \tag{5–46}$$

Ohm's law:
$$j_x = \frac{\sigma_0}{1 + \frac{1}{4}\omega_e^2\tau_e^2}\left[E_x - \frac{1}{2}\omega_e\tau_e(E_y - uB)\right],$$

$$j_y = \frac{\sigma_0}{1 + \frac{1}{4}\omega_e^2\tau_e^2}\left[(E_y - uB) + \frac{1}{2}\omega_e\tau_e E_x\right]. \tag{5–47}$$

These are, of course, considerably simplified when the Hall current is zero or negligible. We shall mention a number of basic limitations to the use of these one-dimensional equations, after a brief discussion of the methods of their solution.

As an illustration, suppose that

$$(\omega_e \tau_e)^2 \ll 1 \quad \text{and that} \quad E_x = j_x \cong 0.$$

Suppose also that the applied electrostatic and magnetic fields are known, and that the magnetic field induced by the current j is small compared to the applied magnetic field. (The conditions under which induced magnetic fields may be ignored will be shown subsequently.)

Under these restrictions, it has been shown [13] that by algebraic rearrangement Eqs. (5–43) through (5–48) yield the following relations for the rates of change of velocity and Mach number with distance downstream:

$$\frac{du}{dx} = \frac{1}{M^2 - 1}\left\{\frac{u}{A}\frac{dA}{dx} - \frac{\sigma B^2}{p}\left(u - \frac{E}{B}\right)\left[u - \left(\frac{\gamma - 1}{\gamma}\right)\frac{E}{B}\right]\right\} \quad (5\text{–}48)$$

and

$$\frac{dM}{dx} = \left\{\frac{1 + [(\gamma - 1)/2]M^2}{M^2 - 1}\right\}\left[\frac{M}{A}\frac{dA}{dx} - \frac{\sigma B^2 M}{pu}\left(u - \frac{E}{B}\right)\left(u - m_1 \frac{E}{B}\right)\right], \quad (5\text{–}49)$$

in which σ is the conductivity, M is the Mach number and

$$m_1 = \frac{(1 + \gamma M^2)(\gamma - 1)}{[2 + (\gamma - 1)M^2]\gamma}.$$

The density distribution can be determined from Eq. (5–44), and the pressure distribution can be determined from a combination of Eqs. (5–43) and (5–44):

$$\frac{dp}{p} = \frac{dT}{T} - \frac{dA}{A} - \frac{du}{u}.$$

The temperature is related to the velocity and Mach number by

$$T = u^2/\gamma R M^2,$$

so that

$$\frac{dp}{p} = +\frac{du}{u} - 2\frac{dM}{M} - \frac{dA}{A}. \quad (5\text{–}50)$$

If E, B, and A are known or prescribed as functions of x, Eqs. (5–48), (5–49), and (5–50) can be integrated simultaneously by a step-by-step numerical procedure to obtain calculated distributions of Mach number, velocity, and pressure along the duct.

Thus numerical solutions for any one-dimensional channel flow (under the given assumptions) may easily be obtained. Only under further restrictions may these equations be integrated algebraically. Let us now consider one of these cases.

Constant-area, constant-temperature acceleration

For this case Eqs. (5–46), (5–44), and (5–43) may be combined to yield

$$u\frac{du}{dx} = \frac{jB}{\rho} + \frac{1}{u}\frac{du}{dx}RT = \frac{jB}{\rho} + u\frac{du}{dx}\frac{1}{\gamma M^2}$$

or

$$u\frac{du}{dx} = \frac{jB}{\rho}\frac{\gamma M^2}{\gamma M^2 - 1}. \tag{5–51}$$

Also Eq. (5–45) becomes

$$u\frac{du}{dx} = \frac{jE}{\rho u}. \tag{5–52}$$

Equating these last two expressions to eliminate $u(du/dx)$ yields

$$\frac{E}{B} = u\left(\frac{\gamma M^2}{\gamma M^2 - 1}\right). \tag{5–53}$$

For constant area, Eq. (5–44) may be integrated to

$$\rho u = \frac{\dot{m}}{A} = \text{const.}$$

Using also the definition of Mach number and the expression $j = \sigma[E - uB]$, Eq. (5–51) may be written

$$\left(\frac{\dot{m}}{A}\right)a\frac{dM}{dx} = \sigma B^2\left(\frac{E}{B} - u\right)\frac{\gamma M^2}{\gamma M^2 - 1}$$

where a is the speed of sound in the plasma. Substituting Eq. (5–53) in this expression yields

$$\left(\frac{\dot{m}}{A}\right)a\frac{dM}{dx} = \sigma B^2\frac{u\gamma M^2}{(\gamma M^2 - 1)^2} \quad \text{or} \quad \frac{(\gamma M^2 - 1)^2\, dM^2}{2\gamma M^4} = \frac{\sigma B^2}{\dot{m}/A}\, dx.$$

This expression may be integrated directly with the result that

$$\int_{x_0}\sigma B^2\, dx = \frac{\dot{m}}{A}\left[\frac{\gamma}{2}(M^2 - M_0^2) - \frac{1}{2\gamma M^2}\left(1 - \frac{M^2}{M_0^2}\right) - 2\ln\frac{M}{M_0}\right], \tag{5–54}$$

where the subscript 0 denotes the beginning of the constant-area, constant-temperature acceleration. Equation (5–54) provides a method for determining the variation of the Mach number along the duct length. The electrostatic field strength E does not appear explicitly, but it may be shown from Eq. (5–53) that, for constant-temperature constant-area acceleration,

$$\frac{E}{B} = \frac{\gamma M^3\sqrt{\gamma RT}}{\gamma M^2 - 1}.$$

Thus the E-field is in fact a dependent variable if σB^2 is regarded as independent. If $(M/M_0) \gg 1$ at $x = L$ and σB^2 is assumed constant, then

$$L \simeq \frac{\gamma}{2} \left(\frac{\dot{m}}{A}\right) \frac{M^2}{\sigma B^2}. \tag{5-55}$$

On the other hand, if E is regarded as constant, it is useful to integrate Eq. (5–52) directly. Ohm's law $[j = \sigma(E - uB)]$ and Eq. (5–53) yield

$$jE = \sigma E^2 \left(1 - \frac{uB}{E}\right) = \frac{\sigma E^2}{\gamma M^2}.$$

Also $\rho u = \dot{m}/A$ and $u = aM$. Thus Eq. (5–52) can be written

$$M^3 \frac{dM}{dx} = \frac{\sigma E^2}{\gamma(\dot{m}/A)a^2}.$$

Integrating, we have,

$$M^4 - M_0^4 = \frac{4\sigma E^2 x}{\gamma(\dot{m}/A)a^2}$$

or

$$L = \frac{\gamma}{4} \left(\frac{\dot{m}}{A}\right) \frac{a^2}{\sigma E^2} (M^4 - M_0^4). \tag{5-56}$$

Magnetic Reynolds number

In the foregoing discussion of one-dimensional flow of a conducting fluid, it has been assumed that the magnetic field B is known and equal to the externally applied field. However, as Eq. (5–12) shows, currents flowing inside the fluid induce an additional magnetic field and it is important to be able to predict when this induced field is negligible relative to the applied field. The criterion for this is the *magnetic Reynolds number* Re_m, which may be interpreted as proportional to the ratio of the induced field B_i due to currents in the plasma to the applied field B_a; thus

$$\text{Re}_m \propto \frac{B_i}{B_a}.$$

The induced field may be treated by writing Eq. (5–12) in the form

$$d\mathbf{B}_i = \frac{\mu}{4\pi} \frac{\mathbf{j} \times \mathbf{r}}{r^2} d\upsilon$$

or

$$\mathbf{B}_i = \frac{\mu}{4\pi} \int \frac{\mathbf{j} \times \mathbf{r}}{r^2} d\upsilon, \tag{5-57}$$

where $\mathbf{r}$ extends from the volume element $d\upsilon$ to the point where B_i is calculated, and the integral extends over the volume of the plasma.

Consider a geometrically similar series of volumes, each having a characteristic dimension L. Suppose all have similar internal distributions of current density, velocity, and conductivity, so that each will also have a characteristic value of these variables. Then it may be seen from Eq. (5–57) that $B_i \propto \mu j L$, where j is the characteristic current density, or $B_i \propto \mu_0 j L$, since $\mu \cong \mu_0$ in a plasma. As shown earlier, in the absence of Hall currents, Ohm's law for a plasma is

$$\frac{\mathbf{j}}{\sigma_0} = \mathbf{E} + \mathbf{u} \times \mathbf{B}.$$

For interesting devices the gradient induced by the plasma motion ($\mathbf{u} \times \mathbf{B}$) will be of the order of $\mathbf{E}$ (just as in conventional motors the "back" EMF is of the order of the terminal voltage). Since we are interested in small induced fields, $\mathbf{B} \approx \mathbf{B}_a$, and in order of magnitude,

$$j = 0(\sigma_0 u B_a),$$

where $=0(\quad)$ signifies "order of magnitude of." Combining this with the previously discussed proportionality, we have

$$\frac{B_i}{B_a} \propto \mu_0 \sigma_0 u L.$$

The magnetic Reynolds number is defined as

$$\mathrm{Re}_m = \mu_0 \sigma_0 u L. \tag{5–58}$$

Experience indicates that induced magnetic fields are negligible if the order of magnitude of Re_m is less than unity. The solutions for one-dimensional flow discussed so far are therefore limited to small magnetic Reynolds numbers.

Plasma boundaries

The assumption of a uniform plasma is somewhat analogous to the assumptions leading to one-dimensional-flow equations for ordinary fluids. That is, while it is convenient to consider one-dimensional flows, it must be remembered that there results only an approximate description of flow processes. In particular the one-dimensional model does not account for conditions adjacent to the fluid boundaries.

As an example of boundary phenomena, consider the potential adjustment of a plasma boundary. For simplicity we restrict our attention to the plasma which is at rest within a grounded conducting container and is not subject to applied electric or magnetic fields. We might expect that the very high thermal velocity of the electrons (relative to that of the ions for equal kinetic energy or temperature) would lead to a rapid depletion of the electrons as they diffuse through the plasma to the walls. This tendency is offset by the space charge that builds up in the remaining positive material. As a result, the plasma comes to an equilibrium

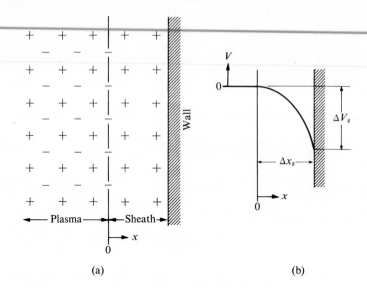

FIG. 5–11.   Approximate configuration of plasma sheath. Neutral atoms are not shown.

potential slightly above that of its container, the potential adjustment occurring in a thin layer or sheath at the edge. The magnitude of the potential difference across the sheath, ΔV_s, is about equal to that required to reflect an average electron. That is,

$$e\,\Delta V_s \approx \frac{\overline{m_e u^2}}{2} = \frac{3}{2}kT_e, \qquad \Delta V_s = \frac{3}{2}\frac{k}{e}T_e = 1.292 \times 10^{-4}\,T_e,$$

in which T_e is the electron temperature in °K.

The approximate thickness of the sheath is also related to the thermal energy of the electrons, since it is only this energy which can bring about charge separation. Suppose the sheath is composed entirely of ions at a density n_e, the same as that of the ions and the electrons in the plasma shown in Fig. 5–11.

Applying Poisson's equation (5–8) to the (one-dimensional) sheath region, we obtain

$$\nabla^2 V = \frac{d^2 V}{dx^2} = -\frac{\rho_q}{\epsilon_0} = -\frac{en_e}{\epsilon_0}.$$

Integrating twice, and setting $V = dV/dx = 0$ at $x = 0$, as in Fig. 5–11, we have

$$V = \frac{en_e}{2\epsilon_0}x^2.$$

Setting V equal to the foregoing expression for ΔV_s, we arrive at

$$x_s = \left(\frac{3\epsilon_0 kT_e}{n_e e^2}\right)^{1/2}.$$

Thus the sheath thickness is of the order of the Debye length, as might be expected, since here again thermal and electrostatic energies are balanced. According to this model the net positive charge necessary to reflect (or reattract) the fast electrons from the walls is actually concentrated in the sheath. Actual sheaths are, of course, not this well defined, and the model indicates only the order of magnitude of the sheath thickness and the potential drop.

When a current is conducted through such a sheath in practical devices, a number of other complications can occur; for example, the space charge contribution of the current-carrying electrons themselves. Then there are complications arising from the behavior of the surface, the availability of electrons at the surface, and from nonequilibrium behavior in the strong potential gradients near surfaces.

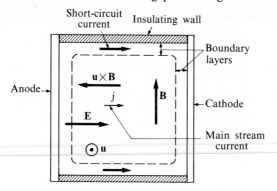

Fig. 5–12.   Boundary layer short-circuiting.

The velocity and the thermal boundary layers can both have an effect on the electrical behavior of a plasma near a wall. The reduction of the $\mathbf{u} \times \mathbf{B}$ portion of the apparent field E' near the wall can be particularly important near the insulating walls parallel to the applied electrostatic field. Figure 5–12 shows a cross section of a plasma accelerator channel. In the boundary layer on the insulating walls the induced field $\mathbf{u} \times \mathbf{B}$ which opposes E may be so much lower than in the main stream that large "short-circuit" currents flow, avoiding the main body of plasma.

The variation of temperature within the thermal boundary layer may have important effects on the conductivity, as may be seen by Fig. 5–3. Since the wall will, in all practical devices, be cooler than the plasma, there may be a significant drop in conductivity near the walls. This results in increased joule heating within the boundary layer which, on the one hand, tends to restore the temperature and conductivity to free-stream values but, on the other hand, can cause significantly increased wall heat transfer. Along the insulating walls this reduced conductivity may to some extent reduce the short-circuiting tendency previously discussed. However, electrically insulating walls are also thermally insulating so that their temperature, and hence that of their boundary layers, may not be far below that of the free stream.

~~Nonequilibrium effects (discussed previously) may significantly modify this~~ picture under certain circumstances.

The general subject of plasma nonuniformities and boundary behavior is very complex, not well understood at the present time, and certainly beyond the scope of this treatment. The foregoing discussion has been included mainly as a reminder that (perhaps to an even greater extent than in ordinary fluid dynamics) the relatively simple one-dimensional flow model may not adequately describe conditions within actual plasma devices.

References

1. SEARS, F. W., and M. V. ZEMANSKY, *University Physics*. Reading, Mass.: Addison-Wesley, 1953, pages 434, 565

2. ROGERS, W. E., *Introduction to Electric Fields*. New York: McGraw-Hill, 1954

3. PUGH, E. M., and E. W. PUGH, *Principles of Electricity and Magnetism*. Reading, Mass.: Addison-Wesley, 1960

4. SEARS, F. W., *An Introduction to Thermodynamics, the Kinetic Theory of Gases, and Statistical Mechanics*, second edition. Reading, Mass.: Addison-Wesley, 1959, Chapters 11, 12, 13

5. DELCROIX, J. L., *Introduction to the Theory of Ionized Gases*, translated from the French by M. Clark, Jr., D. J. BenDaniel, and J. M. BenDaniel. New York: Interscience Publishers, 1960, pages 19, 20, 39

6. LIN, S. C., E. L. RESLER, and A. R. KANTROWITZ, "Electrical Conductivity of Highly Ionized Argon Produced by Shock Waves." *Journal of Applied Physics*, **26**, 1, Jan. 1955, page 95

7. SAHA, M. N., and N. K. SAHA, *A Treatise on Modern Physics*. Allahabad and Calcutta: The Indian Press, Ltd., 1934

8. FRANCIS, GORDON, *Ionization Phenomena in Gases*. London: Butterworth Scientific Publications, 1960, Chapter 2

9. ROSA, R. J., "Physical Principles of MHD Power Generators," *Physics of Fluids*, **4**, 2, Feb., 1961

10. PIERCE, J. R., *Theory and Design of Electron Beams*. New York: D. Van Nostrand, 1954, pages 25–27

11. SUTTON, G. W., and L. STEG, "The Prospects for MHD Power Generation," in *Energy Conversion for Space Power*, edited by N. W. Snyder. New York: Academic Press, 1961

12. COWLING, T. G., *Magnetohydrodynamics*. New York: Interscience Publishers, 1957

13. RESLER, E. L., JR., and W. R. SEARS, "The Prospects for Magneto-Aerodynamics," *Journal of Aeronautical Science*, **25**, 4, April 1958, pages 235–245

14. CAMAC, M., "Plasma Propulsion Devices," *Advances in Space Sciences*, Volume II. New York: Academic Press, 1960

15. KERREBROCK, J. L., "Nonequilibrium Ionization Due to Electron Heating: I, Theory," and KERREBROCK, J. L. and M. A. HOFFMAN, "Nonequilibrium Ionization Due to Electron Heating: II, Experiments." *Journal of Amer. Inst. of Aeronautics and Astronautics*, **2**, 6, June 1964, pages 1072–1087

GENERAL PLASMA REFERENCES

16. CLAUSER, F. H., *Plasma Dynamics*. Baltimore: Johns Hopkins University, Dept. of Aeronautics, 1959

17. CAMBEL, ALI BULENT, *Plasma Physics and Magnetofluid Mechanics*. New York: McGraw-Hill, 1963

18. ROSE, DAVID J., and MELVILLE CLARK, JR., *Plasmas and Controlled Fusion*. Cambridge, Mass.: M.I.T. Press, 1961

19. VON ENGEL, A., *Ionized Gases*. New York: Oxford University Press, 1955

20. BROWN, SANBORN C., *Basic Data of Plasma Physics*. New York: John Wiley & Sons, 1959

Problems

1. Two positive ions, of the same mass, one singly charged and the other doubly charged, are accelerated from zero velocity through the same potential, V_{acc}, after which they travel through a deflection region of length L. Calculate the ratio θ_2/θ_1, where θ_2 refers to the doubly and θ_1 to the singly charged ion for the following cases: (a) Deflection due to a uniform **E** field as shown. (b) Deflection due to a uniform **B** field as shown. (c) Deflection due to both **E** and **B** fields shown. The deflection angles may be assumed small.

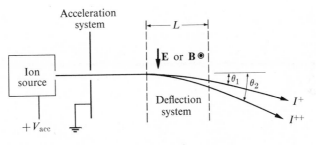

PROBLEM 1

2. Suppose that argon, seeded with 1% potassium by weight, as described in Fig. 5–3, flows in a channel at a Mach number of 10 and a static temperature and pressure of 2500°K and 8.5 atm. (a) Determine the equilibrium electron density in the plasma. (b) If an electric field, $E = 100$ v/cm, is applied across the flow, calculate the resultant current density. (c) For the above conditions calculate the gas translational velocity and the root-mean-square speeds for argon atoms and electrons. (d) Determine the drift velocity of the electrons under the applied electric field.

The molecular weight of argon is 39.944 and the ratio of specific heats is 1.67. The total cross section for electron–atom collision is 6×10^{-17} cm^2 for argon and 3×10^{-14} cm^2 for potassium. The mass of the electron is 2.11×10^{-28} gm and its charge is 1.602×10^{-19} coul. The ionization potential ϵ_i is 4.318 eV for potassium and 15.68 eV for argon.

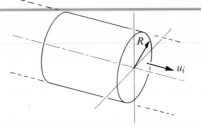

PROBLEM 3

3. A cylindrical positive ion beam of radius R carries a total current I with an ion velocity u_i. The beam is surrounded by chargefree space extending to infinity. Assuming the current density and ion velocity uniform across the beam, calculate potential as a function of r if $V = 0$ at $r = 0$. Is the ion beam likely to expand? For cylindrical coordinates (r, θ, z), Poisson's equation is:

$$\frac{1}{r}\frac{\partial}{\partial r}\left(r\frac{\partial V}{\partial r}\right) + \frac{1}{r^2}\frac{\partial^2 V}{\partial \theta^2} + \frac{\partial^2 V}{\partial z^2} = -\frac{\rho_q}{\epsilon_0}.$$

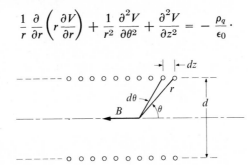

PROBLEM 4

4. Determine the magnetic field strength along the axis of a long coil of constant diameter d with n turns per unit axial distance. A current I flows in the coil. The spacing is very much smaller than d so that the contribution of an incremental length of coil dz to the axial component of B at any point may be written [from Eq. (5–12)]

$$dB_z = \frac{\mu_0(In\, dz)}{4\pi r^2}\pi d \sin\theta.$$

Show that if the coil is infinitely long in both directions,

$$B = \mu_0 In.$$

Suppose the coil has 2000 coils per meter and a current of 100 amps; calculate the magnetic field strength in gauss.

5. Calculate the equilibrium conductivity of potassium-seeded argon at a temperature of 3000°K and pressure of 10 atm for ratios of potassium to argon varying from 0 to 2% by weight. Again, assume $Q_s = 3 \times 10^{-14}\,\text{cm}^2$, $Q_g = 6 \times 10^{-17}\text{cm}^2$, $T_e = T$, and $\epsilon_i = 4.318$ eV for potassium and 15.68 eV for argon.

6. If an electron of mass m_e strikes an initially stationary atom of mass $m_a \gg m_e$, its average deflection angle will be 90° (for isotropic scattering). Show that in this case the loss of electron energy in one elastic collision is approximately $2(m_e/m_a)$ times the difference between the energies of the electron and the neutral particle.

7. In problems 5–4 and 5–5 it was implicitly assumed that the average energies (or temperatures) of the electrons and neutral particles in the plasma were identical. It may be shown, however, that under an applied E-field the electron temperature may be considerably higher than the temperature of the remainder of the plasma. Consider the exchange of energy between electrons and atoms or molecules. In steady state the rate at which the electrons, on the average, gain energy from the electric field must approximately equal the rate at which they lose energy in collisions with neutral particles.

The rate at which, on the average, the electric field does work on the electron is eEu_d. Even if the temperatures of electrons and neutral particles were identical, the average electron velocity would be very much higher than the velocity of the neutrals, which may therefore be neglected as a first approximation. Show that, using the result of Problem 6, the electron energy required to sustain this balance will be approximately given by

$$eEu_d \approx 2\frac{m_e}{m_a}(\tfrac{3}{2}kT_e - \tfrac{3}{2}kT)\nu_e,$$

in which e is the electronic charge, u_d is the drift velocity, T_e is the electron temperature, T is the temperature of the neutral particles, and ν_e is the collision frequency. Show also that, using an approximate average momentum balance, the above expression may be transformed to

$$\frac{T_e}{T} \approx 1 + \frac{2}{3T}\left(\frac{m_a}{k}\right)u_d^2 = 1 + \frac{2}{3}\left(\frac{m_a}{kT}\right)\left(\frac{j}{en_e}\right)^2.$$

According to this very approximate estimate, what would the ratio of electron and neutral temperatures be for $j = 10\ \text{amp/cm}^2$, $T = 3000°\text{K}$, mol wt = 40 and $n_e = 10^{22}/\text{m}^3$?

8. Compare the electron collision and cyclotron frequencies in potassium vapor at 2500 K° and 10 atm under a magnetic field of 1 weber/m^2. The collision cross section is $3 \times 10^{-14}\ \text{cm}^2$ and $\epsilon_i = 4.318$ eV.

Determine the Debye length and mean free path. Could this fluid be considered a plasma if it were flowing in a duct 10 cm in width?

Suppose the fluid Mach number were 2.0. Could induced magnetic fields be neglected in an apparatus of this typical size?

9. An idealized one-dimensional plasma accelerator has a constant cross-sectional area and is designed to operate at constant temperature. At entrance to the channel the Mach number is unity and the stagnation pressure and temperatures are 100 atm and 3000°R. Given that conductivity is 1000 mho/m and the magnetic field strength is 30,000 gauss, both of them being uniform along the channel, estimate the channel length required for a final Mach number of (a) M = 3, and (b) M = 6.

The ratio of specific heats is 1.4 and the molecular weight is 15.44. How must the electric field vary from one end of the channel to the other?

PART 2

Air-Breathing
Engines

Pratt & Whitney JT3C engines in Boeing 707
in flight. (Courtesy the Boeing Company.)

6

Thermodynamics of
Aircraft Jet Engines

6–1 INTRODUCTION

Fluid propulsion devices achieve thrust by imparting momentum to a fluid called the propellant. Two general classes of propulsion devices are considered in this book: air-breathing engines and rockets. An air-breathing engine, as its name implies, uses the atmosphere for most of its propellant. In contrast, a rocket engine carries all its propellant with it. Piston engines, gas turbines, and ramjets are all air-breathing engines, though only the latter two are discussed in this book.

The gas turbine produces hot, high-temperature gas which may be used either to generate power for a propeller or to develop thrust directly by expansion and acceleration of the hot gas in a nozzle. In any case an air-breathing engine continuously draws air from the atmosphere, compresses it, adds energy in the form of heat, and then expands it in order to convert a part of the added energy to shaft work or jet kinetic energy. Thus, in addition to acting as propellant, the air acts as the working fluid in a thermodynamic process in which a fraction of the energy consumed is made available for propulsive purposes.

The main sources of energy for air-breathing engines are the hydrocarbon fuels. Nuclear energy could be used to heat the working fluid, but this idea has not been developed practically.

The purpose of this chapter is to show how the performance of aircraft engines may be understood by means of the laws of thermodynamics. It will be shown that these laws restrict the performance to certain upper limits which depend strongly on the maximum temperature the engine can withstand. A consideration of fundamental thermodynamic principles permits comparison and simple classification of a considerable variety of mechanically complex engines. Only turbine and ramjet engines are discussed in detail. Piston engines, which will no doubt be of continuing importance for certain propulsion applications, are treated in numerous volumes. (See, for example, reference [1].)

139

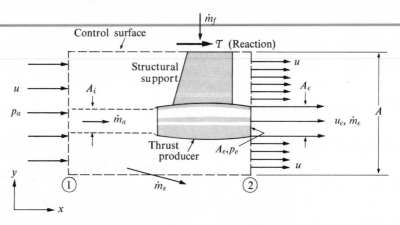

FIG. 6–1. Generalized thrust-producing device.

6–2 THRUST AND EFFICIENCY

Fairly general equations for thrust and efficiency of air-breathing jet engines can be derived from the momentum and energy laws without the need for detailed consideration of the internal mechanisms of particular engines. Consider, for example, the generalized thrust-producing device illustrated in Fig. 6–1, as observed from a position stationary with respect to the device. While this device looks like a pod-mounted ramjet or turbojet, the results obtained will be applicable to the calculation of thrust produced by any air-breathing engine with a single exhaust jet. In Chapter 2 the thrust of a stationary jet engine was derived; we now generalize this result to show the effect of aircraft flight speed on thrust. An additional effect of flight speed—the drag on the external surface of the engine nacelle or pod—is of secondary importance in this discussion, so for simplicity the flow external to the engine will be assumed reversible.

In Fig. 6–1, a control surface is specified which passes through the propellant outlet plane at ② and extends far upstream at ①. The side surfaces of the control volume are parallel to the upstream (flight) velocity $\mathbf{u}$ and far removed from the thrust device. It is assumed that the thrust and conditions at all points within the control volume do not change with time.

The thrust equation

The reaction to the thrust T transmitted through the structural support is indicated in Fig. 6–1. In this sense the engine thrust may be defined as the vector summation of all forces on the internal *and* external surfaces of the engine and nacelle.

The thrust of the generalized thrust-producer may be derived from Eq. 2–4 as it applies to steady flow:

$$\sum \mathbf{F} = \int_{\text{cs}} \mathbf{u}(\rho \mathbf{u}) \cdot \mathbf{n}\, dA.$$

Considering the components of force and momentum flux in the x-direction only, we have

$$\sum F_x = \int_{cs} u_x(\rho \mathbf{u} \cdot \mathbf{n}) \, dA. \tag{6–1}$$

With the assumption of reversible external flow, both the pressure and the velocity may be assumed constant over the entire control surface, except over the exhaust area A_e of the engine. If the exhaust velocity u_e is supersonic, the exhaust pressure p_e may differ from the ambient pressure p_a. The net pressure force on the control surface is therefore $+(p_a - p_e)A_e$. The only other force acting on this control volume is the reaction to the thrust T. Adding up the forces on the control surface which act in the x-direction, we obtain

$$\sum F_x = (p_a - p_e)A_e + T. \tag{6–2}$$

Far upstream at station ①, the air which is drawn into the engine crosses the control surface through capture area A_i at a rate $\dot{m}_a$ given by $\dot{m}_a = \rho u A_i$, in which ρ is the ambient density and u is the flight velocity. The mass flux crossing the exhaust area A_e is $\dot{m}_e = \rho_e u_e A_e$. Taking account of the fuel flow rate $\dot{m}_f$, we have $\dot{m}_e = \dot{m}_a + \dot{m}_f$, or

$$\dot{m}_f = \rho_e u_e A_e - \rho u A_i. \tag{6–3}$$

Now, considering the requirement of continuity for the control volume as a whole, and assuming that the fuel flow originates from outside the control volume, Eq. (2–2) for steady flow is

$$\int_{cs} \rho \mathbf{u} \cdot \mathbf{n} \, dA = 0,$$

which for the present case may be written

$$\rho_e u_e A_e + \rho u (A - A_e) + \dot{m}_s - \dot{m}_f - \rho u A = 0,$$

in which A is the cross-sectional area of the control volume normal to the velocity u, and $\dot{m}_s$ is the mass flow of air through the side surfaces of the control volume (Fig. 6–1). Rearranging, we get

$$\dot{m}_s = \dot{m}_f + \rho u A_e - \rho_e u_e A_e.$$

When we use Eq. (6–3), this equation becomes

$$\dot{m}_s = \rho u (A_e - A_i). \tag{6–4}$$

If the sides of the control volume are sufficiently distant from the thrust producer, it may be assumed that this flow crosses the control surface with a very small velocity in the y-direction and an essentially undisturbed velocity component in the x-direction. Thus the momentum carried out by the control volume with this flow is simply $\dot{m}_s u$, and when we take components only in the x-direction, the right-hand side of Eq. (6–1) may be written

$$\int_{cs} u_x \rho(\mathbf{u} \cdot \mathbf{n}) \, dA = \dot{m}_e u_e + \dot{m}_s u + \rho u(A - A_e)u - \dot{m}_a u - \rho u(A - A_i)u,$$

which is the net outward flux of x-momentum from the control volume. Using Eq. (6–4), we may reduce this to

$$\int_{cs} u_x \rho(\mathbf{u} \cdot \mathbf{n}) \, dA = \dot{m}_e u_e - \dot{m}_a u. \tag{6-5}$$

When we use Eqs. (6–2) and (6–5), the momentum equation (6–1) becomes

$$T = \dot{m}_e u_e - \dot{m}_a u + (p_e - p_a)A_e$$

or, defining the fuel–air ratio $f = \dot{m}_f/\dot{m}_a$, we have

Turbojet

$$T = \dot{m}_a[(1 + f)u_e - u] + (p_e - p_a)A_e. \tag{6-6}$$

The term $(p_e - p_a)A_e$ is not zero only if the exhaust jet is supersonic and the nozzle does not expand the exhaust jet to ambient pressure. Even if it is not zero, it is usually small compared to the momentum-flux term.

It should be borne in mind that in the derivation of Eq. (6–6) the flow external to the engine has been assumed reversible. If this is not so, due to significant boundary layer effects such as separation, the actual force transmitted by the structural support of Fig. 6–1 could be appreciably less than Eq. (6–6) would predict.

For engines which have two distinct exhaust streams, Eq. (6–1) must be applied separately to each stream. As we shall explain in Fig. 6–2 and in more detail in Section 6–4, the additional propellant streams of turbofan and turboprop engines may be distinguished according to whether they pass through or around the combustion chamber and turbine of the engine. The propellant streams may be labeled "hot flow" and "cold flow," respectively. The thrust equation for such an engine, neglecting pressure terms, becomes

$$T = (\dot{m}_{aH} + \dot{m}_f)u_{eH} - \dot{m}_{aH}u + \dot{m}_{aC}(u_{eC} - u),$$

where the subscripts H and C refer to the hot and cold flows, respectively. Basing the definition of the fuel–air ratio on that air which is actually mixed with the fuel, $f \equiv \dot{m}_f/\dot{m}_{aH}$, we obtain the thrust equation

Turboprop

$$T = \dot{m}_{aH}[(1 + f)u_{eH} - u] + \dot{m}_{aC}(u_{eC} - u). \tag{6-7}$$

Engine performance

In describing the performance of aircraft engines, it is helpful if we first define several efficiencies and performance parameters. Various definitions will now be stated and, for simplicity, representative expressions will be presented as they would apply to an engine with a single propellant stream (i.e., a turbojet or ramjet). Turbofan and turboprop engines would require slightly more complex expressions, but the qualitative conclusions regarding comparative engine performance would be similar.

Propulsion efficiency. The product of thrust and flight velocity, Tu, is sometimes called thrust power. One measure of the performance of a propulsion system is the ratio of this thrust power to the rate of production of propellant kinetic energy. This ratio is commonly known as the *propulsion efficiency*, η_p. For a single propellant stream,

$$\eta_p = \frac{Tu}{\dot{m}_a[(1 + f)(u_e^2/2) - u^2/2]}. \tag{6-8}$$

With two reasonable approximations this relationship may be considerably simplified. Firstly, for air-breathing engines in general, $f \ll 1$ and may therefore be ignored in Eqs. (6–6) and (6–8) without leading to serious error. Secondly, the pressure term in Eq. (6–6) is usually much smaller than the other two terms, so that $T \approx \dot{m}_a(u_e - u)$. Thus

$$\eta_p \approx \frac{(u_e - u)u}{(u_e^2/2) - u^2/2} = \frac{2u/u_e}{1 + u/u_e}. \tag{6-9}$$

The thrust equation shows that u_e must exceed u so that the right-hand side of Eq. 6–9 has a maximum value of unity for $u/u_e = 1$. However, for $u/u_e \to 1$, the thrust per unit mass flow is practically zero; and of course for a finite thrust the engine required would be infinitely large. Thus it is not realistic to try to maximize the propulsion efficiency of a jet engine, and other parameters are required to evaluate its overall performance.

It is interesting to note that if the "propellant" of traction vehicles on the earth's surface is considered to be the earth itself, then according to this expression their propulsion efficiency is unity.

Thermal efficiency. Another important performance ratio is the *thermal efficiency*, η_{th}, of the engine. For ramjets, turbojets, and turbofans it is defined as the ratio of the rate of addition of kinetic energy to the propellant to the total energy consumption rate $\dot{m}_f Q_R$, where Q_R is the heat of reaction of the fuel as defined by Eq. (2–31). Thus the thermal efficiency may be written, again for a single propellant stream, as

$$\eta_{th} = \frac{\dot{m}_a[(1 + f)(u_e^2/2) - u^2/2]}{\dot{m}_f Q_R}$$

or

$$\eta_{th} = \frac{[(1 + f)(u_e^2/2) - u^2/2]}{f Q_R}. \tag{6-10}$$

The output of a turboprop or turboshaft engine is largely shaft power. In this case thermal efficiency is defined as for other shaft-power devices by

$$\eta_{th} = \frac{\mathcal{P}_s}{\dot{m}_f Q_R}, \tag{6-11}$$

where $\mathcal{P}_s$ is shaft power.

Propeller efficiency. Shaft power is converted to thrust power of a moving aircraft by a propeller. *Propeller efficiency*, η_{pr}, is customarily defined as the ratio of thrust power to shaft power, or

$$\eta_{pr} = \frac{\mathcal{T}_{pr}u}{\mathcal{P}_s} \tag{6–12a}$$

where, in this case, $\mathcal{T}_{pr}$ is that portion of the thrust due to the propeller. However, since many turboprop engines derive appreciable thrust from the hot exhaust of their turbines, an *equivalent shaft power*, $\mathcal{P}_{es}$, may be defined such that the product of η_{pr} and $\mathcal{P}_{es}$ is equal to the total thrust power (often at some arbitrarily selected flight speed). Then

$$\eta_{pr} = \frac{\mathcal{T}u}{\mathcal{P}_{es}}. \tag{6–12b}$$

Overall efficiency. The product of $\eta_p\eta_{th}$, or $\eta_{pr}\eta_{th}$ as applicable, is called the *overall efficiency*, η_o, and is defined by

$$\eta_o = \eta_p\eta_{th} = \frac{\mathcal{T}u}{\dot{m}_f Q_R}. \tag{6–13}$$

Using Eq. (6–9), it follows that (for $f \ll 1$)

$$\eta_o = 2\eta_{th}\left(\frac{u/u_e}{1 + u/u_e}\right).$$

Thus the overall efficiency depends only on the velocity ratio u/u_e and on the thermal efficiency η_{th}, which depends somewhat on the velocity ratio. The importance of overall efficiency may be demonstrated by a simple aircraft range analysis.

Aircraft range

In many cases the distance or range that an aircraft can travel with a given mass of fuel is an important criterion of the excellence of performance of that aircraft-engine combination. Ignoring the climb to, and descent from, cruise flight conditions and assuming all aircraft and engine characteristics except total mass are constant throughout cruise, an estimate of range is easily obtained.

In level flight at constant speed, engine thrust and vehicle drag are equal, as are lift and vehicle weight. Therefore

$$\mathcal{T} = \mathcal{D} = L\left(\frac{\mathcal{D}}{L}\right) = \frac{mg}{L/\mathcal{D}},$$

where m is the instantaneous vehicle mass, g is the acceleration due to gravity, $\mathcal{D}$ is drag force, and $L/\mathcal{D}$ is its lift–drag ratio. The thrust power is then

$$\mathcal{T}u = \frac{mgu}{L/\mathcal{D}}.$$

By using Eq. (6–13) with this expression, we obtain

$$\dot{m}_f = \frac{mgu}{\eta_o Q_R(L/\mathfrak{D})}.$$ (6–14)

However, since the fuel is part of the total aircraft mass m,

$$\dot{m}_f = -\frac{dm}{dt} \quad \text{or} \quad \dot{m}_f = -u\frac{dm}{ds},$$

where s denotes distance along the flight path. Substituting this expression in Eq. (6–14), we have

$$\frac{dm}{ds} = -\frac{mg}{\eta_o Q_R(L/\mathfrak{D})}.$$ (6–15)

If, as an approximation, the denominator $\eta_o Q_R(L/\mathfrak{D})$ is assumed constant, the range s of the vehicle can be obtained by integrating Eq. (6–15) to obtain Brequet's range formula,

$$s = \eta_o Q_R \left(\frac{L}{\mathfrak{D}}\right) \ln \frac{m_1}{m_2},$$ (6–16)

where m_1 and m_2 are the initial and final masses of the vehicle, the difference being the mass of fuel consumed. Thus the range is directly proportional to the overall efficiency η_o.

The importance of this term prompts an interest in its variation with flight speed and engine variables. If $f \ll 1$ and the pressure term in the thrust equation is neglected, and if we again consider a single propellant stream for simplicity, it follows that

$$\eta_o = \frac{(u_e - u)u}{fQ_R}.$$ (6–17)

For a given fuel–air ratio f and a *given* exhaust jet velocity (i.e., approximately for a given engine), the overall efficiency is maximized when $u \approx u_e/2$. It will be shown that, for a given engine and peak allowable temperature, f and u_e vary somewhat with u, but if this is neglected it follows that the engine should be employed by an aircraft which travels at about half the jet velocity.

Note that it does not follow from Eq. (6–17) that for a given flight speed the overall efficiency is maximized when $u_e \approx 2u$. Since u_e and f are rather strongly interrelated, as will be shown, it is not at this point obvious what exhaust velocity is most desirable in this case.

It may be shown that Eq. (6–17) may be modified to express the overall efficiency of an engine of double propellant streams. It is only necessary to replace the exhaust velocity u_e by the "thrust-averaged" exhaust velocity $\bar{u}_e$, which is defined by

$$\bar{u}_e = \frac{\dot{m}_{aH}(1 + f)u_{eH} + \dot{m}_{aC}u_{eC}}{\dot{m}_{aH}(1 + f) + \dot{m}_{aC}},$$

and to replace the fuel–air ratio f by the term $f/[1 + (\dot{m}_{aC}/\dot{m}_{aH})]$; the fuel–air ratio is defined by $f = \dot{m}_f/\dot{m}_{aH}$. Thus

$$\eta_o = \left(1 + \frac{\dot{m}_{aC}}{\dot{m}_{aH}}\right)\frac{(\bar{u}_e - u)u}{fQ_R} .$$

For a given ratio $\dot{m}_{aC}/\dot{m}_{aH}$ and a given exhaust velocity $\bar{u}_e$, the overall efficiency is approximately maximized for $u = \bar{u}_e/2$, a generalization of the previous result.

Takeoff thrust

One of the most important characteristics of a turbine engine installed in an aircraft is its ability to provide static and low-speed thrust so that the aircraft may take off under its own power (ramjets cannot provide static thrust and so are excluded from this discussion). The static thrust may be derived from Eq. (6–6), ignoring the pressure term and neglecting f relative to unity:

$$\frac{\mathcal{T}}{\dot{m}_a} = u_e \quad \text{(static)}. \tag{6–18}$$

Thus the static thrust per unit mass flow of air is directly proportional to the exhaust jet velocity. However, another important question arises; for a given fuel flow, how does thrust depend on exhaust velocity? The answer to this question illuminates one of the major differences between turbojets, turbofans, and turboprops and it plays a significant role in the choice of propulsion system for many applications. Using Eqs. (6–18) and (6–10) for the static case, we have

$$\mathcal{T} = \frac{2\eta_{\text{th}}Q_R\dot{m}_f}{u_e} \quad \text{(static)}. \tag{6–19}$$

This equation shows that for a given fuel flow $\dot{m}_f$ and thermal efficiency η_{th}, the takeoff thrust is inversely proportional to exhaust velocity. In other words, for a given rate of energy consumption, the takeoff thrust can be increased by accelerating a larger mass flow of air to a smaller exhaust velocity.

The importance of thrust per unit fuel consumption rate, other than fuel economy (which is relatively unimportant during takeoff), is that it roughly determines engine size for a given thrust requirement. Turbine engines which have been developed for high performance have approximately the same peak temperature limit and, therefore, about the same fuel–air ratio. Thus equal fuel flow rates imply approximately equal "hot" air flow rates, and it is the hot air flow that, to a large extent, determines the size of the basic gas generator and engine.

A turbine engine consists basically of a gas generator, that is, a compressor-burner-turbine combination, which provides a supply of hot, high-pressure gases,

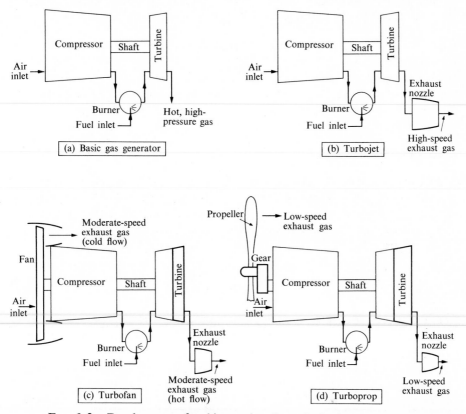

FIG. 6–2.   Development of turbine engines from the basic gas generator.

as illustrated schematically in Fig. 6–2(a). The *turbojet* utilizes this gas generator
with an exhaust nozzle in which the hot gases are accelerated, as in Fig. 6–2(b).
The *turbofan* engine has in addition a fan (shown in Fig. 6–2(c) as an addition
to the front of the compressor) along with a duct for the "cold" air flow and an
enlarged turbine to power the fan. The *turboprop* has a propeller which is driven,
by means of gears, by a substantially enlarged turbine, as in Fig. 6–2(d). Both the
turbofan and turboprop engines accelerate the hot turbine exhaust gases in nozzles,
just as the turbojet does, but to significantly lower velocities, since in a turbofan
and turboprop more of the available energy is extracted by the enlarged turbines
and transferred to the cold flow. It can be seen that the gas generator remains
unchanged. The performance and the actual configuration of these engines are
discussed in detail in Section 6–4 and in Chapters 7 and 8. In general, the turbofan
and turboprop engines accelerate a larger mass of air (per unit of fuel) to a smaller
average exhaust velocity $\bar{u}_e$ than is the case in the turbojet. Thus, according to
Eq. (6–19), their fuel consumption per unit thrust is smaller.

Specific fuel consumption

As shown by Eqs. (6–13) and (6–19) and the discussions on aircraft range and takeoff thrust, one of the important measures of an aircraft engine's performance is its fuel consumption rate per unit thrust. For ramjets, turbojets, and turbofans, the *thrust specific fuel consumption* is defined as

$$\text{TSFC} = \frac{\dot{m}_f}{T}. \tag{6–20}$$

Commonly the TSFC is expressed in units of pounds (mass) per hour per pound (force) of thrust. Using Eq. (6–6) for a turbojet and neglecting the pressure term,

$$\text{TSFC} = \frac{\dot{m}_f}{\dot{m}_a[(1+f)u_e - u]}. \tag{6–21}$$

For turbofan and turboprop engines, Eq. (6–7) may be used to evaluate the thrust T in Eq. (6–20). It can be seen that the thrust specific fuel consumption depends significantly on the flight speed. To avoid ambiguity TSFC is often based on static thrust and fuel consumption for turbojets and turbofans, although such a convention is impossible for ramjets. Typical values of TSFC for modern engines are:

For ramjets: $1.7\,–2.6\,\dfrac{\text{lb}_m/\text{hr}}{\text{lb}_f}$ (at M = 2),

For turbojets: $0.75–1.0\,\dfrac{\text{lb}_m/\text{hr}}{\text{lb}_f}$ (static),

For turbofans: $0.5\,–0.6\,\dfrac{\text{lb}_m/\text{hr}}{\text{lb}_f}$ (static).

For turbine engines which produce shaft power, a *brake specific fuel consumption*, BSFC, is defined by

$$\text{BSFC} = \frac{\dot{m}_f}{\mathcal{P}_s},$$

and expressed usually as pounds (mass) per hour per horsepower.

To account for the thrust of the hot gases, we may define an *equivalent brake specific fuel consumption* as

$$\text{EBSFC} = \frac{\dot{m}_f}{\mathcal{P}_{es}} = \frac{\dot{m}_f}{\mathcal{P}_s + T_e u},$$

in which $\mathcal{P}_s$ is the shaft power supplied to the propeller, T_e is the thrust produced by the turbine engine exhaust, and u is an arbitrarily selected flight speed. Typical values of equivalent brake specific fuel consumption are

$$\text{EBSFC} = 0.45 = 0.60\,\frac{\text{lb}_m/\text{hr}}{\text{hp}}.$$

In this respect the best turboprop engines are as efficient as the best piston engines. In addition, the turboprop engine is considerably lighter and smaller (in frontal area) than a piston engine of equal power, at least in the high-power sizes.

6–3   THE RAMJET

The simplest of all air-breathing engines is the ramjet. As shown schematically in Fig. 6–3, it consists of a diffuser, a combustion chamber, and an exhaust nozzle. Air enters the diffuser where it is compressed before it is mixed with the fuel and burned in the combustion chamber. The hot gases are then expelled through the nozzle by virtue of the pressure rise in the diffuser as the incoming air is decelerated from flight speed to a relatively low velocity within the combustion chamber. Consequently, although ramjets *can* operate at subsonic flight speeds, the increasing pressure rise accompanying higher flight speeds renders the ramjet most suitable for supersonic flight. Figure 6–3 is typical of supersonic ramjets which employ partially supersonic diffusion through a system of shocks. Since the combustion chamber requires an inlet Mach number of about 0.2 to 0.3, the pressure rise at supersonic flight speeds can be substantial. For example, for isentropic deceleration from $M = 3$ to $M = 0.3$, the static pressure ratio between ambient and combustion chamber pressures would be about 34! Only a fraction of the isentropic pressure ratio is actually achieved since, especially at high Mach numbers, the stagnation pressure losses associated with shocks are substantial. After compression the air flows past the fuel injectors, which spray a stream of fine fuel droplets so that the air and fuel mix as rapidly as possible. The mixture then flows through the combustion chamber which usually contains a "flameholder" to stabilize the flame, much as is indicated in Fig. 6–3. Combustion raises the temperature of the mixture to perhaps 4000°R before the products of combustion expand to high velocity in the nozzle. The reaction to the creation of the propellant momentum is a thrust on the engine in accordance with Eq. (6–6). This thrust is actually applied by pressure and shear forces distributed over the surfaces of the engine.

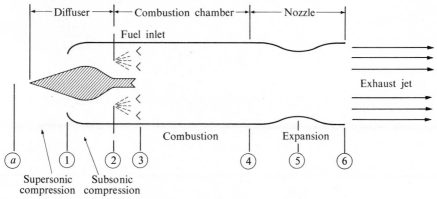

FIG. 6–3.   Schematic diagram of a ramjet engine.

The materials presently used for the walls of combustion chambers and nozzles cannot tolerate temperatures much above 2000°R, but they can be kept much cooler than the main fluid stream by a fuel-injection pattern which leaves a shielding layer of relatively cool air next to the walls. In contrast, the turbine engine cannot be operated at nearly so high a temperature. The turbine blades are subjected to high centrifugal stresses and cannot be cooled so readily. The lower maximum temperature limit of the turbojet greatly affects the relative performance and operating ranges of the two engines, as will be shown.

This relatively high peak temperature limit of the ramjet allows operation at high flight Mach numbers. As the Mach number is increased, however, the combustion chamber inlet temperature also increases, and at some limiting Mach number it will approach the temperature limit set by the wall materials and cooling methods. For example, at a flight Mach number of 6.7 in an atmosphere at −60°F, the stagnation temperature is about 4000°R.

At temperatures above 4000°R, dissociation of the combustion products may be significant. At such temperatures, the major effect of further fuel addition is further dissociation rather than actual temperature rise. Indeed, at extreme flight speeds, it may not be possible to add fuel at all (without exceeding the maximum temperature limit) unless there is considerable dissociation in the combustion chamber. If the dissociated combustion products recombine as they expand in the nozzle, the combustion energy can still be partly transformed to kinetic energy of the propellant. If not, the occurrence of dissociation could severely penalize performance. This question will be considered in the discussion of nozzles in Chapter 13.

A disadvantage of the ramjet is that the pressure ratio is strictly limited by flight speed and diffuser performance. The most serious consequence of this is the fact that the ramjet cannot develop static thrust and, therefore, cannot acceler-

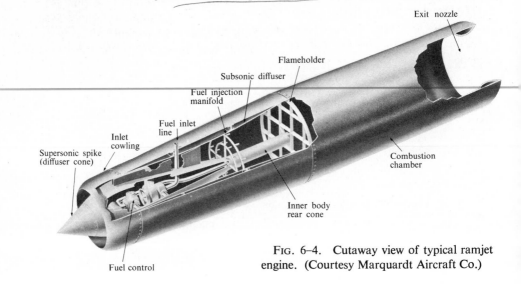

FIG. 6–4. Cutaway view of typical ramjet engine. (Courtesy Marquardt Aircraft Co.)

ate a vehicle from a standing start. Furthermore, the diffuser, whose behavior is so important to the engine as a whole, is difficult to design for high efficiency. This is due to detrimental boundary layer behavior in rising pressure gradients, especially in the presence of shocks which are practically unavoidable during supersonic operation. Supersonic diffusers designed for best efficiency at a given Mach number usually have poor performance at other Mach numbers unless their geometry is variable. The development of a large supersonic diffuser of reasonable performance and operating range requires extensive experimental work and substantial test facilities.

Figure 6–4 is a view of a typical ramjet, showing the geometry of the supersonic and subsonic diffusers, the fuel injector, flameholder, combustion chamber and nozzle. These components will all be discussed in detail in Chapter 7.

The ideal ramjet

In order to understand the performance of the ramjet, it is helpful to perform a thermodynamic analysis of a simplified model. Let us assume that the compression and expansion processes in the engine are reversible and adiabatic, and that the combustion process takes place at constant pressure. These assumptions are not, of course, realistic. In the actual diffuser, there are always irreversibilities due to shocks, mixing, and wall friction. Further, referring back to Chapter 3, it may be noted that, unless the combustion occurs at very low fluid velocity, both static and total pressures will drop, due to heat addition. Departures from isentropic flow in a real nozzle occur due to friction and heat transfer. However, the ideal ramjet is a most useful concept, since its performance is the highest that the laws of thermodynamics will permit, and is the limit which real engines will approach if their irreversibilities can be reduced.

Using the station numbers of Fig. 6–3, Fig. 6–5 shows, on a temperature-entropy diagram, the processes the air goes through in an ideal ramjet. The compression process takes the air from its condition at station a isentropically to its stagnation state 02 at station 2. The combustion process is represented by a constant-pressure heat and mass addition process 02 to 04 up to the maximum temperature T_{04}. The exit nozzle expands the combustion products isentropically to the ambient pressure. (Isentropic expansion demands that exhaust

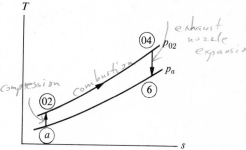

FIG. 6–5.   Thermodynamic path of the fluid in an ideal ramjet.

pressure equal ambient pressure, as will be discussed in Chapter 7.) The ideal engine thrust may be obtained from Eq. 6–6 as

$$\mathcal{T} = \dot{m}_a[(1 + f)u_e - u]. \tag{6-22}$$

With isentropic compression and expansion processes, and constant-pressure heat and mass addition, it follows that the stagnation pressure must be constant throughout the engine. Therefore $p_{0a} = p_{06}$.

If variations in fluid properties (R, γ) through the engine are ignored for this ideal case,

$$\frac{p_{0a}}{p_a} = \left(1 + \frac{\gamma - 1}{2} M^2\right)^{\gamma/(\gamma-1)} \quad \text{and} \quad \frac{p_{06}}{p_e} = \left(1 + \frac{\gamma - 1}{2} M_e^2\right)^{\gamma/(\gamma-1)},$$

in which M is the flight Mach number and M_e is the Mach number in the plane of the exhaust. Therefore, with the condition $p_e = p_a$, it is clear that

$$\frac{p_{0a}}{p_a} = \frac{p_{06}}{p_e} \quad \text{and} \quad M_e = M_a. \tag{6–23}$$

Thus the exhaust velocity may be determined from

$$u_e = \frac{a_e}{a_a} u,$$

where a is the speed of sound. Since $a = \sqrt{\gamma R T}$, then $a_e/a_a = \sqrt{T_e/T_a}$. However, for the case $M_e = M_a$, $T_e/T_a = T_{06}/T_{0a}$ and, since $T_{04} = T_{06}$, then

$$u_e = \sqrt{T_{04}/T_{0a}}\, u. \tag{6–24}$$

The energy equation applied to the idealized combustion process, neglecting the enthalpy of the incoming fuel, is

$$(1 + f)h_{04} = h_{02} + fQ_R, \tag{6–25}$$

where f is the fuel–air ratio and Q_R is the heating value of the fuel. If the specific heat is assumed constant, then Eq. (6–25) may be solved for f in the form

$$f = \frac{(T_{04}/T_{0a}) - 1}{(Q_R/c_p T_{0a}) - T_{04}/T_{0a}} \tag{6–26}$$

Equations (6–22) and (6–24) may be combined to give the thrust per unit mass flow of air,

$$\frac{T}{\dot{m}_a} = M\sqrt{\gamma R T_a}\left[(1 + f)\sqrt{T_{04}/T_a}\left(1 + \frac{\gamma - 1}{2} M^2\right)^{-1/2} - 1\right], \tag{6–27}$$

where f is given by Eq. (6–26). The thrust specific fuel consumption is given by

$$\text{TSFC} = \frac{\dot{m}_f}{T} = \frac{f}{T/\dot{m}_a}, \tag{6–28}$$

with appropriate constants to convert to the desired units, usually pounds (mass) per hour per pound force.

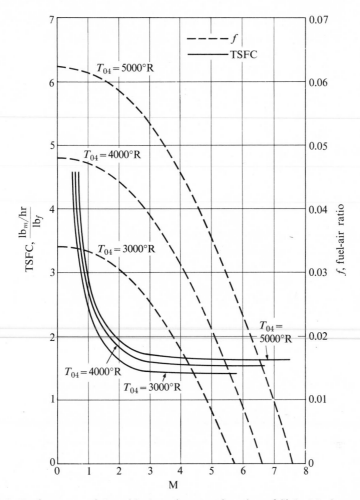

FIG. 6–6. Performance of three ideal ramjets as a function of flight Mach numbers and peak temperature limit for $T_a = -60°F$, $Q_R = 19,000$ Btu/lb$_m$, $c_p = 0.24$ Btu/lb$_m$·°R, and $\gamma = 1.4$.

Figure 6–6 indicates the thrust specific fuel consumption and the required fuel–air ratio of an ideal ramjet as a function of flight Mach number and peak temperature. It can be seen that for any given temperature there is a maximum flight Mach number at which no fuel may be burned in the air. Conversely, for any given flight Mach number it would appear from the figure that operation at low temperature is advantageous, since it results in lower TSFC. However, as shown in Fig. 6–7, this means a relatively low thrust per unit airflow rate and hence a larger engine for a given thrust. But the larger the engine, the greater its mass and drag. As a result, maximum speed operation of an aircraft is often based on maximum tolerable temperature. Cruise operation may then employ

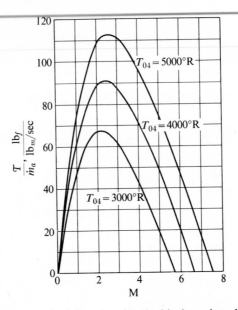

FIG. 6–7.   Thrust per unit airflow rate for the ideal ramjets shown in Fig. 6–6.

a lower engine temperature, which engenders both longer engine life and lower specific fuel consumption. The choice of the best engine size and operating temperature for maximum cruise economy requires a careful analysis of the drag and weight penalty associated with larger engines, as well as their lower fuel consumption.

The effect of aerodynamic losses

The propellant in a real ramjet engine, of course, suffers stagnation pressure losses as it flows through the engine. The most striking effect of such aerodynamic losses is that, as flight Mach number increases, the engine thrust goes to zero long before its fuel consumption does. Hence the TSFC of a real ramjet, rather than decreasing to a relatively low value as in the ideal case, at first decreases and then rises with increasing flight Mach number. The Mach number at which minimum TSFC occurs and that at which TSFC becomes infinite ($T \to 0$) are dependent on the severity of the component stagnation pressure losses, as will be shown.

Figure 6–8 shows, on a T-s diagram, the effect of these irreversibilities on the processes of compression, burning, and expansion. The compression process, $\textcircled{a}$ to $\textcircled{02}$, is no longer isentropic, though the isentropic process is shown for comparison. The stagnation pressure at the end of the process is lower than it would be if the compression were isentropic. The performance of diffusers may be characterized by a stagnation-pressure ratio r_d defined by

$$r_d = \frac{p_{02}}{p_{0a}}. \qquad (6\text{--}29)$$

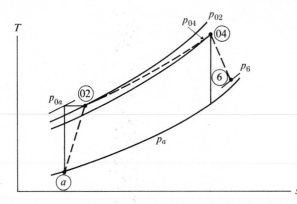

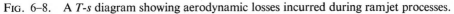

FIG. 6–8.   A T-s diagram showing aerodynamic losses incurred during ramjet processes.

Similarly, stagnation-pressure ratios can be defined for combustors r_c and nozzles r_n as follows:

$$r_c = \frac{p_{04}}{p_{02}}, \tag{6–30}$$

$$r_n = \frac{p_{06}}{p_{04}}. \tag{6–31}$$

The overall stagnation-pressure ratio is therefore

$$\frac{p_{06}}{p_{0a}} = r_d r_c r_n.$$

Further, the actual exhaust pressure p_e or p_6 may not equal the ambient pressure p_a. However, using Eq. (3–11) for constant γ, the exhaust Mach number may be written as

$$\mathrm{M}_e^2 = \frac{2}{\gamma - 1}\left[\left(1 + \frac{\gamma - 1}{2}\mathrm{M}^2\right)\left(\frac{p_{06}}{p_{0a}}\frac{p_a}{p_e}\right)^{(\gamma-1)/\gamma} - 1\right].$$

Thus, in terms of the component stagnation-pressure ratios,

$$\mathrm{M}_e^2 = \frac{2}{\gamma - 1}\left[\left(1 + \frac{\gamma - 1}{2}\mathrm{M}^2\right)\left(r_d r_c r_n \frac{p_a}{p_e}\right)^{(\gamma-1)/\gamma} - 1\right]. \tag{6–32}$$

If heat transfer from the engine is assumed negligible (per unit mass of fluid) then the exhaust velocity is given by $u_e = \mathrm{M}_e\sqrt{\gamma R T_e}$ or, in terms of the exhaust stagnation temperature

$$u_e = \mathrm{M}_e \sqrt{\gamma R T_{04}/\left(1 + \frac{\gamma - 1}{2}\mathrm{M}_e^2\right)}. \tag{6–33}$$

Since irreversibilities have no effect on stagnation temperatures throughout the

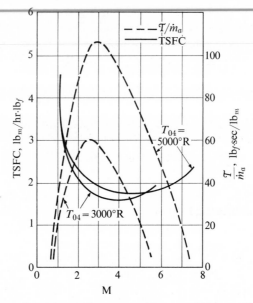

FIG. 6–9.   Estimated ramjet performance with $r_d = 0.7$, $r_c = 0.97$, $r_n = 0.96$, $\gamma = 1.4$, $Q_R = 19{,}000$ Btu/lb$_m$, and $\eta_b = 1.0$.

engine, the fuel–air ratio necessary to produce the desired T_{04} is given by Eq. (6–26)

$$ f = \frac{(T_{04}/T_{0a}) - 1}{(\eta_b Q_R / c_p T_{0a}) - (T_{04}/T_{0a})}, \tag{6–26} $$

where η_b is the combustion efficiency and $\eta_b Q_R$ is the actual heat release per unit mass of fuel. The thrust per unit airflow rate then becomes

$$ \frac{T}{\dot{m}_a} = [(1 + f)u_e - u] + \frac{1}{\dot{m}_a}(p_e - p_a)A_e $$

or, using Eqs. (6–32) and (6–33),

$$ \frac{T}{\dot{m}_a} = (1 + f)\sqrt{\frac{2\gamma R T_{04}(m - 1)}{(\gamma - 1)m}} - M\sqrt{\gamma R T_a} + \frac{p_e A_e}{\dot{m}_a}\left(1 - \frac{p_a}{p_e}\right), \tag{6–34} $$

in which

$$ m = \left(1 + \frac{\gamma - 1}{2}M^2\right)\left(r_d r_c r_n \frac{p_a}{p_e}\right)^{(\gamma - 1)/\gamma}. $$

Again, the thrust specific fuel consumption is given by Eq. (6–28):

$$ \text{TSFC} = \frac{f}{T/\dot{m}_a}. \tag{6–28} $$

The performance of inlets, combustors, and nozzles will be discussed in considerably more detail in Chapter 7. At this point, however, the effect of losses on performance can be appreciated by comparing Fig. 6–6 for the ideal ramjet and Fig. 6–9, which shows calculations of specific fuel consumption and specific thrust for assumed loss coefficients $r_d = 0.7$, $r_c = 0.97$, and $r_n = 0.96$. The specific heat ratio γ has been assumed constant in the derivation of Eq. (6–34), so that it will not have high accuracy. Furthermore, the coefficients (especially r_d) are not constant over large Mach number variations. However, Fig. 6–8 does show that the introduction of moderate losses significantly decreases the ramjet performance, so that it is very important to improve the fluid behavior in the diffuser nozzle and in the combustor. In contrast to Fig. 6–6, Fig. 6–9 shows that there is, at any given peak temperature, a reasonably well-defined minimum TSFC.

6–4 GAS TURBINE ENGINES

The turbojet

It has been mentioned that one of the disadvantages of the ramjet is that its pressure ratio depends on the flight Mach number. It cannot develop takeoff thrust and, in fact, it does not perform well unless the flight speed is considerably above the speed of sound. One way to overcome this disadvantage would be to install a mechanical compressor in the inlet duct, so that even at zero flight speed air could be drawn into the engine, burned, and then expanded through a nozzle. However, this introduces the need for power to drive the compressor. If a turbine is coupled to the compressor and driven by the hot gas passing from the burner on its way to the exhaust nozzle, the ramjet has become a turbojet. The addition of the turbomachinery, however, entirely changes the characteristic performance of the engine.

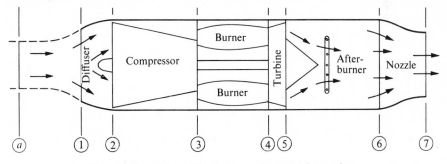

FIG. 6–10. Schematic diagram of a turbojet engine.

The internal arrangement of the turbojet is shown schematically in Fig. 6–10. In flowing through the machine the air undergoes the following processes:

(a)–(1) From far upstream, where the velocity of the air relative to the engine is the flight velocity, the air is brought to the intake, usually with some acceleration or deceleration.

①–② The air velocity is decreased as the air is carried to the compressor inlet through the inlet diffuser and ducting system.

②–③ The air is compressed in a dynamic compressor.

③–④ The air is "heated" by the mixing and burning of fuel in the air.

④–⑤ The air is expanded through a turbine to obtain power to drive the compressor.

⑤–⑥ The air may or may not be further "heated" by the addition and burning of more fuel in an afterburner.

⑥–⑦ The air is accelerated and exhausted through the exhaust nozzle.

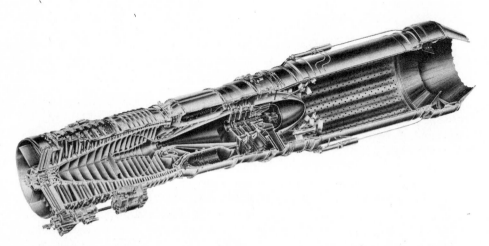

Fɪɢ. 6–11.   General Electric J79 turbojet engine, which has thrust of approximately 16,000 pounds.   (Courtesy General Electric Co.)

Figure 6–11 shows in cutaway a typical afterburner-equipped turbojet, the General Electric J-79 engine.   In this case the compressor is an axial compressor of seventeen stages, a stage being defined as a pair of stationary and rotating blade rows.   Its pressure ratio is 13:1, and the airflow rate is about 170 lb/sec. The combustion chamber consists of 10 distinct "cans" enclosed in a single annular space.   The fuel, a mixture of kerosene and low-vapor-pressure gasoline known as JP-4, is injected within each can.   The hot gases leaving the combustor enter a three-stage turbine.   Figure 6–11 shows the afterburner fuel nozzles and flameholders, as well as a perforated liner which shields the outer casing from the combustion zone.   The exit nozzle area is mechanically variable.

The thermodynamic path of the fluid within a turbojet may be conveniently shown on an enthalpy-entropy or temperature-entropy diagram.   To gain an understanding of the overall process, it is useful at first to study a highly simplified model.   For this reason, let us assume that all components except the burners are reversible and adiabatic, that the burners may be replaced by simple frictionless heaters, and that velocities at sections ② through ⑥ are negligible.   The

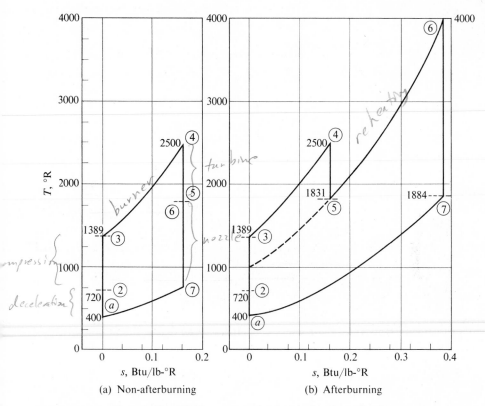

FIG. 6–12.   Fluid processes in the ideal turbojet engine.

T-s diagram for such an engine is shown in Fig. 6–12 for non-afterburning and afterburning engines, assuming the working fluid to be a perfect gas.

In the ideal case the pressure rises from (a) to (1), and still more from (1) to (2) as the air is decelerated relative to the engine. Since the velocity at (2) is assumed zero and the deceleration is isentropic, p_2 is the stagnation pressure for states (a), (1), and (2). Also, T_2 is the stagnation temperature for these states. The power consumed in compression from (2) to (3) must be supplied through the turbine in expansion from (4) to (5). Hence, if the compressor and turbine mass-flow rates are equal, $h_3 - h_2 = h_4 - h_5$, and if the specific heat is constant, the corresponding temperature differences are also equal. In the non-afterburning case, states (5) and (6) are identical and the enthalpy drop from (5) or (6) to (7) is proportional to the square of the exhaust velocity. In the afterburning case the air is reheated between (5) and (6). From the shape of the constant-pressure curves it can be seen that $(T_6 - T_7)$, and hence the exhaust velocity, will be greater in the afterburning case. In fact, as we shall see later, the absence of highly stressed material in the afterburner allows T_6 to be much higher even than T_4, so that the increase in exhaust velocity can be on the order of 50%.

Although this is a greatly simplified model, it illustrates the functions of the various components and the relationships between them. It shows clearly that the output or kinetic energy of the exhaust fluid is, in a sense, a remainder after power has been extracted from the fluid to drive the compressor.

An actual engine differs from this ideal model in several respects. First, and most important, no components are actually reversible, although it is usually reasonable to assume them adiabatic. Second, the burners are not simple heaters and the composition of the working fluid will change during the combustion processes. Third, the fluid velocities within the engine are not negligible. If the fluid velocity in the combustor were actually zero (as constant-pressure combustion requires), it would be impossible to have a stable flame, since the flame propagates relative to the fluid at fairly large velocities. There is a fourth difference, in that the turbine and compressor flow rates may not be equal since, on the one hand, fuel is added between the two and, on the other, air may be extracted at various positions for cooling purposes.

Figure 6–13 shows an enthalpy-entropy diagram for a real engine with reasonable irreversible effects, and typical temperatures, for a compressor pressure ratio of ten. Afterburning and non-afterburning processes are shown, with the exhaust pressure equal to ambient pressure in both cases.

The process begins with atmospheric air at h_a, p_a. By virtue of the relative (flight) velocity between the air and the engine, this air has a stagnation enthalpy h_{0a}, higher than h_a. Further, since no work or heat transfer occurs between (a) and (2), the stagnation enthalpy is constant through station (2). The air is externally decelerated from (a) to (1). For all practical purposes this external deceleration is an isentropic process (unless an external shock occurs), hence state (1) is on an isentrope with state (a) and $p_{01} = p_{0a}$. From (1) to (2) the air is further decelerated, accompanied by an increase in entropy through frictional effects. Note that this results in a decrease in stagnation pressure. From (2) to (3) the air is compressed, again with an increase of entropy due to irreversibilities in the compression process. State $(03s)$ is defined as that state which would exist if the air could be compressed isentropically to the actual outlet stagnation pressure. State (03) is the actual outlet stagnation state. These states will be discussed later in defining compressor efficiency.

From station (3) to station (4) (see Fig. 6–10), some fuel is mixed with the air and combustion occurs. Strictly speaking, the fluid composition changes between these stations, and a continuous path between them should not be shown. However, since the fluid characteristics do not change markedly, there is no difficulty in showing the two substances on different portions of the same diagram. The stagnation pressure at (4) must be less than at (3) because of fluid friction, and also because of the drop in stagnation pressure due to heat addition at finite velocity. As we shall see later, it is advantageous to make T_{04} as high as material limitations will allow. Hence states (04) and (4) are fairly well fixed.

From (4) to (5), the fluid expands through the turbine, providing shaft power equal to the shaft power input to the compressor (plus any mechanical losses or

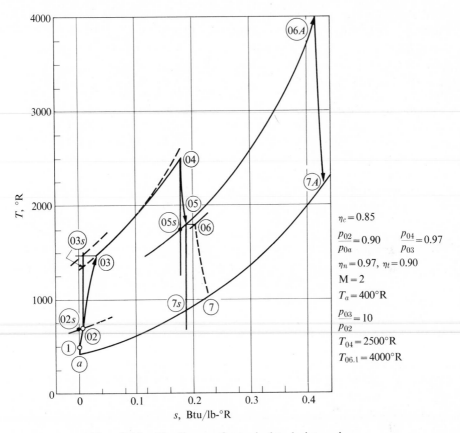

Fig. 6-13. *T-s* diagram for typical turbojet engines.

accessory power). Since no work or heat transfer occurs downstream of station ⑤, the stagnation enthalpy remains constant throughout the rest of the machine. State ⑥ depends on the geometry involved, but p_{06} must be less than p_{05}. The exhaust pressure p_7 generally equals the atmospheric pressure p_a, but it may be different if the exhaust flow is supersonic. If state ⑦ is known, the velocity u_7 can be calculated from the known h_{07} (or h_{05}) regardless of the properties at ⑥. If the afterburner is operative the fluid is raised in temperature to state 06A, after which it expands in the nozzle to state 7A.

Again it can be seen that the exhaust kinetic energy is the relatively small difference between the total available enthalpy drop from state ④ and the compressor work input. For a given compressor-pressure ratio, irreversibilities increase the compressor power requirement while at the same time increasing the necessary turbine pressure drop. Both effects decrease the exhaust kinetic energy, so that overall performance may be expected to be very sensitive to compressor and turbine performance.

Utilizing the fact that compression and expansion processes in turbojet engines are very nearly adiabatic, we can make realistic estimates of engine performance by defining the adiabatic component efficiencies as follows:

For the inlet diffuser, an adiabatic efficiency η_d may be defined as the ratio of the ideal to the actual enthalpy change during the diffusion process (for the same pressure ratio p_{02}/p_1), or

$$\text{Diffuser} \qquad \eta_d = \frac{h_{02s} - h_a}{h_{02} - h_a} . = \frac{ideal}{actual}$$

Alternatively, the stagnation-pressure ratio as defined for the ramjet [Eq. (6–29)] may be used.

Correspondingly, for a compressor, a useful definition of an adiabatic efficiency η_c is the ratio of the work required in an *isentropic* process to that required in the *actual* process, for the same stagnation-pressure ratio and inlet state:

$$\text{Compressor} \qquad \eta_c = \frac{h_{03s} - h_{02}}{h_{03} - h_{02}} . \quad \frac{isentropic}{actual}$$

For the turbine, the adiabatic efficiency may be defined as

$$\text{Turbine} \qquad \eta_t = \frac{h_{04} - h_{05}}{h_{03} - h_{05s}} , \quad \frac{actual}{isentropic}$$

which is the ratio of actual turbine work to that which would be obtained during an isentropic expansion to the same exhaust stagnation pressure.

A nozzle adiabatic efficiency may be defined as

$$\text{Nozzle} \qquad \eta_n = \frac{h_{06} - h_7}{h_{06} - h_{7s}} .$$

In addition to these four adiabatic efficiencies, a fifth kind of efficiency is often employed: the burner efficiency η_b, which is simply the fraction of the chemical energy of the fuel which is released in the combustor.

For well-designed engines, the foregoing efficiencies will generally be in the following range:

$$0.7 < \eta_d < 0.9 \text{ (depending strongly on flight Mach number)},$$
$$0.85 < \eta_c < 0.90, \qquad 0.90 < \eta_t < 0.95,$$
$$0.95 < \eta_n < 0.98, \qquad 0.97 < \eta_b < 0.99.$$

Note that these definitions have utilized stagnation values of enthalpy. It is usually much more convenient experimentally to measure stagnation, rather than static, values of pressure and temperature in a fluid stream. Stagnation values are convenient analytically, since they contain the kinetic-energy terms.

Later studies of components will focus on the questions of component performance in more detail. The immediate objective of this discussion is to show how overall engine performance depends on component performance and, in particular, on maximum pressure ratio and temperature.

Two questions of considerable importance are: how large is the thrust per unit mass flow through the engine, and what is the fuel consumption per unit thrust? Since an engine of given diameter is limited in mass flow rate for given inlet conditions, the answer to the first question largely governs the thrust–size relationship.

Using the thrust equation (6–6) for the case in which the exhaust plane pressure is atmospheric, the thrust per unit airflow is

$$\frac{T}{\dot{m}_a} = [(1 + f)u_e - u], \tag{6–35}$$

and the thrust specific fuel consumption is given by

$$\text{TSFC} = \frac{\dot{m}_f}{T} = \frac{f}{(1 + f)u_e - u}. \tag{6–36}$$

The exhaust velocity is obtained from conservation of energy in the nozzle,

$$\frac{u_e^2}{2} = h_{06} - h_7 \quad \text{or} \quad \frac{u_e^2}{2} = \eta_n(h_{06} - h_{7s}).$$

If the fluid properties are assumed constant,

$$h_{06} - h_{7s} = c_p(T_{06} - T_{7s}) \quad \text{or} \quad h_{06} - h_{7s} = c_p T_{06}\left[1 - \left(\frac{p_7}{p_{06}}\right)^{(\gamma-1)/\gamma}\right],$$

where γ is the ratio of specific heats of the nozzle fluid. Therefore

$$u_e = \sqrt{2c_p T_{06}\eta_n[1 - (p_7/p_{06})^{(\gamma-1)/\gamma}]}. \tag{6–37}$$

To relate the nozzle pressure ratio and the exhaust velocity to flight speed, ambient conditions p_a and T_a, component efficiencies, maximum engine temperature T_{04}, and compressor pressure ratio p_{03}/p_{02}, it is convenient to use the following identity:

$$\frac{p_{06}}{p_7} = \frac{p_{06}}{p_{05}} \cdot \frac{p_{05}}{p_{04}} \cdot \frac{p_{04}}{p_{03}} \cdot \frac{p_{03}}{p_{02}} \cdot \frac{p_{02}}{p_a} \cdot \frac{p_a}{p_7}. \tag{6–38}$$

The ratio p_a/p_7 may be known or assumed, as well as the stagnation pressure ratios p_{04}/p_{03} and p_{06}/p_{05} (which should be fairly close to unity). We now assume that the compressor ratio p_{03}/p_{02} is known, and proceed to evaluate the diffuser ratio p_{02}/p_a and the turbine ratio p_{05}/p_{04}.

For the diffuser, since

$$T_{02} = T_{01} \quad \text{and} \quad \frac{T_{01}}{T_a} = 1 + \frac{\gamma - 1}{2} M^2,$$

then

$$T_{02} - T_a = T_a \frac{\gamma - 1}{2} M^2, \tag{6–39}$$

and, using the definition of diffuser efficiency, we obtain

$$\frac{p_{02}}{p_a} = \left(1 + \eta_d \frac{\gamma - 1}{2} M^2\right)^{\gamma/(\gamma-1)}. \tag{6–40}$$

For a given compressor pressure ratio p_{03}/p_{02} and compressor efficiency η_c, the compressor work input per unit mass is

$$h_{03} - h_{02} = \frac{c_p T_{02}}{\eta_c}\left[\left(\frac{p_{03}}{p_{02}}\right)^{(\gamma-1)/\gamma} - 1\right]. \tag{6–41}$$

The turbine work output per unit mass is

$$h_{04} - h_{05} = \eta_t c_p T_{04}\left[1 - \left(\frac{p_{05}}{p_{04}}\right)^{(\gamma-1)/\gamma}\right]. \tag{6–42}$$

Assuming turbine and compressor mass flow rates about equal,

$$h_{03} - h_{02} = h_{04} - h_{05},$$

and the turbine and compressor pressure ratios are related by

$$\frac{c_p T_{02}}{\eta_c}\left[\left(\frac{p_{03}}{p_{02}}\right)^{(\gamma-1)/\gamma} - 1\right] = \eta_t c_p T_{04}\left[1 - \left(\frac{p_{05}}{p_{04}}\right)^{(\gamma-1)/\gamma}\right],$$

or

$$\left(\frac{p_{05}}{p_{04}}\right)^{(\gamma-1)/\gamma} = 1 - \frac{T_a}{\eta_c \eta_t T_{04}}\left[1 + \frac{\gamma - 1}{2} M^2\right]\left[\left(\frac{p_{03}}{p_{02}}\right)^{(\gamma-1)/\gamma} - 1\right]. \tag{6–43}$$

Hence with p_{02}/p_a and p_{05}/p_{04} given by Eqs. (6–40) and (6–43), the nozzle pressure ratio p_{06}/p_7 can be obtained from Eq. (6–38). Since $T_{05} = T_{06}$, the nozzle inlet temperature T_{06} can then be found from Eq. (6–42); and now with known values of p_{06}/p_7 and T_{06}, the exhaust velocity can be evaluated from Eq. (6–37). Lastly, the fuel–air ratio can be found from an equation similar to Eq. (6–26):

$$f = \frac{(T_{04}/T_{03}) - 1}{(Q_R/c_p T_{03}) - (T_{04}/T_{03})} \tag{6–44}$$

and Eqs. (6–41) and (6–39), which relate T_{03} to T_a.

Thus equations (6–37) through (6–44) allow calculations of thrust specific fuel consumption and thrust per unit airflow rate in terms of component efficiencies, compressor pressure ratio, and maximum engine temperature. Although it is usually adequate to assume that air behaves as a perfect gas within each component —as these equations imply—better accuracy can be achieved if the constants

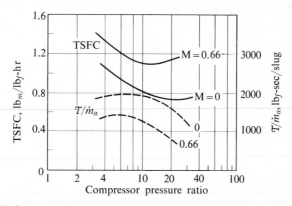

FIG. 6–14. Typical turbojet thrust specific fuel consumption as a function of compressor–pressure ratio for two values of flight Mach number M. Here $T_{04} = 1960°R$, $T_a = 519°R$, $\eta_c = 0.85$, $\eta_b = 0.96$, $\eta_t = 0.90$, $Q_R = 18,900$ Btu/lb$_m$; (after [7]).

c_p and γ are chosen separately for each component. This is particularly true of the compressor and turbine calculations.

A typical calculation is indicated in Fig. 6–14, which shows the dependence of thrust specific fuel consumption on the choice of compressor pressure ratio. Here we observe that at a given flight Mach number, there is one compressor pressure ratio which will produce minimum TSFC. It depends on component efficiencies and maximum temperature T_{04} as well as flight velocity.

Factors other than TSFC must be considered in the choice of compressor pressure ratio. For example, a pressure ratio of 20, which appears desirable for low-speed flight, would require a heavy machine to produce a given thrust, since the thrust per unit mass flow of air is relatively low.

Figure 6–15 shows the effect of peak temperature on thrust specific fuel consumption and thrust per unit airflow rate for the indicated component efficiencies. The pressure ratio for minimum TSFC decreases considerably with T_{04}, and so does the thrust per unit airflow rate. In addition, it may be seen that reducing the turbine inlet temperature from 2460°R to 1960°R leads to a slight reduction in the minimum TSFC. Further reductions in T_{04}, however, lead to an increase in minimum TSFC. It should be borne in mind, of course, that the performance variations illustrated in Fig. 6–15 are quite dependent on component efficiencies as well as on flight Mach number.

In Chapters 7 and 8 the physical factors which explain engine performance limitations are discussed. Since these factors strongly influence the configuration of actual machines, it is appropriate to point them out here before component analyses are made. Basically they are of two kinds: material limits, as expressed by allowable stress and temperature levels, and aerodynamic limits, imposed mainly by the behavior of boundary layers in the presence of rising pressure.

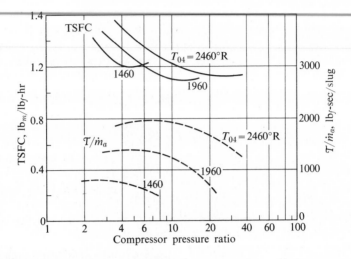

Fig. 6-15. Variation of thrust specific fuel consumption and thrust per unit mass flow of air with compressor–pressure ratio for a typical turbojet engine. Flight Mach number $M = 0.66, T_a = 519°R, \eta_c = 0.85, \eta_b = 0.96, \eta_t = 0.90, Q_R = 18,900$ Btu/lb; (after [7]).

As it does in all flight applications, the desire for minimum engine weight leads to highly stressed engine components as well as high peak temperatures. The largest stresses are the result of centrifugal force. It may be shown that centrifugal stress depends on the tangential velocity of a rotating element; currently this limitation appears to be about 1400 ft/sec in the turbine. Most designs use a tip speed between 900 and 1200 ft/sec. Stresses other than centrifugal ones are present, of course, and cannot be ignored. The allowable operating stress in an element is intimately related to the temperature at which the element must operate, which implies the necessity of compromising the turbine-blade stress and the combustion temperature. With presently available material, the maximum stagnation temperature of the gas is limited to about 1600°F for uncooled blades. If the blades are cooled internally by passing a fluid through small interior passages, it is possible to use stagnation temperatures approaching 2000°F.

Due primarily to the necessity of avoiding boundary layer separation, and (to a lesser degree) to the high losses associated with compressibility, the pressure ratio which can be accommodated in a single stage of a turbomachine is limited. (A *stage* may be defined as a single circumferential row of rotating blades and an associated set of stationary guide vanes, stators, or nozzles.) This limitation is most severe in compressors where the fluid necessarily flows against a rising pressure gradient (see Chapter 4). Primarily for this reason, axial-compressor-stage pressure ratios are much lower than turbine-stage pressure ratios. Hence axial compressors of ten to twenty stages can be driven by turbines of two or three stages.

The turbofan engine

It was shown in Section 6–2 that, for a given energy or fuel-consumption rate, thrust increases with airflow rate. Further, it will be shown that the cruise economy of the engine can, in some cases, be improved by raising its propellant mass flow. Increased airflow for a given thrust could be achieved by simply enlarging the engine while reducing the fuel flow rate, as shown in Fig. 6–15. However, this results in an unnecessarily heavy engine. The turbofan engine achieves the same result in a much lighter configuration.

As shown schematically in Fig. 6–2, and again in greater detail in Fig. 6–16, the turbofan consists of a basic gas generator plus sufficient turbomachinery for acceleration of a separate or "cold" propellant stream. In this way power is mechanically extracted from the "hot" or primary flow, thus lowering the velocity of the exhaust from the gas generator. This power is added through a fan or relatively low-pressure-ratio compressor to a secondary airflow which bypasses the burner and turbine of the gas generator. Thus the fuel–air ratio, peak temperature, and size of the gas generator are similar to that of a conventional turbojet of the same energy consumption rate. Figures 6–17 and 6–18 show, in cutaway views, typical configurations of Pratt & Whitney Aircraft and General Electric engines.

Figures 6–16(a) and (b) are typical of engines which combine the fan with the front stages of the primary flow compressor. In one case the secondary air completely bypasses the remainder of the engine, while in the other it is ducted around the engine and mixed with the primary air before the two flows are expanded through a common nozzle. Both these engines employ compressors consisting of two mechanically independent rotating elements or "spools" driven by separate turbines. This configuration is common in both turbojet and turbofan designs. The JT3D has a bypass ratio (ratio of bypass airflow to primary airflow rate) of about 1.4, while the Conway employs a bypass ratio of about 0.6.

Figure 6–16(c) illustrates the aft-fan configuration employed by General Electric. In this case the fan consists of a single, mechanically independent, rotor downstream from the gas generator. As shown in Fig. 6–18, the fan rotor is made up of blades which act as a turbine on their inner portions and as a compressor on their outer portions. The gas generator in this case is identical to that of the J79 engine of Fig. 6–11, which had an airflow rate of 170 lb/sec. With a bypass flow rate of 250 lb/sec, this engine then has a bypass ratio of about 1.5. The choice of bypass ratio will be discussed shortly.

The effect on performance of adding a secondary stream to a turbojet engine may be estimated by elementary thermodynamic arguments. Let us say that the mass flow through the combustor and turbine, i.e., the primary mass flow, is denoted by $\dot{m}_a$, and that the ratio of secondary to primary mass flows is β. In addition to the nomenclature of Fig. 6–10, let $\boxed{08}$ be the stagnation state of the secondary stream leaving the fan, and η_f be the adiabatic efficiency of the fan. The secondary stream is assumed to expand through a nozzle to ambient pressure,

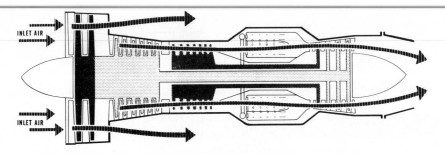

(a) Pratt & Whitney JT3D

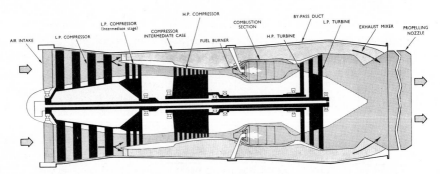

(b) Rolls-Royce Conway R.Co.42

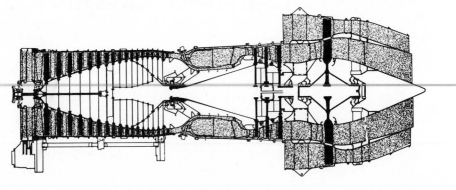

(c) General Electric CJ-805-23

FIG. 6–16.   Schematic diagrams of turbofan engines. (a) Courtesy Pratt & Whitney Aircraft Division of United Aircraft Corp. (b) Courtesy Rolls-Royce. (c) Courtesy Flight International.

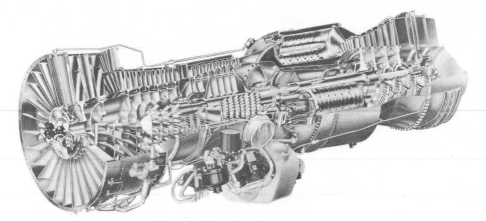

FIG. 6–17. Cutaway view of Pratt & Whitney JT3D engine, shown schematically in Fig. 6–16(a). (Courtesy Pratt & Whitney Aircraft Division of United Aircraft Corp.)

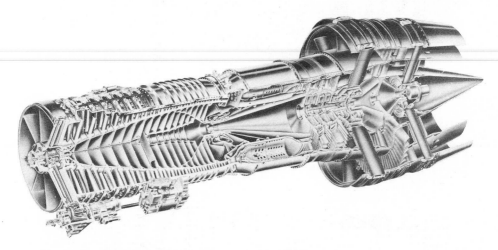

FIG. 6–18. Aft-fan rotor assembly for the General Electric CJ-805-23 engine, shown schematically in Fig. 6–16(c). (Courtesy General Electric Co.)

the adiabatic nozzle efficiency being η_n, the same as for the primary jet nozzle. (It can be shown that there may be a small increase in thrust if the primary and secondary streams are mixed prior to expansion, as they are in the Rolls-Royce RCo-42, though it is difficult to estimate the losses in the mixing process. However, this complication will be ignored in the present discussion.)

The turbine must now supply the power for compression of both primary and secondary airflows, so that the energy balance is

$$\dot{m}_a(1 + f)c_{ph}(T_{04} - T_{05}) = \dot{m}_a c_{pc}(T_{03} - T_{02}) + \beta \dot{m}_a c_{pc}(T_{08} - T_{0a}),$$

in which c_{ph} and c_{pc} are the specific heats of the air after and before combustion, respectively. Rearranging, we have

$$\frac{T_{05}}{T_{04}} = 1 - \frac{[(T_{03}/T_{02}) - 1] + \beta[(T_{08}/T_{0a}) - 1]}{(T_{04}/T_{0a})(c_{ph}/c_{pc})(1 + f)}.$$

Using definitions of adiabatic efficiency which have already been derived, we obtain

$$\frac{T_{03}}{T_{02}} - 1 = \frac{1}{\eta_c}\left[\left(\frac{p_{03}}{p_{02}}\right)^{(\gamma-1)/\gamma} - 1\right], \qquad \frac{T_{08}}{T_{0a}} - 1 = \frac{1}{\eta_f}\left[\left(\frac{p_{08}}{p_{0a}}\right)^{(\gamma-1)/\gamma} - 1\right].$$

We have here assumed that the losses in the inlet to the fan are included in the definition of adiabatic fan efficiency. Also

$$\frac{T_{0a}}{T_a} = 1 + \frac{\gamma - 1}{2}\mathbf{M}^2, \qquad \frac{p_{0a}}{p_a} = \left(1 + \frac{\gamma - 1}{2}\mathbf{M}^2\right)^{\gamma/(\gamma-1)},$$

and

$$\frac{p_{02}}{p_a} = \left(1 + \eta_d\frac{\gamma - 1}{2}\mathbf{M}^2\right)^{\gamma/(\gamma-1)}.$$

If the ratio of specific heats γ_1 of the turbine working fluid is assumed constant, then

$$\frac{T_{05}}{T_{04}} = \frac{T_{06}}{T_{04}} = 1 - \eta_t\left[1 - \left(\frac{p_{05}}{p_{04}}\right)^{(\gamma_1-1)/\gamma_1}\right].$$

With these relations, the pressure ratio across the turbine may be written as

$$\frac{p_{05}}{p_{04}} = \left\{1 - \frac{(1/\eta_c)[(p_{03}/p_{02})^{(\gamma-1)/\gamma} - 1] + (\beta/\eta_f)[(p_{08}/p_{0a})^{(\gamma-1)/\gamma} - 1]}{(1 + f)(T_{04}/T_{0a})\eta_t(c_{ph}/c_{pc})}\right\}^{\gamma_1/(\gamma_1-1)},$$

where the fuel–air ratio f is again determined from Eq. (6–44),

$$f = \frac{(T_{04}/T_{03}) - 1}{(Q_R/c_pT_{03}) - (T_{04}/T_{03})}. \tag{6–48}$$

The pressure ratio across the primary nozzle is

$$\frac{p_{06}}{p_a} = \left(\frac{p_{03}}{p_a}\right)\left(\frac{p_{04}}{p_{03}}\right)\left(\frac{p_{05}}{p_{04}}\right)\left(\frac{p_{06}}{p_{05}}\right). \tag{6–49}$$

The steady-flow energy equation for the nozzles yields

$$u_e' = \sqrt{[2\gamma_1/(\gamma_1 - 1)]RT_{06}\eta_n[1 - (p_a/p_{06})^{(\gamma_1-1)/\gamma_1}]},$$

$$u_e'' = \sqrt{[2\gamma/(\gamma - 1)]RT_{08}\eta_n[1 - (p_a/p_{08})^{(\gamma-1)/\gamma}]},$$

in which u_e' and u_e'' are the exhaust velocities from primary and secondary nozzles, respectively. The foregoing relations show how to relate the variables T_{08}, T_{06},

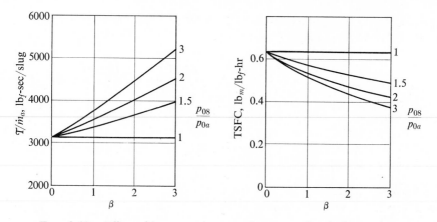

FIG. 6–19. Effect of bypass ratio on performance of turbofan engine.

p_a/p_{06}, and p_a/p_{08} to the flight Mach number, compressor pressure ratio p_{03}/p_{02}, and fan pressure ratio p_{08}/p_{0a}, for assumed efficiencies. The thrust can then be calculated from

$$\frac{T}{\dot{m}_a} = (1 + f)u'_e + \beta u''_e - u(1 + \beta),$$

and the thrust specific fuel consumption from

$$\text{TSFC} = \frac{\dot{m}_f}{T} = \frac{f}{T/\dot{m}_a}.$$

Figure 6–19 shows the result of a particular calculation for flight at a Mach number of 0.9, for different bypass ratios β and pressure ratios p_{08}/p_{0a}, and demonstrates that the thrust of the engine increases with increasing bypass ratio. The thrust specific fuel consumption decreases quite significantly with increasing β.

Two questions of importance in the design of turbofan engines are:

1. What is the best bypass ratio?
2. What is the best pressure ratio of the secondary stream?

The answers to these questions depend on a number of factors, including component efficiencies, structural weights, and the nature of the application.

As shown by the calculations leading to Fig. 6–19, the increase in thrust ΔT gained by adding a fan to a turbojet engine will be of the form indicated in Fig. 6–20. Associated with an increasing mass flow is an increasing drag $\Delta \mathcal{D}$ exerted on the necessarily larger engine installation, as suggested in Fig. 6–20. Thus, for the desired flight condition, the best bypass ratio would be that which corresponded to the maximum difference between the increased thrust and the increased drag, i.e., the net thrust gain.

If, on the other hand, takeoff thrust is a very important consideration, it might be worth while to increase the bypass ratio beyond this value since, on takeoff, the drag would be small. It would not be worth while to increase the airflow indefinitely, because engine weight per unit thrust would tend to rise. It may be shown from Eqs. (6–10) and (6–18) that, for a given energy consumption rate, the takeoff thrust increases approximately as the square root of the airflow. Since the inlet velocity is generally limited (by compressibility effects, as shown in Chapter 8), takeoff thrust must therefore vary approximately with the square root of the inlet area, or the first power of the engine diameter. Engine volume, and hence weight, would at the same time increase with some power greater than the square of the engine diameter, since diameter increases would very likely require length increases as well. It might, for example, be necessary to use additional turbine stages to drive an enlarged fan.

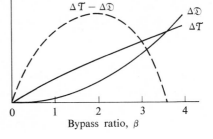

FIG. 6–20.   Effect of bypass ratio on net thrust gain at a given flight Mach number.

Variations in estimate of nacelle drag made by different designers account for differences in bypass ratio advocated by various companies in the field. One factor in this connection is the engine-mounting configuration. Nacelle drag is more serious with wing-pylon installations than aft-fuselage mountings, owing to possible interference, in the former case, with the wing pressure distribution.

An additional advantage of turbofan engines over simple turbojets is reduction of jet noise associated with the lower jet velocity. The intensity of jet noise has been shown to be proportional to the eighth power of the velocity of the jet relative to the ambient air; thus small reductions in velocity may mean significant reductions in noise. On the other hand, a good deal of the noise of a jet engine, especially those noise components with frequencies disagreeable to humans, appears to come from the compressor.

It may be shown by approximate calculations of the mixing process (and it has been verified in tests) that there is a thermodynamic advantage in mixing the cold and hot jets prior to expansion. The thrust gain is of the order of a few percent. Another possible advantage associated with lowering the hot exhaust temperature before the nozzle is that the exhaust stream is then much less susceptible to infrared detection.

Turboprop and turboshaft engines

For relatively high takeoff thrust or for low-speed cruise applications, turboprop engines are employed to accelerate a secondary propellant stream which is much larger than the primary flow through the engine. The relatively low work input per pound of secondary air can be adequately transmitted by a propeller, though a ducted fan could also be used for this purpose. However, a propeller would generally be lighter and, with variable pitch, may be capable of a wider range of satisfactory performance.

Figure 6–21 is a schematic diagram of a typical turboprop engine. Note that the propeller receives its power from a "power turbine" which is mechanically independent of the gas generator rotor elements. This so-called free-turbine configuration, though not the only possibility, has the advantage of flexibility in meeting a range of performance demands.

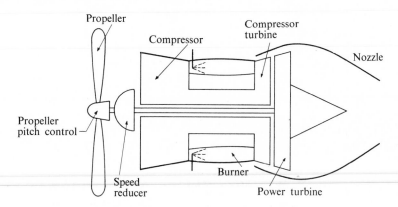

FIG. 6–21. Schematic diagram of a typical free-turbine turboprop engine.

Figure 6–22 shows cross sections of two popular turboprop engines. The Dart engine, perhaps the most widely used of all turboprop engines, is one of the few engines employing a centrifugal compressor. It does not have a free power turbine. The Tyne has a one-stage high-pressure turbine and a three-stage low-pressure turbine. The low-pressure turbine drives the low-pressure compressor and the propeller. The high-pressure turbine drives the high-pressure compressor.

Stress limitations require that the large-diameter propeller rotate at a much lower rate than the relatively small power turbine. Hence a rather large speed-reduction unit, having a speed ratio of perhaps 15 to 1, is required. (A fan of the same diameter would also require a speed reducer.) The propeller, its pitch control mechanism, and the power turbine contribute additional weight, so that a turbo-prop engine may be about 1.5 times as heavy as a conventional turbojet of the same gas-generator size. However, the performance benefits which the turboprop provides on takeoff and at low flight speeds more than compensate for this added weight.

At high subsonic Mach numbers (above M ≈ 0.7), the performance of the turboprop deteriorates compared with that of turbofan or turbojet engines, due to poor propeller efficiency. At such flight speeds the relative velocity of the propeller tip becomes supersonic, and compressibility effects reduce propeller efficiency and create undesirable aerodynamic noise.

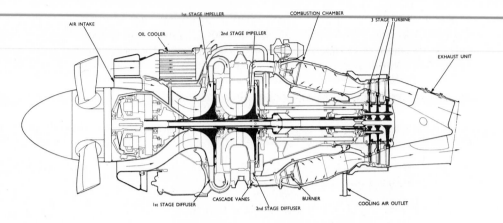

(a) Rolls-Royce Dart RDa-7

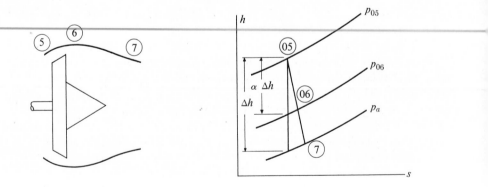

(b) Rolls-Royce Tyne

FIG. 6–22.   Cross sections of two Rolls-Royce turboprop engines.   (Courtesy Rolls-Royce.)

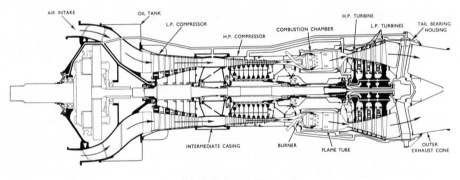

FIG. 6–23.   Enthalpy–entropy diagram for power turbine–exhaust nozzle analysis.

In practice, turboprop engines may exert a significant part of their thrust by means of the hot exhaust jet. Figure 6–23 indicates on an enthalpy-entropy diagram the thermodynamic path of the hot gases expanding through the power turbine and exhaust nozzle. It has been shown [6] that there is an optimum hot-gas exhaust velocity which yields maximum thrust for a given gas generator and flight speed. Referring to Fig. 6–23, let

> Δh = enthalpy drop available in an ideal (isentropic) power turbine and exhaust nozzle,
>
> α = fraction of Δh that would be used by an isentropic turbine having the actual stagnation pressure ratio,
>
> η_{pt}, η_n = adiabatic efficiencies of the power turbine and exhaust nozzle, respectively, and
>
> η_g, η_{pr} = gear and propeller efficiencies, respectively.

The propeller thrust T_{pr} may be determined by considering the energy flux through the free-turbine and propeller shafts:

$$T_{pr}u = \eta_{pr}\eta_g\eta_{pt}\alpha\,\Delta h\dot{m} \quad \text{or} \quad T_{pr} = \frac{\eta_{pr}\eta_g\eta_{pt}\alpha\,\Delta h\dot{m}}{u},$$

where u is the flight velocity and $\dot{m}$ is the hot-gas flow rate. The exhaust nozzle thrust T_n may be written as $T_n = \dot{m}(u_e - u)$, where the exhaust velocity is given by $u_e = \sqrt{2(1 - \alpha)\,\Delta h\eta_n}$. Thus the total thrust is

$$T = \frac{\eta_{pr}\eta_g\eta_{pt}\alpha\,\Delta h\dot{m}}{u} + \dot{m}(\sqrt{2(1 - \alpha)\eta_n\,\Delta h} - u).$$

Maximizing the thrust T for fixed-component efficiencies, flight speed, and Δh yields an optimum value of

$$\alpha = 1 - \frac{u^2}{2\,\Delta h}\left(\frac{\eta_n}{\eta_{pr}^2\eta_g^2\eta_{pt}^2}\right).$$

A turboshaft engine might be defined as a gas turbine engine designed to produce only shaft power. Such engines find application in helicopters, where their light weight and small size compared with piston engines render them attractive. Turboshaft engines are similar to turboprop engines, except that the hot gases are expanded to a lower pressure in the turbine, thus providing greater shaft power and little exhaust velocity.

Figure 6–24 is a cutaway view of a small (1250 hp) turboshaft engine, the General Electric T-58. Power is transmitted from the two-stage free power turbine via a shaft extending through the split exhaust duct. Note the relatively long power-turbine blades and large exhaust duct, indicative of low-pressure (low-density)

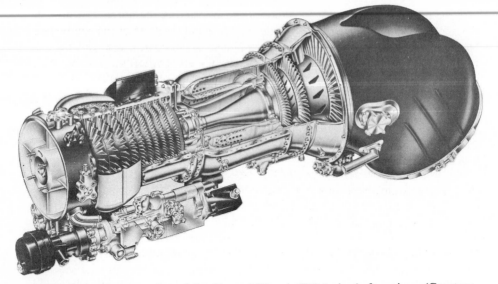

Fig. 6–24. Cutaway view of the General Electric T58 turboshaft engine. (Courtesy General Electric Co.)

low-velocity exhaust. An idea of the size advantage of turbine engines over piston engines may be gained by noting that this engine, producing 1250 hp, is only 16 inches in diameter by 59 inches long, and weighs about 300 pounds. Of course, the speed reducer required to match the output (at 19,500 rpm) to the load adds to these figures.

6–5 TYPICAL ENGINE PERFORMANCE

In previous sections methods have been presented for calculating the overall performance of an aircraft engine in terms of the performances of its components. Calculated performances are accurate, of course, only to the extent that realistic component efficiencies are assumed, and we shall have more to say about actual components in the following three chapters. In this section we shall present typical examples of actual (or estimated) overall engine performance, and in addition show how dimensional analysis permits a correlation of engine performance variables in a form which has minimum complexity and maximum generality. There are many ways to present performance data for an engine, of course, and different forms of several manufacturers will be shown here.

Considering the complexity of some of the engines, it might be appropriate first to determine how many of the engine variables can actually be considered independent. This is perhaps most easily determined by considering the actual variables by which an engine is controlled. If the engine has fixed geometry, there are only two ways to change the thrust or power output: (1) alter the fuel flow, or (2) change the condition of the incoming airstream.

If the engine geometry can be altered (e.g., by varying the nozzle exit area or the angle of pitch of the propeller blades or the angular setting of the compressor blades), then this constitutes a third kind of independent control of engine output. In practice such variations in geometry are usually automatically controlled by a given device as a fixed function of speed, fuel flow, or other variables, and may not therefore be considered capable of independent control by the pilot or test engineer.

The condition of the incoming airstream can be described by three variables: e.g., pressure, temperature, and velocity; or pressure, density, and Mach number. Organizations such as the NACA* have compiled average tables and equations which give the mean dependence of the pressure and temperature (or density) of the ambient atmosphere as a function of altitude only. Using such a "standard day" relationship, the state of the airstream entering an engine can be simply described as a function of two variables, such as altitude and flight speed, or altitude and flight Mach number. To be precise, of course, the angle at which the air enters the engine should also be specified, because this may (at very high angles of attack) have a definite effect on engine performance. For the sake of simplicity, however, this factor will be ignored in the immediate discussion: The air will be assumed to enter the inlet in the direction for which the inlet was designed.

Thus the performance of a given engine can be considered to depend wholly on the fuel flow, flight velocity, and altitude (i.e., ambient pressure and temperature). This set of independent variables is complete but not unique. Another set of three variables could equally well be considered a complete set of independent variables. For example, fuel-flow rate could be replaced by shaft speed or engine-pressure ratio.

With three independent variables, the operating characteristics of an engine can be presented in a series of charts, on each of which one independent variable is held constant. Figure 6–25 indicates the estimated performance of a small turbojet engine, the Pratt & Whitney JT-12, at two different altitudes for an assumed perfect intake system. In this case, lines of constant TSFC and airflow rate are plotted on airspeed–thrust coordinates. Other variables, for example compressor speed, fuel flow, various pressure ratios, etc., could also be plotted on these coordinates.

The various engine ratings shown, such as maximum continuous, maximum climb, and maximum cruise, are performance limitations imposed by the manufacturer to assure reasonable engine reliability and durability for commercial operation. The maximum continuous rating is intended for emergency use only, and although exceeding it would not result in immediate engine destruction, it would result in more frequent overhauls. Higher performance demands, such as takeoff thrust, are permissible if they are of limited duration (for example, five minutes at takeoff rating). For military applications, where immediate perform-

---

* Now known as NASA, the National Aeronautics and Space Administration.

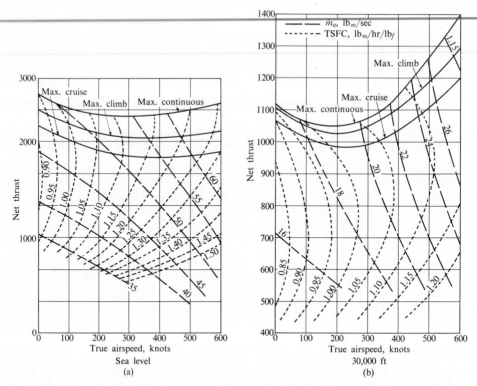

FIG. 6–25. Estimated performance of Pratt & Whitney JT12A-8 turbojet engine for ICAO standard atmosphere, 100% intake efficiency. (Courtesy Pratt & Whitney Aircraft, Division of United Aircraft Corp.)

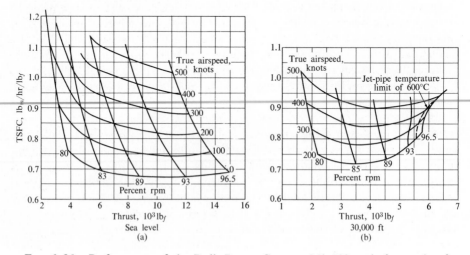

FIG. 6–26. Performance of the Rolls-Royce Conway Mk-509 turbofan engine for ISA standard atmosphere, 100% intake efficiency. (Courtesy Rolls-Royce.)

ance is often of greater importance than time between overhauls, engines may be operated at higher ratings.

Figure 6–26 indicates typical performance for a large turbofan engine, the Rolls-Royce Conway, for 100% intake efficiency. In this case airspeed and low-pressure compressor speed, expressed as percentage of maximum, are plotted on thrust–TSFC coordinates. Note in the altitude case that a line of constant jet-pipe temperature is also plotted, since this particular temperature is the maximum allowable value.

The turboprop engine, as normally equipped with a variable-pitch propeller, possesses four independent variables, and hence a complete presentation of performance would require a much larger set of charts. Since such an engine usually operates with an automatic pitch-control mechanism which maintains constant propeller speed, it is convenient to take engine speed as the fourth independent variable. Figure 6–27 presents the maximum power performance of a large turbo-prop engine, the Rolls-Royce Tyne R. Ty. 11 (Fig. 6–22), as limited by the allowable turbine inlet temperature (1173°K in this case) or the maximum available fuel flow, at the cruising speed of 13,500 rpm. Thus the two variables which are held constant (or independently specified) are speed and peak temperature (or fuel flow if applicable). Obviously, with the flexibility of a variable-pitch propeller, this engine could, at the same rpm, provide less power at any given altitude and speed by simply reducing the fuel flow rate while adjusting the propeller (decreasing the blade pitch) to absorb less power.

The Tyne is a particularly efficient turboprop engine, which compares favorably in fuel consumption with highly developed piston engines. For comparison, Fig. 6–28 indicates the performance of a large 28-cylinder air-cooled radial piston engine with gear-driven supercharger, the Pratt & Whitney Wasp Major B13. It is possible by carburetor adjustment to vary the fuel–air ratio, and these curves are for an automatic "normal" carburetor setting. Flight speed has much less importance for a positive-displacement engine because the pressure rise in the inlet diffuser is generally much less than the pressure rise in the cylinder. At high altitude the power output of the piston engine and the thrust of the jet engine are both greatly reduced, primarily because at low ambient density the mass flow of air into the engine is low.

Information on ramjet performance is much less readily available in the literature than is information on other engines. Hence we must content ourselves with calculated performances based on realistic assumptions of component efficiencies. Figure 6–29 is the result of such calculations for ramjets whose geometry was assumed adjustable for best operation at any given Mach number and altitude. At a given altitude, the thrust rises rapidly with Mach number due to the increase in engine pressure ratio. Due primarily to decreasing density, the thrust decreases sharply with increase in altitude. The specific fuel consumption is essentially independent of altitude at a given flight Mach number. Specific fuel consumption is shown both for the maximum-thrust condition and at a lower thrust cruise condition for which the thrust specific fuel consumption is minimum. At high

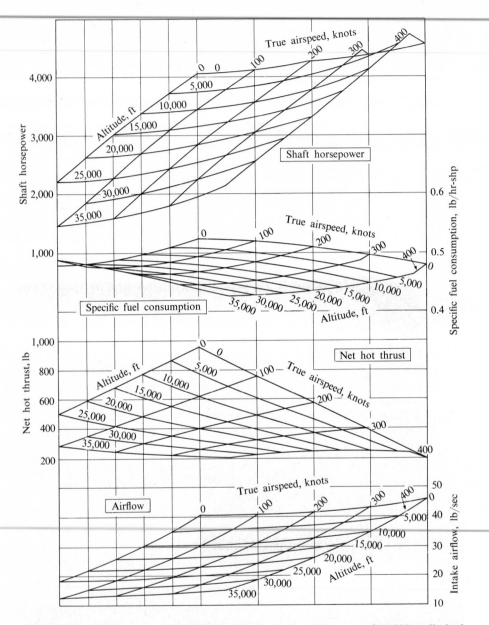

Fig. 6–27.   Performance of Rolls-Royce R. Ty. 11 at cruising rpm of 13,500, as limited by turbine inlet temperature or fuel-flow rate; ICAO standard atmosphere. (Courtesy Rolls-Royce.)

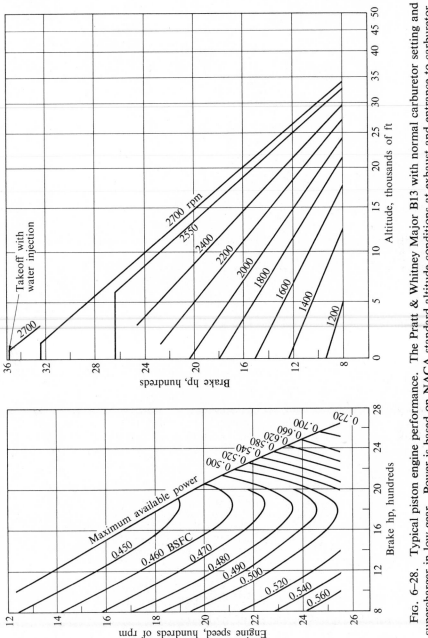

Fig. 6–28. Typical piston engine performance. The Pratt & Whitney Major B13 with normal carburetor setting and supercharger in low gear. Power is based on NACA standard altitude conditions at exhaust and entrance to carburetor, with a mixture strength corresponding to fuel-consumption curves. (Courtesy Pratt & Whitney Aircraft, Division of United Aircraft Corp.)

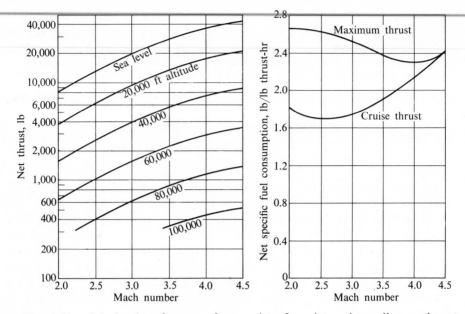

FIG. 6–29. Calculated performance for a series of ramjet engines, all operating at design point. (Courtesy Bristol-Siddeley Engines, Ltd.)

Mach numbers the high inlet stagnation temperature limits the maximum fuel addition so that the maximum possible thrust and the thrust for minimum specific fuel consumption approach each other.

The performance of an engine can be presented in much more compact form by using appropriate dimensionless groups of variables. To derive a general set of such groups it is necessary to consider, in addition to the independent variables already mentioned, all the fluid properties which could be considered important in the operation of the engine. Thus the air viscosity μ_a and the gas constants R and γ must also be considered, in addition to the local pressure and temperature p_a, T_a, and the flight speed u. Taking thrust as a typical dependent variable, we may express it as some function of the independent variables:

$$\mathcal{T} = f_1(p_a, T_a, u, R, \gamma, \mu_a, T_{04}, D, \text{design}), \qquad (6\text{--}50)$$

where D, the engine diameter, and "design," the engine shape (expressed in terms of a very large number of dimension ratios), entirely determine the engine geometry. The combustion temperature, T_{04}, has been taken as a more convenient independent variable, replacing the physical variable $\dot{m}_f$ for reasons which will be discussed subsequently.* A result of the "π theorem" [1] may be stated in the

---

* This is permissible so long as $\dot{m}_f$ is small compared to $\dot{m}_a$; i.e., so long as the fuel–air ratio is much less than one since, in this case, the only appreciable effect of the fuel addition is the rise in temperature to T_{04}.

following form. Any function of n variables involving q fundamental quantities (such as mass, length, time, temperature) can be reduced to a function of $(n - q)$ dimensionless groups. The form of the $(n - q)$ groups is arbitrary, of course, but as in most compressible flow problems, the above variables can be conveniently expressed in terms of a Mach number, a Reynolds number, a temperature ratio, and the design ratios. Thus, it may be shown that the functional relationship (6–50) can be simplified to the equally general form

$$\frac{T}{p_a D^2} = f_2\left(\frac{u}{\sqrt{\gamma R T_a}}, \frac{\rho_a u D}{\mu_a}, \frac{T_{04}}{T_a}, \gamma, \text{design}\right). \tag{6–51}$$

The convenience of this form is not readily apparent, since it would appear that the dimensionless thrust variable is still a function of four variables, not counting the large number of design ratios. However, for a *given* engine (or for a series of geometrically similar engines), the design ratios need not be considered, since they remain constant. Further, it has been found that, for ordinary operation, the effect of change in Reynolds number is insignificant. (At extreme altitudes, Reynolds number effects may be significant, as will be discussed in Chapter 8 with regard to compressor performance.) Thus Eq. (6–51) may be reduced to the approximation,

$$\frac{T}{p_a D^2} \cong f_3\left(M, \frac{T_{04}}{T_a}\right). \tag{6–52}$$

Equation (6–52) is generally valid for both ramjet and turbine engines, although for turbine engines an alternate form is often chosen. The rotor speed N of a turbine compressor is an additional dependent variable which can be conveniently expressed in the dimensionless form $ND/\sqrt{\gamma R T_a}$. Just as for thrust,

$$\frac{ND}{\sqrt{\gamma R T_a}} = f_4\left(M, \frac{T_{04}}{T_a}\right).$$

However, since N is an easily measured quantity and T_{04} is not, it is often convenient to treat N as independent and T_{04} as dependent, writing

$$\frac{T_{04}}{T_a} = f_5\left(M, \frac{ND}{\sqrt{\gamma R T_a}}\right).$$

Using this expression, Eq. (6–52) can be written

$$\frac{T}{p_a D^2} \cong f_6\left(M, \frac{ND}{\sqrt{\gamma R T_a}}\right). \tag{6–53}$$

The validity of an expression of this form is shown by test data in Fig. 6–30(a), where the "thrust parameter" $T/\gamma p_a A_x$ is plotted as a function of the engine-speed parameter ND/a_a for three altitude conditions (and thus for three different Rey-

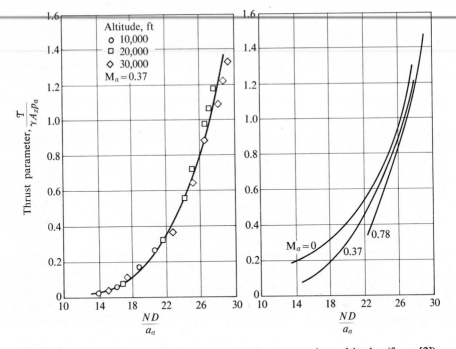

FIG. 6–30.   Nondimensional turbojet engine thrust at various altitudes (from [2]).

nolds numbers, since the Reynolds number is proportional to the local pressure level). In these expressions, D is a characteristic size of the engine (or of the geometrically similar series of engines) under discussion and the symbol $a_a = \sqrt{\gamma R T_a}$ is the speed of sound in the ambient air. The frontal area A_x is of course proportional to D^2 for a given engine. It can be seen that the Reynolds number variation has little or no influence over this operating range. The influence of Mach number can be seen in Fig. 6–30(b).

Figure 6–30 is naturally only one example of the representation of turbojet engine performance data in dimensionless form. A function of five independent variables, $T = f(p_a, T_a, U, N, D)$, requiring many pages of graphs for adequate representation, has been shown to be essentially a function of two independent variables,

$$\frac{T}{p_a D^2} = f\left(M, \frac{ND}{\sqrt{\gamma R T_a}}\right),$$

which can be entirely displayed on one graph. In a similar way many other dimensionless variables can be shown as unique functions of the two arguments shown on the right-hand side. Curves such as Fig. 6–30 are very useful for illustrating in simple form the characteristic behavior of an engine. For the engineer who requires the absolute magnitude of certain variables, these curves have the dis-

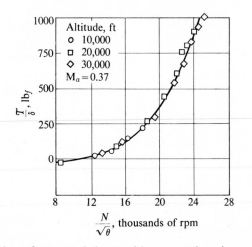

FIG. 6–31. Variation of corrected thrust with corrected engine speed for static test conditions for a given engine.

advantage that the dimensionless numbers are not immediately meaningful to him. He therefore may prefer to modify the relationship (6–53) as follows.

For a given engine, D is constant and can therefore, like γ and R, be omitted. Further, by use of pressure and temperature ratios,

$$\delta = \frac{p_a}{p_{std}}, \qquad \theta = \frac{T_a}{T_{std}}$$

in which p_{std} and T_{std} are constant or "standard" values of pressure and temperature, the engineer may legitimately write

$$\frac{T}{\delta} = f\left(M, \frac{N}{\sqrt{\theta}}\right) \qquad \text{for a given engine.}$$

The thrust and shaft speed variables are now no longer dimensionless, but (as applied to a single engine) this expression is just as general as Eq. (6–53). Figure 6–31 gives an example in which the scale of the thrust variable is the thrust the engine would develop in pounds at "standard" ambient pressure ($\delta = 1$). Similarly, the abscissa is the actual shaft speed at standard ambient temperature ($\theta = 1$). For nonstandard operating conditions it is only necessary to use the actual values of δ and θ to find the actual engine thrust. The variables T/δ and $N/\sqrt{\theta}$ are usually called "corrected" variables.

A special problem arises, however, in the correlation of the fuel flow $\dot{m}_f$ with these variables. The energy equation for the combustor is

$$\dot{m}_f \eta_b Q_R = \frac{\gamma}{\gamma - 1} R(T_{04} - T_{03})\dot{m}_a \qquad \text{or} \qquad \frac{\gamma - 1}{\gamma} \frac{\dot{m}_f Q_R \eta_b}{\dot{m}_a R T_a} = \frac{T_{04}}{T_a} - \frac{T_{03}}{T_a}.$$

Following Sanders [2], we transform this relationship as follows: It may be shown by dimensional analysis (see page 183) that the temperature ratios T_{03}/T_a and T_{04}/T_a may be written as functions of M and ND/a_a only. Further, it may be shown that the engine thrust may be expressed as

$$\frac{\mathcal{T}}{\dot{m}_a a_a} = g_1\left(\text{M}, \frac{ND}{a_a}\right).$$

Using these functional relationships the combustor energy equation may be written

$$(\gamma - 1)\,\frac{\eta_b \dot{m}_f Q_R}{\mathcal{T} a_a} = g_3\left(\text{M}, \frac{ND}{a_a}\right).$$

As a rule, γ and Q_R are essentially constant and could therefore be omitted from this relationship. The combustion efficiency, however, is not constant. At high altitude, for example, the absolute fuel–flow rate is very much lower than it is at sea level. The combustor drop spray pattern may be so seriously altered that a significant part of the fuel passes through the combustor unburned. Figure 7–37 shows how the combustion efficiency varies with altitude for three typical combustors. The relationship between combustor efficiency η_b and other characteristic variables is complex and not well understood. At any rate it is not simply a function of variables like M and ND/a_a. Thus it is not reasonable to expect that

$$\frac{\dot{m}_f Q_R}{\mathcal{T} a_a} = g\left(\text{M}, \frac{ND}{a_a}\right).$$

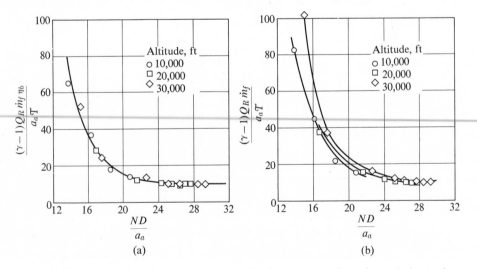

FIG. 6–32.   Effect of altitude on thrust specific fuel consumption of a turbojet engine (from [2]).   Given M.

Figure 6–32 demonstrates that the dimensionless engine fuel-flow variable may be expressed as a function of ND/a_a, $\dot{m}_a$ (or T_{04}/T_a and M) only if it is multiplied by the actual combustor efficiency.

Sanders [2] develops a much more complete set of nondimensional and corrected performance parameters, many of which find widespread use. These parameters are discussed more fully by Zucrow [3].

References

1. BRIDGMAN, P., *Dimensional Analysis*. New Haven: Yale University Press, 1931

2. SANDERS, N. D., "Performance Parameters for Jet-Propulsion Engines," NACA TN 1106, July 1946

3. ZUCROW, M. J., *Aircraft and Missile Propulsion*, Volume II. New York: John Wiley & Sons, 1958

4. SMITH, C. W., *Aircraft Gas Turbines*. New York: John Wiley & Sons, 1956

5. FOA, T. V., *Elements of Flight Propulsion*. New York: John Wiley & Sons, 1960

6. LANCASTER, O. E. (editor), *Jet Propulsion Engines*, Volume XII, "High-Speed Aerodynamics and Jet Propulsion." Princeton, N. J.: Princeton University Press, 1959

7. PINKEL, B., and I. M. KARP, "A Thermodynamic Study of the Turbojet Engine," NACA Report No. 891, 1947

Problems

1. Show that for the ideal turbojet (all efficiencies equal unity), the exhaust Mach number M_e is given by

$$M_e^2 = \frac{2}{\gamma - 1}\left[\left(\frac{p_{03}}{p_{02}}\right)^{(\gamma-1)/\gamma}\left(1 + \frac{\gamma - 1}{2}M^2\right)\right.$$

$$\left.\times\left\{1 - \frac{1 + \frac{\gamma - 1}{2}M^2}{T_{04}/T_a}\left[\left(\frac{p_{03}}{p_{02}}\right)^{(\gamma-1)/\gamma} - 1\right]\right\} - 1\right],$$

in which   M is the flight Mach number,

   $\dfrac{p_{03}}{p_{02}}$ is the compressor pressure ratio,

   $\dfrac{T_{04}}{T_a}$ is the ratio of turbine inlet and ambient temperatures, and

   γ is the ratio of specific heats (assumed constant).

The mass flow rates through the compressor and turbine may be assumed equal.

2. For a given work input per unit mass of fluid and given adiabatic efficiency η_c, how does the actual pressure ratio for a compressor differ from the ideal (isentropic) pressure

ratio? Derive an expression which relates the actual pressure ratio to the ideal one and η_c. As mentioned in Chapter 6, the adiabatic efficiency η_c is usually defined by

$$\eta_c = \frac{h_{03s} - h_{02}}{h_{03} - h_{02}}.$$

3. Consider two versions of a jet engine. The first is a standard engine run with a turbine-inlet temperature of 2000°R. The second is identical, except that its turbine is cooled by bleeding air from the compressor; in this way the allowable turbine-inlet temperature may be raised to 2500°R. The basic engine and the bleed air line are indicated in the figure. To make an estimate of the effect of this modification on the thrust, suppose that the engine is flying at Mach 2 at an altitude where the ambient temperature is −60°F, and that the compressor pressure ratio is 9:1. Suppose also that for the second engine 10% of the airflow is bled from the compressor at a point where the pressure ratio is 3:1. After cooling the turbine, the bleed air is exhausted from the engine with no appreciable velocity. For simplicity assume all components of both engines to be reversible. Let $\gamma = 1.4$ and $c_p = 0.24$ Btu/lb$_m$·°R.

(a) Determine the thrust of the two engines per unit mass flow of air entering the compressor.

(b) What is the ratio of the thrust specific fuel consumption of the second engine to that of the first?

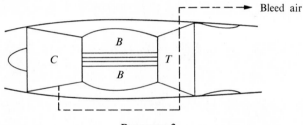

PROBLEM 3

4. Estimate the propulsion and thermal efficiencies of a turbojet engine during subsonic cruise. The flight Mach number is 0.8 and the ambient temperature is −60°F. The compression ratio is 12 and the turbine inlet temperature is 1960°F. The respective adiabatic efficiencies of the diffuser, compressor, turbine, and nozzle are 0.92, 0.85, 0.85 and 0.95. The burner stagnation pressure ratio is 0.97 and the average specific heat during and after combustion is 0.27 Btu/lb$_m$·°F.

5. Compare the specific fuel consumption of a turbojet and a ramjet which are being considered for flight at M = 1.5 and 50,000 feet altitude (ambient conditions of pressure and temperature: 242 psf and −69.7°F respectively).

The turbojet pressure ratio is 8 and the maximum allowable temperature is 2000°R. For the ramjet the maximum temperature is 4000°R. For simplicity ignore aerodynamic losses in both engines. Conventional hydrocarbon fuels are to be used (heating value 18,000 Btu/lb$_m$). Assume $\gamma = 1.4$ and $c_p = 0.24$ Btu/lb$_m$.

6. The performance of a series of ramjet engines, each operating under design conditions, is to be calculated as a function of flight Mach number. The engines are to fly at an altitude of 50,000 ft, where $T_a = -70°F$, $p_a = 242$ lb$_f$/ft^2. The fuel heating value

is 19,000 Btu/lb$_m$ and the peak temperature is limited to 4000°R. According to the Aircraft Industries Association, a reasonable estimate of ramjet diffuser losses is given by

$$\frac{p_{02}}{p_{0a}} = 1 - 0.1(M - 1)^{1.5}, \qquad M > 1$$

in which M is the flight Mach number.

The stagnation pressure ratio across the flameholders, ② to ③, is 0.97, and the stagnation pressure loss in the combustor, ③ to ④, may be estimated roughly by the methods of Chapter 3. The nozzle and combustion efficiencies are 0.95 and 0.98, respectively, and the combustion chamber exit Mach number is 0.5. For the nozzle $A_6 \leqq A_4$. Assume the propellant is a perfect gas having $R = 53.3$ ft lb$_f$/lb$_m$·°R throughout, but having $\gamma = 1.4$ from ⓐ to ③ and $\gamma = 1.3$ from ③ to ⑥. (η_N is based on conditions upstream of shock in this case.) Also, check to see that the fuel–air ratio does not exceed stoichiometric (about 0.0667). Calculate TSFC and specific thrust for flight Mach numbers of 1.0, 2.0, 3.0, 4.0.

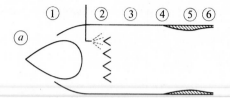

PROBLEM 6

7. A ramjet has low-speed stagnation pressure losses given by $r_d = 0.90$, $r_c = 0.95$, $r_n = 0.96$. The ambient and maximum allowable temperatures are 400°R and 4000°R, respectively. A very high energy fuel is used, and $f \ll 1$. What is the minimum flight Mach number at which the engine develops positive thrust? What will physically limit the maximum flight Mach number at which it can produce positive thrust? (Assume $\gamma = 1.4$.)

8. The idling engines of a landing turbojet produce forward thrust when operating in a normal manner, but they can produce reverse thrust if the jet is properly deflected. Suppose that, while the aircraft rolls down the runway at 100 mph, the idling engine consumes air at 100 lb$_m$/sec and produces an exhaust velocity of 450 ft/sec.

(a) What is the forward thrust of this engine?

(b) What is the magnitude and direction (i.e., forward or reverse) if the exhaust is deflected 90° without affecting the mass flow?

(c) What is the magnitude and direction of the thrust (forward or reverse) after the plane has come to a stop, with 90° exhaust deflection and an airflow of 75 lb$_m$/sec?

7

Aerothermodynamics of Inlets, Combustors, and Nozzles

7-1 INTRODUCTION

In Chapter 6 we used the laws of thermodynamics and fluid mechanics to explain the behavior of aircraft jet engines. The several engine components were treated as "black boxes," in the sense that discussion was confined to the inlet and outlet conditions of the propellant, without regard to the internal mechanisms which produce its change of state. Where necessary, the actual performance was related to some easily calculated or "ideal" performance by the definition of an appropriate component efficiency or stagnation-pressure ratio. The purpose of this and the following chapter is to examine the internal mechanisms of the various components in order to describe the factors which impose practical limits on performance. Conditions required for high performance of components will be considered and, in some cases, methods for quantitative prediction of their behavior are given.

For the ramjet, Eq. (6–34) indicates that a given percentage loss in stagnation pressure has the same effect on engine performance wherever it occurs through the engine. For turbine engines, the same conclusion holds although it is not so easily seen, since component performances are usually stated in terms of adiabatic efficiencies rather than stagnation-pressure ratios. Hence high performance is of equal importance for all engine components. The *attainment* of high performance is generally more difficult in those regions requiring a rise in static pressure than in those where the pressure falls. This, as pointed out in Chapter 4, may be attributed to boundary layer behavior and the tendency for separation in the presence of a rising static pressure. Thus inlets (which generally have rising pressure gradients even for turbojets) are more difficult to design for efficient operation than nozzles. Similarly, as will be seen in Chapter 8, compressors are more diffi-cult to design (and have lower ultimate efficiency) than turbines.

A somewhat arbitrary division has been made between the material discussed in this chapter and that in the following one. This chapter deals with flow in sta-tionary ducts, while Chapters 8 and 9 are concerned primarily with the flow in rotating fluid machines. Only Chapter 7 is relevant to ramjets; but all three are relevant to the flow in a variety of turbine engines.

7–2 SUBSONIC INLETS AND DIFFUSERS

An engine installed in an aircraft must be provided with an air intake and a ducting system. As shown in Chapter 6, it is generally necessary that the intake of a ramjet engine provide a static pressure rise. It will be shown in Chapter 8 that, for turbojet engines, the airflow entering the compressor must be of low Mach number, of the order of 0.4 or less. Thus for either engine the entrance duct usually acts as a diffuser which, at high flight Mach numbers, should provide a high pressure rise. It is important that the stagnation-pressure loss in the inlet be small. It is also important that the flow leaving the inlet system be uniform, since distortions in the velocity profile at the compressor inlet can severely upset the compressor aerodynamics and may lead to failure of the blades due to vibra-tions. One of the main engineering difficulties in designing a good inlet is that the space available is usually quite restricted.

Flow patterns

Depending on the flight speed and the mass flow demanded by the engine, the inlet may have to operate with a wide range of incident stream conditions. Figure 7–1 shows the streamline patterns for two typical subsonic conditions, and the corresponding thermodynamic path of an "average" fluid particle. During level cruise the streamline pattern may include some deceleration of the entering fluid external to the inlet plane (Fig. 7–1a). During low-speed high-thrust operation (e.g., during takeoff and climb), the same engine will demand more mass flow and the streamline pattern may resemble Fig. 7–1(b), which illustrates external acceler-ation of the stream near the inlet. In both cases there is an external change of state which is essentially isentropic, since there are no walls on which friction may act. For given air velocities at stations (a) and (2), external acceleration raises the inlet velocity and lowers the inlet pressure, therefore increasing the internal pressure rise across the diffuser. If this pressure increase is too large, the diffuser may stall, due to boundary layer separation, and stalling usually reduces the stagnation pressure of the stream as a whole. Conversely, external deceleration requires less internal pressure rise, and hence a less severe loading on the boundary layer. Therefore, the inlet area is often chosen so as to minimize external acceleration during takeoff, with the result that external deceleration occurs during level-cruise operation. Under these conditions the "upstream capture area" A_a is less than the inlet area A_1, and some flow is "spilled over" the inlet.

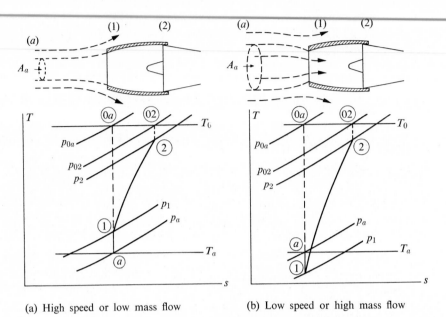

(a) High speed or low mass flow   (b) Low speed or high mass flow

FIG. 7–1. Typical streamline patterns for subsonic inlets.

These circumstances apply primarily to low-speed flight, where the inlet area may be increased without the external surface of the engine mounting suffering excessive drag. For higher flight speeds, a drag penalty may offset the benefit to internal flow of increasing inlet area. For supersonic flight, in fact, the "spilling" action is necessarily accompanied by a shock system which reduces the relative velocity at inlet to subsonic values so that the air may sense the presence of the inlet and flow around it. This usually results in an intolerable increase in drag on the nacelle.

Internal flow

The flow within the inlet is usually required to undergo further diffusion. Because of wall friction such diffusion will be irreversible, and in fact it may be accompanied by boundary layer separation or stall. In a diffuser the term "stall" is used to denote backward motion of some part of the fluid stream. When this takes place the whole diffuser performance usually suffers seriously. The static pressure gradient decreases and the mixing losses increase rapidly.

Figure 7–2 denotes four types of flow in diffusers. The type to be expected in a given diffuser is dependent, largely, on the rate at which the flow area increases. If the flow area does not increase rapidly in the flow direction, the boundary layer will remain well behaved. As explained in Chapter 4, the pressure gradient tends to retard the slow-moving fluid near the wall more than it retards the main stream. However, if the pressure gradient is not too large, viscous and turbulent shear

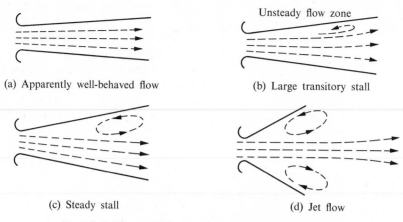

(a) Apparently well-behaved flow   (b) Large transitory stall

(c) Steady stall   (d) Jet flow

FIG. 7–2. Types of flow in diffusers. (After Kline [1].)

stresses will act on the slow-moving fluid and keep it from stagnating. Under these conditions the flow is relatively well behaved, as shown in Fig. 7–2(a).

If the divergence or rate of area increase is larger, the flow may be characterized by unsteady zones of stall (Fig. 7–2b). The shear forces are no longer able to overcome the pressure forces at all points in the flow, and local separation takes place at some point. Due to separation, the main flow pattern changes and relieves the pressure gradient. The boundary of the separation zone is typically quite unstable and the flow as a whole may become unsteady. The average stagnation pressure decreases markedly due to irreversible mixing of fairly large portions of semistagnant fluid with the main stream. Under some circumstances (Fig. 7–2c), the separation pattern becomes steady. If the diffuser walls diverge rapidly, the flow will separate completely and behave much like a jet, as shown in Fig. 7–2(d). In this case practically no static pressure rise takes place in the diffuser.

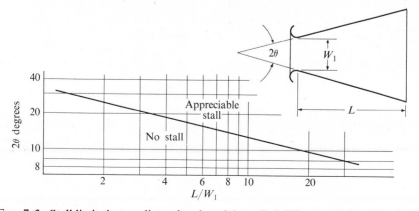

FIG. 7–3. Stall limits in two-dimensional straight-walled diffusers. (After Kline [1].)
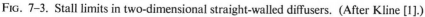

Kline [1] has carried out a systematic study of two-dimensional straight-walled diffusers of variable divergence angle, and has shown the limiting divergence angle for good flow in the diffuser. His results are indicated in Fig. 7–3 (which applies to incompressible fluid). Measurements were made at Reynolds numbers (based on inlet width and velocity) varying from 6000 to 800,000, and surprisingly little variation was noted in the stall limit indicated. While these results apply to incompressible flow in a particular geometry, they do give a qualitatively valid indication of the sensitivity of any diffuser to rapid divergence of flow area.

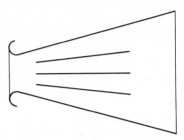

FIG. 7–4. Splitter vanes used to prevent stall in a diffuser (incompressible flow).

One might think that boundary layer theory could be used to predict the onset of separation and stall in such a diffuser. Up to now, however, such attempts have not been successful. Apparently the behavior of the diffuser cannot be predicted by examining each of its boundary layers as though it were isolated from the others. The boundary layers and the main stream interact; near the point of stall the flow may be quite unstable. Kline has shown that the flow in the diffuser may be stabilized by the addition of "splitter vanes" as shown in Fig. 7–4. The addition of these vanes can prevent stall in a diffuser and cause the overall pressure rise to increase so that the wall boundary layers which were stalled are more highly loaded than before, yet they do not separate. Thus it appears that prediction of diffuser stall cannot be made solely on the basis of boundary layer theory.

As one would expect, stall in a diffuser decreases the average stagnation pressure of the stream, since energy is consumed in mixing slow and fast portions of the stream. Kline describes the relative diffuser performance in his experiments by means of a pressure recovery coefficient c_{PR}, defined by

$$c_{\mathrm{PR}} = \frac{1/A_2 \int_{A_2} p \, dA - (1/A_1) \int_{A_1} p \, dA}{1/A_1 \int_{A_1} \tfrac{1}{2}\rho u^2 \, dA}, \tag{7–1}$$

in which A_1 and A_2 are the entrance and exit areas, respectively, of the diffuser. For the ideal one-dimensional diffuser, it follows from Bernoulli's equation that

$$c_{\mathrm{PR}} = c_{\mathrm{PR}i} = 1 - \left(\frac{u_2}{u_1}\right)^2, \quad \text{or} \quad c_{\mathrm{PR}i} = 1 - \left(\frac{A_1}{A_2}\right)^2. \tag{7–2}$$

Typical results of measurements [2] of pressure recovery in two-dimensional

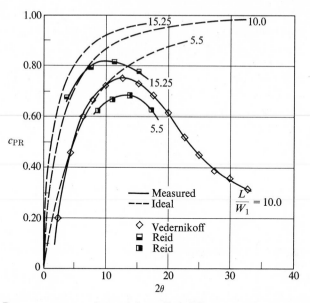

FIG. 7–5. Pressure recovery in straight-walled diffusers. (After Kline, *et al.* [2].)

diffusers are shown in Fig. 7–5, for straight-walled diffusers of given length-width ratio L/W_1 and various divergence angles. It may be seen that even for divergence angles for which Fig. 7–3 shows no stall, the ratio of actual to ideal pressure recovery is significantly less than unity. However, when stall takes place, the ratio c_{PR}/c_{PRi} decreases very rapidly. Thus very large stagnation pressure losses can be incurred in diffusers.

In the actual engine inlet, separation can take place in any of the three zones shown in Fig. 7–6. Separation of the external flow in Zone 1 may result from local high velocities and subsequent deceleration over the outer surface. This possibility, which leads to high nacelle drag, will be discussed subsequently.

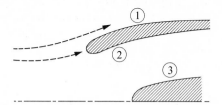

FIG. 7–6. Possible locations of boundary layer separation.

Separation on the internal surfaces may take place either in Zones 2 or 3, depending on the geometry of the duct and the operating conditions. Zone 3 may be the scene of quite large adverse pressure gradients, since the flow accelerates around the nose of the center body, then decelerates as the curvature decreases. In some installations it has not been possible to make the exit area of the intake more than about 30% greater than the inlet area without the incidence of stall and large losses. The pressure recovery measurements cited in Fig. 7-5 can be only an approximate guide for this case, due to the effects of wall curvature and compressibility. Reynolds number effects may also be important for large inlets and high-speed flow. At high angles of attack, all three zones could be subjected to unusual pressure gradients.

External flow   Skip

In view of the difficulties of compression of the airflow in an internal diffuser, one might think that purely external diffuson would be the best solution. This solution would, in fact, provide good propellant flow and engine performance. However, it would very likely be accompanied by poor flow over the external surfaces of the engine mounting, resulting in high drag and reduced overall aircraft performance. The optimum solution thus requires a compromise between internal and external misfortunes.

A typical streamline pattern for large external deceleration is shown in Fig. 7-7. In flowing over the lip of the inlet, the external flow is accelerated to high velocity, much as the flow is accelerated over the suction surface of an airfoil. This high velocity and the accompanying low pressure can adversely affect the boundary layer flow in two ways: For entirely subsonic flow, the low-pressure region must be followed by a region of rising pressure in which the boundary layer may separate. Hence one might expect a limiting low pressure p_{min} or, equivalently, a maximum local velocity u_{max}, beyond which boundary layer separation can be expected downstream. For higher flight velocities (or higher local accelerations), partially supersonic flow can occur. Local supersonic regions usually end abruptly in a shock, and the shock-wall intersection may cause boundary layer separation.

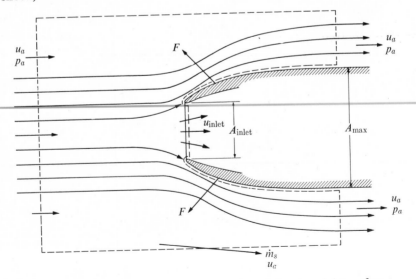

FIG. 7-7. Control volume for the calculation of thrust on inlet surface.

One might expect a limiting local Mach number which should not be exceeded. Whatever the cause, boundary layer separation is to be avoided, since it results in poor pressure recovery in the flow over the after portions of the aircraft or engine housing. This, of course, results in a net rearward force or drag on the body.

To illustrate the major features of the external flow near the inlet, consider the simplified problem of an inlet on a semi-infinite body.* Küchemann and Weber [3] have shown how to relate the external flow over such a body to the extent of external deceleration of the flow entering the inlet. Suppose (see Fig. 7–7) that the external cross section of the inlet grows to a maximum area A_{max} and that the body remains cylindrical from this point downstream. A control surface is indicated which extends far from the inlet on the sides and upstream end, crosses the inlet at its minimum area A_i, passes over the inlet surface, and extends downstream far enough for the external fluid velocity to return essentially to the upstream or flight velocity u_a (neglecting boundary layer effects). Thus all the external flow enters and leaves the control volume with an axial velocity component u_a, it being assumed that the sides of the control volume are sufficiently removed from the inlet. The internal flow enters the control volume with velocity u_a and leaves with velocity u_i (assuming, for simplicity, one-dimensional flow in the inlet). The net momentum flux out of the control volume is then, ignoring changes in the air density,

$$\dot{m}_s u_a + \rho u_i^2 A_i - \rho u_a^2 A_{\mathrm{max}}.$$

From continuity, the side flow rate is $\dot{m}_s = \rho u_a A_{\mathrm{max}} - \rho u_i A_i$, so that the net momentum flux can be expressed as $\rho A_i(u_i^2 - u_i u_a)$. The net force in the axial direction on the control volume is .

$$p_a A_{\mathrm{max}} - p_i A_i - F_x,$$

where F_x is the axial component of the force on the control volume due to the forces on the external surface of inlet. If friction is neglected,

$$F_x = - \int_{\mathrm{inlet}} p \mathbf{x} \cdot \mathbf{n} \, dA = \int_{A_i}^{A_{\mathrm{max}}} p \, dA_x$$

where p = pressure on surface,

          $\mathbf{x}$ = unit vector along axis in the flow direction,

          $\mathbf{n}$ = outward (from the inlet surface) pointing unit vector,

      dA = increment of external surface area,

     dA_x = increment of external surface area normal to $\mathbf{x}$ ($2\pi r \, dr$ for an axisymmetric inlet).

---

* An inlet followed by a cylindrical portion several diameters in length would behave similarly, but few practical engine housings are actually cylindrical over any appreciable length.

Combining these expressions, the momentum equation requires

$$p_a A_{\max} - p_i A_i - \int_{A_i}^{A_{\max}} p \, dA_x = \rho A_i(u_i^2 - u_i u_a)$$

or

$$\int_{A_i}^{A_{\max}} (p_a - p) \, dA_x = \rho A_i(u_i^2 - u_i u_a) + (p_i - p_a)A_i. \tag{7-3}$$

The equation is arranged in this form since the integral can be considered a component of thrust ΔT_i which acts on the front external surface of the inlet due to reduction of local surface pressure. Applying Bernoulli's equation to the external deceleration of internal flow, we obtain

$$p_i - p_a = \rho \left(\frac{u_a^2 - u_i^2}{2}\right).$$

Thus

$$\int_{A_i}^{A_{\max}} (p_a - p) \, dA_x = \rho A_i(u_i^2 - u_i u_a) + \rho A_i \left(\frac{u_a^2}{2} - \frac{u_i^2}{2}\right)$$

or

$$\frac{\Delta T_i}{\frac{1}{2}\rho u_a^2 A_i} = \left(1 - \frac{u_i}{u_a}\right)^2. \tag{7-4}$$

Before proceeding, let us point out that large ΔT_i ($u_i \ll u_a$) is not, in itself, desirable. Although this force contributes to thrust, it must be remembered that for any body of finite length (i.e., double-ended), large ΔT_i would be accompanied by a proportionately large force in the negative direction over the after portions of the body. For real flows, including boundary layer effects, the negative force could easily exceed the positive force, thus creating a net drag.

To return to Eq. (7–4): this expression indicates the magnitude of the force, but it says nothing of its distribution. In order to avoid boundary layer separation, the pressure distribution must be such that the surface pressure is nowhere less than the allowable value $p_{\min}$. Clearly, within this limitation, the *minimum* frontal area for a given $\Delta T_i/\frac{1}{2}\rho u_a^2 A_i$ (or u_i/u_a) will be presented by an inlet shaped so as to experience the minimum allowable pressure over the entire front external area. In this case,

$$\Delta T_i = (p_a - p_{\min})(A_{\max} - A_i)$$

or, using Bernoulli's equation again,

$$\Delta T_i = \frac{\rho(u_{\max}^2 - u_a^2)}{2} (A_{\max} - A_i).$$

Using Eq. (7–4), we can express the minimum area ratio as a function of the desired external deceleration ratio and the maximum allowable velocity ratio:

$$\left(\frac{A_{\max}}{A_i}\right)_{\min} = \frac{[1 - (u_i/u_a)]^2}{(u_{\max}/u_a)^2 - 1} + 1. \tag{7-5}$$

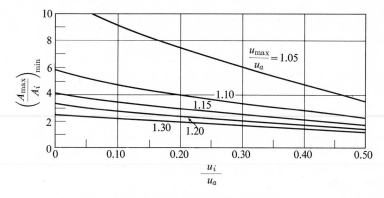

FIG. 7–8. Minimum frontal area ratio as a function of deceleration ratio and maximum surface velocity ratio, from Eq. (7–5).

Figure 7–8 is a plot of this expression, indicating that large external decelerations ($u_i/u_a \ll 1$) are accompanied by rather large area ratios for a given value of u_{max}/u_a. An estimate of a reasonably safe upper limit for the ratio u_{max}/u_a may be obtained by considering the pressure rise during deceleration. A maximum pressure coefficient may be defined by

$$C_{p\,max} = \frac{p_a - p_{min}}{\frac{1}{2}\rho u_{max}^2} = 1 - \left(\frac{u_a}{u_{max}}\right)^2.$$

Section 4–1 pointed out that the turbulent boundary layer is liable to separate if the pressure coefficient exceeds a value in the range $0.4 < C_p < 0.8$. Taking the lower limit for the pressure coefficient, the above equation then restricts the velocity ratio u_{max}/u_a to a value of about 1.3. Of course, this estimate has ignored the effects of compressibility. If the flight Mach number M_a is close to unity, the local Mach number of the flow adjacent to the surface could significantly exceed unity, so that a shock wave strong enough to separate the external flow could be established.

As an example, consider a flight Mach number of 0.9, and suppose that the inlet Mach number is to be reduced entirely externally to 0.4 (typical compressor inlet value), and that the maximum local external Mach number is not to exceed 1.2. Then u_i/u_a is 0.47, and (for a limiting Mach number of 1.2) $u_{max}/u_a = 1.28$. From Eq. (7–5), or Fig. 7–8, one can see that the minimum area ratio A_{max}/A_i is about 1.44.

This discussion gives a considerably simplified picture of the real flow around inlets. Nevertheless it demonstrates that large external deceleration can require large engine-housing diameters. The use of partial internal diffusion is, of course, doubly effective in reducing maximum diameter, since it permits a reduction in both A_i and $(A_{max}/A_i)_{min}$.

The foregoing analyses are strictly applicable only to incompressible flow, of course. Corresponding results which take compressibility into account and are

in qualitative agreement with the conclusions of this discussion have been obtained
by Küchemann and Weber [3]. They also present experimentally verified methods
to calculate inlet shapes having constant pressure over the surface, as required for
minimum frontal area. Such inlets require a rather sharp corner near the leading
edge so that their performance drops substantially with off-design operation,
and they have not found extensive application.

To summarize: This simplified analysis shows that the performance of an inlet
depends on the pressure gradient on both internal and external surfaces. The
external pressure rise is fixed by the external compression and the ratio of maximum to inlet areas A_{max}/A_i. The internal pressure rise depends on the reduction of velocity between entry to the inlet diffuser and entry to the compressor
(or burner, for a ramjet).

Diffuser performance

In the literature on this subject, several performance parameters have been
used to describe the relative excellence of diffusers. Two of these will be discussed
in the following. The first is the adiabatic diffuser efficiency used in the discussion
of engine performance.

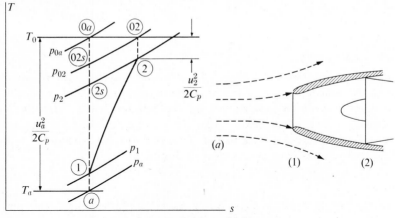

FIG. 7–9. Definition of inlet states.

(a) *Isentropic efficiency* η_d: Referring to Fig. 7–9, we can define the isentropic
efficiency of a diffuser as follows:

$$\eta_d = \frac{h_{02s} - h_a}{h_{0a} - h_a} \approx \frac{T_{02s} - T_a}{T_{0a} - T_a}.$$

State (02s) is defined as that state which would be reached by isentropic compression to the *actual* outlet stagnation pressure. Since

$$\frac{T_{02s}}{T_a} = \left(\frac{p_{02}}{p_a}\right)^{(\gamma-1)/\gamma} \quad \text{and} \quad \frac{T_{02}}{T_a} = 1 + \frac{\gamma - 1}{2}M^2,$$

the diffuser efficiency η_d is also given by

$$\eta_d = \frac{(p_{02}/p_a)^{(\gamma-1)/\gamma} - 1}{[(\gamma - 1)/2]M^2}.$$ (7–6)

(b) *Stagnation-pressure ratio, r_d*: The stagnation-pressure ratio,

$$r_d = p_{02}/p_{0a},$$ (7–7)

is widely used as a measure of diffuser performance. Diffuser efficiency and stagnation-pressure ratio are, of course, related. In general,

$$\frac{p_{02}}{p_a} = \frac{p_{02}}{p_{0a}} \cdot \frac{p_{0a}}{p_a} = \frac{p_{02}}{p_{0a}}\left(1 + \frac{\gamma - 1}{2}M^2\right)^{\gamma/(\gamma-1)},$$

and, with Eqs. (7–6) and (7–7),

$$\eta_d = \frac{\left(1 + \dfrac{\gamma - 1}{2}M^2\right)(r_d)^{(\gamma-1)/\gamma} - 1}{[(\gamma - 1)/2]M^2}.$$ (7–8)

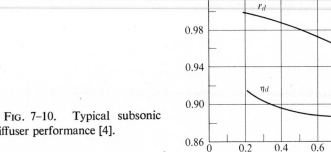

FIG. 7–10. Typical subsonic diffuser performance [4].

As the preceding discussion shows, it is unfortunate that these criteria are based on overall diffusion rather than on internal diffusion. The relationship between internal and external diffusion depends on engine mass flow as well as flight Mach number M, since this determines the amount of "spilling," and hence external diffusion. Figure 7–10 gives typical values of stagnation-pressure ratio r_d. The diffuser efficiency η_d was calculated from r_d, using Eq. (7–8).

7–3 SUPERSONIC INLETS

Even for supersonic flight it remains necessary, at least for present designs, that the flow leaving the inlet system be subsonic. Compressors capable of ingesting a supersonic airstream could provide very high mass flow per unit area and, theoretically at least, very high pressure ratio per stage. However, the difficulty of diffusing the supersonic stream in blade passages without excessive shock losses (especially at off-design conditions) has so far made the development of multistage supersonic compressors a possibility that is somewhat remote. Similarly, there are certain advantages to be gained in the use of supersonic velocities within ramjet combustors, as discussed in Section 7–4. Again, although supersonic combustion is perhaps attainable, there remain formidable problems which must be solved before such combustors are a reality. For the present, then, both ramjets and turbojets require diffusers which provide subsonic airstreams. For both cases, outlet Mach numbers of the order of 0.4 are typical.

Reverse-nozzle diffuser

For low supersonic Mach numbers, it is obvious that subsonic velocities can be attained without excessive stagnation pressure loss through a simple normal shock. However, at higher Mach numbers the desire for low losses makes isentropic diffusion an increasingly attractive goal. From a simple one-dimensional analysis, it might appear that a supersonic nozzle operated in reverse, as will be indicated in Fig. 7–13, would be the ideal device to produce isentropic deceleration. While such deceleration is not theoretically impossible, it does seem impractical, at least for inlet Mach numbers high enough to make shockfree deceleration especially desirable. The difficulty can be traced to the behavior of the boundary layer in the presence of the adverse pressure gradient along the walls. At best, the wall contour would have to be very carefully designed and constructed to avoid "stacking up" of the compression process in the form of a shock. The design of an appropriate contour would be considerably complicated by the effect of the boundary layer which, under the desired free-stream pressure gradient, would at least be rather thick, and might easily separate in the absence of shocks. Due to boundary layer thickening, shocks would generally be formed, and it is unlikely that isentropic flow could be established. Any shock in a diffuser designed for

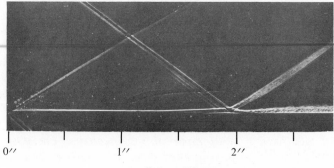

$\theta = 6°, R_i = 630,000$

Fig. 7–11.   Interaction of an oblique shock and a boundary layer. (Courtesy M.I.T. Gas Turbine Laboratory.)   Flow from left to right.

isentropic deceleration to subsonic velocities would be accompanied by a drastic alteration of the flow pattern.

The effect of a shock on a boundary layer can be seen in Fig. 7–11, which is a Schlieren photograph of the interaction between an oblique shock and a boundary layer on a flat plate. Schlieren photographs indicate regions of varying density gradient [5] and hence, in this figure, both the shock and the boundary layer appear as rather distinct lines, while regions of gradual compression or expansion appear less distinctly. One can see that the effect of the shock-induced pressure rise propagates upstream in the boundary layer (which is of course partially subsonic), causing thickening ahead of the shock. Unlike the simple shock reflection which would occur in the absence of the boundary layer, this interaction includes gradual compression due to the upstream thickening of the boundary layer, expansion around the "bump" in the boundary layer, and recompression as the flow is again turned parallel to the wall. The net effect on the boundary layer is seen to be at least substantial thickening and increased turbulence. In fact, the downstream boundary layer is very likely separated.

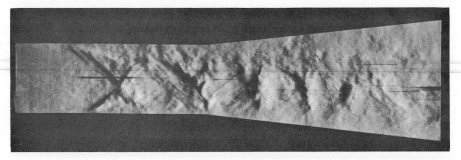

FIG. 7–12. Shock boundary layer interactions in a duct. (Courtesy M.I.T. Gas Turbine Laboratory.) Flow from left to right.

Within a closed duct, shock intersections with adjacent walls and the accumulation of appreciable stagnant or back-flowing fluid set up sizable obstructions to the main stream flow. This grossly distorts the entire flow field, and is a source of losses from irreversible mixing in addition to losses from the shocks themselves. Figure 7–12 is a Schlieren photograph of a shock system within a duct, showing the effects of separation and mixing.

Even if one could design a converging-diverging inlet that actually gave isentropic deceleration, it would not be a practical diffuser for a flight vehicle. A fixed-geometry diffuser would be isentropic for only one operating condition, although it might have a small range of near-isentropic operation near the design condition. Since proper operation demands shockfree flow, at least upstream of the minimum area, such a device would be exceedingly sensitive to variations in angle of attack. Finally, as we shall discuss next, a converging-diverging supersonic diffuser would be difficult to start.

The starting problem

Internal supersonic deceleration in a converging passage (of nonporous walls) is not easy to establish. In fact, as we shall now show, design conditions cannot be achieved without momentarily overspeeding the inlet air or varying the diffuser geometry. This difficulty is due to shocks which arise during the deceleration process, and it need not be related to boundary layer behavior. Therefore, let us neglect boundary layer effects for the moment, while we examine the starting behavior of a converging-diverging diffuser which is one-dimensional and isentropic except for losses which occur in whatever (normal) shocks may be present. This simplified analysis contains all the essential features of the phenomena, and it could be a valid representation of a real flow from which the wall boundary layer fluid was carefully removed by suction through porous walls.

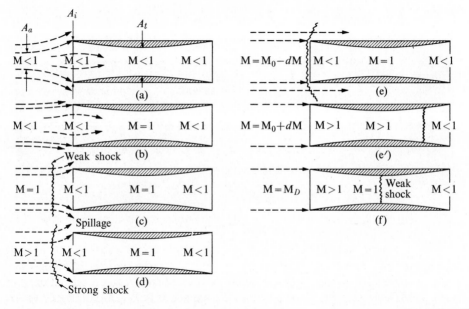

Fɪɢ. 7–13. Successive steps in the acceleration and overspeeding of a simplified (one-dimensional) supersonic inlet.

Figure 7–13 illustrates successive steps in the acceleration of a fixed-geometry converging-diverging inlet. To isolate the inlet behavior from that of the rest of the propulsion device, we assume that whatever is attached to the diffuser exit is always capable of ingesting the entire diffuser mass flow. Thus mass flow rate is limited only by choking at the minimum diffuser area A_t.

Condition (a) illustrates low subsonic speed operation, for which the inlet is not choked. In this case the airflow through the inlet, and hence the upstream capture area A_a, is determined by conditions downstream of the inlet. In condition (b), although the flight velocity is still subsonic, the flow is assumed to be acceler-

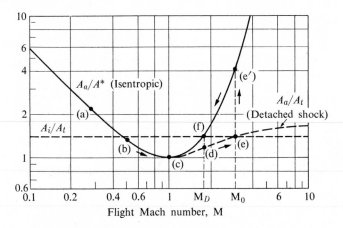

FIG. 7–14. Acceleration and overspeeding of a one-dimensional supersonic inlet.

ated to sonic velocity at the minimum area A_t, and the inlet mass flow rate is limited by the choking condition at A_t. Since the flow is assumed isentropic, then $A_t = A^*$ and the upstream capture area A_a is given by

$$\frac{A_a}{A^*} = \frac{A_a}{A_t} = \frac{1}{M}\left[\frac{2}{\gamma+1}\left(1 + \frac{\gamma-1}{2}M^2\right)\right]^{(\gamma+1)/2(\gamma-1)}.$$

For sufficiently high subsonic values of M [see point (b) in Fig. 7–14], we have

$$\frac{A_a}{A^*} = \frac{A_a}{A_t} < \frac{A_i}{A_t}.$$

Thus, for these conditions, the capture area A_a must be less than the inlet area A_i, and there will therefore be spillage around the inlet.

For sonic or supersonic flight speeds the spillage mechanism is necessarily nonisentropic. That is, in order to "sense" the presence of the inlet and flow around it, the spilled air must be reduced to subsonic velocity upstream of the inlet plane. The mechanism for this deceleration is a detached "bow wave" which stands sufficiently far upstream to allow the required spillage. The process of establishing the detached shock wave can be imagined as follows: Suppose that when the air first reached supersonic velocity there were no shock. Then flow would have to enter, without deviation, the entire inlet area, in effect making A_i act as the capture area A_a. But for low supersonic Mach numbers (see Fig. 7–14 where $A_i/A_t > A_a/A^*$) the *allowable* capture area as limited by choking at A_t is *less* than A_i. Hence there would be an accumulation of mass and a rise in pressure in the inlet. This pressure rise would build up rapidly until a shock of sufficient strength moved upstream against the supersonic flow and became established at a position which would allow the required spillage.

Once the shock is established, the flow entering the inlet is no longer isentropic. Hence, when the design Mach number of the aircraft is first reached, as at (d) in

Fig. 7-13, the "reversed isentropic nozzle" mass flow cannot pass through the throat area A_t. This follows from Eq. (3-14), which indicates that the choked mass flow through a given area (A_t) is proportional to p_0, and from the fact that the fluid suffers a stagnation-pressure loss in traversing the shock. For the flow which *does* enter the inlet (assuming isentropic flow from a point just downstream of the shock to the throat), Eq. (3-15) gives

$$\frac{A_2}{A_t} = \left(\frac{A}{A^*}\right)_2 = \frac{1}{M_2}\left[\frac{2}{\gamma+1}\left(1 + \frac{\gamma-1}{2}M_2^2\right)\right]^{(\gamma+1)/2(\gamma-1)},$$

where the subscript 2 refers to conditions just downstream of a normal shock. It may be assumed that the slightly curved shock may be approximated over the capture area by a normal shock, so that M_2 may be expressed as a function of the upstream Mach number M by the normal shock relation

$$M_2 = \sqrt{\frac{[2/(\gamma-1)] + M^2}{[2\gamma/(\gamma-1)]M^2 - 1}}.$$

These two expressions may be combined to give the ratio of capture area to throat area, A_a/A_t, as a function of flight Mach number for the flow which includes a detached normal shock. The result, plotted in Fig. 7-14, indicates that the inlet area A_i will *remain* too large and spillage will continue even beyond the design Mach number M_D, unless the inlet can be *overspeeded* to a Mach number M_0. At this Mach number [or just below it, as at (e)], the inlet is capable of ingesting the entire incident mass flow without spillage. The shock position will be just on the lip of the inlet, as in Fig. 7-13(e), and a slight increment in speed, as to (e'), will cause the shock to enter the convergence. Since a shock cannot attain a stable position within the convergence,* it will move quickly downstream to

---

* Suppose that conditions are such that a stationary shock exists someplace within the convergence. Two conditions are possible for the flow downstream of the shock: either it is choked at A_t, or it is not choked. If it is choked, the mass flow through the throat is given by Eq. (3-14),

$$\dot{m}_t = p_0\left[\frac{A_t}{\sqrt{RT_0}}\sqrt{\gamma}\left(\frac{2}{\gamma+1}\right)^{(\gamma+1)/2(\gamma-1)}\right] = \mathcal{C}p_{02},$$

where, for constant inlet velocity (flight speed), $\mathcal{C}$ is a constant and p_{02} is the stagnation pressure *downstream* of the shock. Now suppose that a small disturbance moves the shock slightly upstream where the shock Mach number is slightly higher. A greater stagnation-pressure loss will occur across the shock, lowering the downstream stagnation pressure and the throat mass flow. Since conditions upstream of the shock are not affected, the mass flow through the shock remains constant and there results an accumulation of mass and a rise in pressure behind the shock. As the static pressure ratio across the shock increases, its propagation speed relative to the fluid increases and the shock moves further upstream. This further increases the stagnation-pressure loss, of course, so that the mass-flow imbalance increases and the shock continues to move upstream and out of the inlet. The final position of the shock is far enough outside the inlet area come to rest within the divergence, at a position determined by downstream conditions. The flow to the throat is now isentropic, and the area ratio A_a/A^* (now greater than A_a/A_t, since $A_t > A^*$) would be given by (e′) of Fig. 7–14. The incoming flow is decelerated from A_i to A_t, whereupon it is reaccelerated supersonically in the divergence. Having thus attained isentropic flow in the inlet, the Mach number may be reduced from M_0 to M_D, as at (f). At exactly the design speed, the throat Mach number would be just unity and isentropic deceleration from supersonic to subsonic flow would exist. Even for this simplified model, however, this condition would be unstable. A slight decrease of flight speed or increase of back pressure would require spillage, and a shock would move rapidly out through the inlet to reestablish condition (d). Thus it might be better to maintain the throat Mach number slightly greater than unity while reaching subsonic velocities through a very weak shock just downstream of the throat.

This simplified description contains the essential feature of the starting problem associated with an internally contracting passage. That is, an inlet having A_i/A_t greater than one will *always* require spillage upon reaching supersonic flight velocities, since A_a/A_t will always pass through a minimum of one just as sonic flight velocity is attained. It is necessary to perform some operation other than simply accelerating to the design speed in order to "swallow" the starting shock and establish isentropic flow. Overspeeding is one such operation, but there are others.

If overspeeding is not feasible (note that, except for very modest design Mach numbers, substantial overspeed would be required), it might be possible to swallow the shock by a variation of geometry at constant flight speed. The principle is easily seen in terms of the simple one-dimensional analysis. Suppose the inlet is accelerated to the design Mach number M_D with the starting shock present, as at point (d) in Figs. 7–14 and 7–15. Now, if the actual area ratio can be *decreased*

---

to allow the correct subsonic "spilling" of air. By similar arguments it can be seen that slight motion downstream by the shock is continued until the shock is carried through the throat.

If the flow downstream of A_t is not choked, an appropriate downstream boundary condition might be that at some exit area A_e a constant static pressure p_e is maintained. In this case, for steady flow,

$$\dot{m}_t = \dot{m}_e = \rho_e u_e A_e = \frac{A_e p_e}{RT_e}\sqrt{2c_p(T_0 - T_e)}.$$

Assuming isentropic flow downstream of the shock, this can be expressed as

$$\dot{m}_t = \frac{A_e p_e}{\sqrt{RT_0}}\sqrt{\frac{2\gamma}{\gamma-1}\left(\frac{p_{02}}{p_e}\right)^{(\gamma-1)/\gamma}\left[\left(\frac{p_{02}}{p_e}\right)^{(\gamma-1)/\gamma}-1\right]}.$$

Thus it can be seen that, since $\dot{m}_t$ is dependent only on p_{02}, changes in shock position result in a mass flow imbalance similar to that for the choked case, and the same conclusions hold: Since any shock would exist in the convergence by *moving* into it, it follows that no shock can reach a stable position within the convergence.

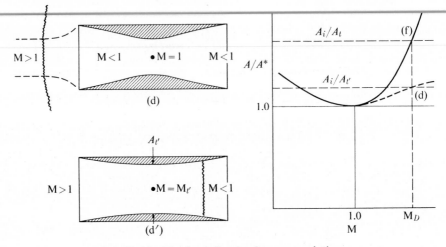

FIG. 7–15. Shock swallowing by area variation.

from A_i/A_t to that value which can ingest the entire inlet flow behind the shock, the shock will be swallowed to take up a position downstream of the throat. This variation would normally involve a momentary *increase* of throat area from A_t to a new value which we shall call A_t' (see Fig. 7–15). Having thus achieved isentropic flow within the convergence, the throat Mach number $M_{t'}$ is greater than one, and a relatively strong shock occurs further downstream. Complete isentropic flow can then be achieved by returning the area ratio to its original value, while the operating conditions move from (d) to (f).

A geometric variation such as that shown schematically in Fig. 7–15 would, of course, be exceedingly difficult mechanically. However, geometries which permit the axial motion of a center plug between nonparallel walls can be used. A somewhat similar effect can be had through the use of porous convergent walls. With the shock external, the high static pressure within the convergence causes considerable "leakage" through the walls, thus effectively increasing the throat area to permit shock swallowing. Once this occurs, the lower static pressure within the convergence somewhat decreases the flow loss. However, since the total porosity required is even greater than the throat area [6], there remains a high mass flow loss under operating conditions. This is desirable to the extent that it eliminates boundary layer fluid, but the lost flow is more than is needed for this purpose. Flow which is decelerated but not used internally contributes substantially to the drag of the propulsion device.

From this discussion it would seem that the starting problem could be avoided altogether by using a simple divergent inlet, i.e., one for which $A_i/A_t = 1$, as shown in Fig. 7–16. For supersonic flight speeds and sufficiently low back pressure it is possible to accelerate the internal flow within the divergence before decelerating it in a shock. To reduce stagnation-pressure losses, it is desirable to have the shock occur at the minimum possible Mach number which, for this

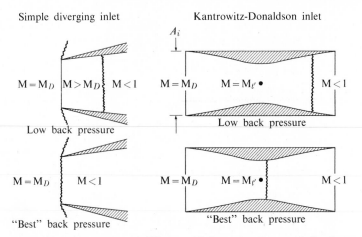

Simple diverging inlet   Kantrowitz-Donaldson inlet

$M = M_D$ $M > M_D$ $M < 1$ $M = M_D$ $M = M_{t'}$ • $M < 1$

Low back pressure   Low back pressure

$M = M_D$ $M < 1$ $M = M_D$ $M = M_{t'}$ • $M < 1$

"Best" back pressure   "Best" back pressure

Fɪɢ. 7–16. Fixed-geometry diffuser with intentional normal shocks at the design Mach number M_0. (Refer to Fig. 7–15.)

geometry, is the flight Mach number. This condition can be achieved by adjusting the back pressure (by varying the engine exhaust area, for example) so that the shock is positioned just on the inlet lip.

A slight improvement, called the Kantrowitz-Donaldson inlet by Foa [7], is also illustrated in Fig. 7–16. This configuration uses the maximum internal convergence which will just permit shock swallowing at the design flight Mach number. Referring to Fig. 7–15, we can see that this is just $A_i/A_{t'}$ for a design Mach number of M_D. As in the simple divergent inlet, it is necessary to adjust the back pressure to assure that the shock occurs at the minimum possible Mach number. The advantage of the internal convergence is, of course, that this minimum Mach number is less than the free-stream value. A disadvantage is that there is an abrupt change in performance just at the design Mach number, since the shock will not be swallowed below this value and it will be immediately disgorged if the speed falls off slightly from the design value. Further, since shocks of any origin are unstable within the convergence, such an inlet would be quite sensitive to changes in angle of attack.

The condition in which a shock just hangs on the inlet lip is called the *critical condition*. Operation with the shock swallowed is called *supercritical*, while that with a detached shock and spillage is called *subcritical*. Note that subcritical operation may occur as the result of choking in the inlet, as discussed here for inlets alone; or, for a complete engine, as the result of any downstream flow restriction which cannot accept the entire mass flow $\rho_a u_a A_i$.

External deceleration

The absence of a starting problem for the normal shock inlet is offset by the accompanying stagnation-pressure loss at all but low flight Mach numbers (less than about 1.5). For example, Fig. 3–8 indicates a 28% stagnation-pressure loss at M = 2 (γ = 1.4). If we wish to obtain reasonable performance while maintaining the starting characteristics of a simple divergent inlet, then it is clear that some *external* deceleration must occur upstream of the inlet plane in order that the Mach number of the normal shock be reduced to a suitable value.

The simplest and most practical external deceleration mechanism is an oblique shock or, in some cases, a series of oblique shocks. While such shocks are not isentropic, the stagnation-pressure loss in reaching subsonic velocity through a series of oblique shocks followed by a normal shock is less than that accompanying a single normal shock at the flight velocity. The losses decrease as the number of oblique shocks increases, especially at high flight Mach numbers.

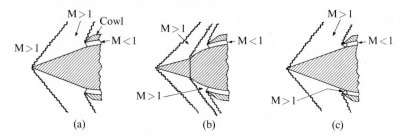

FIG. 7–17. Typical configurations for oblique shock diffusers. (Adapted from Oswatitsch [8].)

In the external compression process, shocks and boundary layers may interact strongly, so that it is highly desirable to locate the oblique shocks at points where boundary layers are absent. This can be arranged easily if a center body (primarily for axisymmetric flow) is used, as in Fig. 7–17. These schematics, taken from Oswatitsch [8], illustrate the typical single oblique shock system and two double oblique shock systems. Although the double shock systems theoretically give better performance, one can see that several problems may arise in their use. In the configuration of Fig. 7–17(b), the second shock, generated by a turn of the wall, occurs at a point where the boundary layer has had time to develop, and separation may result. The configuration of 7–17(c) avoids this at the point of shock generation, but the second shock still intersects the boundary layer on the center body. Deceleration by a series of shocks arranged as in 7–17(b) introduces a net turning of the flow that may result in excessive cowling size and drag. The arrangement of 7–17(c) avoids this difficulty to some extent, but it introduces a region of internal convergence. This would be accompanied by some starting difficulty but, since the convergence is not intended to reduce the Mach number to unity, the starting problem need not be difficult, and it could be even simpler than the Kantrowitz-Donaldson inlet behavior. In all cases, of course, the normal shock–boundary layer intersection can cause considerable difficulty, since it is desirable to have a well-behaved boundary layer at the inlet to the subsonic diffuser.

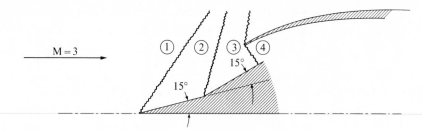

FIG. 7–18. Two-dimensional diffuser.

The performance gain to be expected through the use of multiple oblique-shock deceleration can be appreciated by looking at a simple two-dimensional example. Consider the diffuser in Fig. 7–18, in which the flow is deflected through two 15-degree angles before entering a normal shock. Using Fig. 3–12, we can show that the Mach numbers in regions (2), (3), and (4) are 2.26, 1.65, and 0.67, respectively. Then, using Fig. 3–11, we can obtain the stagnation-pressure ratios as follows:

$$\frac{p_{02}}{p_{01}} = 0.895, \qquad \frac{p_{03}}{p_{02}} = 0.945, \qquad \frac{p_{04}}{p_{03}} = 0.870.$$

Thus, the overall stagnation-pressure ratio is approximately $p_{04}/p_{01} = 0.735$. If the deceleration had been achieved by a single normal shock, the overall stagnation-pressure ratio would have been only 0.33. It should be remembered that these estimates do not include losses due to boundary layer effects, which may be especially important in the subsonic diffuser.

This example does not necessarily employ the best arrangement of three shocks, of course, since a variation of their relative strengths might provide a higher overall stagnation-pressure ratio. For the simple two-dimensional case, in which conditions downstream of each shock are uniform, Oswatitsch [8] has shown by theoretical analysis that, for a given flight Mach number and a given number of oblique shocks followed by one normal shock, the overall stagnation-pressure ratio will be maximized if all the oblique shocks have equal strength. The oblique shocks may be said to have equal strength when the Mach numbers of the velocity components normal to each shock, and incident to it, are equal. It follows, of course, that the stagnation-pressure ratios will also be equal. This result will not seem unreasonable if we recall that for a normal shock the stagnation-pressure loss rises very quickly with incident Mach number. It is interesting to note that the Mach number of the final normal shock should be *less* than the normal component of the oblique shocks (about 0.94 times as great [9]).

For a diagram of the best performance using *n* oblique shocks of equal strength followed by one normal shock, see Fig. 7–19.

For the more common case of axisymmetric inlets and conical shocks, the best arrangement is not so easily determined, since the downstream flow of a conical shock is not uniform. In fact, fluid properties are constant along conical surfaces

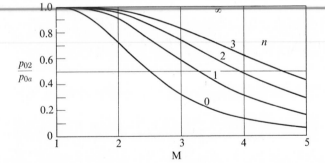
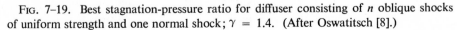

FIG. 7–19. Best stagnation-pressure ratio for diffuser consisting of n oblique shocks of uniform strength and one normal shock; $\gamma = 1.4$. (After Oswatitsch [8].)

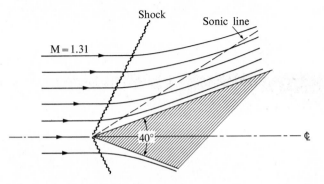

FIG. 7–20. Typical conical-flow streamlines. (After Shapiro [10].)

emanating from the shock vertex, and the streamlines downstream of the shock are curved [10]. The effect of the streamline curvature is further diffusion, which can be very nearly isentropic, so that subsequent shocks occur at reduced (but nonuniform) Mach numbers. It is even possible to achieve deceleration to subsonic velocities behind a conical shock without subsequent shocks, as shown for a rather low Mach number in Fig. 7–20.

The performance advantage of multiple conical shocks is qualitatively similar to that of the multiple-plane shocks shown in Fig. 7–19. Going to the limit (for either axisymmetric or two-dimensional geometries) of an infinite number of infinitesimal shocks, generated by continuous wall curvature, one could, theoretically at least, achieve isentropic external deceleration to sonic velocity. Such an inlet, indicated qualitatively in Fig. 7–21, would seem to provide the ideal geometry to achieve low losses, while at the same time avoiding the starting problems of an internal convergence. However, several practical difficulties would be encountered in the operation of such an inlet. This geometry, like that of the isentropic internal flow diffuser, would function properly at only one Mach number, and would be very sensitive to angle of attack. Furthermore, the boundary layer along the curved surface, unlike that along plane or conical surfaces, would

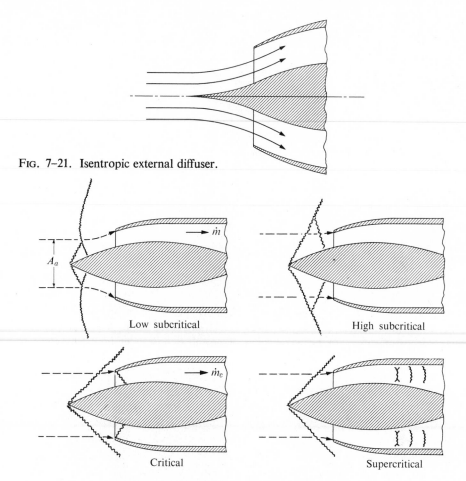

FIG. 7–21. Isentropic external diffuser.

FIG. 7–22. Typical modes of inlet operation.

be subject to a high adverse pressure gradient which might cause separation. Finally, for high flight Mach numbers it would be necessary that the flow turn through large angles before reaching sonic velocity. The resultant large cowl angle would at least exhibit high drag, and perhaps even interfere with the inlet flow.

The operation of external shock diffusers is divided into subcritical, critical, and supercritical modes, which depend on the external and internal shock configuration. The three modes are shown schematically for a typical case in Fig. 7–22. With entirely diverging internal flow such as this, the normal shock position is determined by a downstream flow restriction rather than by the inlet geometry. Hence the operating mode is sensitive to variations in exhaust-nozzle area and fuel-flow rate. Subcritical operation entails "spilling" of flow and a normal shock upstream of the inlet. "Low" and "high" subcritical operation differ only in the extent of spilling. Supercritical operation occurs at the same mass flow as critical operation, but with increased losses, since the normal shock occurs at a higher Mach number.

The actual performance of supersonic diffusers has been tested by a great number of investigators (References [8, 9, 11–17] are typical), owing to the substantial importance of inlet performance, especially at high Mach numbers. A wide variety of geometries has been considered, each depending on a particular application. Special attention is paid to the off-design performance of an inlet, since it is of obvious importance in an actual flight application and yet not so amenable to analyses as the design conditions. As the reference titles indicate, various adjustable inlets have been considered to extend the favorable operating range of an inlet.

The data of Dailey [12] are typical of many investigations of an important instability which occurs during the subcritical operation of most supersonic inlets. This phenomenon, known as "buzz," consists of a rapid oscillation of the inlet shock and flow pattern, and the resultant internal disturbance is very detrimental to engine performance. In a ramjet, for instance, the onset of buzz usually extinguishes combustion. Although the pulses of the shock system are similar, the interval between pulses is not constant [12]; hence buzz cannot be considered a periodic phenomenon. Although it is not thoroughly understood at present, buzz has been shown to be a function of conditions only at, and immediately downstream of, the inlet [12, 14]. In some cases boundary layer bleed from the center body can delay the onset of buzz. In other cases the use of a length of nondivergent subsonic passage just downstream of the inlet can have a beneficial effect. This latter case could be attributed either to the establishment of a healthy boundary layer (as compared with a separated one just downstream of the normal shock) or to the establishment of more uniform free-stream conditions (see below) before subsonic diffusion is attempted.

Figure 7–23 is a plot of typical diffuser performance as a function of mass flow, expressed as the ratio of actual to critical (or supercritical) mass flow rate $\dot{m}_c$, or the equivalent ratio of actual to critical capture area. Ratios of $\dot{m}/\dot{m}_c$ less than one signify subcritical operation. It can be seen that for mass flows slightly less than critical (the so-called "high subcritical" flow) the stagnation-pressure ratio increases slightly with mass flow. This is explained by the presence of lower aerodynamic losses in the internal passages due to reduced mass flow and velocity. *Overall* performance, of course, does not increase with subcritical operation, since spilling is accompanied by both increased drag and decreased thrust. As the flow is further reduced the normal shock is pushed further upstream and stagnation pressure losses increase. A portion of the air entering the cowl may travel through a single strong shock (see "low subcritical," Fig. 7–22) rather than the two weaker shocks it would pass through during critical operation. The entrance of this low-stagnation-pressure (or high-entropy) air causes a reduction in performance. In supercritical operation the stagnation-pressure ratio drops rather rapidly at constant mass flow rate.

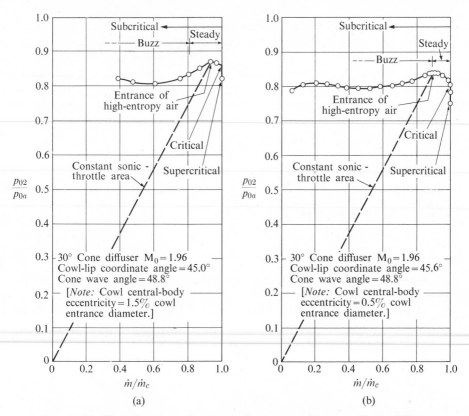

Fig. 7–23. Typical supersonic inlet performances. (Courtesy Dailey [12].)

It can be seen by comparing Figs. 7–23(a) and (b) that the ingestion of low-stagnation-pressure air may or may not correspond to the onset of buzz (although it usually does).

7–4 COMBUSTORS

The working fluid in the engine is "heated" by an internal combustion process. Before this chemical reaction can occur, the liquid fuel must be injected into the airstream, atomized, vaporized, and the vapor must be mixed with the air. All this takes time and space. Space is of course at a premium in aircraft applications, so that great effort is made to reduce the size of the combustion chamber by hastening completion of the above processes. Very high combustion intensity is achieved in aircraft combustion chambers as compared with conventional combustion devices: 12,000 Btu/ft^3·sec in turbojets as compared to 8 Btu/ft^3·sec in a typical steam power plant. This intensity must be achieved without sacrificing other equally important performance characteristics. Some of the desirable features of combustion chambers are as follows: (1) completeness of combustion,

(2) proper temperature distribution at exit, (3) low stagnation-pressure loss, (4) absence of "hot spots," (5) stability, (6) freedom from flameout, (7) relight ability, (8) absence of deposits on surface, and (9) small cross section and length.

Many of these requirements are incompatible. For example, high efficiency and low pressure loss are in direct opposition to small size. Satisfying these requirements to a reasonable degree over a wide range of operating conditions is not easy.

Combustion limits and flame speed

In general the performance of air-breathing jet engines is strongly dependent on the mass flow rate per unit cross-sectional area of the engine. A large area for a given flow rate implies not only large engine mass per unit thrust, but also large nacelle drag for an externally mounted engine or large volume for an engine placed inside the fuselage. For this reason it is desirable for the velocity of the working fluid to be as high as possible without incurring excessive losses due to shocks, wall friction, and viscous mixing. A limitation is imposed by the combustor since it is necessary to maintain a stationary flame within a moving airstream.

For simplicity let us neglect the complications of injection, etc., and imagine the flame as propagating through a combustible mixture at the "flame speed," while the mixture is carried downstream. To avoid flame travel out of the combustion chamber, it is necessary to maintain the mixture velocity within certain limits: if the velocity is too high the flame will be "blown out" the exit; if too low, the flame will travel upstream and be extinguished. It is not necessary, however, that the mixture velocity be exactly equal to the flame speed since, as we shall see, the flame front need not be normal to the flow. In fact, if it were, an enormous cross-sectional area would be required.

In addition to limitations imposed by flame speed, there are, under some conditions, difficulties associated with the combustibility of the fuel–air mixture. This is always true in turbojet combustors which are limited to low turbine inlet temperature and thus to very low fuel–air ratios, and it can also occur during some phases of ramjet operation.

For a turbojet, the allowable fuel addition can be determined in terms of the compressor outlet temperature T_{03}, the turbine inlet temperature T_{04}, the fuel "heating value" Q_R, and the air properties by

$$\dot{m}_f Q_R = \dot{m}_a c_p (T_{04} - T_{03}),$$

where $\dot{m}_f$ = fuel flow rate, $\dot{m}_a$ = air flow rate, and c_p = specific heat at constant pressure. In terms of the fuel–air ratio f,

$$f = \frac{\dot{m}_f}{\dot{m}_a} = \frac{c_p}{Q_R}(T_{04} - T_{03}).$$

As an example, an engine flying in a $-60°F$ atmosphere at Mach 0.8 with a twelve-to-one compressor–pressure ratio and 85% efficiency will have a compressor outlet temperature of about 1000°R. If the turbine inlet temperature is 2000°R, Q_R is 18,000 Btu/lb, and c_p averages 0.26 Btu/lb·°R, the fuel–air ratio is

$$f = \frac{0.26}{18,000}(2000 - 100) = 0.0145,$$

which will now be compared with the stoichiometric fuel–air ratio. Assuming octane as a representative hydrocarbon, and writing the stoichiometric chemical reaction:

$$2C_8H_{18} + 25(O_2 + \tfrac{79}{21}N_2) \rightarrow 16CO_2 + 18H_2O + 25(\tfrac{79}{21})N_2,$$

the fuel–air ratio is found to be

$$f = \frac{2(96 + 18)}{25(32 + \tfrac{79}{21}28)} = 0.0667.$$

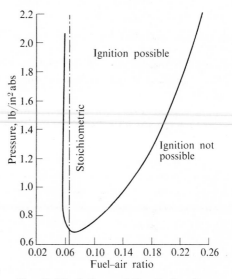

FIG. 7–24. Inflammability limits of gasoline–air mixtures. (Courtesy Olson, *et al.* [18].)

Thus, to prevent excessive temperature at the exit of the combustion chamber, the fuel–air ratio must be much less than stoichiometric. The difficulty of lean mixtures in a combustion chamber is illustrated in Fig. 7–24, taken from Reference [18].

From the figure it can be seen that if all the air were to be initially mixed with all the fuel, the resulting mixture would be much too lean for combustion. For this reason the fuel is initially mixed and burned with a small amount of "primary" air, after which the combustion products are diluted and cooled to T_{04} by the remaining air.

Let us now return to flame speed: A combustion zone will travel into a mixture of reactants at a rate which is dependent on complex chemical kinetics and the state of the reactants. Conversely, a flame remains stationary in a traveling mixture of reactants if the speed of the flame relative to the reactants is just equal to the reactant velocity. Figure 7–25 shows the flame velocity observed by holding a stationary flame front in a mixture of flowing reactants. The lower curve represents undisturbed (probably laminar) flow and the upper curve depicts combustion downstream of a perforated plate. These curves were obtained for one particular inlet temperature and pressure. Again we see that typical turbojet fuel–air mixtures are much too lean. Ramjet fuel–air ratios can vary from very lean to very

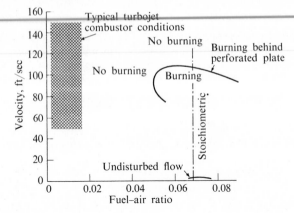

FIG. 7–25. Dependence of flame speed on fuel–air ratio. (Courtesy Olson, *et al.* [18].)

rich values, owing to the wide variance in operating conditions of ramjets and to higher allowable temperatures than for turbojets.

Figure 7–25 points out two important things. First, it reemphasizes the necessity of initial burning in primary air for lean mixtures. Second, it shows the tremendous influence of turbulence on apparent flame velocity. In this case the perforated plate created local low-velocity regions which "held" the flame, and at the same time the resulting turbulence greatly increased the area of the flame front, accounting for the much higher gross flame speed. The turbulence level varies considerably from the level exhibited in this test, for various combustion chambers and among the combustion chambers themselves. Thus, while this figure is of great qualitative significance, it has little quantitative value.

Combustion at near-stoichiometric fuel–air ratios, while alleviating some of the above difficulties, is not without problems. Provided stable combustion is attained, complete combustion in the case of lean mixtures is virtually assured since, with excess oxygen, local fuel-rich areas are unlikely. On the other hand, combustion of a near-stoichiometric mixture obviously requires an essentially uniform distribution of constituents to avoid wasting some fuel in local fuel-rich regions.

Ramjet combustors and afterburners

Ramjet combustors and turbojet afterburners are essentially similar devices and they are, physically at least, simpler than turbojet combustors. Usually they consist of a straight through-flow region containing several bluff bodies which act as "flameholders," as shown in Figs. 6–3, 6–4, and again in 7–26(a). In some cases perforated liners similar to those in turbojets might be used, as shown in Fig. 7–26(b). For very high flight speeds, flameholders may not be needed if the combustor inlet temperature is above that for spontaneous ignition of the fuel.

The performance of a bluff body as a flameholder may be seen in terms of a simplified model (Fig. 7–27). Suppose that a combustible mixture travels with

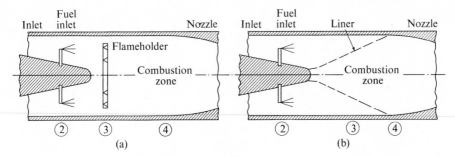

FIG. 7–26. Schematic ramjet (or afterburner) combustion chambers.

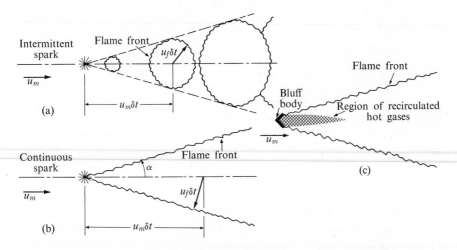

FIG. 7–27. Simplified picture of the operation of a flameholder.

uniform and parallel velocity u_m, and that the flame speed within the mixture is
u_f, where $u_f < u_m$. Combustion initiated by an intermittent spark in such a
flow would proceed into the unburned mixture, producing a series of spherical
regions of burned material, as in Fig. 7–27(a). Each sphere's radius would increase at the rate u_f (with the simplifying assumption that reactants and products
are of the same density), while its center travels downstream at the rate u_m. If
the intermittent spark is replaced by a continuous one, a conical region of com-
busted material will result, as in 7–27(b). The cone angle would obviously be
$\sin^{-1}(u_f/u_m)$. The reader will recognize the similarity of flame propagation to
that of the propagation of a small disturbance in supersonic flow. The flame-
holding effect of a bluff body results from the recirculation of hot and combusting
gases in its wake, as in Fig. 7–27(c). This region acts as a continuous ignition
source just as a spark does; and, neglecting velocity distortion and turbulent
effects, a conical region of burned material will result, much as in the case of
the spark. Actually the wake of the body distorts the flame front, while the turbulence in the wake has a favorable effect on flame speed. It is obvious that by this

means one can maintain a stationary flame within a stream whose average velocity is greater than the flame speed, and the combustion chamber velocity need not be as low as the flame speed. For given flame and mixture speeds, the length of a combustion chamber can be reduced if a larger number of flameholders can be installed. However, a compromise must be reached in the flameholder arrangement. If the total cross-sectional area of the obstructions is too large, an excessive stagnation-pressure loss will occur across the flameholders.

A useful yet simple one-dimensional analysis of a ramjet combustion chamber may be made by treating the friction and heating effects as though they occurred in separate regions. The flow is not actually one-dimensional, nor do frictional and heating effects occur separately. However, these approximations do yield some insight into the overall effects of the combustion process.

First the flow through the flameholders [(2) to (3) in Fig. 7–26] is assumed adiabatic (ignoring local effects of fluid vaporization and combustion) with the friction losses that are appropriate to the chamber as a whole. The flow at entrance to and exit from the flameholder is assumed one-dimensional. Then the combustion process is considered to occur in a frictionless, one-dimensional, constant-area flow following the friction process. In this way the overall effects of the combustion process may be estimated with reasonable accuracy.

The pressure loss due to friction can be written in the form

$$\frac{p_{02} - p_{03}}{q_2} = K, \tag{7–9}$$

where q_2 is the "dynamic pressure" at station (2) defined by $q_2 = \frac{1}{2}\rho_2 u_2^2$, and K is an empirical number which may lie in the range $1 < K < 4$ in practical combustion chambers.

Equation (7–9) can be written

$$\frac{p_{03}}{p_{02}} = 1 - K\frac{q_2}{p_{02}}$$

or, using the definition of Mach number and stagnation pressure, we can write

$$\frac{p_{03}}{p_{02}} = 1 - \frac{K}{2}\frac{\gamma M_2^2}{\{1 + [(\gamma - 1)/2]M_2^2\}^{\gamma/(\gamma-1)}}. \tag{7–10}$$

Since, for convenience, we have assumed that the flow at the end of the friction process is one-dimensional, with the same cross section as at entrance, we write

$$\rho_3 u_3 = \rho_2 u_2.$$

Using the perfect gas law, we may write this as

$$\frac{p_3}{p_2} = \frac{u_2}{u_3}\frac{T_3}{T_2} \quad \text{or} \quad \frac{p_3}{p_2} = \frac{M_2}{M_3}\sqrt{\frac{T_3}{T_2}}.$$

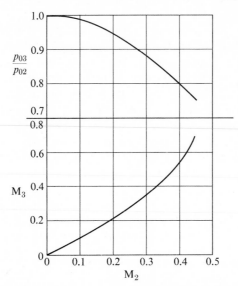

FIG. 7-28.   Flow through an idealized flameholder; $K = 2$, $\gamma = 1.4$. (Courtesy Wyatt [19].)

If the flow in the friction process is taken to be adiabatic, then $T_{02} = T_{03}$. Thus

$$\frac{p_3}{p_2} = \frac{M_2}{M_3} \left\{ \frac{1 + [(\gamma - 1)/2]M_2^2}{1 + [(\gamma - 1)/2]M_3^2} \right\}^{1/2}$$

and

$$\frac{p_{03}}{p_{02}} = \frac{M_2}{M_3} \left\{ \frac{1 + [(\gamma - 1)/2]M_3^2}{1 + [(\gamma - 1)/2]M_2^2} \right\}^{(\gamma+1)/2(\gamma-1)}. \qquad (7\text{-}11)$$

The conditions at the exit of the hypothetical adiabatic friction process can then be determined by solving for p_{03}/p_{02} and M_3 from Eqs. (7-10) and (7-11). A typical solution for the case $K = 2$ is given in Fig. (7-28).

Neglecting the enthalpy of the liquid fuel, the enthalpy of the gas at the end of the combustion process can be determined from

$$h_{04} = \frac{h_{03} + f\eta_b h_f}{1 + f}, \qquad (7\text{-}12)$$

where η_b is the combustion efficiency and f is the fuel-air ratio.

For the combustion process the lower heating value of hydrocarbon fuels may be written [19]

$$Q_R = 15,900 + 15,800\frac{H}{C} \text{ Btu/lb}_m,$$

where H/C is the hydrogen-carbon ratio given by H/C $= 1.008m/12.01n$ for a fuel represented by C_nH_m. It may be seen that most hydrocarbon fuels have approximately the same heating value.

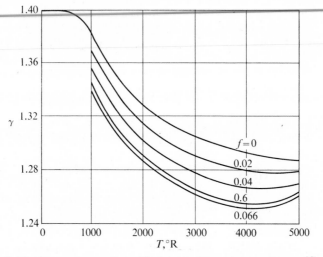

FIG. 7–29. Dependence of specific heat ratio on temperature. (Courtesy Wyatt [19].)

The gas constant R depends on the composition of the gas phase [19]:

$$R = \left[1716 + \frac{12,300f}{1 + (H/C)} \left(\frac{H}{C} \right) \right] \left(\frac{1}{1 + f} \right) \frac{\text{ft·lb}}{\text{lb·mole}} .$$

The ratio of specific heats γ depends on temperature as well as composition. The dependence is illustrated in Fig. 7–29 for C_8H_{18} burned in air.

From continuity, we have

$$\rho_3 u_3 (1 + f) = \rho_4 u_4 .$$

Using the perfect gas relation, we obtain

$$\frac{p_4}{p_3} = \frac{u_3}{u_4} \frac{R_4}{R_3} \frac{T_4}{T_3} (1 + f)$$

or

$$\frac{p_4}{p_3} = \sqrt{\frac{\gamma_3}{\gamma_4}} \frac{M_3}{M_4} \sqrt{\frac{R_4 T_4}{R_3 T_3}} (1 + f).$$

And hence we can write

$$\frac{T_{04}}{T_4} = 1 + \frac{\gamma_4 - 1}{2} M_4^2, \qquad \frac{T_{03}}{T_3} = 1 + \frac{\gamma_3 - 1}{2} M_3^2,$$

$$\frac{p_4}{p_3} = \sqrt{\frac{\gamma_3}{\gamma_4}} \frac{M_3}{M_4} \sqrt{\frac{R_4 T_{04}}{R_3 T_{03}}} (1 + f) \left\{ \frac{1 + [(\gamma_3 - 1)/2] M_3^2}{1 + [(\gamma_4 - 1)/2] M_4^2} \right\}^{1/2} . \qquad (7–13)$$

Another expression for the pressure ratio p_4/p_3 can be derived from the momentum equation for the combustion zone. Ignoring wall friction forces,

$$(p_3 - p_4)A_3 = (\rho_4 A_3 u_4)u_4 - (\rho_3 A_3 u_3)u_3$$

or

$$1 - \frac{p_4}{p_3} = \frac{p_4}{p_3}\frac{\rho_4 u_4^2}{p_4} - \frac{\rho_3 u_3^2}{p_3}.$$

Thus

$$\frac{p_4}{p_3} = \frac{1 + \gamma_3 M_3^2}{1 + \gamma_4 M_4^2}. \tag{7-14}$$

Equating the right-hand sides of Eqs. (7–13) and (7–14) yields an expression for the exit Mach number

$$\frac{\sqrt{\gamma_4}\, M_4\{1 + [(\gamma_4 - 1)/2]M_4^2\}^{1/2}}{1 + \gamma_4 M_4^2}$$
$$= (1 + f)\sqrt{\frac{R_4 T_{04}}{R_3 T_{03}}}\, \frac{\sqrt{\gamma_3}\, M_3\{1 + [(\gamma_3 - 1)/2]M_3^2\}^{1/2}}{1 + \gamma_3 M_3^2}. \tag{7-15}$$

The stagnation temperature T_{04} may be determined from Eq. (7–12), using an average value of specific heat to relate enthalpy change to temperature rise:

$$T_{04} - T_{03} = \frac{h_{04} - h_{03}}{c_p},$$

where c_p is the average specific heat of the combustion products over the temperature interval T_{03} to T_{04}.

When the exit Mach number has been determined from Eq. (7–15), the stagnation-pressure ratio can be obtained from Eq. (7–14) by using the relations

$$\frac{p_{04}}{p_4} = \left(1 + \frac{\gamma_4 - 1}{2}M_4^2\right)^{\gamma_4/(\gamma_4-1)}, \qquad \frac{p_{03}}{p_3} = \left(1 + \frac{\gamma_3 - 1}{2}M_3^2\right)^{\gamma_3/(\gamma_3-1)},$$

with the result that

$$\frac{p_{04}}{p_{03}} = \frac{1 + \gamma_3 M_3^2}{1 + \gamma_4 M_4^2}\left[\frac{\left(1 + \frac{\gamma_4 - 1}{2}M_4^2\right)^{\gamma_4/(\gamma_4-1)}}{\left(1 + \frac{\gamma_3 - 1}{2}M_3^2\right)^{\gamma_3/(\gamma_3-1)}}\right]. \tag{7-16}$$

Equations (7–15) and (7–16) permit solution of all conditions at the exit of the chamber when p_{03}, M_3 and T_{04} are known. A typical solution of the equations is drawn in Fig. (7–30).

Figures 7–28 and 7–30 suggest that considerable changes in stagnation pressure and average Mach number may take place in the combustor, in addition to large changes in the stagnation temperature. However, these results do not teach us how to design a combustion chamber. The behavior of the combustion zone is

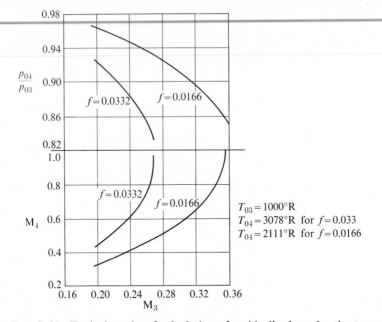

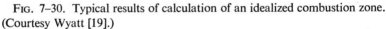

FIG. 7–30. Typical results of calculation of an idealized combustion zone. (Courtesy Wyatt [19].)

so complex that the designer relies almost wholly on experience and further experiment. Few analytical results are quantitatively useful to him in predicting flame speed, rates of turbulent mixing, atomization and mixing of fuel sprays, chemical kinetics, and other important matters.

Ramjet operation at very high ($M > 5$) Mach numbers is accompanied by a different sort of combustion problem, stemming from the very high stagnation temperatures encountered, and it is in this application that some of the features of supersonic combustion appear attractive [21, 22, 23]. With ordinary subsonic combustion the problems associated with efficient diffusion at such velocities are, of course, severe. Aside from these problems, however, the unavoidably high static temperatures (practically equal to stagnation temperatures) at the inlet to a subsonic combustion chamber substantially alter the effect of fuel injection and "combustion." At such temperatures an appreciable amount of the energy that might have been released by combustion is tied up in dissociated air and product molecules, so that the temperature rise is much less than would occur at lower inlet temperatures. Figure 7–31(a) shows the approximate effect of burning hydrocarbon fuels with a stoichiometric fuel–air ratio at a chamber pressure of 2 atmospheres [21]. The energy tied up in dissociated material is not available for the expansion and acceleration of the propellant, unless recombination can occur within the exhaust nozzle as a result of the static temperature drop of the propellant. Even if recombination does occur within the nozzle, the energy release occurs at a lower pressure and hence does not contribute as effectively to

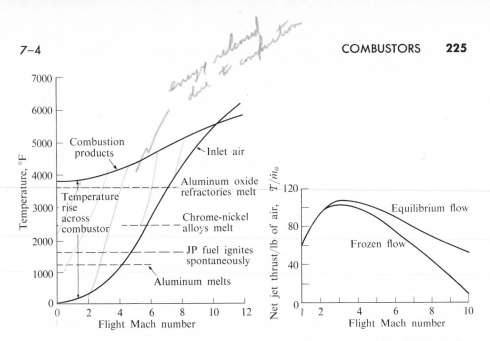

FIG. 7–31. The effect of dissociation on ramjet performance at high flight speeds. (a) Combustion temperature. (Courtesy Breitweiser [21].) (b) Thrust. (Courtesy Olson [24].)

propellant acceleration as does that released in the combustion chamber. The most favorable possibility, called *equilibrium flow*, is that for which recombination occurs as fast as temperature drop allows, so that the propellant remains in thermodynamic equilibrium throughout expansion. The least favorable possibility, called *frozen flow*, is that for which practically no recombination occurs, so that the propellant remains "frozen" at the combustion chamber composition.* Figure 7–31(b) shows the performance difference between these two extremes for the case of a stoichiometric hydrocarbon fuel–air ratio, for flight at an altitude of 100,000 feet with a constant diffuser efficiency of 0.875. Although the results are probably optimistic due to the rather high diffuser efficiency assumption (especially at very high Mach numbers), the substantial difference between equilibrium and frozen flow can be seen. In fact, it would appear that high flight speeds are impossible without flow conditions approaching equilibrium. At present, both calculation and experiment [24] indicate that, for reasonable nozzle lengths, nearly frozen flow prevails for hydrocarbon fuels and probably even for hydrogen fuel.

The main advantage of supersonic combustion is that, for a given flight Mach number, supersonic velocities within the combustion chamber are accompanied by lower static temperatures. Since the composition of combustion products is a

---

* The concepts of frozen and equilibrium flow will be discussed in greater detail later, with regard to chemical rocket applications, where they are currently of greater practical significance.

function of static temperature, this decreases the amount of dissociation. It is also true that supersonic combustion requires less inlet diffusion; hence better inlet performance can be expected. This apparent advantage is largely offset, however, by the larger stagnation-pressure loss occurring at supersonic combustion velocities, as may be seen by examination of Fig. 3–4. The existence of lower static temperature is of little significance to the wall heat-transfer problem, since this is largely a function of stagnation temperature. It may be seen in Fig. 7–31(a) that any ramjet of very high velocity will require some form of chamber wall cooling.

At present, the use of supersonic combustion must await the solution to the very difficult problem of efficient (i.e., shockless or weak-shock) fuel injection and mixing. A step in the right direction is that flames have been stabilized in supersonic streams of premixed combustibles.

Turbojet combustion chambers

As previously stated, the turbojet combustion chamber normally operates at a very low fuel–air ratio. Its configuration must be such that a small portion of the air (the primary air) is initially mixed and burned with the fuel, while the remainder of the air is mixed in further downstream. Several configurations are used to satisfy the many requirements of the combustion chamber. Early models (e.g., the Whittle type engines $W1$, $W2B$, $WR1$) employed the reverse-flow type of chamber, shown in Fig. 7–32, mainly in order to keep the engine shaft short by maintaining a small distance between the turbine and compressor. At the time this was deemed necessary in order to prevent severe whirling vibrations of the shaft. Difficulties were experienced with combustion in the reversed-flow burners, and later improvements in the design of shafting made it possible to eliminate whirling in relatively long shafts so that straight-through flow combustion chambers could be used. These chambers were easier to develop for acceptable flow distribution and combustion efficiency. Figure 7–33 shows several modern combustion chambers.

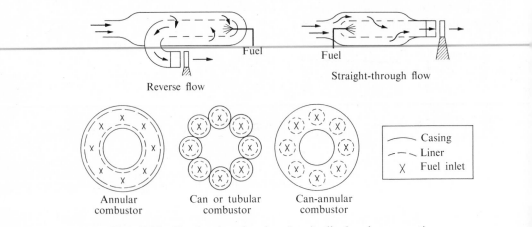

Fig. 7–32. Combustion chamber, longitudinal and cross sections.

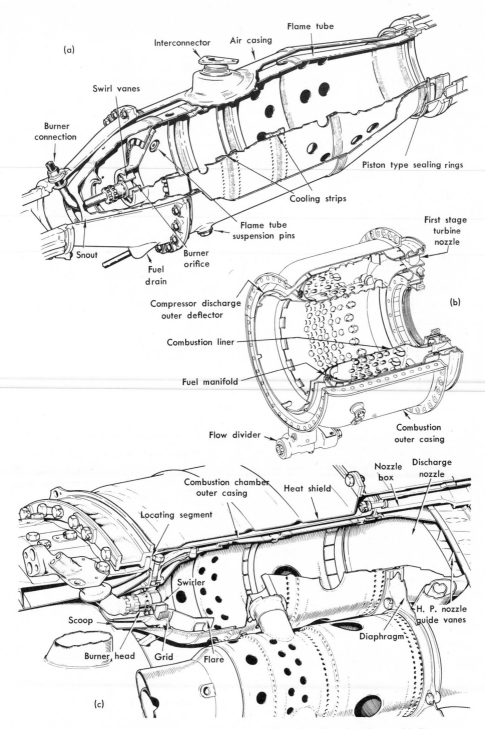

Fig. 7–33. Cutaway sketches of various types of combustion chambers. (a) Can type, Rolls-Royce Dart. (b) Annular type, General Electric T58. (Courtesy General Electric Co.) (c) Can-annular type, Rolls-Royce Tyne. [(a) and (c) courtesy Rolls-Royce.]

In every type the primary combustion is seen to occur within a perforated sheet-metal liner mounted within the outer casing. The arrangement may consist of a number of "cans" or tubular sections. Later designs have tended toward the annular arrangement, in which both casing and liner are continuous annular pieces. A compromise employing an annular casing but retaining tubular liners is presently being used. The annular arrangement obviously embodies the most efficient use of frontal area, and the can type is the least efficient in this respect. On the other hand, the can type is much more easily developed, since the single can requires much smaller test facilities. The acquisition of large test facilities is responsible for the trend toward the more attractive annular type.

Combustion chambers of these types have been developed to rather remarkable performance levels. Under favorable conditions, nearly complete combustion is attained with little pressure loss in combustion chambers of reasonable size. Operation is possible under widely varying ambient conditions, although performance of course suffers under extreme conditions.

Analysis

The techniques used in the development of a combustion chamber are largely empirical, but by approximate analysis it is possible to indicate qualitatively the effect of variations in the size of liner and casing, and in the distribution of liner holes. Grobman *et al.* [25] have performed such an analysis for a tubular combustion chamber with constant casing and liner diameters, as shown in Fig. 7–34(a).

Figure 7–34(b) is an idealized picture of the flow through a typical hole in the combustor liner. With the aid of empirical data it is possible to estimate this flow by idealizing it as a jet whose discharge coefficient and direction depend not only on the hole geometry, but also on the velocities u_A and u_L. The subscripts A and L refer to the annular and liner flow, respectively.

Considering the holes to be small, and uniformly and closely spaced, the flow in both annulus and liner may be predicted by approximate application of momentum and continuity equations. Figure 7–35 is a typical result of Grobman, *et al.*, for the case in which combustion is absent. It may be seen that, due to wall friction, the stagnation pressure in the annulus p_{0A} drops below the inlet stagnation pressure value p_{01}. The static pressure p_A in the annulus rises as fluid is transferred to the interior of the liner and the fluid remaining in the annulus decelerates.

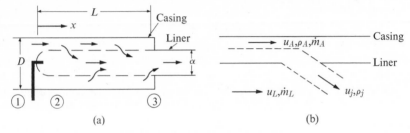

FIG. 7–34. Idealized combustor.

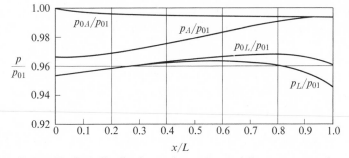

FIG. 7–35. Typical distributions of annulus and liner pressure, for cold flow. (Courtesy Grobman, *et al.* [25].)

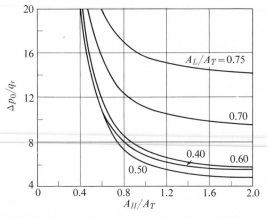

FIG. 7–36. Typical stagnation-pressure loss for cold flow. (Courtesy Grobman, *et al.* [25].)

Due to the irreversibility of the jet discharge, the value of the stagnation pressure inside the liner is significantly below p_{0A} at $x = 0$. However, p_{0L} varies somewhat with x, as the result of assumptions about the speed and magnitude of discharge from the liner holes. Initially the velocity inside the liner is low enough that the stagnation and static pressures p_{0L} and p_L are nearly identical. Toward the exit of the liner, however, the velocity is significant.

The presence of combustion would, of course, have a considerable effect on these results. The effect could be estimated by assuming an axial temperature distribution, or by more detailed assumptions concerning the rate of chemical reaction within the chamber.

Confining our attention to the aerodynamics of the cold flow only, we turn next to Fig. 7–36, which gives typical losses in stagnation pressure for the combustion chamber of Fig. 7–34. Figure 7–36 can be derived directly from the result of calculations (typically shown in Fig. 7–35) of pressure distributions for a variety of combustor area ratios. Two area ratios are significant: the ratio of the cross-sectional area A_L of the liner to the total cross section A_T, and the ratio of the total hole area A_H to the total cross section A_T. The stagnation-pressure loss Δp_0 is normalized in Fig. 7–36 by the upstream dynamic pressure q_r, which may be expressed as

$$q_r = \frac{\dot{m}^2}{2\rho_1(A_T)^2},$$

in which $\dot{m}$ is the mass flow through the combustor and ρ_1 is the inlet density.

From this figure it can be seen that Δp_0 can be greatly reduced if A_H/A_T is made as large as 1.5, but that relatively little is gained by further increase. It also appears that there is a value of A_L/A_T (lying between 0.6 and 0.4) which will produce minimum stagnation-pressure loss.

Analytical results of this kind are often useful in providing a general picture of pressure and mass flow distributions. In addition they may help the designer to decide the relative liner, casing, and hole areas. A much more complete discussion is given in Reference [25].

Altitude performance

Under favorable conditions, very intense and efficient combustion can be achieved in turbojet combustion chambers. Unfavorable conditions such as low inlet temperature and pressure, and high air velocity, are likely to occur at high altitude, where ambient temperature and pressure are low. Reference [18] presents experimental results obtained with typical combustion chambers. Figure 7–37 shows the effect of altitude on three combustion chambers. Combustion efficiency is a measure of combustion completeness, and may be defined as the actual temperature rise divided by the theoretical (complete combustion) temperature rise.

Although, as mentioned above, very high efficiency of combustion is possible under favorable conditions, at some altitude limit a given combustor will cease operating altogether. It has been proved possible, however, to build jet engine combustors capable of operation at altitudes up to 80,000 feet or more.

The separate effects of pressure, temperature, and air velocity have been investigated. The effects of pressure and temperature on combustion can be appreciated by considering the nature of the chemical process. The probability of chemical

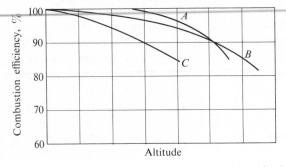

FIG. 7–37. Variation of combustion chamber efficiency with altitude, for three combustors. (Courtesy Olson, *et al.* [18].)

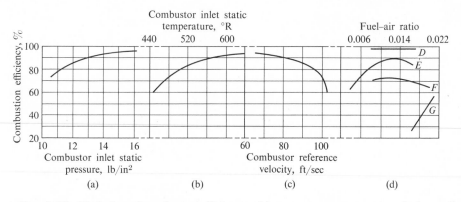

FIG. 7–38. Variation of combustor efficiency with pressure, temperature, velocity, and fuel–air ratio. (Courtesy Olson, *et al.* [18].)

combination of atoms or molecules depends not only on the frequency of molecular collisions but on the energy of the colliding molecules. Below a certain temperature (ignition temperature), practically no reaction occurs. A certain "threshold" energy level is necessary to initiate significant reaction. Combustion proceeds as a chain reaction in which energy released from one combination is sufficient to raise neighboring atoms above the threshold energy, thereby inciting further chemical combinations so that the process continues. For a given temperature, high pressure corresponds to high density and this means greater frequency of collision of fuel and oxidant atoms. Raising the temperature decreases the energy which, on the average, must be added to a particle before the reaction threshold is reached.

High combustion chamber velocity simply means that residence time within the chamber is reduced, and hence there is less chance for completion of the combustion process. Figure 7–38 shows the gross effects of these factors on a particular combustion chamber. In each part of the figure the effect of one variable is presented with fixed values of the other variables. The combustor was generally operated under conditions which produce low efficiency, so that the effects of each variable on efficiency would be obvious.

The last portion of Fig. 7–38 shows the variation of combustion efficiency with fuel–air ratios for four different combustors. A satisfactory combustor would perform as shown by curves D and E; that is, peak efficiency at the design value of f accompanied by a gradual fall-off at higher or lower values.

Owing to the great complexity of the fluid dynamic, chemical kinetic, and heat transfer problems encountered in the study of typical jet engine combustors, it has not been possible to develop satisfactory theoretical methods for designing combustors or predicting their performance. Combustion chamber design therefore relies strongly on empirical methods, and on a considerable experimental development program. This discussion has attempted only to present certain factors which aid understanding of typical combustor performance.

7–5 EXHAUST NOZZLES

From the standpoint of fluid dynamics, at least, the ramjet or turbojet exhaust nozzle is an engine component which is relatively easy to design, since the pressure gradient can be favorable all along the wall.   Considering mechanical design, however, the exhaust nozzle can be rather complex if it is necessary that the area of the nozzle be adjustable, especially if a converging-diverging (i.e., supersonic) configuration is desired.   Adjustable nozzles are often constructed of a series of pie-shaped segments which can be moved to form a conical fairing of variable outlet area.   Figure 6–11 illustrates such a construction.   In this case a kind of converging-diverging geometry is achieved by utilizing a cushion of secondary air which bypasses the engine and flows through a second nozzle to form the diverging nozzle section.   At present, converging-diverging nozzles are used only in highly supersonic applications, since at lower flight velocities the exhaust velocity would probably not be sufficiently supersonic to justify the added weight of these nozzles. Furthermore, the desire for simplicity in ramjet applications often precludes adjustable nozzles.   Even for afterburning turbojet engines, the adjustment is often just a two-position arrangement which is open for afterburning operation and closed for nonafterburning.   This adjustment is necessary in order to accommodate the very large variation in exhaust stagnation temperature, since the mass flow rate per unit area of nozzle is inversely proportional to the square root of the stagnation temperature.   [See Eq. (7–18).]

The process in the nozzle, like that in the diffuser, can be very close to adiabatic, since the heat transfer per unit mass of fluid is much smaller than the difference of enthalpy between inlet and exit.   Unlike the diffuser process, however, the nozzle process can be described quite well by the equations of one-dimensional isentropic flow.   Friction may reduce the nozzle adiabatic efficiency (Fig. 7–39), defined by

$$\eta_n = \frac{h_{04} - h_6}{h_{04} - h_{6s}}$$

to values in the range $0.95 < \eta_n < 0.98$, provided the nozzle is well designed.

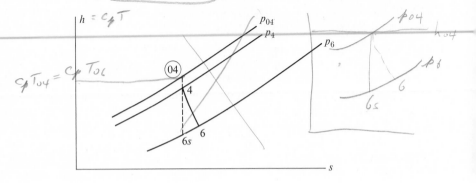

FIG. 7–39.   Definition of fluid states in a nozzle.   The numbers correspond to the turbojet nomenclature of Chapter 6.

If the expansion (of a divergent section) is too rapid, it is possible for boundary layer separation to occur, which considerably lowers the efficiency.

Since the flow is approximately adiabatic, $h_{06} = h_{04}$. Also $h_{06} = h_6 + (u_6^2/2)$. Thus the exhaust velocity may be calculated from

$$u_6 = \sqrt{2(h_{04} - h_6)} \quad \text{or} \quad u_6 = \sqrt{2(h_{04} - h_{6s})\eta_n}.$$

If the specific heat ratio γ is taken to be constant, then

$$u_6 = \sqrt{2[\gamma/(\gamma - 1)]\eta_n R T_{04}[1 - (p_6/p_{04})^{(\gamma-1)/\gamma}]}. \tag{7-17}$$

By introducing the Mach number M_6, we have $M_6 = u_6/\sqrt{\gamma R T_6}$. Equation (7–17) may be transformed to

$$M_6^2 = \frac{2}{\gamma - 1}\left\{\frac{\eta_n[1 - (p_6/p_{04})^{(\gamma-1)/\gamma}]}{1 - \eta_n[1 - (p_6/p_{04})^{(\gamma-1)/\gamma}]}\right\}.$$

If the flow, up to the throat, is assumed isentropic with nozzle entrance conditions M_4, A_4, p_{04}, T_{04}, the throat area A_t will be given by

$$\frac{A_t}{A_4} = \frac{M_4}{M_t}\frac{[1 + (\gamma - 1)/2M_t^2]^{(\gamma+1)/2(\gamma-1)}}{[1 + (\gamma - 1)/2M_4^2]^{(\gamma+1)/2(\gamma-1)}}.$$

For most cases of interest to ramjets the throat will be choked, so that $M_t = 1$ and

$$\frac{A_t}{A_4} = \frac{M_4[(\gamma + 1)/2]^{(\gamma+1)/2(\gamma-1)}}{[1 + (\gamma - 1)/2M_4^2]^{(\gamma+1)/2(\gamma-1)}}.$$

The mass flow which will pass through the choked throat is given by

$$\dot m = \frac{A_t p_{04}}{\sqrt{R T_{04}}}\sqrt{\gamma}\left(\frac{2}{\gamma + 1}\right)^{(\gamma+1)/2(\gamma-1)}. \tag{7-18}$$

If the flow is to be expanded ideally so that the pressure in the exhaust plane equals the ambient pressure, the required exhaust area may be obtained from

$$\frac{A_6}{A_4} = \frac{p_{04}}{p_{06}}\frac{M_4}{M_6}\left\{\frac{1 + [(\gamma - 1)/2]M_6^2}{1 + [(\gamma - 1)/2]M_4^2}\right\}^{(\gamma+1)/2(\gamma-1)}. \tag{7-19}$$

The ratio p_{04}/p_{06} may be obtained from the pressure ratio and the adiabatic nozzle efficiency. Writing

$$\eta_n = \frac{1 - (T_6/T_{04})}{1 - (T_{6s}/T_{04})} = \frac{1 - (T_6/T_{06})}{1 - (T_{6s}/T_{04})}$$

and recognizing that

$$\frac{T_6}{T_{06}} = \left(\frac{p_6}{p_{06}}\right)^{(\gamma-1)/\gamma}, \quad \frac{T_{6s}}{T_{04}} = \left(\frac{p_6}{p_{04}}\right)^{(\gamma-1)/\gamma},$$

it may be shown that

$$\frac{p_{04}}{p_{06}} = \frac{p_{04}}{p_6} \left\{ 1 - \eta_n \left[1 - \left(\frac{p_6}{p_{04}}\right)^{\frac{(\gamma-1)}{\gamma}} \right] \right\}^{\frac{\gamma}{\gamma-1}}.$$ (7–20)

In a practical case it may easily happen that the ratio A_6/A_4 calculated from Eq. (7–19) will exceed unity by a considerable margin. Considerations of external drag on the nacelle or of structural limitations may then suggest the desirability of reducing A_6/A_4 to near one, even though this will mean incomplete expansion.

References

1. KLINE, S. J., "On the Nature of Stall." *Jour. of Basic Engineering*, **81,** Series D, No. 3, Sept. 1959, pages 305–320

2. KLINE, S. J., D. E. ABBOTT, and R. W. FOX, "Optimum Design of Straight-walled Diffusers." *Jour. of Basic Engineering*, **81,** Series D, No. 3, Sept. 1959, pages 321–331

3. KÜCHEMANN, D., and J. WEBER, *Aerodynamics of Propulsion.* New York: McGraw-Hill, 1953

4. HANSEN, E. A., and E. A. MESSMAN, "Effects of Pressure Recovery on the Performance of a Jet-Propelled Airplane." NACA *TN*-1695, Nov. 1948

5. SHAPIRO, ASCHER H., *The Dynamics and Thermodynamics of Compressible Fluid Flow*, Volume I. New York: The Ronald Press Company, 1953

6. WU, J. H. T., "On a Two-Dimensional Perforated Intake Diffuser." *Aerospace Engineering*, **21,** No. 7, July 1962, page 58

7. FOA, J. V., *Elements of Flight Propulsion.* New York: John Wiley & Sons, 1960

8. OSWATITSCH, K., "Pressure Recovery for Missiles with Reaction Propulsion at High Supersonic Speeds" (translation). NACA *TM*-1140, 1947

9. HERMANN, R., "Supersonic Diffusers and Introduction to Internal Aerodynamics." Minneapolis-Honeywell Regulator Co., Aeronautical Div., Minneapolis, Minn.

10. SHAPIRO, ASCHER H., *The Dynamics and Thermodynamics of Compressible Fluid Flow*, Volume II, Chapter 17. New York: The Ronald Press Company, 1954

11. FERRI, A., and L. M. NACCI, "Preliminary Investigation of a New Type of Supersonic Inlet." NACA Report 1104, 1952

12. DAILEY, C. L., "Supersonic Diffuser Instability." *Jour. of the Aeronautical Sciences*, **22,** No. 11, Nov. 1955, pages 733–749

13. GORTON, G. C., "Investigation at Supersonic Speeds of a Translating Spike Inlet Employing a Steep-Lip Cowl." NACA *RM* E54G29, Oct. 11, 1954

14. TRIMPI, R. L., and N. B. COHEN, "Effect of Several Modifications to Center Body and Cowling on Subcritical Performance of a Supersonic Inlet at Mach Number of 2.02." NACA *RM* L55C16, May 20, 1955

15. OBERY, L. J., and L. E. STITT, "Investigation at Mach Numbers 1.5 and 1.7 of Twin-Duct Side Air-Intake System with 9° Compression Ramp Including Modifications to Boundary-Layer-Removal Wedges and Effects of a Bypass System." NACA *RM* E53H04, Oct. 12, 1953

16. NETTLES, J. C., and L. A. LEISSLER, "Investigation of Adjustable Supersonic Inlet in Combination with *J*34 Engine." NACA *RM E*54*H*11, Oct. 15, 1954

17. CONNORS, J. F., and R. R. WOOLLETT, "Preliminary Investigation of an Asymmetric Swept-Nose Inlet of Circular Projection at Mach Number 3.85." NACA *RM E*54626, Oct. 11, 1954

18. OLSON, W. K., J. H. CHILDS, and E. R. JONASH, "The Combustion Efficiency Problem of the Turbojet at High Altitude." *Trans. ASME,* **77,** 1955, pages 605–615

19. WYATT, D. DEM., "The Ramjet Engine," Section *E,* Chapter 1 in "Jet Propulsion Engines," Volume XII, *High-Speed Aerodynamics and Jet Propulsion,* edited by O. E. Lancaster. Princeton, N. J.: Princeton University Press, 1959

20. HENRY, J. R., and J. B. BENNETT, "Method for Calculation of Ramjet Performance." NACA *TN*-2357, 1951

21. BREITWEISER, R., "Ramjet Combustion." *Astronautics,* **4,** No. 4, April 1959, page 44

22. DUGGER, G. L., "A Future for Hypersonic Ramjets." *Astronautics,* **4,** No. 4, April 1959, page 38

23. PERCHONOCK, E., "Ramjet Trends." *Astronautics,* **4,** No. 4, April 1959, page 40

24. OLSON, W. T., "Recombination and Condensation Processes in High-Area-Ratio Nozzles," *Jour. Amer. Rocket Society,* **32,** No. 5, May 1962, page 672

25. GROBMAN, J. S., C. C. DILTRICH and C. C. GRAVES, "Pressure Drop and Air Flow Distribution in Gas Turbine Combustors." *Trans. ASME,* **79,** 1957, pages 1601–1609

Problems

1. A two-dimensional or wedge-type diffuser consists of two external oblique shocks, a normal shock, and a subsonic internal section, as shown in the figure. Assuming no external losses other than shock losses, and assuming the subsonic section efficiency is given by Fig. 7–10, what is the overall total pressure ratio at a flight Mach number of 2.5? Sketch the inlet shock system roughly to scale and label all angles.

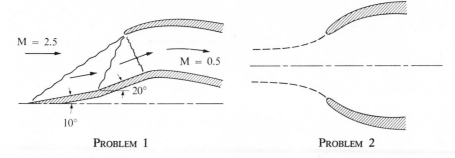

PROBLEM 1   PROBLEM 2

2. A jet engine in an aircraft flying at $M = 0.9$ ingests an airflow of 100 lb_m/sec through an inlet area of 15 ft^2. The adiabatic efficiency of the (internal) diffuser is 0.9 and the Mach number of the flow entering the compressor is 0.4. The ambient temperature and pressure are 400°R and 200 psf, respectively.

(a) What is the ratio of the inlet static pressure (at entrance to the engine intake) to the ambient pressure?

(b) What is the static pressure ratio across the internal diffuser?

(c) What fraction of the inlet dynamic pressure is converted to static pressure in the intake?

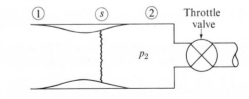

PROBLEM 3

3. The figure indicates a hypothetical one-dimensional supersonic inlet installed in a wind tunnel and equipped with a throttle valve by which the downstream static pressure p_2 might be varied. Suppose that the inlet is designed for a Mach number $M = 3.0$ and that with this flight Mach number the shock has been swallowed and an internal shock exists, as at ⑤. Neglecting all losses except those occurring in the shock, calculate and plot the shock Mach number M_s and the stagnation pressure ratio p_{02}/p_{0a} as a function of the static pressure ratio p_2/p_a (for $\gamma = 1.4$). Let p_2/p_a range from unity to well beyond that value which disgorges the shock.

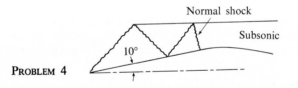

PROBLEM 4

4. The figure indicates a two-dimensional diffuser which produces no net turning of the flow. For the geometry shown, calculate the overall stagnation pressure ratio for a flight Mach number of 3.0. Neglect all losses except those occurring in the shocks. Would this diffuser be easy to start?

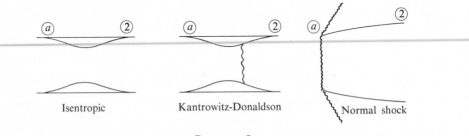

PROBLEM 5

5. Sketched are three supersonic inlets: an isentropic inlet, the Kantrowitz-Donaldson inlet of Fig. 7–16, and a simple normal shock inlet. For flight Mach numbers M from 1 to 4, calculate and plot p_{02}/p_{0a} as a function of M, with each inlet operating with best back pressure.

6. A ramjet flies at a Mach number of 2.5 in an atmosphere of 400°R. Its maximum temperature is 4000°R. Show how the overall stagnation pressure ratio of the combustor, and its exit Mach number, vary with inlet Mach number M_2 (as M_2 varies from 0 to 0.3 or until the chamber chokes). Due to friction, the static pressure loss is about twice the inlet dynamic pressure. To facilitate calculations, it may be assumed that $\gamma = 1.4$ (although Fig. 7–29 indicates that this is not a very good assumption).

7. It was shown that during starting an isentropic diffuser would experience a detached shock and consequent losses. In order to swallow the shock, a fixed-geometry diffuser must be overspeeded. However, as shown in Fig. 7–14, as the design Mach number increases, the required overspeeding increases very rapidly, so that even if the aircraft could be infinitely overspeeded, the design Mach number would be limited to a finite value. Assuming one-dimensional flow and constant γ (1.4), what is the absolute maximum *design* Mach number for which an otherwise isentropic diffuser of fixed geometry may be expected to start, assuming any amount of overspeed is possible?

8. In an afterburning jet engine, the fluid entering the afterburner is at a temperature $T_0 = 1600°R$, whether or not the afterburner is operating. The fluid temperature rises to 3000°R when the burner is operating, and the stagnation pressure is about 5% less than it is without burning. If the average specific heat in the nozzle rises from 0.27 to 0.30 Btu/lb$_m$·°R due to burning, how much relative change in exhaust nozzle area is required as the burner is ignited? It may be assumed that the nozzle is choked in both cases and that the mass flow is unchanged.

9. A ramjet has a constant-diameter combustion chamber followed by a nozzle whose throat diameter is 0.94 of the chamber diameter. Air enters the combustion chamber with $T_0 = 800°R$ and $M = 0.3$. How high may the temperature be raised in the combustion chamber without necessarily changing the chamber inlet conditions? For simplicity, neglect frictional losses in the chamber and nozzles and assume that the working fluid has the properties of air ($\gamma = 1.4$).

10. A hypothetical isentropic diffuser is being considered for deceleration of an airstream from $M = 2.0$ to $M = 0.4$. Is the boundary layer likely to separate?

8

Jet Engine
Turbomachines:
Axial Compressors

8-1 INTRODUCTION

In Chapter 6 we discussed the behavior of turbojet engines without giving
attention to the internal details of the turbomachinery. It was shown, however,
that the pressure ratio and efficiency of the turbine and compressor do have very
important effects on the performance of the engine as a whole. The purpose of
this chapter is to discuss the internal fluid mechanics of these machines, to show
how they work and what their basic limitations are. In a sense these devices are
exceedingly complex, and their behavior cannot be accurately predicted with
purely theoretical methods. However, simple analyses can provide very useful
insight and, when combined discretely with empirical information, they can
provide a satisfactory basis for design.

Fundamentally the axial compressor is limited by boundary layer behavior in
adverse (positive) pressure gradients. Each blade passage of a compressor may
be thought of as a diffuser, so that the boundary layers on all its walls are subject
to a pressure increase. Unless this pressure gradient is kept under control, separation or stall will occur. Chapter 4 showed that the two-dimensional turbulent
boundary layer in an adverse pressure gradient may be predicted theoretically
(with difficulty) for those cases in which the boundary layer remains reasonably
well behaved (i.e., not separated). However, in turbomachine blade passages the
boundary layers are generally three-dimensional, and almost no analytical techniques are available for calculating their behavior on surfaces whose geometries
are as complex as these. Thus boundary layer theory is not sufficient for the
design of such machines, and we are at present forced to rely on experimental
measurements of the performance of blade rows in cascade wind tunnels. This
necessary combination of empiricism and theory in the analysis and design of

axial compressors need not be considered an unwholesome feature. In fact, it is an essential aspect of almost all engineering activity; thus the study of how formalized theory and experience can work together in a problem of this kind is of very general interest.

It will be shown that, while the behavior of a single-stage machine is relatively easy to understand, the combination of many similar stages to make a multistage compressor or turbine raises a number of problems. Still more problems are introduced when one multistaged machine is driven by another, e.g., the turbine–compressor combination in a jet engine. Others occur when the combination is called on to operate under conditions different from those for which it was designed. A number of these effects will be taken up in the following pages.

Turbomachines are often referred to as "dynamic" fluid machines, since they cannot operate unless their working fluid has finite momentum. The volumetric or positive-displacement type of compressor, by contrast, can transfer energy to its fluid without changing its momentum. For the mass flows and pressure ratios required in aircraft jet engines, the dynamic machine is generally very much lighter than an equivalent volumetric one. For this reason the next two chapters are restricted entirely to turbomachines.

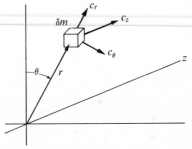

FIG. 8–1. Motion of a single particle in a cylindrical coordinate system.

8–2 ANGULAR MOMENTUM

The concept of angular momentum is very useful for describing the dynamics of turbomachinery. Consider the motion of single particle of fluid of mass δm, shown in Fig. 8–1. For convenience a cylindrical coordinate system (r, θ, z) is specified in which radial, tangential, and axial particle velocity components are c_r, c_θ, and c_z, respectively. The angular momentum of the particle about the z-axis is simply $\delta m r c_\theta$. Newton's law for the particle may be written

$$\mathbf{F} = \delta m \frac{d\mathbf{c}}{dt}, \tag{8–1}$$

in which

$$\mathbf{F} = \mathbf{r}F_r + \boldsymbol{\theta}F_\theta + \mathbf{z}F_z, \qquad \mathbf{c} = \mathbf{r}c_r + \boldsymbol{\theta}c_\theta + \mathbf{z}c_z,$$

and $(\mathbf{r}, \boldsymbol{\theta}, \mathbf{z})$ are unit vectors in the radial, tangential, and axial directions. (The use of $\mathbf{c}$ for velocity will be clarified later.) The unit vectors $\mathbf{r}$ and $\boldsymbol{\theta}$ may change direction so that their time derivatives are not zero, even though their magnitudes are constant. In fact,

$$\frac{d\mathbf{r}}{dt} = \mathbf{\theta}\frac{d\theta}{dt} \quad \text{and} \quad \frac{d\mathbf{\theta}}{dt} = -\mathbf{r}\frac{d\theta}{dt}.$$

Thus Eq. (8-1) can be resolved into three component equations:

$$\mathbf{r}: \quad F_r = \delta m \left(\frac{dc_r}{dt} - c_\theta \frac{d\theta}{dt} \right), \qquad (8\text{-}2)$$

$$\mathbf{\theta}: \quad F_\theta = \delta m \left(\frac{dc_\theta}{dt} + c_r \frac{d\theta}{dt} \right), \qquad (8\text{-}3)$$

$$\mathbf{z}: \quad F_z = \delta m \left(\frac{dc_z}{dt} \right). \qquad (8\text{-}4)$$

Multiplying by r, we can write Eq. (8-3) as

$$rF_\theta = \delta m \left(r\frac{dc_\theta}{dt} + c_r r\frac{d\theta}{dt} \right).$$

Since $c_r = dr/dt$, and $c_\theta = r(d\theta/dt)$, this expression may be written

$$rF_\theta = \delta m \frac{d}{dt}(rc_\theta). \qquad (8\text{-}5)$$

The product rF_θ is called the *torque* acting on the particle with respect to the z-axis. The right-hand side of Eq. (8-5) is the time rate of change of the angular momentum of the particle. If no torque acts on the particle, then the product rc_θ must obviously be constant. The angular momentum rc_θ per unit mass of fluid about any given axis may be considered a property of the fluid, just as are velocity and kinetic energy.

As we mentioned in Chapter 2, when one is dealing with fluid flows it is usually much more convenient to write Newton's laws as they apply to control volumes than as they apply to particular portions of matter. The next step, then, is to transform Eq. (8-5) into a form which is applicable to a control volume. Consider the general control volume shown in Fig. 8-2, with a fixed z-axis which may, or may not, intersect the control surface. By using reasoning exactly analogous to that which led to the derivation of Eq. (2-4), we can show that the total torque $\sum\tau$ acting on the control volume is given by

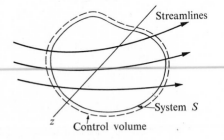

FIG. 8-2. General control volume.

$$\sum\tau = \frac{d}{dt}\int_{cv} \rho r c_\theta \, dv + \int_{cs} \rho r c_\theta (\mathbf{c} \cdot \mathbf{n}) \, dA,$$

where $\mathbf{n}$, as before, is a unit vector pointing outward from the control surface. The first integral, extending throughout the control volume, represents the rate of change of angular momentum stored within the control volume. The second integral, extending over the entire control surface, is the net outward flux of angular momentum for the control volume.

The internal conditions of turbomachines are actually unsteady, but generally the unsteadiness is periodic and of high frequency so that, on the average,

$$\sum \tau = \int_{cs} \rho r c_\theta (\mathbf{c} \cdot \mathbf{n}) \, dA. \tag{8–6}$$

Since flow through rotating machinery is generally axisymmetric as well as steady, a further simplification (and specialization) of this result is convenient for our purposes. Consider a cylindrical coordinate system set up along the axis of a rotating machine, as shown in Fig. 8–3. The control volume illustrated is intended to enclose one rotating element (i.e., one row of blades or vanes) of a turbomachine including the blades, disc, and shaft within the volume. The control surface contains inlet and outlet flow areas along with sides, usually chosen at the inner surface of the rotor housing, and ends chosen to isolate a single machine element. The control surface intersects the shaft, through which torque may act on the control volume.

The cylindrical coordinates are axial position z, radius r, and angle θ (which is here defined as positive in the direction of rotation of the machine Ω).

Consider an incremental inlet area of radius r_1 and radial width dr, through which mass flows at the rate $d\dot{m}$. The conditions of steady, axially symmetric flow enable us to trace an incremental control volume which encloses the fluid

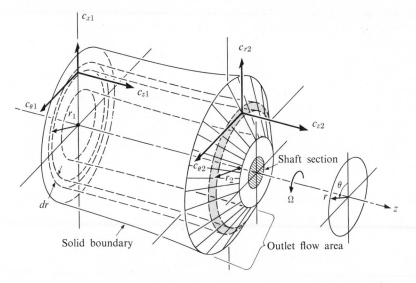

FIG. 8–3. Special control volume for application to axisymmetric flow through a rotor.

entering through this area, until it emerges at r_2. Furthermore, the axial symmetry means that all fluid entering at r_1 will have the same c_{z_1}, c_{r_1}, c_{θ_1}, and similarly, the outlet velocity is independent of θ.

The equation of angular momentum [Eq. (8-6)] for steady flow through this incremental control volume requires that the incremental torque acting on it be proportional to the incremental mass flow rate times the change in the fluid's angular momentum per unit mass:

$$d\tau = d\dot{m}(r_2 c_{\theta 2} - r_1 c_{\theta 1}),$$

where $d\tau$ is the sum of the torques acting on the inner and outer surfaces of the angular stream tube.

The total torque acting on the control volume is the sum of these incremental torques, which may be expressed in terms of integrals over the inlet and exit flow areas, in the form

$$\sum \tau = \int_{A_2} (rc_\theta)\rho c_n \, dA - \int_{A_1} (rc_\theta)\rho c_n \, dA, \tag{8-7}$$

where c_n is the component of velocity normal to the area dA. Alternatively, Eq. (8-7) may be obtained directly from Eq. (8-6) without the above intermediate step.

If the flow is such that rc_θ is a constant (so-called "free-vortex" flow) at each area, or if a "suitable average" is taken, the equation may be integrated to

$$\sum \tau = \dot{m}[(rc_\theta)_2 - (rc_\theta)_1]. \tag{8-8}$$

For preliminary work, the conditions at the mean inlet and outlet radii often represent suitable averages. Actual design procedure very often begins with a mean radius analysis, after which there is consideration of radial variations from which refined estimates of torque and power requirements are made.

The axial compressor derives its name from the fact that the air being compressed has very little motion in the radial direction. In contrast, the radial motion

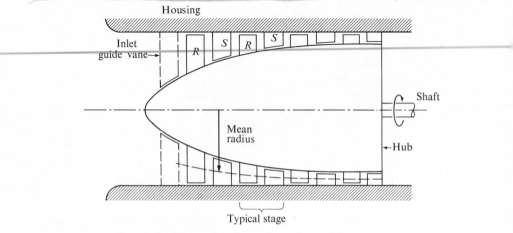

FIG. 8-4. Schematic of a section of an axial compressor.

of the air in a centrifugal compressor is much larger than the axial motion. These machines have quite different characteristics, as the following chapter will show. In general the axial machines have much greater mass flow but much less pressure ratio per stage, assuming two machines with about the same overall diameter. The reason why the axial machine has smaller pressure rise per stage will be explained by reference to the boundary layer behavior. The reason why it may have greater mass flow for a given maximum diameter is simply that its inlet area can be a much greater fraction of its total frontal area.

Figure 8–4 shows a section along the axis of a typical axial compressor. An axial compressor consists of rotating elements of rather closely spaced blades mounted between similar stationary elements. Each rotating element is called a rotor (R), while each stationary element is called a stator (S). One rotor together with the following stator makes up a single stage of the machine. Before the first stage (farthest upstream), some machines employ inlet guide vanes, for reasons which will be discussed later.

This particular sketch shows a machine of constant housing diameter. Actual machines may be made this way, or with constant mean blade diameter, or with constant hub diameter, or with all three varying. However, the mean blade radius does not usually change very much. The blade length usually varies in order to accommodate the variation in air density, so that the axial velocity component will be approximately uniform.

Figure 8–5 is a cutaway sketch of the compressor in the General Electric J85 turbojet engine. This particular compressor consists of eight stages and has a

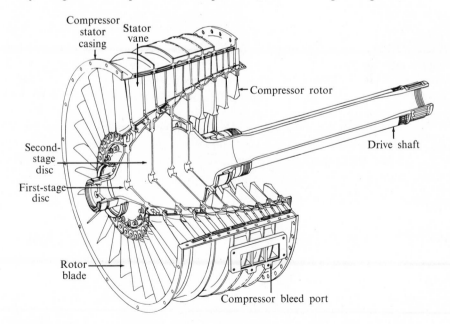

FIG. 8–5. Cutaway sketch of a typical axial compressor assembly: the General Electric J85 compressor. (Courtesy General Electric Co.)

Fig. 8–6. Rolls-Royce Conway RCo12 low-pressure and high-pressure compressor–turbine rotating assembly. (Courtesy Rolls-Royce.)

pressure ratio of about 6.5:1. Each rotor blade row is mounted on a separate disc. Figure 8–6 is a photograph of the complete compressor-turbine shaft assembly for the Rolls-Royce Conway RCo12 engine. It consists of two distinct compressors mounted on independent coaxial shafts and powered by separate turbines. The fact that one compressor can be operated at speeds different from the other has distinct advantages. This design variation and others, including the bleed port of Fig. 8–5 and variable-geometry stator configurations, are necessitated by certain off-design (especially starting) compressor characteristics. We shall discuss these in greater detail after we consider the dynamics of a simple single stage.

8–3 SINGLE-STAGE AXIAL COMPRESSORS

The velocity of an object, such as a fluid "particle," for example, is dependent on the motion of the observer as well as on that of the object. Within a turbomachine, observations may be considered from two reference frames of importance: the *absolute* reference frame, which is fixed to the frame of the machine (test bed or aircraft frame, for instance), and the *relative* reference frame, which rotates with the rotor of the machine. Fluid velocities observed from the absolute reference frame are denoted by c and are called *absolute velocities*, while those viewed from the relative reference frame are denoted by w and are called *relative velocities*. The vector difference between c and w at any point is simply the local blade velocity, which will here be denoted as U, and the three quantities are, by definition, related by the vector addition:

$$c = U + w.$$

The graphical solution of this equation results in a triangular figure called the *velocity triangle*.

Suppose we begin our study of the dynamics of a single axial compressor stage with an analysis of the velocity triangles at the mean radius. If an approxi-

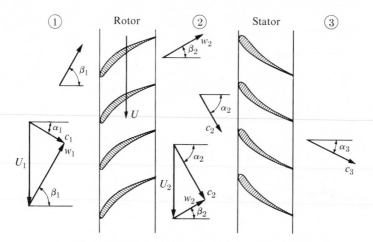

FIG. 8-7. Mean radius section of a compressor stage.

mately cylindrical surface at the mean radius of the typical stage of Fig. 8-4 is "unwrapped," part of it would appear as in Fig. 8-7, which indicates both absolute and relative fluid velocity vectors at the entrance and exit of each blade row of the stage. Absolute flow angles are denoted by α, while β denotes the direction of the flow relative to the rotor. Velocity triangles are constructed at ① and ② to show the relation of absolute and relative velocities.

The angular momentum of the fluid is changed, as it travels through a compressor blade row, as the result of pressure forces on the blade surface. The pressure on the convex surface is generally relatively low and the pressure on the concave side is generally relatively high. For this reason the convex and concave surfaces are customarily called the *suction* and *pressure* sides, respectively, of the blade.

The absolute velocity of the flow entering the rotor is $\mathbf{c}_1$. The velocity relative to the rotor $\mathbf{w}_1$ is then derived by vector subtraction of the rotor velocity, as indicated by the velocity triangle. The flow turns in the blade passage to the new relative velocity vector $\mathbf{w}_2$. The absolute velocity $\mathbf{c}_2$ at the rotor exit or stator entrance is obtained by adding this relative velocity to the rotor speed, as shown by the second vector triangle. It may be seen that in this case the rotor imparts angular momentum to the air, since the component of the absolute velocity in the tangential direction increases. In both rotor and stator, the velocity inside the blade passage (relative to the blades) usually decreases; hence the pressure rises in each blade row. Since U_1 and U_2 are equal in an axial machine (because there is negligible radial displacement of a given streamline), the velocity triangles may be superimposed as in Fig. 8-8, in a form which is convenient for visualizing the change in absolute tangential velocity, Δc_θ. The axial velocity component c_z usually changes very little through the stage. If it were assumed that each fluid particle passing through the stage had the velocity vectors shown in this figure, then the torque and power required to drive the rotor could be easily determined. Actually the blade and flow angles usually vary in the radial direction. This assumption therefore will not generally lead to accurate computation unless the inner and outer radii of the flow annulus are quite close to each other; or unless, of course, the product $r \, \Delta c_\theta$ actually remains constant. But, assuming (as an approximation) that all the mass flow through the stage undergoes the velocity changes indicated in Fig. 8–8, the torque acting on the fluid within the rotor may be determined from Eq. (8–8) applied to a control volume which contains only the rotor. That is,

$$\tau = \dot{m}[(rc_\theta)_2 - (rc_\theta)_1].$$

The power required to drive the stage is $\tau\Omega$, and according to our thermodynamic convention (Chapter 2), the power output from the machine is negative:

$$\mathcal{P}_s = \tau\Omega = -\dot{m}\Omega[(rc_\theta)_2 - (rc_\theta)_1].$$

Since Ωr equals the blade speed U, and since U is constant, we have

$$\mathcal{P}_s = -\dot{m}U(c_{\theta 2} - c_{\theta 1}). \qquad (8\text{–}9)$$

The work per unit mass w_c done on the fluid by the rotor is

$$w_c = -\frac{\mathcal{P}_s}{\dot{m}} = U(c_{\theta 2} - c_{\theta 1})$$

or

$$w_c = U \, \Delta c_\theta. \qquad (8\text{–}10)$$

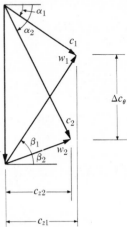

FIG. 8–8. Velocity triangles for a compressor stage.

The work done by the stator on the fluid is zero, of course (even though $\Delta c_\theta \neq 0$), because $U = 0$. There is, however, a torque on the stator (opposite to that on the fluid) of magnitude

$$\tau = \dot{m}[(rc_\theta)_3 - (rc_\theta)_2].$$

The flow at stations ①, ②, and ③ is not truly steady, since the wake of each passing blade will cause a fluctuation in velocity at those points. However, this unsteadiness is periodic (and of high frequency) so that, on the average, conditions are steady. The flow between those stations is close to adiabatic, since even at the high-temperature end of the compressor the heat transfer from the air (per unit mass of air) is negligible relative to the work transfer. Assuming uniform stagnation enthalpy in the radial direction, we apply Eq. (2–8) over the rotor and find that

$$0 = \dot{m}(h_{02} - h_{01}) + \mathcal{P}_s,$$

where $\mathcal{P}_s$ is shaft power. With Eq. (8–9), then, we obtain

$$h_{02} - h_{01} = U(c_{\theta 2} - c_{\theta 1}). \qquad (8\text{–}11)$$

If the specific heat c_p is assumed constant, a nondimensional stagnation-temperature rise across the rotor may be derived from Eq. (8–11)

$$\frac{\Delta T_0}{T_{01}} = \frac{U \, \Delta c_\theta}{c_p T_{01}}.$$

(8–12)

Using the blade angles defined in Fig. 8–8, we have

$$\Delta c_\theta = c_{\theta 2} - c_{\theta 1} = (U - c_{z2} \tan \beta_2 - c_{z1} \tan \alpha_1).$$

Thus

$$\frac{\Delta T_0}{T_{01}} = \frac{U^2}{c_p T_{01}} \left[1 - \frac{c_{z1}}{U} \left(\frac{c_{z2}}{c_{z1}} \tan \beta_2 + \tan \alpha_1 \right) \right].$$

(8–13)

Since the flow is adiabatic and no work is done on the fluid as it passes through the stator row, $T_{03} = T_{02}$, so that this ΔT_0 represents the entire stagnation-temperature rise across the stage.

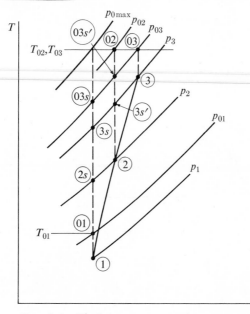

Fig. 8–9. Single-stage compression process.

The isentropic efficiency of the stage does not enter into Eq. (8–13), since the work done on the fluid depends only on the kinematic changes in the flow. The stagnation-pressure rise across the blade stage does depend strongly on the efficiency, however. Figure 8–9 shows on the temperature-entropy plane the path of the compression process through the stage. If the whole process were isentropic, the final stagnation pressure for the same work input would be $p_{0 \, \text{max}}$ as shown. However, due to losses in the rotor and stator, $p_{03} < p_{02} < p_{0 \, \text{max}}$.

If the stage efficiency η_{st}, defined by

$$\eta_{st} = \frac{h_{03s} - h_{01}}{h_{03} - h_{01}},$$ (8–14)

were known, the stagnation-pressure ratio across the stage could be determined, using the isentropic relationships, from

$$\frac{p_{03}}{p_{01}} = \left(1 + \eta_{st}\frac{\Delta T_0}{T_{01}}\right)^{\gamma/(\gamma-1)}.$$ (8–15)

It is, of course, desirable to maintain a high stage efficiency in order that the stage work input, represented by ΔT_0, be accompanied by a suitable stagnation-pressure rise. It is entirely possible to achieve a very high work input per stage, but it is *not* possible to achieve a high $\Delta T_0/T_{01}$ while maintaining high efficiency. It will be shown that in a single-stage axial compressor the temperature rise for efficient operation is always limited to values such that $(\Delta T_0/T_{01}) \ll 1$. Thus it is quite acceptable to approximate Eq. (8–15) by

$$\frac{p_{03}}{p_{01}} = 1 + \frac{\gamma}{\gamma - 1}\,\eta_{st}\frac{\Delta T_0}{T_{01}}.$$ (8–16)

We shall present an approximate method of estimating stage efficiency after we have shown how the stage pressure rise of an axial compressor is limited by the behavior of the boundary layer.

Within the limitations posed by deteriorating efficiency, it is, of course, highly desirable to achieve high work input per compressor stage in order to minimize the total number of stages necessary to provide the desired overall compressor-pressure ratio. Since the axial velocity c_z is proportional to the mass flow rate $\dot{m}$, Eq. (8–13) shows how the work input of a given stage varies with mass flow rate and blade speed U. As will be shown, the relative angle of the velocity *leaving* any blade row is reasonably constant over a range of compressor operating conditions. Thus, in Eq. (8–13), the angles β_2 and α_1 may be treated as constants, leaving only blade speed and axial velocity as variables influencing the work input. Obviously high blade speeds and low axial velocity (or mass flow rate) contribute to high work input. Blade speed may be limited by stress considerations or, more likely, by Mach number difficulties, as we shall see.

Equation (8–13) shows that for constant U the work per pound of air decreases linearly with increasing mass flow rate. Thus a slight increase of mass flow rate would be accompanied by decreased work and, probably, decreased pressure ratio, tending to restore the mass flow rate to a lower value. The opposite is true for a slight decrease of mass flow rate.

It is also instructive, in order to see the effect of rotor turning, to derive the expression for stagnation-temperature rise in terms of rotor blade angles alone.

For this purpose we write:

$$c_{\theta 2} - c_{\theta 1} = (U - c_{z2} \tan \beta_2) - (U - c_{z1} \tan \beta_1) = c_{z1} \tan \beta_1 - c_{z2} \tan \beta_2,$$

and

$$\frac{\Delta T_0}{T_{01}} = \frac{U^2}{c_p T_{01}} \frac{c_{z1}}{U} \left(\tan \beta_1 - \frac{c_{z2}}{c_{z1}} \tan \beta_2 \right). \qquad (8\text{–}17)$$

rotor blade angles

Thus for *fixed* β_1, β_2, and c_{z2}/c_{z1}, the stagnation-temperature rise is proportional to c_{z1}/U. Careful distinction should be made between the seemingly contradictory observations from Eqs. (8–13) and (8–17) on the effect of c_{z1}/U on stage work. Equation (8–13) refers to the variation of work for a *given machine* (i.e., α_1 and β_2 are constant) as a function of axial velocity. Equation (8–17), on the other hand, treats a *design variation* in which β_1 and β_2 are held constant (and for which α_1 must therefore vary as differential axial velocities are considered). Thus Eq. (8–17) shows that for high stage work it would be well to consider high c_z/U in the *design* of the machine, whereas Eq. (8–13) shows how the stage work will change with c_z/U for *fixed* design geometry.

Equation (8–17) also points out that a large turning ($\beta_1 - \beta_2$) within the rotor leads to high work per stage, but we shall show that this possibility is limited by boundary layer behavior.

Pressure-rise limitations

Before applying these equations in an attempt to design a machine with large stage work, we must realize the limitations on the various terms. It has already been mentioned that each blade passage in the stage usually behaves as a diffuser. Therefore the boundary layer on the end walls will be subjected to an adverse (positive in the flow direction) pressure gradient which could cause separation. A two-dimensional turbulent boundary layer on a flat surface will separate if the pressure increase in the flow direction exceeds a certain value of the pressure coefficient C_p defined by

$$C_p = \frac{\Delta p}{\frac{1}{2}\rho w_i^2},$$

where w_i is the free-stream velocity (relative to the surface in question) at the point where the boundary layer begins to grow, and Δp is the static pressure rise from point (i) to the point where C_p is evaluated. The value of the pressure coefficient corresponding to separation lies in the range $0.4 < C_p < 0.8$, depending largely on whether the pressure rise is sudden (as in a shock) or gradual, and on the initial condition of the boundary layer.

The boundary layers in turbomachine blade passages are generally considerably more complex than the two-dimensional one on a flat plate. Nevertheless, they too will be subject to separation and, in the absence of better information, many designers limit the static pressure rise in a given blade row to $C_p < 0.6$. For the typical rotor of Fig. 8–4, the pressure coefficient is given by

$$C_p = \frac{p_2 - p_1}{\frac{1}{2}\rho_1 w_1^2} \cdot = 1 - \frac{w_2^2}{w_1^2} \tag{8–18}$$

Since Bernoulli's equation will hold for fluid on a streamline in the blade passage, we write $C_p = 1 - (w_2/w_1)^2$. Similarly, for the stator, we have $C_p = 1 - (c_3/c_2)^2$. Thus a limitation on C_p directly implies a limitation on the relative velocity reduction within a blade row.

Another limitation stems from the effects of compressibility on the flow over the blade surfaces. Compressibility limits the relative velocity at the inlet to any blade row, and this limitation is especially important for the first stage of the machine. As the Mach number of the flow incident to a blade increases, the Mach number at some point near the blade surface will exceed unity. With further increase in the incident Mach number, shock waves will form on the blade surface, inducing separation of the boundary layer, raising the airfoil drag and decreasing the lift. If the blade is part of a blade row, still further increases in inlet Mach number will lead to choking of the flow. In this way compressibility at least limits the mass flow through a given stage. However, to avoid excessive aerodynamic losses, it is usually desirable to keep the incident Mach number well below the choking limit. The incident Mach number for which drag losses begin to increase rapidly depends strongly on the shape of the airfoil. It usually lies in the range of

$$0.6 < \mathrm{M}_{1\,\mathrm{rel}} < 0.85, \qquad \text{where} \qquad \mathrm{M}_{1\,\mathrm{rel}} = \frac{w_1}{a_1}$$

and a is the sonic speed. Efficient operation in the upper limit of this range requires thin airfoils. Some compressors—called *transonic compressors*—are designed to operate with supersonic flow over part of the blade surface, in spite of compressibility effects.

The advantage of high Mach number operation (provided reasonable efficiency can be maintained) is that the mass flow per unit area and the pressure ratio per stage will both be high. The pressure coefficient and inlet Mach number limitations together limit the stage pressure ratio. Using the definition of Mach number and the perfect gas law, we have $\mathrm{M}_{1\,\mathrm{rel}}^2 = \rho_1 w_1^2/\gamma p_1$. With this value, we can write Eq. (8–18) as

$$\frac{p_2}{p_1} = 1 + C_p \frac{\gamma \mathrm{M}_{1\,\mathrm{rel}}^2}{2}.$$

Using $C_p = 0.6$, $\mathrm{M}_1 = 0.8$ and $\gamma = 1.4$, we obtain $p_2/p_1 = 1.27$. If the same values hold for the stator, the stage pressure ratio will be

$$\frac{p_3}{p_1} = \left(1 + C_p \frac{\gamma \mathrm{M}_1^2}{2}\right)^2 = 1.6.$$

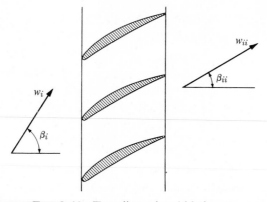

FIG. 8–10. Two-dimensional blade row.

Most compressor stages have pressure ratios much less than this limit. Turbine stages, on the other hand, operate with pressure ratios as high as three without encountering boundary layer losses because the pressure is generally falling in the flow direction. In practice, the pressure ratio across rotor and stator is considerably less than this limit because of the problems of operating the compressor at flows and speeds for which it was not designed. The Mach-number limit is likely to be reached only at the tip of the first-stage rotor blade, where the relative velocity is high and the speed of sound is low. As the air is compressed, the temperature, and thus the speed of sound, increases and, unless the blade speed or the axial velocity components are increased from stage to stage, the Mach number decreases. Thus the actual pressure ratio per stage is quite low, and real machines require many stages for high pressure ratios.

In addition to high work input, high axial velocity c_z means high flow per unit area, and hence small engine diameters for given mass flows. However, as we have seen, high losses result from high velocities relative to the blades, so that some sort of compromise must be made between size and efficiency. At present, designers specify axial velocities of the order of 500 ft/sec. It is advantageous to have low velocity at the compressor outlet in order to reduce outlet diffuser losses. However, in most compressors the axial velocity does not decrease a great deal from stage to stage.

The limiting pressure coefficient can be rather easily expressed for a simple case in terms of blade angles. Consider a two-dimensional incompressible flow, as in Fig. 8–10, in which the blades may be either rotor or stator blades.* For this flow, continuity requires that the velocity component c_z normal to the cascade be the same at inlet and outlet.

---

* The inlet and outlet of a general blade row will be denoted as i and ii, respectively, to distinguish them from the inlet and outlet of a compressor stage. To emphasize the fact that it is the flow *relative to the blades* that is of importance, w and β will be used for velocity and flow angle, with the understanding that they are automatically c and α if the expressions are applied to a stator.

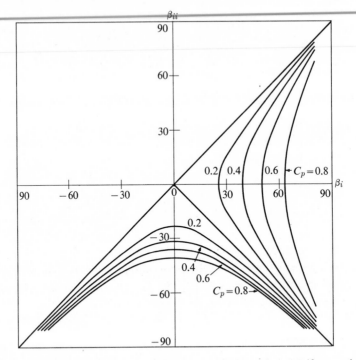

FIG. 8–11. Pressure coefficient as a function of blade angles. Uniform axial velocity. (Courtesy M.I.T. Gas Turbine Laboratory.)

The assumptions are not as restrictive as they may sound, since the axial velocity c_z varies little in an actual machine, and the pressure ratio per blade row is very low so that density change in the single stage is small. Also, by definition, the radial velocity must be very low in an axial compressor. Since the axial velocity is constant, $w_i \cos \beta_i = w_{ii} \cos \beta_{ii}$, and therefore

$$C_p = 1 - \frac{\cos^2 \beta_i}{\cos^2 \beta_{ii}}.$$

Curves of constant C_p may be plotted as a function of β_i and β_{ii}, as in Fig. 8–11. This analysis is good for either stator or rotor if w and β are taken to be the flow velocity and angle *relative to the blade*. In Fig. 8–11 the region of interest for axial compressors is limited, for positive values of β_i, to positive values of β_{ii}, in order to avoid excessive absolute velocities at station (*ii*).

To reduce compressibility effects, the relative Mach number at the inlet to the first rotor can be reduced by the addition of inlet guide vanes to give the air swirl in the direction of rotation, as shown by Fig. 8–12. In addition to reducing w_1 however, the inlet guide vanes impart a velocity component $c_{\theta 1}$ in the direction of rotation which, as Eq. 8–9 shows, reduces the stage work. Figure 8–12 shows

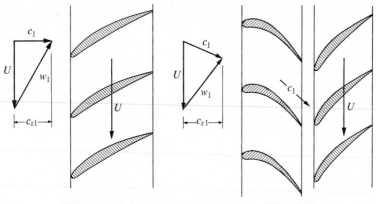

Without guide vanes   With guide vanes

FIG. 8–12. Effect of inlet guide vanes with constant mass flow and blade speed.

that the inlet guide vanes act as nozzles rather than diffusers, since the flow in their passages accelerates. Hence there should be no boundary layer difficulty within the inlet guide vanes.

Cascade aerodynamics

If the actual fluid angles are known for any given blade row, as has been assumed in the above derivations, it is a rather simple matter to determine the stagnation-temperature rise per stage. The problem is to relate the fluid angles to the blade geometry. At one time the aerodynamic performance of blade rows was calculated from plentifully available data on single airfoils. More recently the results of measurements in cascade wind tunnels have become available, and are used widely in design. A cascade is an array of blades similar to the blade row in the machine, except that it is usually two-dimensional, as shown by Fig. 8–13. Measurements from two-dimensional cascades can be directly useful for rotating machines, though, as we shall see later, corrections are often made for radial effects. The variables which define the cascade geometry are the following: (1) blade shape, (2) solidity C/s, (3) stagger angle λ; in which the blade chord C, spacing s, and stagger angle λ are defined by Fig. 8–13.

Specification of the blade shape requires that the camber (or mid-thickness) line be specified, as well as the distribution of blade thickness along the camber line. Often the camber line is a circular arc or part of a parabola. In the following discussion, reference will be made to results of two-dimensional cascade tests on a series of blades developed by the National Advisory Committee for Aeronautics (now NASA) and known as the 65 Series Cascade Data. Each blade in the series is specified by a design lift coefficient and a maximum thickness–chord ratio. A convention for identification of the blades is as follows:

$$65(x)y \qquad \text{or} \qquad 65 - x - y,$$

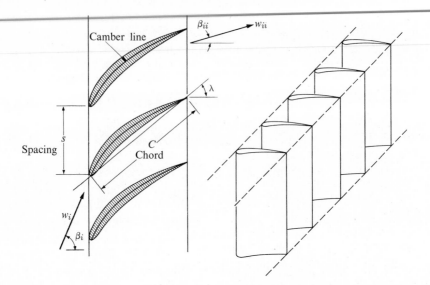

FIG. 8–13. Geometry of a rectilinear cascade.

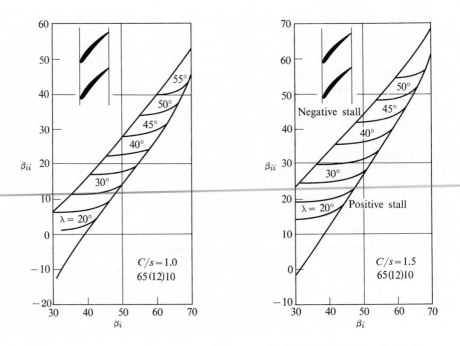

FIG. 8–14. Typical results of cascade tests on the NACA 65 Series of airfoils.

where x is ten times the design lift coefficient and y is the maximum thickness–chord ratio. The lift coefficient will be defined subsequently.

If the blade geometry is fixed, it is usually assumed that the mean exit angle β_{ii} and the stagnation-pressure loss coefficient ξ, defined by

$$\xi = \frac{\Delta p_0}{\frac{1}{2}\rho w_i^2},$$

are dependent only on β_i. That is, Mach-number and Reynolds-number effects can be neglected over suitable ranges of both these parameters.

The number of variables present in a series of tests for even one type of airfoil means that the results must be presented in a series of curves. Mellor [1] has correlated data for the NACA 65 Series of airfoils in the following way. For a given blade and a particular solidity (C/s), curves of β_{ii} versus β_i are plotted for a series of values of stagger angle λ. Then, for other values of solidity, the plot is repeated, giving a series of plots covering all possible variables for that particular blade shape. Similar series are found for other blade shapes. Further family relations may be presented if the blade shape is considered to be made up of separately specified camber line, thickness distribution, and maximum thickness, but this is just an organizational refinement. Typical results are shown in Fig. 8–14, where blade shape is specified by the array of numbers appearing on each curve, and approximate appearance is indicated in the inset. With the proper curve of this type, β_{ii} can be found as a function of β_i. It may be seen that, within a certain range of β_i, β_{ii} is nearly constant for fixed C/s and λ. However, the range of possible angles β_i is restricted by "stall" limits. Stall is the result of boundary layer separation on the blade surface.

Consider, for example, a cascade operating under normal conditions, as in Fig. 8–15. For normal operation, increasing β_i does not change β_{ii} substantially, but

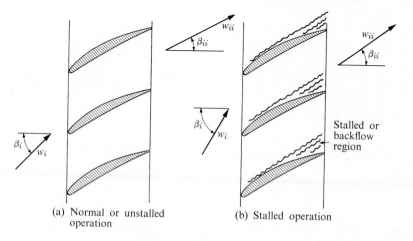

(a) Normal or unstalled operation (b) Stalled operation

FIG. 8–15. Cascade stall.

it does raise the pressure coefficient. When this is raised to an intolerable value, boundary layer separation takes place. The separated region is the scene of a large amount of turbulent mixing action which leads to marked stagnation-pressure loss and an increase in β_{ii} as the streamlines cease to follow the convex surface of the blades. For operation in and beyond this range, β_{ii} varies considerably with β_i and the flow losses are prohibitively high. The same can be said for decreases of β_i beyond reasonable values, decreases for which separation occurs on the pressure side of the blade. Separation resulting from an increase of β_i is termed "positive stall." "Negative stall" is separation due to drastically reduced β_i. The losses usually become prohibitively high before the flow is fully stalled, and the stall limit is commonly defined as that condition for which losses are some arbitrary multiple of their minimum value. For the NACA 65 Series blades, Mellor has arbitrarily defined stall as a 50% increase in the blade profile drag coefficient. It is important to recognize that in a compressor blade row the profile drag is only one of the sources of stagnation-pressure loss. Other losses to be discussed subsequently include losses due to the end wall boundary layer and also those which have been commonly ascribed to secondary flows.

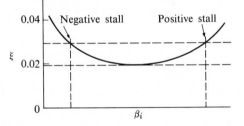

FIG. 8–16. Typical stagnation-pressure loss measurement for a cascade of NACA series airfoils.

Figure 8–16 shows a typical measurement of the stagnation-pressure loss coefficient: $\xi = \Delta p_0 / \frac{1}{2}\rho w_i^2$ for one of the 65 Series of blades.

Efficiency

The efficiency of a single stage, defined with reference to Fig. 8–9 as

$$\eta_{\text{st}} = \frac{h_{03s} - h_{01}}{h_{03} - h_{01}}, \tag{8–14}$$

is less than one because of frictional effects on the flow channel surface (blades, hub, and housing). Frictional effects are also responsible for the stagnation-pressure loss which occurs in the flow relative to each blade row. Therefore, the stage efficiency may be expressed in terms of stagnation-pressure losses which, in turn, may be estimated from cascade tests. However, to include the effects of friction along hub and shroud surfaces, the cascade data must be modified somewhat. The following derivation of stage efficiency in terms of stagnation-pressure losses follows somewhat that of Horlock [2], who also derives more general expressions applicable to compressible flow.

With reference to the fluid states defined by Fig. 8–9, the ideal work input can be expressed in terms of the identity

$$h_{03s} - h_{01} \equiv (h_{03} - h_{01}) - (h_{03s'} - h_{03s}) - (h_{03} - h_{03s'}). \qquad (8-19)$$

The first term on the right-hand side will be recognized as the *actual* work input, while the last two terms will be related to the loss of stagnation pressure occurring in the rotor and stator, respectively. Examination of Fig. 8–9 (or, more rigorously, the equations of a perfect gas) reveals that for small pressure changes as encountered within a single stage,

$$h_{03s'} - h_{03s} \approx h_2 - h_{2s} \quad \text{and} \quad h_{03} - h_{03s'} \approx h_3 - h_{3s'}. \qquad (8-20)$$

From a reference frame rotating with the rotor, there is *no* work done on or by the fluid within the rotor passages, since the fluid-rotor forces are *not moving* within this reference frame. Thus the *relative* stagnation enthalpy remains constant, or,

$$h_1 + \frac{w_1^2}{2} = h_2 + \frac{w_2^2}{2}. \qquad (8-21)$$

For any process involving a pure fluid substance,

$$T\,ds = dh - \frac{1}{\rho}\,dp.$$

Thus, for an *isentropic* process, as from ①to ②s

$$h_{2s} - h_1 = \int_1^{2s} \frac{1}{\rho}\,dp \approx \frac{1}{\rho}(p_2 - p_1), \qquad (8-22)$$

where the integrated expression is sufficiently accurate for the small density variations which occur within axial compressor blade rows. Combining Eqs. (8–21) and (8–22), we have

$$h_2 - h_{2s} = \left(\frac{p_1}{\rho} + \frac{w_1^2}{2}\right) - \left(\frac{p_2}{\rho} + \frac{w_2^2}{2}\right).$$

The combination $(p/\rho) + (w^2/2)$ is, again for nearly incompressible flow, just $p_{0\,\text{rel}}/\rho$, where $p_{0\,\text{rel}}$ is the stagnation pressure as observed from the relative (rotor) reference frame. Thus, for the rotor,

$$h_2 - h_{2s} = \frac{1}{\rho}\Delta p_{0R}, \qquad (8-23a)$$

the subscript R signifying relative or rotor. Similarly, for the stator,

$$h_3 - h_{3s'} = \frac{1}{\rho}\Delta p_{0S}. \qquad (8-23b)$$

Combining Eqs. (8–20), (8–23), and (8–19) with the definition of stage efficiency,

$$\eta_{st} \approx 1 - \frac{\Delta p_{0R} + \Delta p_{0S}}{\rho(h_{03} - h_{01})}. \qquad (8\text{–}24)$$

This formulation of efficiency in terms of stagnation-pressure losses relative to each blade row is convenient, since it is just such loss information that may be obtained from cascade performance tests. Since $h_{03} - h_{01}$, the actual work input, is known from the stage velocity diagram, it only remains to estimate the stagnation-pressure loss in each of the blade rows in order to estimate the stage efficiency. If the actual machine behaved exactly as the corresponding (two-dimensional) cascade, data as presented in Fig. 8–16 would be adequate. However, a large portion of the actual Δp_0 may be due to end-wall (hub and shroud) boundary layers, so that these effects must be considered. This may be done through a somewhat roundabout process which considers modified lift and drag forces on the blades in the cascade.

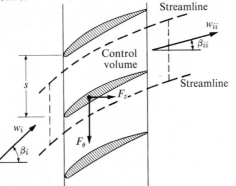

FIG. 8–17. Forces on a two-dimensional cascade.

To relate stagnation-pressure loss to lift and drag forces on a blade in a simple rectilinear cascade, consider a control volume enclosing one blade of a two-dimensional cascade, as indicated in Fig. 8–17. It may be assumed that the wakes of the blades mix quickly with faster-moving fluid. Thus the flow through the end surfaces of the indicated control volume may be assumed uniform. The side surfaces are chosen to be at all points parallel to streamlines, so that no fluid can cross them. Since the flow is taken to be two-dimensional and nearly incompressible, the continuity requirement for flow through the control volume is

$$w_i \cos \beta_i = w_{ii} \cos \beta_{ii}. \qquad (8\text{–}25)$$

It is convenient to consider the two components of blade force per unit length normal and parallel, respectively, to the cascade axis. These forces are actually applied to the control volume as shear forces where the control surface intersects the blade (as in the plane of the figure).

The force F_θ acting on the fluid per unit blade length may be determined by applying the general momentum equation (2–4) to the control volume of Fig. 8–17. From symmetry, the pressure forces on the sides of the control volume will cancel, so that

$$F_\theta = \dot{m}(w_{\theta i} - w_{\theta ii}).$$ (8–26)

In this expression, F_θ is taken to be positive in the direction shown, while $w_{\theta i}$ and $w_{\theta ii}$ represent the magnitude only of the tangential velocity. [The difference in direction of F_θ and w_θ accounts for the apparent sign change in the use of Eq. (2–4).] The mass flow rate per unit length is given by

$$\dot{m} = \rho w \cos \beta s.$$ (8–27)

Similarly, the momentum equation in the z-direction is

$$F_z + (p_i - p_{ii})s = \dot{m}(w_{zii} - w_{zi}),$$

in which F_z is the axial component of the force on the fluid. However, since from continuity $w_{zi} = w_{zii} = w_z$, therefore

$$F_z = s(p_{ii} - p_i).$$ (8–28)

Equation (8–28) may be rewritten using stagnation pressure (relative to the blade row in question), recognizing that in incompressible flow,

$$p_{ii} - p_i = p_{0ii} - p_{0i} + \frac{\rho}{2}(w_i^2 - w_{ii}^2).$$

Since the axial velocity component w_z is constant,

$$p_{ii} - p_i = p_{0ii} - p_{0i} + \frac{\rho}{2}(w_{\theta i}^2 - w_{\theta ii}^2)$$

or

$$p_{ii} - p_i = p_{0ii} - p_{0i} + \rho\left(\frac{w_{\theta i} + w_{\theta ii}}{2}\right)(w_{\theta i} - w_{\theta ii}).$$

Defining $w_{\theta m} = (w_{\theta i} + w_{\theta ii})/2$ and $\Delta p_0 = p_{0i} - p_{0ii}$, Eq. (8–28) may be written

$$F_z = s[\rho w_{\theta m}(w_{\theta i} - w_{\theta ii}) - \Delta p_0].$$ (8–29)

The forces exerted on the blades by the fluid, which are equal and opposite to the forces given by Eqs. (8–26) and (8–29), may be expressed in terms of lift and drag components normal and parallel, respectively, to the mean velocity vector $\mathbf{w}_m$ indicated in Fig. 8–18(a). The forces as they act on the blade are given in Fig. 8–18(b). It may be seen that the lift and drag forces (per unit length), L and $\mathfrak{D}$, are

$$L = F_\theta \cos \beta_m + F_z \sin \beta_m, \qquad \mathfrak{D} = F_\theta \sin \beta_m - F_z \cos \beta_m, \qquad (8\text{–}30)$$

where

$$\sin \beta_m = \frac{w_{\theta m}}{w_m} \quad \text{and} \quad \cos \beta_m = \frac{w_z}{w_m}.$$

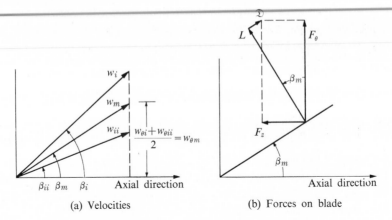

(a) Velocities   (b) Forces on blade

FIG. 8–18. Definition of blade force, velocities, and angles.

Using Eqs. (8–26) and (8–29), we can write these in the form

$$L = \rho s w_m (w_{\theta i} - w_{\theta ii}) - s \Delta p_0 \frac{w_{\theta m}}{w_m}, \qquad \mathfrak{D} = s \Delta p_0 \frac{w_z}{w_m}. \qquad (8\text{--}31)$$

Then, defining conventional lift and drag coefficients as follows,

$$C_L = \frac{L}{(\rho/2) w_m^2 C}, \qquad C_{\mathfrak{D}} = \frac{\mathfrak{D}}{(\rho/2) w_m^2 C}, \qquad (8\text{--}32)$$

it may be shown that

$$C_L = 2 \frac{s}{C} \cos \beta_m (\tan \beta_i - \tan \beta_{ii}) - \frac{\Delta p_0}{(\rho w_m^2)/2} \frac{s}{C} \sin \beta_m,$$

$$C_{\mathfrak{D}} = \frac{\Delta p_0}{\rho w_m^2 / 2} \frac{s}{C} \cos \beta_m. \qquad (8\text{--}33)$$

Using Eq. (8–33), we can write the lift coefficient as

$$C_L = 2 \frac{s}{C} \cos \beta_m (\tan \beta_i - \tan \beta_{ii}) - C_{\mathfrak{D}} \tan \beta_m. \qquad (8\text{--}34)$$

Equation (8–33) relates the drag coefficient to the stagnation-pressure loss. One method [4] of calculating the actual loss in the machine is to use an "effective" drag coefficient, $C_{\mathfrak{D}}'$, which attempts to account for the additional losses occurring in the rotor. The rotor (or stator) stagnation-pressure loss is then calculated from the effective drag coefficient. The effective drag coefficient may be estimated as follows:

$$C_{\mathfrak{D}}' = (C_{\mathfrak{D}})_{\text{cascade}} + (C_{\mathfrak{D}})_{\text{walls}} + (C_{\mathfrak{D}})_{\text{secondary flow}}, \qquad (8\text{--}35)$$

where $(C_{\mathfrak{D}})_{\text{cascade}}$ is just the two-dimensional result calculated from Eq. (8–33) and cascade data as given in Fig. 8–16.

The losses due to the end walls (as distinct from the losses due to the drag on the cascade profiles) may be estimated very approximately in the following manner. For turbulent flow at a bulk mean velocity w_m in a constant-area channel of length X, height h, and width s, the stagnation-pressure loss would be

$$\Delta p_0 = 4 \frac{C_f X}{D_H} \frac{\rho w_m^2}{2},$$

in which C_f is the wall friction coefficient (defined here by $C_f = \tau_0 / \frac{1}{2}\rho w_m^2$, where τ_0 is the wall shear stress), and D_H is the hydraulic diameter, defined as four times the area of the channel divided by the "wetted perimeter." To account for friction on the end walls only, one might consider a hypothetical channel with frictionless side walls, in which case the wetted perimeter would be just $2s$ and $D_H = 4(hs)/2s = 2h$. For fully developed flow in a constant-area channel, the total drag force due to this pressure drop would be $\mathcal{D}_{\text{walls}} = \Delta p_0 (hs)$. The drag coefficient $(C_\mathcal{D})_{\text{walls}}$ per unit blade length might be written in the form

$$(C_\mathcal{D})_{\text{walls}} = \frac{\mathcal{D}_{\text{walls}}/h}{\rho(w_m^2/2)C}.$$

Since, in the blade row $X \approx C$, we can write this, using the above expression for D_H, as

$$(C_\mathcal{D})_{\text{walls}} = 2C_f \left(\frac{s}{h}\right),$$

in which h is the blade height. It has been suggested [3] that a suitable value of $C_f \approx 0.01$, so that

$$(C_\mathcal{D})_{\text{walls}} = 0.02 \frac{s}{h}.$$

This is of course a very crude method of estimating frictional losses due to the end walls. Nevertheless it is of some use in the absence of a more detailed and reliable procedure.

The secondary flow losses appear to be mainly due to the development of patterns of circulation normal to the main flow direction in curved passages. A tangential pressure gradient is established between the blades to turn the fluid in the blade passage. This tangential gradient turns the slow-moving fluid near the end walls much more than the fast-moving fluid in the main stream. Thus patterns of "secondary flow" circulation are established, as suggested by Fig. 8–19. It is not fully understood at present how these secondary circulations are related to the losses in a curved passage. Nevertheless, experimental data of Howell [3] appear to correlate the losses unaccounted for by $(C_\mathcal{D})_{\text{cascade}}$ and $(C_\mathcal{D})_{\text{walls}}$ by $(C_\mathcal{D})_{\text{secondary flow}} = 0.018 C_L^2$. The total effective drag coefficient for the cascade may then be written

$$C_\mathcal{D}' \approx (C_\mathcal{D})_{\text{cascade}} + 0.02 \left(\frac{s}{h}\right) + 0.018 C_L^2. \tag{8–36}$$

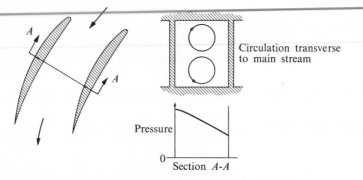

FIG. 8–19. Typical secondary flow pattern.

The term $(C_{\mathcal{D}})_{\text{cascade}}$ is usually in the range from 0.018 to 0.025, which may be less than half the total effective drag coefficient.

An equivalent drag coefficient, calculated from Eq. (8–36), may be used to calculate a modified stagnation-pressure loss from Eq. (8–33). This stagnation-pressure loss, calculated for the rotor and stator, may then be used in Eq. (8–24) to obtain an estimate of the stage efficiency. It cannot be said that this method of estimating stage losses is really reliable or satisfactorily founded on basic principles. It is discussed here to show how our lack of understanding of complex three-dimensional turbulent boundary layers prevents a more rational method of designing axial compressors. The principal justification for using the method is that it enables fairly realistic prediction of stage performance from cascade data, in the absence of a deeper understanding of the actual fluid behavior.

Off-design performance

Equation (8–13) may be written in terms of the temperature rise parameter, $c_p \, \Delta T_0 / U^2$, as

$$\frac{c_p \, \Delta T_0}{U^2} = 1 - \frac{c_z}{U} (\tan \alpha_1 + \tan \beta_2), \qquad (8\text{–}37)$$

for constant axial velocity. The corresponding stage pressure ratio is given by Eq. (8–16) as

$$\frac{p_{03}}{p_{01}} = 1 + \frac{\gamma}{\gamma - 1} \, \eta_{\text{st}} \frac{\Delta T_0}{T_{01}}. \qquad (8\text{–}16)$$

It has been shown by typical cascade data (Fig. 8–14) that over a considerable range of inlet angle the outlet angle of the flow through a cascade changes very little. This holds as long as the solidity is reasonably high and the stage remains unstalled. Thus the angles α_1 and β_2 may be said to be approximately constant and the ideal stage temperature rise parameter $c_p \, \Delta T_0 / U^2$ would be a linear function of the velocity ratio c_z/U, as shown by Fig. 8–20. The axial velocity c_z is averaged in the radial direction and U is the mean blade speed. The actual

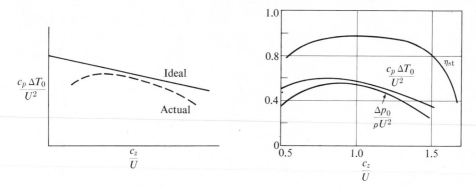

FIG. 8–20. Temperature rise of a typical   FIG. 8–21. Typical characteristics of a
single-stage axial compressor.   single-stage axial compressor.

temperature rise will be less than the ideal, for reasons already considered in
the discussion of radial variations in the flow annulus. The ratio c_z/U is directly
related to the angle of incidence for given blade geometry and, as Fig. 8–20 suggests, the actual stage temperature rise is linear over quite a range of values of
c_z/U. For very high values (negative incidence) and very low values of c_z/U
(positive incidence), the blade will stall, and therefore the work done and the pressure ratio will decrease.

Figure 8–21 shows typical characteristics of a single-stage axial compressor.
Equation (8–37) indicates that the shape of the characteristic curve may be easily
altered by changing the blade angles α_1 and β_2.

By combining Eqs. (8–13) and (8–16), we can write the single-stage pressure
ratio as

$$\frac{p_{03}}{p_{01}} = 1 + \eta_{st}\frac{U^2}{RT_{01}}\left[1 - \frac{c_z}{U}(\tan\beta_2 + \tan\alpha_1)\right]. \qquad (8\text{–}38)$$

The ratio c_z/U, a function of the flow through the compressor and the
rotor speed, can be written in terms of a Mach number and a wheel speed ratio
$U/\sqrt{\gamma RT_{01}}$. Since the Mach number at the inlet to the compressor is

$$\mathrm{M}_1 = \frac{c_1}{a_1} = \frac{c_z/\cos\alpha_1}{\sqrt{\gamma RT_1}} \quad\text{or}\quad \frac{c_z}{\sqrt{\gamma RT_{01}}} = \frac{\mathrm{M}_1\cos\alpha_1}{\{1 + [(\gamma - 1)/2]\mathrm{M}_1^2\}^{1/2}}.$$

Therefore

$$\frac{c_z}{U} = \frac{\mathrm{M}_1}{\{1 + [(\gamma - 1)/2]\mathrm{M}_1^2\}^{1/2}}\frac{\sqrt{\gamma RT_{01}}}{U}\cos\alpha_1.$$

Owing to the uncertainties in evaluating η_{st} and to the approximations of a mean
radius analysis, Eq. (8–38) does not provide a very accurate method of computing
the pressure ratio of an actual single stage compressor. However, it does show that

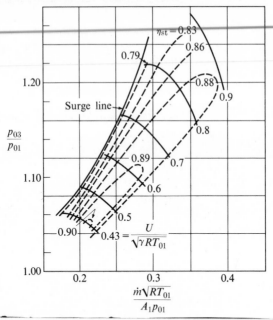

FIG. 8–22.   Characteristics of a single-stage axial-flow air compressor.   (Courtesy M.I.T. Gas Turbine Laboratory.)

for a given machine (that is, α_1 fixed), the pressure ratio depends on the following dimensionless groups:

$$\frac{p_{03}}{p_{01}} = f\left(\frac{U}{\sqrt{\gamma R T_{01}}}, M_1, \gamma, \eta_{st}\right).$$

However, from continuity, the mass flow may be written

$$\frac{m}{A_1}\frac{\sqrt{R T_{01}}}{p_{01}} = \sqrt{\gamma} M_1 \bigg/ \left(1 + \frac{\gamma - 1}{2} M_1^2\right)^{(\gamma+1)/2(\gamma-1)}.$$

Further, an examination of the method of calculating η_{st} will show that, for a given machine, it is just a function of blade incidence angle or c_z/U.   Therefore

$$\eta_{st} = f\left(\frac{U}{\sqrt{\gamma R T_{01}}}, M_1\right).$$

Thus the pressure ratio across the single stage may be written

$$\frac{p_{03}}{p_{01}} = f\left(\frac{U}{\sqrt{\gamma R T_{01}}}, \frac{\dot{m}}{A_1}\frac{\sqrt{R T_{01}}}{p_{01}}, \gamma\right).$$

Figure 8–22 shows a typical variation of both pressure ratio and stage efficiency, with mass flow and wheel speed parameters for a single stage axial air compressor (γ is not a significant variable here).

It may be noticed that the viscosity does not appear in these relationships. A statement of the dependence of the pressure ratio of any turbomachine on a set of independent variables may take the form

$$p_{02} = f(\dot{m}, p_{01}, T_{01}, \gamma, \Omega, R, \nu, \text{design}, D),$$

where (1) is the inlet state and (2) the outlet state. All the geometrical description of the machine is given by its design and its scale of size D. It is of course possible to include other variables. For example, if heat transfer is a necessary element of the operation of the machine, the thermal conductivity of the fluid would also need to be included. However, under most circumstances the variables in the parentheses may be assumed to comprise a complete set of independent variables. Since R and T are the only variables which include temperature in their dimensions, they must always appear together as RT. Then, for a particular design of machine,

$$p_{02} = f(\dot{m}, p_{01}, RT_{01}, \gamma, \Omega, \nu, D).$$

By dimensional analysis this statement can be reduced to

$$\frac{p_{02}}{p_{01}} = f\left(\frac{\dot{m}\sqrt{RT_{01}}}{p_{01}D^2}, \frac{\Omega D}{\sqrt{\gamma RT_{01}}}, \gamma, \frac{\Omega D^2}{\nu}\right),$$

in which the first three independent dimensionless variables are essentially the same as those already given. The fourth, which is a Reynolds number, is additional. Thus the question of when viscosity is important may be transformed to the question of the importance of Reynolds number in turbomachines.

In an axial compressor, for example, the most important effect of Reynolds number concerns transition from laminar to turbulent boundary layer behavior. If the Reynolds number is not sufficiently high the boundary layer will remain laminar, and be much more liable to separate and generate stall regions within the blade passages. Thus one would expect that below a certain level of Reynolds number the performance of the machine would deteriorate very rapidly. This does in fact happen with axial compressors, as demonstrated by their performance at very high altitudes, where the density and thus the Reynolds number are much less than at sea level. For such applications blades of enlarged chord (i.e., increased Reynolds number) may be necessary to attain suitable performance. Cascade tests usually show that as long as the Reynolds number (based on the mean velocity and the blade chord) is larger than about 2×10^5, viscous effects are unimportant.

Degree of reaction

The degree of reaction R of a compressor stage may be defined as the static pressure rise in the rotor divided by the static pressure rise in the whole stage. If the flow is taken to be (approximately) isentropic in the rotor,

$$T \, ds = dh - (dp/\rho) = 0.$$

For nearly incompressible flow this integrates to

$$h_2 - h_1 = (1/\rho)(p_2 - p_1).$$

Combining this with the energy equation relative to the rotor [Eq. (8–21)], we have

$$\frac{p_2 - p_1}{\rho} = \frac{w_1^2}{2} - \frac{w_2^2}{2}.$$

Writing the energy equation of the entire stage using Eq. (8–11), we have, since $h_{02} = h_{03}$,

$$h_3 - h_1 + \frac{c_3^2}{2} - \frac{c_1^2}{2} = U(c_{\theta 2} - c_{\theta 1}),$$

in which, as before, c_θ signifies the absolute velocity in the tangential direction. If, as is often the case for a single stage within a multistage machine, the initial and final absolute velocities are nearly identical, this reduces to $h_3 - h_1 = U(c_{\theta 2} - c_{\theta 1})$.

Again assuming approximately incompressible and isentropic flow, this expression can be written

$$\frac{p_3 - p_1}{\rho} = U(c_{\theta 2} - c_{\theta 1}).$$

Thus the degree of reaction R is

$$R = \frac{p_2 - p_1}{p_3 - p_1} = \frac{w_1^2 - w_2^2}{2U(c_{\theta 2} - c_{\theta 1})}. \tag{8–39}$$

If the axial velocity is constant,

$$w_1^2 - w_2^2 = w_{\theta 1}^2 - w_{\theta 2}^2. \qquad \text{Also} \qquad w_{\theta 1} - w_{\theta 2} = c_{\theta 2} - c_{\theta 1}.$$

Thus

$$R = \frac{w_{\theta 1} + w_{\theta 2}}{2U} = \frac{c_z}{U} \tan \beta_m \tag{8–40}$$

or

$$R = \frac{1}{2} - \frac{c_z}{U}\left(\frac{\tan \alpha_1 - \tan \beta_2}{2}\right). \tag{8–41}*$$

Now if the exit flow angles α_1 and β_2 (Figs. 8–7 and 8–8) are equal (still assuming that c_1 and c_3 are identical), the degree of reaction is $\frac{1}{2}$ and the velocity triangles are symmetrical. This is the 50% reaction stage, in which half the static pressure rise takes place in the stator and half in the rotor. Estimations of stage efficiency have shown that best efficiency is attained with a stage of about this configuration,

---

* If, as is sometimes done [4], the degree of reaction is defined as the ratio of rotor to overall static enthalpy rise, these results require only the assumption of constant axial velocity c_z.

although stage efficiency is not strongly sensitive to variations in the degree of reaction around this optimum. This conclusion seems plausible, since stage losses occur as the result of boundary layer effects, and well-behaved flow in an axial compressor is therefore essentially limited by static pressure rise. It is clear that if the degree of reaction is much different from 50%, one of the blade rows must tolerate greater pressure rise than the other. This means that its boundary layer will be more highly loaded and more likely to stall or at least create higher losses. Greater stage pressure rise can therefore be obtained if both blade rows are loaded to about the same pressure rise. As will be shown later, however, the designer may choose to vary the degree of reaction in the radial direction by adjusting the blade angles.

Solidity

Equation (8–36) suggests that the spacing of the blades in the annulus has an effect on efficiency. The greater the number of blades the larger the losses due to frictional effects. This consideration by itself would suggest the desirability of wide spacing between blades, that is, low solidity (C/s). However, there are other important considerations. For a given fluid turning angle, Eq. (8–34) shows that the lift on a given airfoil increases as the solidity decreases. Increase of blade lift means that the pressure gradient along the suction surface of the blade becomes more adverse so that the flow is more liable to separate. Closer blade spacings mean that a given blade can operate over a considerably wider range of inlet flow angles without separation, and this factor is of great importance to multistage axial compressors. Also, closer spacing means that the flow angle leaving a blade row (which in all our calculations has been assumed a constant or average value) exhibits smaller variation across the blade passage.

The designer who wishes to achieve the best balance of all these effects has various empirical rules to aid him in his choice of solidity. It may be seen from Fig. 8–6 that a typical value is not greatly different from unity.

Radial variations

In the derivation of Eq. (8–13) we have assumed that radial variations in the flow through the compressor annulus are negligible. This is not really true, and designers usually take into account radial variations in

(1) Blade speed $U = \Omega r$,

(2) Axial velocity c_z,

(3) Tangential velocity c_θ,

(4) Static pressure.

The axial velocity at the inlet to a multistage machine may be quite uniform radially, but large variations can develop after the fluid has passed through two or three stages. The radial variation in tangential velocity depends on the way

~~in which the blade angles vary with radius.~~ The radial pressure gradient depends directly on the absolute tangential velocity component as follows.

Consider a small element of fluid δm with tangential velocity component c_θ, as in Fig. 8–23. It has a centripetal acceleration toward the axis of $-c_\theta^2/r$. If the radial velocity does not change ($\partial c_r/\partial r \approx 0$), this is the total radial acceleration of the fluid and the radial pressure force must be equal to

$$F_r = -\delta m(c_\theta^2/r)$$

where

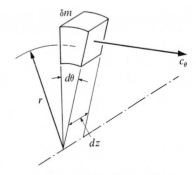

$$F_r = p(r\,d\theta\,dz) - \left(p + \frac{\partial p}{\partial r}\,dr\right)(r + dr)\,d\theta\,dz$$

$$\simeq -\frac{\partial p}{\partial r}\,dr\,r\,d\theta\,dz.$$

Also, $\delta m = \rho r\,dr\,d\theta\,dz$. Combining these expressions, we have

$$\frac{\partial p}{\partial r} = \rho\,\frac{c_\theta^2}{r}, \qquad (8\text{–}42)$$

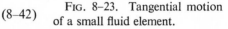

FIG. 8–23. Tangential motion of a small fluid element.

which is the equation of simple radial equilibrium. Actual velocity distributions must satisfy this relationship at the entrance and exit of each blade row.

In order to provide a reasonably uniform flow at the exit of a compressor, it is desirable to maintain a uniform work input along the radial length of a rotor. As has been shown, the change in stagnation enthalpy along a streamline through any stage in an axial compressor is $\Delta h_0 = U\,\Delta c_\theta$. The radial gradient is

$$\frac{\partial}{\partial r}(\Delta h_0) = \Omega\,\frac{\partial}{\partial r}(r\,\Delta c_\theta). \qquad (8\text{–}43)$$

Thus to keep the fluid at constant stagnation enthalpy the product $r\,\Delta c_\theta$ must be constant with radius.

~~One configuration which satisfies this requirement,~~ and at the same time the requirements of radial equilibrium, is the "free vortex" design, in which the product rc_θ is held constant across the exit of each blade row. Unfortunately the blading of a free-vortex compressor usually requires excessive blade twist.

Consider a rotor with hub, mean, and tip radii related by

$$r_t = 1.25\,r_m, \qquad r_h = 0.75\,r_m,$$

and suppose that the mean radius velocity triangles have been specified, as in Fig. 8–24. For free vortex at (1) and (2), the hub and tip velocities can be determined from

$$c_{\theta h} = c_{\theta m}\frac{r_m}{r_h} \qquad \text{and} \qquad c_{\theta t} = c_{\theta m}\frac{r_m}{r_t}.$$

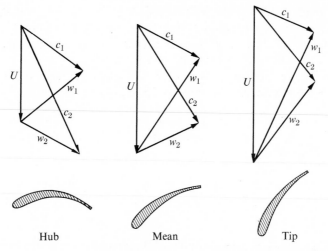

Fig. 8–24. Free-vortex velocity diagrams and blade profiles for $r_h/r_m = 0.75$, $r_t/r_m = 1.25$.

With constant c_z and with blade speed, of course, proportional to radius, the complete velocity triangles may be determined as shown. Also shown are blade cross sections as they might appear at the various radii. Here it can be seen that rather large blade twist is required, as mentioned above.

Aside from the undesirable blade twist, note the very high absolute velocity at the exit of the rotor hub. This follows from the high turning and the relatively small amount of diffusion within the rotor, and results in intolerably large pressure rise in the stator at this radius.

Both the blade twist and the unequal rotor-to-stator diffusion distribution can be improved while constant work is maintained if it is specified that only $r\Delta c_\theta$ instead of rc_θ be constant with radius. Starting with the same mean radius diagram and assuming symmetric diagrams at hub and tip to equalize diffusion, we arrive at the set of diagrams of Fig. 8–25.

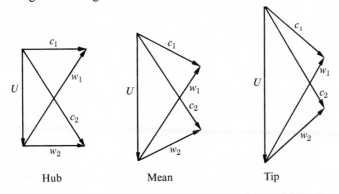

Fig. 8–25. Symmetric diagrams for $r\Delta c_\theta = $ constant; $r_h/r_m = 0.75$, $r_t/r_m = 1.25$.

Such a blade system is desirable in that it avoids the drawbacks encountered by the free-vortex blading. However, it can be shown that these velocity triangles are incompatible because they violate the conditions of radial equilibrium. Radial equilibrium is simply the satisfaction of the *three-dimensional* equations of fluid motion rather than the two-dimensional equations which have been considered. Except for a few fortunate cases which are automatically satisfied, the velocity variations in the radial direction cannot be arbitrarily specified. An attempt to produce incompatible velocity diagrams such as these would be met by radial adjustments in the flow pattern to something which would satisfy all equilibrium conditions. The actual pattern may be reasonably similar to the intended pattern or it may be quite different.

In order to account very approximately for the effect of radial variations, designers sometimes use a "work-done" factor K_w to modify Eq. (8-13) to

$$\frac{\Delta T_0}{T_{01}} = K_w \frac{U^2}{c_p T_{01}} \left[1 - \frac{c_{z1}}{U} \left(\frac{c_{z2}}{c_{z1}} \tan \beta_2 + \tan \alpha_1 \right) \right].$$

This factor is the actual work done by the rotor on the fluid divided by the work calculated, using a constant axial velocity and the blade angles at the mean radius. Values of 0.85 are quoted in the literature, though some designers prefer to use a value of 0.95 for the first stage, decreasing the value through the compressor to about 0.85. This factor is not related to the stage efficiency, but simply expresses the degree to which the velocity triangles and work done at the mean radius are a good average for the annulus as a whole.

In a serious design method, radial variations are taken into account by using computation methods.

8-4 MULTISTAGE AXIAL COMPRESSORS

The very low pressure ratio attainable in a single axial stage requires that, for most aircraft engine applications, a multiple stage compressor be used. In principle the preceding treatment of single stage performance is adequate for the study of a multistage compressor which is, after all, just a series of single stages. In practice, however, it is found that the stage pressure-rise limitations within a multistage compressor must be more modest than those quoted for a single stage. The J79 engine of Fig. 6-10, for instance, has a seventeen-stage compressor producing an overall stagnation-pressure ratio of about 13. The stage pressure ratio, if all stages were the same, would be only $13^{1/17}$, or about 1.16. This lower pressure-rise limitation is a function of interactions between stages, the possibility of cumulative loss effects, and certain starting difficulties associated with high pressure ratio compressors. For example, the outlet flow from a highly loaded stage, which may be partially stalled, can be quite nonuniform and yet produce reasonable efficiency in a single stage. However, the effect of such a velocity distortion on subsequent stages could produce very poor overall efficiency in a multistage compressor.

The overall adiabatic efficiency of a compressor as defined in Chapter 6 is

$$\eta_c = \frac{h_{03s} - h_{02}}{h_{03} - h_{02}}, \tag{8–44}$$

where states (2) and (3) are the compressor inlet and outlet, respectively (Fig. 6–9). The numerator represents the work required for isentropic compression between the actual inlet and outlet stagnation pressures, while the denominator is the actual work. Assuming a perfect gas with constant specific heat, the overall stagnation-pressure ratio can be written

$$\frac{p_{03}}{p_{02}} = \left(1 + \eta_c \frac{\Delta T_0}{T_{01}}\right)^{\gamma/(\gamma-1)}, \tag{8–45}$$

where ΔT_0 is the overall stagnation-temperature rise $T_{03} - T_{02}$ and $T_{02} = T_{01}$.

The compressor efficiency can be calculated from the estimated efficiencies of the single stages. Obviously, for a large number of stages, considerable empirical data is required and the complete calculation is sufficiently lengthy to warrant the use of a computer. Mellor [5] has developed an approximate performance calculation which treats the compression process as a continuous path rather than as a series of discrete steps. The compressor in this case might be imagined to consist of an infinite number of stages, each of infinitesimal pressure rise. The efficiency of such an infinitesimal pressure rise process, called *polytropic efficiency*, is quite similar to that of the conventional single stage efficiency. Hence an examination of the relationship between polytropic and overall compressor efficiency will be of value in illustrating the relationship between stage and overall efficiency in a multistage compressor.

Consider an incremental pressure rise from p_0 to $p_0 + dp_0$, as shown on a T-s diagram in Fig. 8–26. The temperature rise accompanying this pressure rise is dT_0, while the temperature rise for an isentropic process would have been dT_{0s}. The polytropic compression efficiency, η_{pc}, is defined for this incremental process just as the overall efficiency is defined for a finite compression:

$$\eta_{\text{pc}} = \frac{dT_{0s}}{dT_0}. \tag{8–46}$$

For the isentropic process,

$$dT_{0s} = T_0\left[\left(\frac{p_0 + dp_0}{p_0}\right)^{(\gamma-1)/\gamma} - 1\right].$$

Using the binomial expansion for $dp_0/p_0 \ll 1$,

$$\left(1 + \frac{dp_0}{p_0}\right)^{(\gamma-1)/\gamma} = 1 + \frac{\gamma - 1}{\gamma}\frac{dp_0}{p_0}.$$

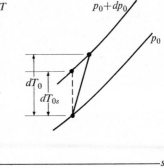

Fig. 8–26. Definition of terms for an incremental compression of a perfect gas.

Thus Eq. (8–46) can be written

$$\eta_{pc} \frac{dT_0}{T_0} = \frac{\gamma - 1}{\gamma} \frac{dp_0}{p_0}. \tag{8–47}$$

For overall compression from state i to state ii, this may be integrated (for constant η_{pc}) to give

$$\frac{T_{03}}{T_{02}} = \left(\frac{p_{03}}{p_{02}}\right)^{(\gamma-1)/\eta_{pc}\gamma}. \tag{8–48}$$

Combining Eqs. (8–48) and (8–45), we see that the relationship between overall adiabatic and polytropic compression efficiency depends on overall pressure ratio:

$$\eta_c = \frac{(p_{03}/p_{02})^{(\gamma-1)/\gamma} - 1}{(p_{03}/p_{02})^{(\gamma-1)/\eta_{pc}\gamma} - 1}. \tag{8–49}$$

Figure 8–27, a plot of η_c versus pressure ratio for various values of η_{pc}, shows that η_c is less than η_{pc} and that the difference increases as the pressure ratio increases. For pressure ratios as low as those encountered in a single stage it can be seen that $\eta_c = \eta_{st} \approx \eta_{pc}$; hence the polytropic and stage efficiencies are about the same. Thus an overall efficiency of 0.9 at a pressure ratio of 10 would require a stage efficiency of about 0.93, a rather high value. With the reasonable assumption that maximum attainable stage efficiency is independent of the total number of stages (or at least that

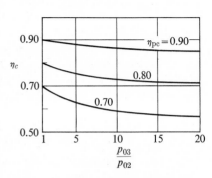

FIG. 8–27. Compressor efficiency as a function of polytropic efficiency and pressure ratio.

η_{pc} does not *increase* as the number of stages increases), it can be seen that overall compressor efficiency decreases as compressor pressure ratio increases.

The dimensional analysis applied to the single-stage compressor could obviously just as well be applied to a multistage machine, so that the compressor performance could be plotted on a "map" similar to Fig. 8–22. The only difference would be that the pressure ratios would be much higher, while the efficiencies would be somewhat lower.

8–5 INSTABILITIES AND UNSTEADY FLOW

Surge

The line marked "surge" in Fig. 8–22 denotes the locus of unstable operation of the compressor. The possibility of this instability, which may easily lead to pressure oscillation violent enough to destroy the compressor blades, can be explained as follows. Consider a single stage whose constant speed (U) pressure-

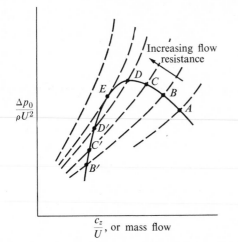

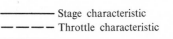

———— Stage characteristic
— — — Throttle characteristic

FIG. 8–28. Pressure rise characteristic compared with throttle characteristic of downstream resistance.

rise characteristic might be as shown in Fig. 8–28 if surge were not to occur. Suppose that the stage exhausts into a duct in which flow is controlled by a throttle whose pressure-drop mass-flow characteristics for a series of throttle settings are also shown in Fig. 8–28. An operating point for a particular throttle setting is determined by the intersection of that throttle characteristic with the stage characteristic. If the slope of the throttle characteristic is greater than that of the stage, the operating point will be stable. For example, at A, B, C, or D, a momentary increase of mass flow will be met by a throttle demand for pressure rise which is greater than that provided by the compressor so that mass flow will return to the original value. The converse is true for a momentary mass flow decrease. By a similar argument the intersections B', C', D', are unstable and any shift from the operating point would be followed by a transition either toward B, C, or D (for increased flow) or to zero flow (for decreased flow). The point of neutral stability is E, where the throttle characteristic is just tangent to the stage characteristic, and small departures from the operating point are neither opposed nor reinforced.

These considerations all have to do with the static stability of the system. Suppose now that the system is operating at B and the throttle is gradually closed. Stable operation should persist through C and D until point E is reached. A further closing of the throttle causes a very rapid reduction in flow, since a reduction in mass flow causes the stage operation to travel along its characteristic (assuming this line describes its transient behavior), while the throttle characteristic moves away from the compressor characteristic altogether. Once air delivery is stopped, or perhaps even reversed, the pressure in the discharge region drops until the compressor can start again, rapidly refill the exhaust chamber, and repeat the cycle. The possibility of cyclic behavior depends on the filling and emptying characteristics of the discharge system as well as on its throttling characteristics.

This simple model would predict that stable operation is possible for any mass flow and throttle setting greater than that for which the characteristics are just tangent, as at E. In fact, however, the region between E and D of positive compressor characteristic slope is not always stable, although it has been demonstrated to be so on low pressure ratio machines. The behavior of multistage machines is much more complex, and the assumption that surge begins at the peak pressure point is usually nearly correct [2, 4]. The model is oversimplified even for single stage operation, of course, since it is not necessarily correct to describe transient behavior in terms of a series of steady state operating points.

It may be seen from Fig. 8–28 that the difficulty occurs at low values of c_z/U, i.e., high positive blade incidence (Fig. 8–7). For a multistage compressor the behavior is complicated by stage interactions. As the mass flow decreases, the value of c_z/U at the first stage decreases. As the first stage value of c_z/U decreases, pressure rise in that stage increases [Eq. (8–13)]. This increases the fluid density at the entrance to the second stage, further reduces the axial velocity in that stage (to satisfy continuity), and leads to progressively higher values of density and progressively lower values of c_z/U through the compressor. At some point, as the compressor mass flow is reduced, the last stage or two is liable to become unstable by the mechanism just described. In an engine this unstable condition must be carefully avoided by automatic governing controls.

Rotating stall

Another important phenomenon which can lead to compressor damage is rotating stall. If a blade row is on the verge of stalling, it is likely that one particular blade will stall before its neighbors do.

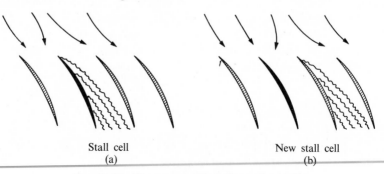

Stall cell   New stall cell
(a)   (b)

FIG. 8–29. The propagation of rotating stall.

As can be seen in Fig. 8–29(a), a single stalled blade results in a flow diversion which tends to overload one adjacent blade and unload the other adjacent blade. The overloaded blade will then stall and the resulting flow diversion will unload the originally stalled blade, as in Fig. 8–29(b). This process is repeated, and the stall cell propagates forward in the direction indicated. In actual cases there may be more than one stall cell rotating on a single blade row. Generally speaking, the zones of stall, which may cover several blade spacings, rotate at about half the engine speed. The phenomenon is also observed on rectilinear cascades where the motion is, of course, linear rather than circular. The alternate loading and unloading of the blades sets up an alternating stress on each blade. This stress in itself is not large unless the forcing frequency happens to match a blade vibrational frequency. In this case large stresses are present and fatigue failure can occur, resulting in complete destruction of an entire blade row. Little can be done to prevent this failure, other than avoiding operation at speeds corresponding to resonance.

8–6 AERODYNAMICS OF STARTING

Starting a high-pressure-ratio compressor can be a major problem. Under design conditions such a machine operates with a large change in density from entrance to exit. There is a corresponding change in blade length to accommodate this density variation, as suggested by Fig. 8–4. When the compressor is started, the density ratio is unity, and it remains low until the compressor is actually working properly. On the other hand, as will be shown, the compressor cannot work properly until the density ratio is reasonably near the design value. Obviously some modification must be made during starting to overcome this impasse.

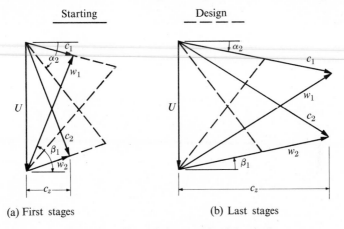

(a) First stages   (b) Last stages

Fig. 8–30. Effect of change of axial velocity.

Consider the effect of operation at design rotor speed with a density ratio below design conditions. If the axial velocity in the first stages is to be correct, the mass flow rate must be about the design value. Hence, with low density, the axial velocity in the last stages must be much too high. Conversely, if the axial velocity is to be correct in the last stages, the mass flow rate must be reduced and the axial velocity in the first stages will then be low. The actual situation will be somewhere between these extremes, with low axial velocity in the first stages and high axial velocity in the last stages. Under these conditions, the high downstream velocity requirements would appear as an excessive flow restriction to the first stages. If the fluid angles were to equal blade angles at blade row outlets, the operational requirements would appear as in Fig. 8–30.

Considering the first stages, we can see that decreased c_z with α_1 and β_2 constant results in increased α_2 and β_1 or *increased loading on both rotor and stator blades*. Considering the last stages, we can see that the opposite is true, that increased c_z results in *decreased loading of both rotor and stator blades*. The starting difficulty can now be restated in more specific terms: operation at below design-density ratio causes variations in c_z which tend to overload the leading stages, thus causing stall and preventing proper compression, which would correct the poor density ratio.

Several solutions are used to allow compressors with high pressure ratios to be self-starting. Variation of blade speed is of little help, since the two ends of the compressor require opposite variations. External assistance such as an air blast to increase c_z in the leading stages is inconvenient for aircraft use,* although it is used in some stationary power plants.

The simplest device to aid starting is a blow-off valve located at about the middle of the compressor, which allows some of the incoming air to bypass the second half of the compressor. Such a valve is shown in Fig. 8–5 as a compressor bleed port. In this way the axial velocity is reduced in the last stages, allowing these stages to do more work, while at the same time the apparent downstream restriction encountered by the first stages is reduced, allowing an increase in c_z. In effect this solution temporarily creates two low-pressure compressors in series with a variable mass-flow relationship between them. The variation is accomplished by a rather rapid or on-off type control, which is acceptable for starting purposes but undesirable for in-flight application. Since in-flight conditions might involve off-design operation similar to that encountered in starting, the blow-off valve is employed only as an auxiliary device to be used with other continuously variable devices.

Another solution is suggested by the above statement that the two ends of the compressor demand opposite variations of blade speed. If the compressor can be broken into two mechanically independent units, the first or low-pressure stages being driven by one turbine and the last or high-pressure stages being driven by another, these demands can be met. With this arrangement the lightly loaded high-pressure stages will automatically run faster, improving their velocity triangles and picking up a greater share of the starting load. This has the effect of reducing the downstream restriction encountered by the low-pressure stages, while at the same time the overloaded low-pressure stages will automatically run more slowly, also improving their velocity triangles and avoiding stall. This is the "two-spool" configuration shown in Figs. 6–16 and 8–6.

Noting that the overloading of the first stages is the result of an increased angle of attack on the blades, one can suggest the solution of variable blade angles. While it would be desirable to vary all blade angles, it is mechanically feasible to vary only stator blade angles, by rotating the stator blades about their radial axis.

---

* Aircraft have occasionally used the exhaust blast from another aircraft to assist in starting.

Figure 8–31 shows how much change in stator angle $\Delta\alpha_1$ is necessary to restore β_1 to approximately the design value for a given c_z/U. Note that this does not change α_2, but if the following stator is rotated in the same manner as the preceding one, the angle of attack on the following stator will be reduced by the same quantity. This is the variable-stator configuration employed by General Electric engines in this country.

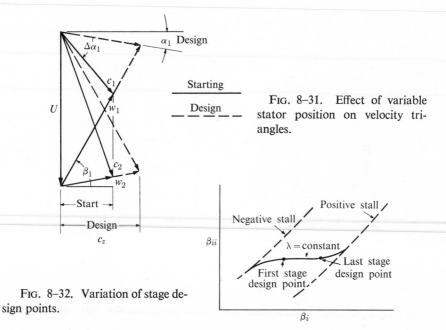

FIG. 8–31. Effect of variable stator position on velocity triangles.

FIG. 8–32. Variation of stage design points.

There is some benefit to be had from varying the design point loading throughout the stages. Realizing that the first stages will be overloaded during starting, an engineer would be wise to design them for light loading. Conversely, the last stages should be heavily loaded. In terms of cascade data, first and last stage design points should appear as shown in Fig. 8–32. Note that this choice leaves a relatively large range of increased β_1 before stall occurs in the first stages. The light loading of the first stages is somewhat compensated for by the heavy loading of the last stages. This technique alone may be sufficient to provide starting ability in moderate pressure ratio machines, but it is not adequate for high pressure ratio machines. Hence one or more of the mechanical methods just described finds application on all modern high pressure ratio engines.

References

1. MELLOR, G. E., "The 65 Series Cascade Data."   Cambridge, Mass., Gas Turbine Laboratory, Massachusetts Institute of Technology (unpublished)

2. HORLOCK, J. H., *Axial Flow Compressors*.   London: Butterworth's Scientific Publications, 1958

3. HOWELL, A. R., "Design of Axial Compressors" and "Fluid Dynamics of Axial Compressors."   *Proc. Inst. Mech. Engrs.*, London, 153, 1945

4. SHEPHERD, D. G., *Principles of Turbomachinery*.   New York: The Macmillan Company, 1956

5. MELLOR, G. L., "The Aerodynamic Performance of Axial Compressor Cascades with Application to Machine Design."   Cambridge, Mass.: Massachusetts Institute of Technology, Gas Turbine Laboratory, Report No. 38, 1957

Problems

1. Shown schematically in part (a) of the figure is the blading of a single-stage axial turbomachine. What kind of machine is represented by Cases 1 and 2? What would happen in Case 3? Would it be desirable to build a compressor stage as in part (b) of the figure?

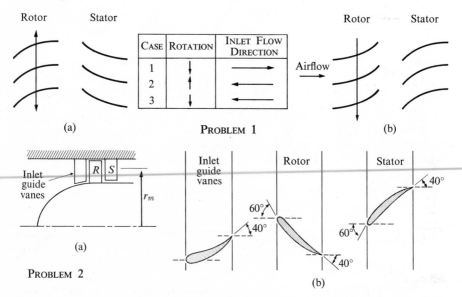

CASE	ROTATION	INLET FLOW DIRECTION
1	↓	→
2	↑	←
3	↓	←

PROBLEM 1

PROBLEM 2

2. Estimate the power required to drive a single-stage compressor shown schematically as in parts (a) and (b) of the figure.

At the mean radius (r_m = 12 in.), the blade configuration is as shown in part (b). For simplicity it may be assumed that the air angles and blade angles are identical.

The overall adiabatic efficiency of the stage is 90%. The hub-tip radius ratio is 0.8, high enough so that conditions at the mean radius are a good average of conditions from root to tip.

The axial velocity component at design flow rate is uniformly 400 ft/sec, and the inlet air is at 1 atm and 20°C. What should the shaft speed be under these conditions?

3. The first stage of an axial compressor is to have blades with constant outlet angles β_2 and α_3 ($\beta_2 \neq \alpha_3$ necessarily). For the dimensions and velocities shown in the figure specify velocity triangles and α_3 at the hub, mean, and tip radii in order that the maximum pressure coefficient be no greater than 0.6. In this problem, calculate the work input per pound of air at the hub, mean, and tip radii.

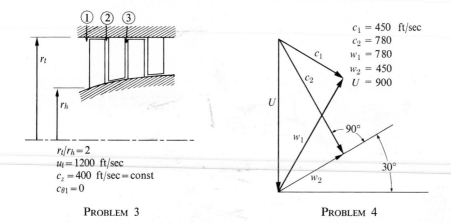

$r_t/r_h = 2$
$u_t = 1200$ ft/sec
$c_z = 400$ ft/sec = const
$c_{\theta 1} = 0$

$c_1 = 450$ ft/sec
$c_2 = 780$
$w_1 = 780$
$w_2 = 450$
$U = 900$

Problem 3   Problem 4

4. At design flow and speed the mean velocity triangle for the second stage of a high pressure ratio axial compressor is as shown in the figure. (a) What is the total pressure ratio of this stage if the stage efficiency is 0.85 and the inlet temperature is 500°R? (b) During starting, the axial velocity rises to 200 ft/sec and no more after the rotor has reached design speed. To prevent stalling in the first stages, variable stator angles are to be employed. How far and in what direction must the stator upstream of this stage be rotated to bring β_1 to the design value when c_z is only 200 ft/sec?

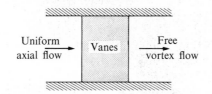

Uniform axial flow   Vanes   Free vortex flow

Problem 5

5. An incompressible fluid flows down a straight pipe with uniform stagnation pressure and passes through vanes producing a free-vortex ($rc_\theta = k$) tangential velocity distribution on the downstream side. Neglecting friction, show that it is proper to assume a uniform axial velocity on the downstream side of the vanes.

6. Determine the axial and tangential components of the aerodynamic forces exerted on a rotor blade if the fluid velocities and angles are as given in Problem 4. Con-

sider two cases $C'_{\mathcal{D}} = 0$ and $C'_{\mathcal{D}} = 0.10$. The average air density is 0.1 lb_m/ft^3. The blade chord, length, and spacing are 1, 2, and 1 in., respectively.

7. Using the typical cascade performance data of Fig. 8–14, show how, for a stagger angle of 40°, the turning angle and pressure rise depend on the inlet flow angle. Plot $\beta_1 - \beta_2$ and $(p_2 - p_1)/\frac{1}{2}\rho w_1^2$ as functions of β_1 for solidities of 1.0 and 1.5, assuming no stagnation pressure loss. Qualitatively, how would you expect these results to change with Reynolds number and Mach number?

8. Show that, at sufficiently high Reynolds number, for any given actual single-stage axial compressor the pressure rise at all speeds and flows may be represented by a single curve relating the dimensionless variables ϕ and ψ which are defined by

$$\phi = \frac{Q}{\Omega D^3} \quad \text{and} \quad \psi = \frac{\Delta p}{\rho \Omega^2 D^2},$$

in which Q is the volume flow rate, Ω is the shaft speed, D is a characteristic dimension, and Δp is the stage presssure rise. Show that ϕ is uniquely related to rotor and stator incidence angles, and thus that ϕ determines the flow pattern regardless of flow and speed. Why are at least three dimensionless variables required to describe the performance of a multistage axial air compressor?

9

Jet Engine
Turbomachines:
Centrifugal Compressors
and Axial Turbines

9-1 INTRODUCTION

In this chapter, starting again from the basic angular momentum equation derived in Section 8–2, we shall discuss the performance characteristics of centrifugal compressors, and compare them with corresponding features of axial compressors. We shall then discuss the axial turbine, and show how compressors and turbines must interact to provide satisfactory engine performance.

One of the important factors which constrains turbine and compressor performance is mechanical stress. We shall deal with this very significant subject, though only briefly, at the close of the chapter.

9-2 THE CENTRIFUGAL COMPRESSOR

The first two independently developed turbojet engines each employed a centrifugal compressor. At that time (1935–40), the centrifugal compressor was the only type sufficiently developed for application. Since that time, with intensive development work, the axial compressor has largely replaced the centrifugal compressor in aircraft gas turbines. However, the centrifugal compressor still finds application in small gas turbine engines. The centrifugal pump is used a great deal in rocket engines. For these reasons, and because the centrifugal compressor may ultimately be capable of much better performance than it has shown up to the present, the centrifugal machine will now be discussed. A comparison of centrifugal and axial compressors will follow this description of the centrifugal machine.

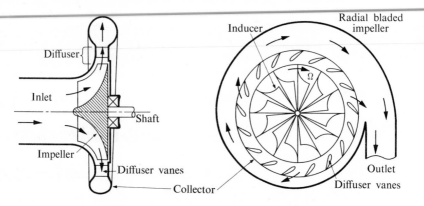

F<small>IG</small>. 9–1. Typical centrifugal compressor.

Figure 9–1 shows a common type of centrifugal compressor. The velocity of the air leaving this impeller, in contrast to that of the air leaving the axial impeller or rotor, has no axial component. The inlet velocity of the air is normally axial, but it may have a tangential component in some cases.

Fluid enters the impeller through the inducer, whose function is to turn the relative flow as it enters the impeller passages. The blades in aircraft centrifugal compressors are usually straight and radial, as shown in this figure. Other blade forms will be considered herein only briefly. After leaving the impeller, the air passes through a radial diffuser in which momentum is exchanged for pressure. Radial diffusers usually consist of a vaneless portion, as shown, followed by a series of stator vanes. The air leaving the diffuser is collected and delivered to the outlet. Figure 9–1 shows a single outlet, but aircraft engine applications in which the air is to be delivered to multiple combustion chambers may employ multiple outlets. These components will be considered separately.

Impeller and inducer

We must consider the inducer, or impeller inlet region, together with the radial portions of the impeller, since the two rotate together and the boundary between inducer and impeller is a matter of arbitrary definition.

Consider a control volume containing the impeller shown in Fig. 9–2. Equation (8–8) may be employed to relate the torque on the rotor to the changes in angular momentum of the fluid. Thus, when we assume that inlet and outlet angular momentum rc_θ are each uniform across the respective flow areas, the torque τ on the rotor is

$$\tau = \dot{m}[(rc_\theta)_2 - (rc_\theta)_1].$$

The power input to the rotor, $\mathcal{P} = \Omega\tau$ (with $U_1 = \Omega r_1$, $U_2 = \Omega r_2$), is

$$\mathcal{P} = \dot{m}(U_2 c_{\theta 2} - U_1 c_{\theta 1}),$$

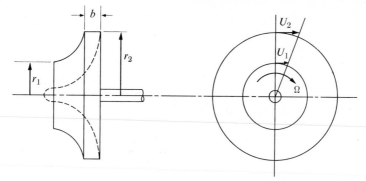

FIG. 9–2. Centrifugal compressor rotor.

and thus the stagnation-temperature rise ΔT_0 of the fluid is given by

$$\frac{T_{02} - T_{01}}{T_{01}} = \frac{U_2 c_{\theta 2} - U_1 c_{\theta 1}}{c_p T_{01}},$$

in which T_{01} is the stagnation temperature of the incoming fluid. Rearranging this expression, we obtain

$$\frac{T_{02} - T_{01}}{T_{01}} = \frac{U_2^2}{c_p T_{01}}\left[\frac{c_{\theta 2}}{U_2} - \left(\frac{r_1}{r_2}\right)^2 \frac{c_{\theta 1}}{U_1}\right]. \qquad (9\text{--}1)$$

The equation is presented in this way since the quantities within the brackets are simply functions of geometry over a fairly wide range of operating conditions. The equation is sometimes further simplified by the introduction of an arbitrarily defined "Mach index," Π_M:

$$\Pi_M = \frac{U_2}{a_{01}},$$

where a_{01} is the velocity of sound at the inlet stagnation state; $a_{01} = \sqrt{\gamma R T_{01}}$. Combining these, we have

$$\frac{T_{02} - T_{01}}{T_{01}} = (\gamma - 1)\Pi_M^2\left[\frac{c_{\theta 2}}{U_2} - \left(\frac{r_1}{r_2}\right)^2 \frac{c_{\theta 1}}{U_1}\right].$$

Blade shape may be one of three types; straight radial, as in Fig. 9–1, backward leaning, or forward leaning. Figure 9–3 shows the three types schematically, together with typical velocity triangles in the radial plane for the outlet of each type. The three rotors are shown with the same tip speed U_2 and the same radial velocity component $w_r = c_r$. As a first approximation, the *relative* fluid velocity leaving the impeller, w_2, is assumed parallel to the blade. The angle of the blade at exit with respect to the radial direction is given the symbol β_2, and is positive in the direction of rotation. Because the rotor speed U is entirely tangential, the radial

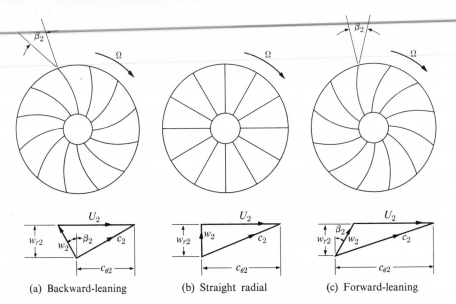

(a) Backward-leaning   (b) Straight radial   (c) Forward-leaning

FIG. 9-3. Shapes of centrifugal impeller blades.

components of relative and absolute velocity, w_r and c_r, respectively, are equal.

Consider the typical machine with purely axial inlet velocity; that is, $c_{\theta 1} = 0$; Eq. (9-1) becomes

$$\frac{T_{02} - T_{01}}{T_{01}} = (\gamma - 1)\Pi_M^2 \frac{c_{\theta 2}}{U_2}.$$

Since, for all three impellers, $c_{\theta 2} = U_2 + w_{r2} \tan \beta_2$, then

$$\frac{T_{02} - T_{01}}{T_{01}} = (\gamma - 1)\Pi_M^2 \left(1 + \frac{w_{r2}}{U_2} \tan \beta_2\right).$$

From continuity, we see that

$$w_{r2} = \frac{\dot{m}}{2\pi r_2 b \rho_2},$$

where b is the blade height at exit and ρ_2 is the exit density.

For any fixed inlet condition and tip speed (Π_M), these expressions state that the three machines will produce the same stagnation-temperature rise at zero mass flow.* For an increase in mass flow the stagnation-temperature rise will decrease

---

* Actually, cases of near-zero through-flow usually involve reverse flow out of at least a part of the inlet. Hence the simple assumptions of this analysis only hold at flow rates reasonably near the design flow rate.

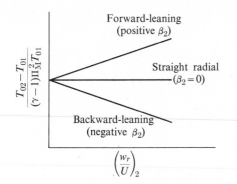

FIG. 9–4. Relative temperature rise characteristics of centrifugal compressors.

for backward-leaning, remain constant for straight radial, and increase for forward-leaning blades. This is shown graphically in Fig. 9–4.

For the centrifugal compressor with no swirl in the entering fluid, the ratio $(w_r/U)_2$ has the same significance as the ratio (c_z/U) for the axial compressor. In both cases the relative stage temperature rise is a linear function of this variable.

For the same tip speed and mass flow, forward-leaning blades do more work per unit mass of fluid, and thereby tend to produce higher pressure ratios, provided their blade boundary layers are well behaved. There is some danger, as shown by the discussion in Chapter 8 of the surge of a single stage compressor, that the rising temperature and pressure characteristic may lead to unstable flow. Another disadvantage of this type of impeller is relatively high blade stress, since the centrifugal forces on curved blades induce bending stresses. Also the absolute impeller exit velocity is higher than that of a straight radial bladed impeller (at the same w_{r2} and U_2) and this increases the pressure rise necessary in the diffuser. For these reasons forward-leaning blades are generally unsuitable for high-speed compressors.

Backward-leaning blades often have higher efficiency than straight radial blades, since the absolute velocity leaving the impeller can be slightly lower than that of a straight radial bladed impeller. However, for the same work output, backward-leaning blades must be run faster. The necessary increase in tip speed only aggravates the already serious stress problem encountered with curved blades, so that the efficiency advantage cannot be realized in highly loaded aircraft compressors. Not only are straight radial blades free from bending stresses; they may also be somewhat easier to manufacture than curved blades. These reasons are sufficient to restrict the choice to straight radial blades for aircraft engine compressors. Pumps are limited by cavitation to speeds which are low enough that stress considerations are secondary and, therefore, backward-curved blading is more likely to be used.

So that the flow may enter the impeller smoothly, it is essential that an inducer be installed. Centrifugal fans that do not have inducers are usually very noisy. Apparently the noise results from separation and fairly violent mixing near the leading edge of the impeller vane. Figure 9–5 shows velocity triangles for an axial flow entering a centrifugal impeller, showing the flow angles for root, mean, and tip

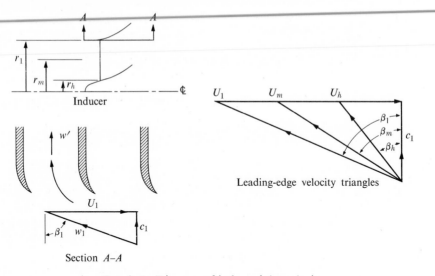

Fɪɢ. 9–5. Diagram of inducer inlet velocity.

radii. Here we may note that at the impeller tip the angle between the relative velocity vector and the flow may be very large. Thus, near the tip, the vane needs to be set at an angle large with respect to the axial direction if the flow is to enter the inducer without leading-edge separation. However, there is a limit to the amount of turning that can occur within the inducer, if separation on the curved inducer surface is to be avoided.

If it is assumed that the inducer accomplishes the turning of the relative velocity vector without a change in radius, then these velocities can be shown in a cylindrical section of radius r_1, shown "unwrapped" in Fig. 9–5. If there is no radial velocity component in the inducer, continuity for incompressible flow requires that the axial component of velocity remain constant. Thus

$$w' = w_1 \cos \beta_1,$$

where w' denotes the relative velocity at the inducer outlet. The fact that $w' < w_1$ indicates diffusion in the inducer. In order to estimate the maximum permissible turning angle which would not produce separation on the inducer surface, it is necessary to know the pressure coefficient $C_{p\,\text{max}}$ corresponding to separation. As defined previously, the pressure coefficient is

$$C_p = \frac{p' - p_1}{\frac{1}{2}\rho w_1^2} = 1 - \left(\frac{w'}{w_1}\right)^2.$$

Then the maximum turning angle is given by

$$\cos \beta_1 = \sqrt{1 - C_{p\,\text{max}}}.$$

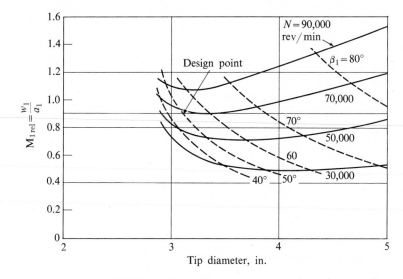

FIG. 9–6. Optimization of centrifugal compressor inlet ($D_h \approx 1$ in., $T_{01} = 690°$R, $\dot{m} = 2.4$ lb$_m$/sec).

If $C_{p\,max}$ is again taken to be 0.6, as seems reasonable for turbulent boundary layers, then the maximum turning angle is around 50°. Unfortunately, however, the behavior of the turbulent boundary layer can be markedly affected by the curvature of the surface over which it is traveling, and it may separate earlier on a curved surface than a flat one for the same variation of pressure with distance in the flow direction.

The inducer is also subject to the effects of compressibility. At high Mach numbers shocks may appear near the tip and cause the performance to deteriorate, as is the case for the axial compressor stage. For a given mass flow and rotor speed a small inlet diameter leads to a low tip speed U_1, but a high absolute inlet velocity, c_1. Conversely, a large inlet diameter leads to a high U_1, and a low c_1. Within these extremes lies an inlet diameter giving a minimum tip relative Mach number $M_{1\,rel}$. Figure 9–6 shows the results of a sample calculation on an inducer to handle 2.4 lb$_m$/sec. The relative tip Mach number is shown as a function of tip diameter and rotor speed for fixed values of hub diameter and inlet stagnation temperature.

Tip velocities U_1 and c_1 determine the inducer turning angle β_1, of course. The value of β_1 is therefore known at all points along any constant-speed line in Fig. 9–6. A series of constant-speed curves can be calculated and, from these, lines of constant β_1 can be plotted. Having these two families of curves, the design point can be located to satisfy both the relative Mach number and inducer turning limits. For the design point of Fig. 9–1, $M_{1\,rel} = 0.9$ and $\beta_1 = 50°$. Knowing the rotor speed, we can determine the rotor diameter for a given work input (U_2) requirement. Thus conditions near the inducer tip and the required stage work are sufficient to determine the inlet and outlet diameters and the speed of a centrifugal compressor having straight radial blades.

The flow area in the impeller passage is usually controlled to avoid excessive divergence of the fluid streamlines in these passages. If the fluid diffuses very rapidly the boundary layer on the end walls or on the vane surfaces may separate. In fact, little is known at the present time about the effective resistance to separation of these layers. Thus the groundwork for a really sound method of designing centrifugal compressors remains to be laid.

Diffuser

Although the relative velocity within the impeller is low, the high impeller tip speed results in a high absolute velocity leaving the impeller. Absolute exit Mach numbers of the order of 1.3 are not uncommon. This high velocity is reduced (while the pressure rises) in a diffuser in which the fluid flows radially outward from the impeller, first through a vaneless region, and then through vanes, as shown by Fig. 9–7.

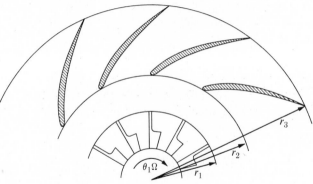

FIG. 9–7. Radial diffuser.

For incompressible flow in a vaneless region of constant axial width h, continuity requires that

$$\dot{m} = \rho(2\pi r h)c_r = \text{constant} \qquad \text{or} \qquad r c_r = \text{constant}.$$

At the same time, the absence of any torque-transmitting mechanism (except wall friction, which is neglected) requires that angular momentum be conserved throughout the vaneless region. Thus $r c_\theta = \text{constant}$. Combining these two equations, we have $c_\theta/c_r = \text{constant} = \tan \alpha$, where α is the angle between the velocity and the radial direction, as indicated in Fig. 9–8.

Thus the streamline makes a constant angle with the radial direction, and its shape is sometimes called a logarithmic spiral. Since α is constant, it follows that velocity is inversely proportional to radius, or $c_2/c_1 = r_1/r_2$.

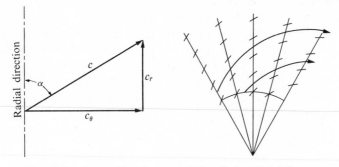

F<small>IG</small>. 9–8. Streamlines in a parallel-walled radial diffuser.

Compressible flow in a vaneless diffuser is also easily analyzed if we assume that the flow is reversible. As in most compressible-flow problems, it is convenient to relate fluid properties at any point to their values at a point where the Mach number is one. Thus, from continuity, $\rho r c_r = \rho^* r^* c_r^*$, where the * denotes the value attained in reversible adiabatic flow from the local condition to sonic velocity.

The conservation of angular momentum requires that $r c_\theta = r^* c_\theta^*$. These two equations can also be written

$$\rho r c \cos \alpha = \rho^* r^* c^* \cos \alpha^*, \tag{9–2}$$

$$r c \sin \alpha = r^* c^* \sin \alpha^*, \tag{9–3}$$

and combined to yield

$$\frac{\tan \alpha^*}{\tan \alpha} = \frac{\rho^*}{\rho}.$$

Applying the equations of one-dimensional reversible adiabatic flow along any streamline, we have

$$\frac{\tan \alpha^*}{\tan \alpha} = \left[\frac{2}{\gamma + 1} \left(1 + \frac{\gamma - 1}{2} \mathrm{M}^2 \right) \right]^{1/(\gamma-1)}. \tag{9–4}$$

With this relation for α as a function of M, Eq. (9–3) can be used to find r as a function of M. Thus

$$\frac{r^* \sin \alpha^*}{r \sin \alpha} = \frac{c}{c^*} = \mathrm{M}\sqrt{T/T^*},$$

and, using the isentropic equations for T/T^*, we obtain

$$\frac{r^* \sin \alpha^*}{r \sin \alpha} = \mathrm{M} \left\{ \frac{(\gamma + 1)/2}{1 + [(\gamma - 1)/2]\mathrm{M}^2} \right\}^{1/2}. \tag{9–5}$$

The known inlet conditions serve to determine the * quantities, after which we can find α from Eq. (9–4) and r from Eq. (9–5), as a function of M. Rather than using these consecutive equations, it is convenient to plot their results for any fixed γ,

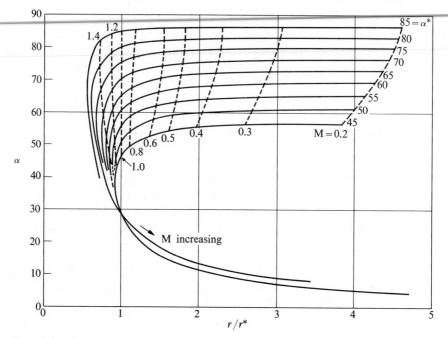

Fɪɢ. 9–9. Compressible flow in a vaneless diffuser; $\gamma = 1.4$. (Courtesy M.I.T. Gas Turbine Laboratory.)

as in Fig. 9–9. In this diagram, we determine the inlet state from the known angle and Mach number, α_1 and M_1. From this we can determine $(r/r^*)_1$ along with α_1^*. Since the value of α^* is constant, the state of the fluid follows a line of constant α^* until the desired outlet Mach number, M_2, is reached. Thus we find $(r/r^*)_2$, and the actual radius ratio of the diffuser is

$$\frac{r_2}{r_1} = \frac{(r/r^*)_2}{(r/r^*)_1}.$$

Along a line of constant α^*, two directions are possible: one corresponding to diffusion and the other to supersonic acceleration. Only the diffusion is of interest for compressors. For example, if $M_1 = 1.3$ and $\alpha_1 = 70°$, then $\alpha^* = 75°$ and $(r/r^*)_1 = 0.8$. If it is desired to diffuse to $M_2 = 0.4$ before entering a combustion chamber, then $(r/r^*)_2 = 2.3$ and

$$\frac{r_2}{r_1} = \frac{2.3}{0.8} = 2.9.$$

The vaneless diffuser is a very efficient device with an essentially limitless operating range, since the angle of attack sensitivity of vanes is absent. In the case of supersonic inlet flow the axial symmetry of the device, coupled with the fact that the radial component of velocity is subsonic, means that there can be no shock.

Unfortunately, however, the desired diffusion usually requires an excessively large radius ratio across the vaneless diffuser, as shown in the above example. This is especially critical in aircraft applications, where large engine diameters are to be avoided. Some additional diffusion can be achieved by increasing the diffuser width h, but this affects only the relatively small radial component of velocity. To save space, vanes as shown in Fig. 9–7 may be employed to reduce the fluid's angular momentum and thereby reduce c_θ more rapidly.

Radial-vaned diffusers are no different from other diffusers, in that boundary layer behavior limits the possible pressure rise. Hence the vane shape must be carefully chosen.

The entire diffuser must decelerate the flow from high Mach numbers (about 1.3) to low Mach numbers (about 0.4) and deliver the air to a collector. It has been shown that a vaneless diffuser to do this job is excessively large. However, to avoid shock effects on the vanes, the vaneless portion is often made large enough to avoid supersonic flow within the vanes, so that the Mach number at the entrance to the vanes is about 0.8 or 0.9, and about 0.4 at the exit.

For aircraft applications the air is usually delivered to multiple combustion chambers. Thus, rather than collect all the air in one pipe, the air is usually collected separately for each combustion chamber. This has a great influence on the spacing of diffuser vanes, since it is easiest to make each collector just an extension of two vanes, with perhaps two or three channels between them. There may be additional diffusion in each collector before delivery to the combustion chamber. If the air is to be delivered to one outlet, one collector which encircles the entire diffuser is used. In this case there is usually no diffusion within the collector, since this would result in an asymmetric pressure distribution around the machine axis. This "scroll collector," as it is called, would then increase linearly with angle θ in area to maintain constant velocity. Further diffusion, if necessary, may occur in the outlet pipe downstream from the diffuser outlet.

Performance

The foregoing sections outline some of the principles of centrifugal compressors. The work-transfer or temperature-rise characteristics of the complete machine can be reasonably well predicted in terms of the tip speed and a "slip factor."

The *ideal* stagnation-temperature rise of a straight-radial-bladed impeller is obtained by assuming that the absolute tangential velocity $c_{\theta 2}$ of the fluid at the impeller outlet is equal to the tip speed (i.e., that the relative velocity is parallel to the blades). Then,

$$\frac{T_{02} - T_{01}}{T_{01}} = (\gamma - 1)\Pi_M^2.$$

Figure 9–10 shows typical values of actual centrifugal-compressor performance for a number of rotor speeds. In two cases the theoretical stagnation-temperature rise has been calculated from the above formula, and we can see that it agrees with the measured results within about 10%. The actual temperature rise is less

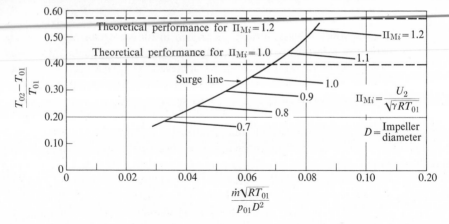

FIG. 9–10. Typical centrifugal compressor characteristics. (Courtesy M.I.T. Gas Turbine Laboratory.)

than predicted because the actual average value of $c_{\theta 2}$ is somewhat less than U_2. That is, there is some slip between the impeller and the fluid. For the impeller with straight radial blades, the ratio of fluid tangential velocity to blade velocity is called the *slip factor*, ζ:

$$\zeta = \frac{c_{\theta 2}}{U_2}. \tag{9–6}$$

Thus

$$\frac{T_{02} - T_{01}}{T_{01}} = (\gamma - 1)\Pi_M^2 \zeta. \tag{9–7}$$

The slip factor is nearly constant for any machine and is related to the number of blades on the impeller. Various theoretical and empirical studies have shown that, for more than 10 blades,

$$\zeta \approx 1 - \frac{2}{z}, \tag{9–8}$$

where z is the number of blades.

The range of possible mass flows* shown in Fig. 9–10 is limited by two phenomena. At the upper value, the flow is choked; the mass-flow rate cannot be increased without changing the inlet state. At the lower end, unstable flow or surging has begun, for reasons given previously.

Figure 9–11 shows typical overall pressure ratios p_{02}/p_{01} and efficiencies η_c of a centrifugal compressor stage. The pressure ratio for a given speed, unlike the temperature ratio, is strongly dependent on mass flow, since the machine

---

* The dimensionless mass flow variable used in Fig. 9–10 is the same as the one used in Chapter 8 for axial compressors. The characteristic size D could, for example, be taken as the tip diameter of the rotor. Figure 9–10 could describe the performance of a series of geometrically similar machines, Reynolds number effects being unimportant.

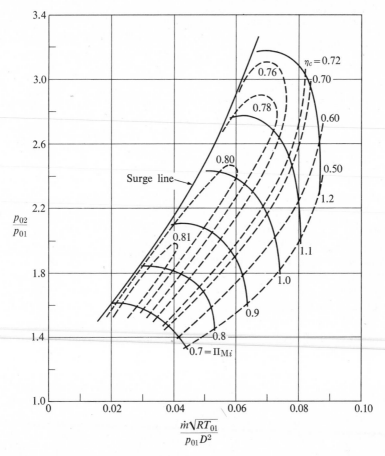
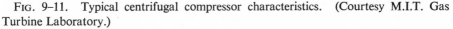

FIG. 9–11.   Typical centrifugal compressor characteristics.   (Courtesy M.I.T. Gas Turbine Laboratory.)

efficiency is usually at its peak value for a narrow range of mass flow. No method, other than actually testing the machine, is available to predict efficiency accurately.

Comparison of centrifugal and axial compressors

For aircraft applications one would be interested in comparing compressors on the basis of efficiency, weight, size, and ease of obtaining the desired pressure ratio.

(1) *Efficiency:* With much greater development effort behind it, today's axial compressor is about 5% more efficient than a centrifugal compressor of equal pressure ratio.

(2) *Weight:* Within the range of application of single stage centrifugal compressors, and for given overall diameters, the weight of the two types probably does not differ greatly. The axial machine has heavier shafts and discs but the centrifugal machine has a heavier housing and diffuser.

(3) *Size:* For aircraft application, the most important aspect of the compressor size is its cross-sectional area per unit mass-flow rate. The axial machine can accept a very much higher mass flow than a centrifugal one of the same maximum diameter, since a much larger fraction of its frontal area can be inlet.

(4) *Ease of obtaining desired pressure ratio:* For any pressure ratio up to about four, axial and centrifugal compressors are comparable in this respect. Centrifugal compressors can reach this ratio with one stage, and axial compressors are easily multistaged to reach this pressure ratio. For higher pressure ratios the difficulty of multistaging the centrifugal machine makes the axial machine much more attractive.

From these facts we can see why axial compressors have virtually eliminated centrifugal compressors from turbojet application. It is possible (with considerable development effort) that the centrifugal compressor may some day be capable of surpassing the axial compressor in efficiency, but this possibility in the face of the other drawbacks seems insufficient to warrant development of centrifugal compressors for turbojets. They have been successfully used for turboprop engines.

9–3   THE AXIAL TURBINE

Turbines, like compressors, can be classified into radial, axial, and mixed-flow machines. In the axial machine the fluid moves essentially in the axial direction through the rotor. In the radial type the fluid motion is mostly radial. The mixed-flow machine is characterized by a combination of axial and radial motion of the fluid relative to the rotor. The choice of turbine type depends on the application, though it is not always clear that any one type is superior.

Comparing axial and radial turbines of the same overall diameter, we may say that the axial machine, just as in the case of compressors, is capable of handling considerably greater mass flow. On the other hand, for small mass flows the radial machine can be made more efficient than the axial one. The radial turbine is capable of a higher pressure ratio per stage than the axial one. However, multistaging is very much easier to arrange with the axial turbine, so that large overall pressure ratios are not difficult to obtain with axial turbines.

In order to minimize the turbojet engine nacelle drag and the engine weight per unit thrust, the mass flow per unit cross-sectional area must be as large as possible. The limiting mass flow of turbojet engines depends mainly on the maximum permissible Mach number of the flow entering the compressor. It is generally large enough to require the mass-flow capability of an axial turbine, even with engines which have centrifugal (radial-flow) compressors. Thus the following study concerns axial turbines only.

By way of illustration, the rotors of the two turbines from the Rolls-Royce Conway RCo12 engine are shown in Fig. 9–12. Unlike the blades on axial compressors, these blades have, at their tips, platform surfaces which together form a circumferential band or shroud around the blade row at its tip diameter. This shroud

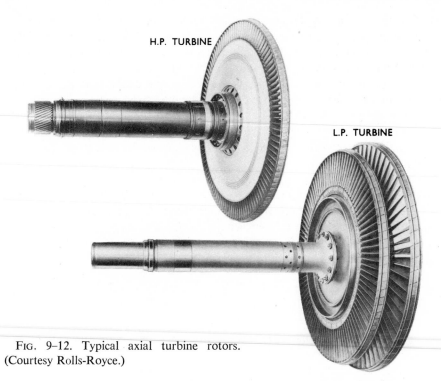

H.P. TURBINE

L.P. TURBINE

FIG. 9–12. Typical axial turbine rotors. (Courtesy Rolls-Royce.)

generally serves two purposes: it reduces blade vibration, and controls the leakage of air past the blade tips and across the blade row. Not all turbines have shrouds, but they are more common on turbines than compressors, due to higher stage pressure ratios (hence greater leakage) and stress conditions which are generally more critical because of high operating temperatures.

Generally the efficiency of a well-designed turbine is higher than the efficiency of a compressor. Moreover, the design process is often much simpler. The principal reason for this fact is that the fluid undergoes a pressure *drop* in the turbine and a pressure *rise* in the compressor. The pressure drop in the turbine is sufficient to keep the boundary layer fluid well behaved, and separation problems which are often serious in compressors can be easily avoided. Offsetting this advantage is the much more critical stress problem, since turbine rotors must operate in very high-temperature gases. Actual blade shape is often more dependent on stress considerations than on aerodynamic considerations, beyond the satisfaction of the velocity-triangle requirements.

An axial turbine stage consists of a row of stationary blades, called nozzles in this case, followed by the rotor, as in Fig. 9–13. Because of the high pressure drop per stage, the nozzle and rotor blades may be of increasing length, as shown, to accommodate the rapidly expanding gases, while holding the axial velocity to something like a uniform value through the stage.

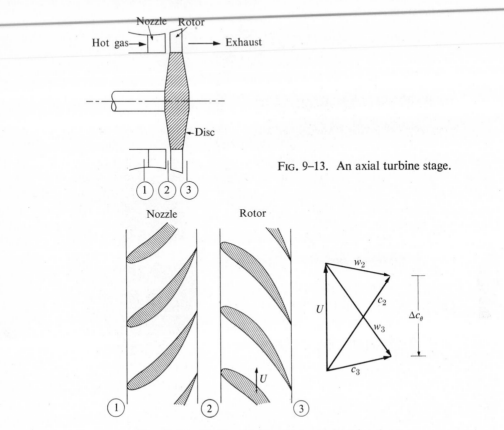

FIG. 9–13. An axial turbine stage.

FIG. 9–14. Turbine blading and velocity triangles.

A section through the mean radius would appear as in Fig. 9–14. It is seen that the nozzles accelerate the flow, imparting an increased tangential velocity component. Note that the velocity diagram of the turbine differs from that of the compressor, in that the change in tangential velocity in the rotor, Δc_θ, is in the direction *opposite* to the blade speed U. The reaction to this change in the tangential momentum of the fluid is a torque on the rotor in the direction of motion. Hence the fluid does work on the rotor. Again applying the angular momentum relationship as in Eq. 8–8, we may show that the power output is

$$\mathcal{P} = \dot{m}(U_2 c_{\theta 2} - U_3 c_{\theta 3}).\qquad (9\text{–}9)$$

In an axial turbine, $U_2 \simeq U_3 = U$. The turbine work per unit mass is

$$W_T = U(c_{\theta 2} - c_{\theta 3}) \qquad \text{or} \qquad W_T = c_p(T_{01} - T_{03}).$$

When we write

$$\Delta T_0 = T_{01} - T_{03} = T_{02} - T_{03},$$

it follows from these relations that

$$\frac{\Delta T_0}{T_{01}} = \frac{U(c_{\theta 2} - c_{\theta 3})}{c_p T_{01}}.$$ (9–10)

We can see that the work output (or ΔT_0) per turbine stage is increased by high blade speed and large turning of the fluid. When we consider the fact that the turbine must operate with a limited pressure ratio, it also becomes obvious that high inlet temperature is desirable. Inlet temperature and blade speed are, of course, limited by stress considerations. While large turning of the fluid can occur within turbine blade rows (as compared to compressors), it is still true that good efficiency places some limits on the fluid-turning, and hence power output of a turbine stage.

The boundary layer limitation on compressor blading points to the desirability of approximately equal pressure change per blade row. Turbine blade row performance is not nearly so sensitive to the condition of the boundary layer (so long as the pressure is not rising through the passage). Thus the designer has considerably more freedom to distribute the total stage pressure drop between rotor and stator. Nonetheless turbine blade row losses are adversely affected by a pressure rise (or velocity decrease), as suggested by Fig.

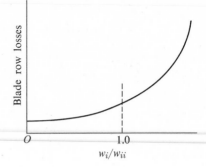

FIG. 9–15. Turbine blade row losses.

9–15. Blade losses are relatively low for accelerating flows ($w_i/w_{ii} < 1$; see footnote page 251). They exceed their minimum value by a factor of two or three when neither acceleration nor deceleration occurs ($w_i/w_{ii} = 1$), and they increase very rapidly when the pressure rises across the blade row ($w_i/w_{ii} > 1$).

Stage dynamics

Turbine stages in which the entire pressure drop occurs in the nozzle, called *impulse* stages, are quite common. Stages in which a portion of the pressure drop occurs in the nozzle and the rest in the rotor are described by their degree of reaction. The degree of reaction may be defined for a turbine as the fraction of overall static enthalpy drop occurring in the rotor. Hence a 50%-reaction machine has equal static enthalpy drops for nozzle and rotor. An impulse turbine would be a zero reaction machine. Attention will here be directed to the relative stage performance of the impulse and 50%-reaction stages.

An impulse turbine stage is shown in Fig. 9–16, along with a typical diagram for the common case of constant axial velocity. Since there is no static enthalpy change within the rotor, the energy equation within the rotor requires that $w_2 = w_3$.

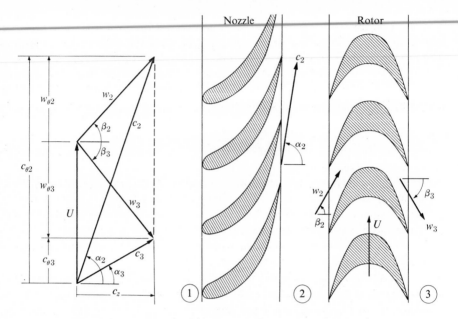

Fɪɢ. 9–16. Impulse turbine stage and constant axial velocity.

If the axial velocity component is held constant, then this requirement can be satisfied by $\beta_2 = -\beta_3$, since c_z is constant.

The impulse turbine work can be expressed in terms of various blade angles in a variety of forms of Eq. (9–10). One useful approach illustrates the effect of nozzle outlet angle α_2. From the velocity diagram,

$$w_{\theta 2} = -w_{\theta 3}, \qquad c_{\theta 2} - c_{\theta 3} = 2(c_{\theta 2} - U) = 2(c_z \tan \alpha_2 - U).$$

With these, Eq. (9–10) can be written

$$\frac{\Delta T_0}{T_{01}} = \frac{2U^2}{c_p T_{01}} \left(\frac{c_z}{U} \tan \alpha_2 - 1 \right). \qquad (9\text{–}11)$$

It is evident, then, that for large power output the nozzle angle should be as large as possible. However, two difficulties are associated with very large α_2. For reasonable axial velocities (i.e., reasonable flow per unit frontal area), it is evident that large α_2 creates very large absolute and relative velocities throughout the stage. High losses are associated with such velocities, especially if the relative velocity w_2 is supersonic. In practice, losses seem to be minimized for values of α_2 around 70°. In addition, it can be seen that for large α_2 [$\tan \alpha_2 > (2U/c_z)$], the absolute exhaust velocity will have a swirl in the direction opposite to U. While we have not introduced the definition of turbine efficiency as yet, it is clear that, in a turbojet engine where large *axial* exhaust velocity is desired, the kinetic energy associated with the tangential motion of the exhaust gases is essentially a loss. Furthermore, application of the angular momentum equation over the entire engine indicates that exhaust swirl is the result of an (undesirable) net torque acting on the propellant (from the aircraft). Thus the desire is for axial or near-axial absolute exhaust velocity (at least for the last stage if a multistage turbine is used). For the special case of constant c_z, and axial exhaust velocity, $c_{\theta 3} = 0$ and $c_{\theta 2} = 2U$. So Eq. (9–10) becomes

$$\frac{\Delta T_0}{T_{01}} = \frac{2U^2}{c_p T_{01}} . \tag{9–12}$$

For a given power and rotor speed, and for a given peak temperature, Eq. (9–12) is sufficient to determine approximately the mean blade speed (and hence radius) of a single stage impulse turbine having axial outlet velocity. If, as is usually the case, the blade speed is too high (for stress limitations), or if the mean diameter is too large relative to the other engine components, it is necessary to employ a multistage turbine in which each stage does part of the work.

If the ratio of blade length to mean radius is small, this mean radius analysis is sufficient for the entire blade length. If the blade is relatively long, however, radial variations should be considered; and we shall see that impulse design at the mean radius leads to difficulty at the hub. It was shown previously that the 50%-reaction compressor stage (with constant c_z) has symmetrical velocity triangles. The same is true for the 50%-reaction turbine stage. Since the changes in static enthalpy are the same in both blade rows, the change in kinetic energy relative to each blade row must be the same. Thus for constant axial velocity the velocity triangles are as shown in Fig. 9–17. Since the diagram is symmetrical,

$$c_{\theta 3} = -(c_z \tan \alpha_2 - U)$$

for constant axial velocity. Therefore, Eq. (9–10) for this case becomes

$$\frac{\Delta T_0}{T_{01}} = \frac{U^2}{c_p T_{01}} \left(2 \frac{c_z}{U} \tan \alpha_2 - 1 \right) . \tag{9–13}$$

Again the desirability of large α_2 is indicated and the same limitations are encountered, so that typical values of α_2 are near 70°. For the special case of axial outlet velocity and constant c_z, α_3 and β_2 are zero and the velocity diagram becomes a rectangle. The stage work output is then

$$\frac{\Delta T_0}{T_{01}} = \frac{U^2}{c_p T_{01}} . \tag{9–14}$$

Thus, for the same blade speed and for axial outlet velocities, the impulse stage work is twice that of the 50%-reaction stage. The impulse stage can be expected to have somewhat greater loss, however, since the average fluid velocity in the stage is higher and since the boundary layer on the suction side of the rotor blades may

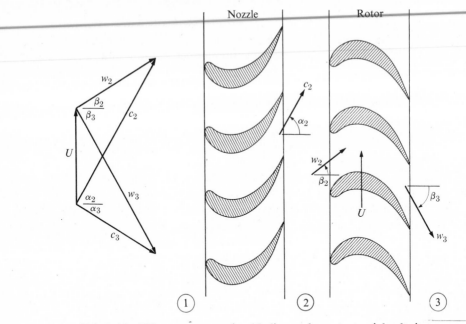

Fig. 9–17. Fifty-percent reaction blading and constant axial velocity.

be significantly thicker and closer to separation, depending on the turning angle and blade spacing.

The 50%-reaction stage is not uniquely desirable, of course. Any degree of reaction (greater than zero) can be used to design a turbine of acceptable performance. As we shall see, the mean-radius velocity triangles may be largely dependent on radial variations and problems encountered near the hub.

Radial variations

In turbines, as in compressors, it is necessary to consider radial variations in flow pattern brought about by radial variations of blade speed. Free-vortex tangential velocity distributions are useful. As shown in the discussion of compressor stages, the free-vortex distribution satisfies radial equilibrium requirements. A simpler example will be discussed here, however: a design with a constant outlet angle. For this machine the relative outlet angles of the nozzle and rotor α_2 and β_3 will be assumed constant.

Consider radial variations with an impulse design and no outlet swirl at the mean radius, as in Fig. 9–18. Holding α_2, β_3 and c_z constant, the hub and tip velocity diagrams may be calculated. We can immediately see that this particular design introduces exhaust swirl, at both hub and tip. Of possibly greater significance is the fact that, at the hub, $w_3 < w_2$, so there is danger of the relatively high losses indicated by Fig. 9–15 for $w_{ii}/w_i < 1$. The magnitude of this effect would of course depend on the particular design.

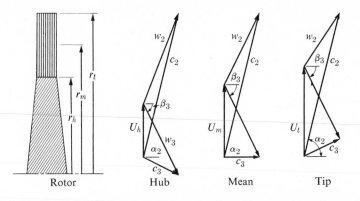

FIG. 9–18. Radial variations in velocity triangles with constant outlet angles.

Deviation

The flow at the exit of a turbine rotor or nozzle blade, just as is the case in compressors, does not leave at exactly the blade exit angle. However, the generally well-behaved flow in a turbine blade row, unlike compressor behavior, is more amenable to simple correlations.

Figure 9–19 shows the conditions between two turbine blades. It has been found by experience that the actual exit flow angle at the design pressure ratio of the machine is fairly well approximated by $\alpha_2 = \cos^{-1}(d/s)$, so long as the nozzle is not choked. If the nozzle is choked, then a supersonic expansion may take place at the blade exit and radically alter the flow direction, providing the exit pressure is sufficiently low.

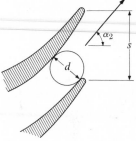

FIG. 9–19. Flow at nozzle exit.

Efficiency

Thus far it has been unnecessary to define or specify turbine efficiency, since we have been interested only in changes in momentum of the working fluid. To describe the actual process of expansion in a turbine, and thus be able to relate work output to fluid pressure, it is convenient to define an ideal turbine, in which the expansion is reversible, and to relate this ideal turbine to the actual turbine through an empirical *efficiency*. The actual turbine work is expressed in Eq. (9–10) as the drop in stagnation enthalpy $\Delta h_0 = c_p \Delta T_0$. Since turbines are essentially adiabatic devices, the ideal turbine would be an isentropic one. The relationships we shall define between actual and ideal processes are shown in the *T-s* diagram of Fig. 9–20.

Two turbine efficiencies are in common usage; the choice between them depends on the application for which the turbine is used. For many conventional turbine applications, useful turbine output is in the form of shaft power, and exhaust

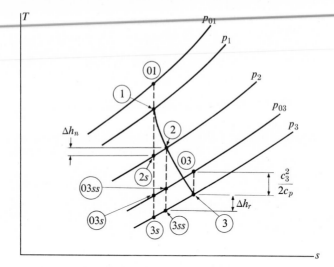

FIG. 9–20. Expansion in a turbine.

kinetic energy, $c_3^2/2$, is considered a loss. In this case, the ideal turbine would then be an isentropic machine with no exhaust kinetic energy, and

$$W_{T,\text{ideal}} = c_p(T_{01} - T_{3s}).$$

If the actual turbine work is compared to this ideal work, the efficiency is called *total-to-static turbine efficiency*, since the ideal work is based on total (or stagnation) inlet conditions and static exit pressure.

The total-to-static turbine efficiency η_{ts} is defined by

$$\eta_{ts} = \frac{T_{01} - T_{03}}{T_{01} - T_{3s}},$$

or

$$\eta_{ts} = \frac{T_{01} - T_{03}}{T_{01}[1 - (p_3/p_{01})^{(\gamma-1)/\gamma}]} = \frac{1 - (T_{03}/T_{01})}{1 - (p_3/p_{01})^{(\gamma-1)/\gamma}}. \qquad (9\text{--}15)$$

In some applications, particularly turbojets, the exhaust kinetic energy is not considered a loss since the exhaust gases are intended to emerge at high velocity. The ideal work in this case is then $c_p(T_{01} - T_{03s})$ rather than $c_p(T_{01} - T_{3s})$. This requires a different definition of efficiency, the *total-to-total turbine efficiency* η_{tt}, defined by

$$\eta_{tt} = \frac{T_{01} - T_{03}}{T_{01} - T_{03s}} = \frac{1 - (T_{03}/T_{01})}{1 - (p_{03}/p_{01})^{(\gamma-1)/\gamma}}. \qquad (9\text{--}16)$$

It is this latter efficiency which is of the most interest for turbojet turbines. The two may be compared by making the approximation (see Fig. 9–20) that

$$T_{03s} - T_{3s} \simeq T_{03} - T_3 = c_3^2/2c_p,$$

and using Eqs. (9–15) and (9–16) to show that

$$\eta_{tt} = \frac{\eta_{ts}}{1 - c_3^2/2c_p(T_{01} - T_{3s})}.$$

Thus

$$\eta_{tt} > \eta_{ts}.$$

Regardless of these definitions, the energy associated with the *tangential component* of the turbine exhaust velocity must be considered a loss, since it will not be transformed to axially directed momentum at the exit of the turbojet exhaust nozzle.

Note that these definitions also serve for multistage turbines if the states (1) and (3) correspond to inlet and outlet, respectively. Thus, with the turbojet engine numbering scheme of Chapter 6, the turbine work, total-to-total efficiency, pressure ratio, and inlet temperature are related by Eq. (6–42).

$$W_T = \eta_{tt} c_p T_{04} \left[1 - \left(\frac{p_{05}}{p_{04}} \right)^{(\gamma-1)/\gamma} \right]. \qquad (6\text{–}42)$$

In the following a method for estimating turbine losses and efficiency will be discussed.

Suppose we are interested in the total-to-total turbine efficiency. Then the ideal turbine work is the difference in stagnation enthalpies $h_{01} - h_{03s}$. The isentropic drop in enthalpy may be written, using the points of Fig. 9–20, as

$$h_{01} - h_{03s} \equiv (h_{01} - h_{03}) + (h_{03} - h_{03ss}) + (h_{03ss} - h_{03s}). \qquad (9\text{–}17)$$

The following approximations may be used for convenience in determining nozzle and rotor losses, Δh_n and Δh_r, respectively:

$$\Delta h_n = h_2 - h_{2s} \simeq h_{03ss} - h_{03s}, \qquad \Delta h_r = h_3 - h_{3ss} \simeq h_{03} - h_{03ss}.$$

With these approximations, Eq. (9–2) can be written

$$h_{01} - h_{03s} = (h_{01} - h_{03}) + \Delta h_n + \Delta h_r.$$

Thus the turbine efficiency η_{tt} may be written

$$\eta_{tt} = \frac{h_{01} - h_{03}}{(h_{01} - h_{03}) + \Delta h_n + \Delta h_r},$$

in which $h_{01} - h_{03}$ is the actual turbine work per unit mass of fluid.

Nozzle and rotor losses in actual machines have been correlated in terms of loss coefficients based on the kinetic energy of the blade row outlet. Loss coefficients ξ_n and ξ_r for nozzle and rotor, respectively, may be defined by

$$\Delta h_n = \xi_n \frac{c_2^2}{2}, \qquad \Delta h_r = \xi_r \frac{w_3^2}{2}. \qquad (9\text{–}18)$$

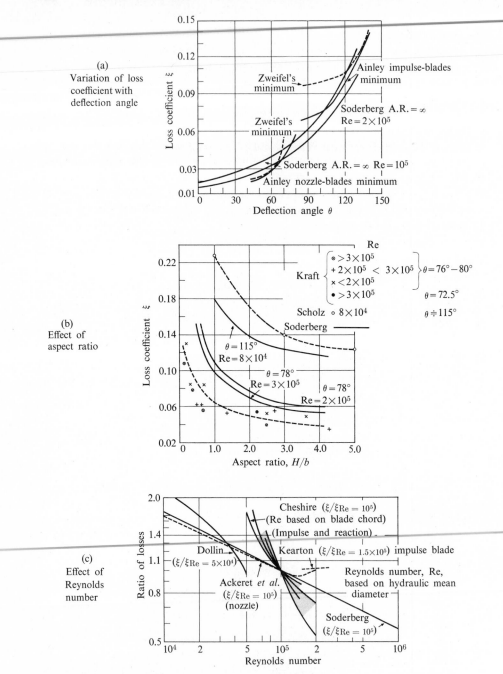

FIG. 9–21. Estimates of loss coefficients for turbine blading (after Horlock [7]). A. R. = blade aspect ratio H/b, H = blade height, b = axial width of blade row, Re = Reynolds number, θ = the fluid turning angle.

The loss coefficients have been found to depend on blade shape, aspect ratio (ratio of blade length to spacing), and Reynolds numbers as well as the fluid incidence angle. Figure 9–21(a) shows a comparison of losses for turbine blading based on the work of Soderberg, Ainley, and Zweifel. (The loss coefficients of Fig. 9–21(a), (b), and (c) apply to either rotor or stator.) The designation "minimum" signifies that the blades are so spaced as to produce minimum loss. The effect of the blade aspect ratio is illustrated in Fig. 9–21(b). Figure 9–21(c) shows various estimates of the effects of variation in Reynolds number. Both these effects are quite important and their influence is much larger than that of changes in incident angle (so long as the incidence is less than 30° or 40°). In Fig. 9–21(c) the Reynolds-number variations in loss coefficient are all normalized by the loss coefficient for a Reynolds number of 10^5. It may be seen that there is considerable disagreement on the magnitude of the effect.

Turbine performance

In the foregoing discussion the performance of turbine stages has been discussed in terms of blade geometry and gas and wheel speeds. If the mass flow and speed of the turbine are such that the blade incidence angles are quite different from the design values, it may nonetheless be a good approximation to say that the relative flow angles α_2 and β_3 of the air leaving each blade row are unchanged. This provides a simple method for estimating the ideal off-design performance of the single stage. For example, for a constant-axial-velocity turbine, we can combine Eq. (9–10) with the statements $U = c_z(\tan \alpha_3 + \tan \beta_3)$ and $c_{\theta 2} - c_{\theta 3} = c_z(\tan \alpha_2 - \tan \alpha_3)$, which may be verified from Fig. 9–16. This yields

$$\frac{\Delta T_0}{T_{01}} = \frac{U^2}{c_p T_{01}}\left[\frac{c_z}{U}(\tan \alpha_2 + \tan \beta_3) - 1\right].$$

For constant values of α_2 and β_3, the temperature rise is thus a function of c_z/U and $(U/\sqrt{\gamma R T_{01}})^2$.

It has already been shown (Chapter 8) that for any given type of turbomachine the overall pressure ratio, for example, may (by dimensional analysis) be written

$$\frac{p_{02}}{p_{01}} = f\left(\frac{\dot{m}\sqrt{RT_{01}}}{p_{01}D^2}, \frac{\Omega D}{\sqrt{\gamma R T_{01}}}, \gamma, \frac{\Omega D^2}{\nu}\right),$$

in which the subscript 01 denotes inlet conditions and D is the scale or size of a series of geometrically similar machines. The Reynolds number $\Omega D^2/\nu$ is not a very important variable so long as it is large enough, as discussed earlier. Thus for a given design of turbine operating with a given fluid at sufficiently high Reynolds number,

$$\frac{p_{04}}{p_{05}} = f\left(\frac{\dot{m}\sqrt{RT_{01}}}{p_{04}D^2}, \frac{\Omega D}{\sqrt{\gamma R T_{04}}}\right),$$

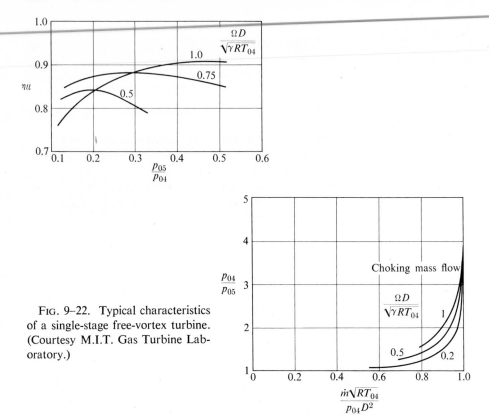

FIG. 9–22. Typical characteristics of a single-stage free-vortex turbine. (Courtesy M.I.T. Gas Turbine Laboratory.)

where stagnation states 04 and 05 are at the turbine inlet and outlet, respectively, in keeping with the scheme of Chapter 6 for numbering turbojet engines. Figure 9–22 shows the overall performance of a particular single stage turbine. It may be seen that pressure ratios much greater than those for compressor stages can be obtained with satisfactory efficiency. The peak efficiency of this machine is not outstandingly high; larger machines may have efficiencies approaching 95%.

The performance of turbines is limited principally by two factors: compressibility and stress. Compressibility limits the mass flow which can pass through a given turbine and, as we shall see, stress limits the wheel speed U. The work per stage, for example, depends on the square of the wheel speed. However, as shown in Chapter 6, the performance of the engine depends very strongly on the maximum temperature. Of course, as the maximum temperature increases the allowable stress level diminishes; hence in the design of the engine there must be a compromise between maximum temperature and maximum rotor tip speed U.

For given pressure ratio and adiabatic efficiency, Eq. (6–42) shows that the turbine work per unit mass is proportional to the inlet stagnation temperature. Since, in addition, the turbine work in a jet or turboshaft engine is commonly three or four times the useful energy output of the engine, a 1% increase in turbine inlet temperature can produce a 3 or 4% increase in engine output. This considerable advantage has supplied the incentive for the adoption of fairly elaborate methods for cooling the turbine nozzle and rotor blades.

Two basic methods of cooling the blades have been used. One method utilizes the flow of liquids or air in thin radial passages internal to the blade. Another, porous cooling, uses the latent heat of an evaporating fluid to cool the surface. The latter method is not successful for rotor blades, however, because the porous blade does not have sufficient strength to withstand the centrifugal stresses.

The advantages gained by operating with higher turbine inlet temperatures are offset to some extent, of course, by the mechanical complexity of the fluid-cooling passages through the blades, rotor, and supporting structure. Also there are performance losses due to the possible necessity of bleeding air from the compressor to accomplish the cooling task.

However, experience has shown that turbine blade cooling is feasible and tends to improve performance appreciably by raising allowable turbine inlet temperatures by up to 300 or 400°F.

9–4 TURBINE AND COMPRESSOR MATCHING

The problem of matching turbine and compressor performance has great importance for jet engines, which must operate under conditions involving large variations in thrust, inlet pressure and temperature, and flight Mach number.

Essentially the matching problem is simple, though the computations can be lengthy. The steady-state engine performance at each speed is determined by two conditions: continuity of flow and a power balance. The turbine mass flow must be the sum of the compressor mass flow and the fuel flow, minus any compressor bleed flow. Also the power output of the turbine must be equal to that demanded by the compressor.

For given flight Mach number, ambient conditions, diffuser and nozzle efficiencies, and flow areas, the performance of a jet engine can be determined from the "maps" of compressor and turbine performance indicated in Fig. 9–23. These diagrams are similar to those of Figs. 8–22 and 9–22 except that, for a given machine, the constants D, R, γ have been omitted and the speed U is replaced by the rotor rpm, N. In principle, the matching computations could proceed as follows:

(1) Select operating speed.
(2) Assume turbine inlet temperature T_{04}.
(3) Assume compressor pressure ratio.
(4) Calculate compressor work per unit mass.
(5) Calculate turbine pressure ratio required to produce this work.
(6) Check to see (Fig. 9–23) if compressor mass flow plus fuel flow equals turbine mass flow; if it does not, assume a new value of compressor pressure ratio and repeat steps (4), (5), and (6) until continuity is satisfied.

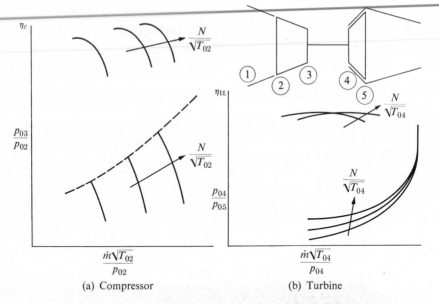

(a) Compressor   (b) Turbine

FIG. 9–23. Typical compressor and turbine performance maps.

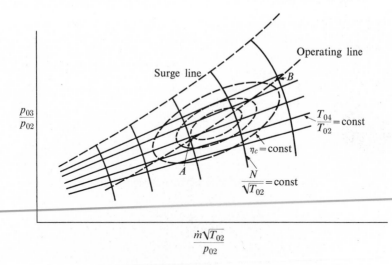

FIG. 9–24. Operating line on a compressor map.

(7) Now calculate the pressure ratio across the jet nozzle from the pressure ratios across the diffuser, compressor, combustor, and turbine.

(8) Calculate the area of the jet nozzle outlet necessary to pass the turbine mass flow calculated in step (6) with the pressure ratio calculated in step (7) and the stagnation temperature calculated. If the calculated area does not equal the actual exit area, assume a new value of T_{04} [step (2)] and repeat the entire procedure.

That the procedure is lengthy may be inferred from the fact that high-speed digital computers may require several minutes to complete the foregoing iteration.

The designer will try to match turbine and compressor so that the compressor is operating near its peak efficiency through the entire range of operation, as indicated in Fig. 9–24, where the operating line (i.e., the locus of steady-state matching conditions) runs through the centers of the islands defined by the constant efficiency lines. It may happen that an operating line located in this manner will be dangerously close to the surge line. If this is the case, it might be possible to adjust the stage velocity triangles so that the maximum efficiency of the compressor occurs farther from the surge line.

Figure 9–24 is useful in a discussion of the important problem of acceleration of gas turbine engines. If an attempt is made to accelerate the engine quickly by increasing the fuel flow very rapidly, there is danger of the compressor surging, which usually causes flameout of the combustor. Also it is possible that the violent changes in aerodynamic load on the compressor blades during the surge period may cause them to fail.

It is convenient to note the form of constant T_{04} lines on the compressor map. In general (referring to Fig. 9–23),

$$\frac{\dot{m}\sqrt{T_{02}}}{p_{02}} = \frac{\dot{m}\sqrt{T_{04}}}{p_{04}} \cdot \frac{p_{04}}{p_{03}} \cdot \frac{p_{03}}{p_{02}} \cdot \sqrt{\frac{T_{02}}{T_{04}}}.$$

The value of p_{04}/p_{03} is practically constant (with a value close to unity). If the turbine nozzles are choked, as they are over a good part of the operating range of turbine engines, then

$$\frac{\dot{m}\sqrt{T_{04}}}{p_{04}} = \text{const.}$$

Thus

$$\frac{p_{03}}{p_{02}} \propto \sqrt{\frac{T_{04}}{T_{02}}} \cdot \frac{\dot{m}\sqrt{T_{02}}}{p_{02}},$$

and lines of constant T_{04}/T_{02} may be plotted as shown in Fig. 9–24: straight lines radiating from the origin. Of course, at lower speeds when the turbine nozzle ceases to be choked, the lines are no longer straight.

Suppose we wish to accelerate the engine between equilibrium operating points A and B. If the fuel flow is suddenly increased, the first effect will be a sudden rise in T_{04}/T_{02}. Thus, before the rotor has time to accelerate, the compressor operation will move along a line of constant $N/\sqrt{T_{02}}$ toward the new T_{04}/T_{02} condition; that is, *toward* the surge line. Thus the fuel control system must limit the rate of additional fuel flow during the acceleration period. As a result, the acceleration process may be quite slow, although some engines (with variable stator blades in the compressor) can accelerate from idle to takeoff power in about six seconds, or from landing approach to full power in about three seconds.

9–5 STRESSES IN TURBINES AND COMPRESSORS

The subject of this text is primarily the thermodynamic and fluid dynamic behavior of propulsion devices. However, the performance demands of propulsion engines, as of all practical devices, must be within limitations set by the materials used in construction. Hence a brief mention of the more important material limitations is necessary.

In general, the geometry, force distribution, and temperature distribution within turbomachine elements are so complex that "exact" stress analyses are not feasible or even possible. In addition, the appropriate material behavior is usually not known with great precision, especially within the complex conditions of an actual application. For example, rotating elements are seldom designed simply to avoid outright failure but, rather, to avoid failure within a specified time. As a result, approximate stress calculations are often all that may be justified. Obviously experience, expressed in the form of adequate safety factors, is a valuable asset in the design of turbomachine elements. Reference [2] contains an excellent qualitative discussion of the methods and limitations of stress analyses.

There are three types of stress which may be of importance in various turbomachine elements; centrifugal, bending, and thermal. Centrifugal stresses are important both in blades and in the discs that support them. Bending stresses arise in the blades from the aerodynamic forces acting on them. Further, as the moving blades pass through wakes from upstream blades, the aerodynamic forces are nonsteady and vibratory bending stresses are set up. The vibratory stresses are usually not large, so long as the forcing frequency does not coincide with one of the natural frequencies of the blade. The steady bending stress can be serious, since it can create a tensile stress in the sharp edges of the blade at the blade root where failure is likely to occur. However, this can be offset, at least at the design speed, by tilting the blade to induce a centrifugal bending stress in the opposite direction. In fact the stress due to centrifugal bending may be deliberately made larger than that due to aerodynamic forces to assure compressive stress near the blade edges.

Turbine discs are in effect heated at the rim by conduction from the hot blades, while they are cooled over their inner portions by air bled from the compressor. As a result, large temperature gradients and thermal stresses may result even during steady operation. Smith [2] states that thermal stresses are usually not important unless other stresses are marginal, although Manson [3] points out that in some instances thermal stresses may be appreciable.

Assuming that bending stresses can be minimized by properly balancing aerodynamic and centrifugal forces and that thermal stresses are important only when added to other stresses, we then encounter centrifugal stresses as a major operating limitation. This limitation can be expressed in the form of a limiting blade speed, as stated earlier.

Consider the simplest possible case, that of a thin bar of length $2L$ rotating about its center, as in Fig. 9–25. The stress at some position r can be calculated in terms of the centrifugal force acting on the mass of material m between r and the end of the bar:

$$F = ma = m\Omega^2 \left(\frac{L + r}{2}\right) = \sigma A,$$

where σ = stress at r, and A = cross-sectional area of bar. The mass m may be expressed as $\rho A(L - r)$, where ρ is the density of the material. We note that ΩL is the tip speed U_t, and then

$$\sigma(r) = \frac{\rho U_t^2}{2}\left[1 - \left(\frac{r}{L}\right)^2\right].$$

Thus the stress at any position r varies as the square of the tip speed. This simple expression may be used directly to calculate the stress at any radius r within an untapered blade by replacing L with the tip radius. The effect of blade taper is relatively easily treated, as in References [2] and [6].

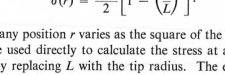

FIG. 9–25. Stress in a rotating bar.

The conclusion that stress is proportional to tip speed squared (in geometrically similar objects) can be demonstrated much more generally with the aid of a simple dimensional analysis: Listing all the quantities which might influence the stress at a particular position in a body, one can say (in the absence of thermal stress),

$$\sigma = f_1(\Omega, \rho, L, \nu, \text{shape, position}),$$

where L = characteristic dimension of body and ν = Poisson's ratio of material. "Shape" can be expressed in terms of dimensionless ratios with L, as can "position," the point at which σ is measured. For a particular point in a series of geometrically similar bodies (shape and position constant), this can be reduced to

$$\frac{\sigma}{\rho\Omega^2 L^2} = f_2(\nu).$$

Hence, taking diameter as a characteristic dimension, we can see that stress is proportional to tip speed squared.

Turbine engines do not, of course, have geometrically similar rotating elements. However, they are roughly similar and, with equally good design (i.e., the use of appropriate materials and the avoidance of stress concentrations), they will be subject to nearly the same limiting tip speed. As stated in Chapter 6, this is about 1400 ft/sec, although most designs use a slightly lower tip speed.

References

1. TAYLOR, E. S., "The Centrifugal Compressor," to be published in *High-Speed Aerodynamics and Jet Propulsion*, Volume X, Section J. Princeton, N. J.: Princeton University Press

2. SMITH, C. W., *Aircraft Gas Turbines*. New York: John Wiley & Sons, 1956

3. MANSON, S. S., "Determination of Elastic Stresses in Gas Turbine Disks," *NACA TR* 871, 1947

4. MANSON, S. S., and M. B. MILLENSON, "Determination of Stresses in Gas Turbine Disks subjected to Plastic Flow and Creep," *NACA TR* 906, 1948

5. MANSON, S. S., "Direct Method of Design and Stress Analysis of Rotating Disks with Temperature Gradient," *NACA TR* 952, 1950

6. VINCENT, E. T., *Theory and Design of Gas Turbines and Jet Engines*. New York: McGraw-Hill, 1950

7. HORLOCK, J. H., "Losses and Efficiencies in Axial-Flow Turbines." *Int. J. Mech. Sci.*, **2**, 1960, pages 48–75 (Pergamon Press)

8. SHEPHERD, D. G., *Principles of Turbomachinery*. New York: The Macmillan Company, 1956

Problems

1. Show that in any ideal turbomachine the total pressure rise in the rotor is given by

$$\int_1^2 \frac{dp_0}{\rho} = \frac{1}{2}(w_1^2 - w_2^2 + U_2^2 - U_1^2 + c_2^2 - c_1^2),$$

if frictional effects are negligible. In this expression the subscripts 1 and 2 denote rotor inlet and exit, respectively. Why are centrifugal compressors generally capable of much greater pressure ratios than axial ones?

2. A centrifugal compressor with 30 straight radial blades operates at a tip speed of 1200 ft/sec. Estimate the pressure ratio for air if the inlet conditions are 1 atm and 70°F, and if the overall adiabatic efficiency is (a) 0.70, and (b) 0.85.

3. Air enters a radial vaneless diffuser at Mach number 1.4 and at an angle 75° from the radial direction. Estimate the pressure ratio across the diffuser if its radius ratio is 1.5.

4. In an attempt to reduce aerodynamic losses associated with the turning of the relative velocity vector in the inducer, it is proposed that stationary inlet guide vanes be installed upstream of the rotor, as shown in the figure. The purpose of these vanes is to give the incoming fluid swirl in the direction of the rotor motion. If the swirl is solid body rotation, $c_\theta = r\Omega$, where Ω is the rotor speed, the inlet turning can be completely eliminated.

(a) Show that if $c_{\theta 1} = r\Omega$, $c_{\theta 2} = U_2$ and the axial velocity c_z is constant over the inlet plane (1), the average work done per unit mass of fluid is

$$\Omega^2 r_2^2 \left[1 - \frac{1}{2}\left(\frac{r_1}{r_2}\right)^2 \right].$$

(b) Is the assumption of constant axial velocity at plane (1) reasonable? (c) Would the inlet guide vanes present boundary layer problems? (d) Could a smaller inlet radius r_1 be used if these inlet guide vanes are installed?

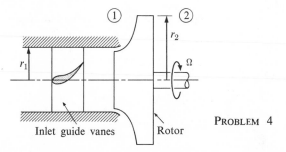

PROBLEM 4

Inlet guide vanes   Rotor

5. Shown in part (a) of the figure is a cross-sectional sketch of a turbojet engine, using a single-stage centrifugal compressor. The compressor tip diameter is 30 in. and the turbine mean diameter is 24 in., as shown. There is no swirl in the compressor inlet and the fuel–air ratio, f, is 0.015 lb fuel/lb air. The axial velocity throughout the turbine (nozzle and rotor) is to be held constant at 400 ft/sec. This axial velocity and the stator outlet velocity have set the nozzle design as in part (b) of the figure. The turbine is to be designed to give *minimum exhaust kinetic energy*, and a slip factor of unity may be assumed for the compressor.

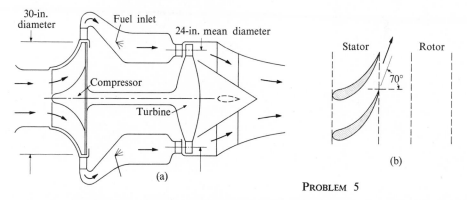

PROBLEM 5

(a) At how many rev/min will the turbine and compressor run? (b) Determine the inlet and outlet relative flow angles for the rotor, and sketch a row of turbine blades to show these angles at the mean diameter.

6. The figure shows velocity diagrams for impulse and 50% reaction turbines which produce the same work (at different speeds). The subscripts 2 and 3 denote conditions before and after the rotor, respectively. In both cases the absolute velocity c_2 is

1200 ft/sec at an angle $\alpha_2 = 70$, and the absolute exhaust velocity c_3 is axial. If the blades are uncooled and well insulated from the turbine disc, approximately what would the equilibrium temperature of the blades be in each case, for an inlet gas temperature $T_{02} = 2000°R$?

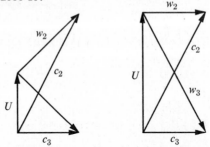

PROBLEM 6

7. A turbine is to be designed with a "free vortex" velocity distribution. This means that the absolute tangential velocity varies with radius according to

$$(rc_\theta)_2 = K_2, \qquad (rc_\theta)_3 = K_3,$$

where K_2 and K_3 are constants, and the subscripts 2 and 3 signify the rotor entrance and exit, respectively.

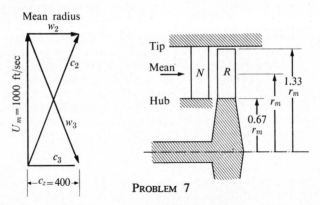

PROBLEM 7

(a) Show that, with this velocity distribution, the work extracted from any unit mass of fluid flowing through the turbine is independent of radial position. (b) Construct velocity triangles (drawn roughly to scale) for the hub (r_h) and tip (r_t) velocities if the mean radius diagram and variation of radius are as shown in the figure.

First complete the following table. Draw approximate *rotor* blade shapes at hub, mean, and tip radii.

Variable	Hub	Mean	Tip
r	$0.667 r_m$	r_m	$1.33 r_m$
c_z			
$c_{\theta 2}$			
$c_{\theta 3}$			

(c) Calculate the percent reaction $[\Delta h_{\text{rotor}}/\Delta h_{\text{total}}]$ at hub, mean, and tip radii. Note that the nozzle inlet velocity is axial, so $c_1 = c_{z1} = 400$ ft/sec. (d) What features of this design would be undesirable from boundary layer considerations?

8. An aft-fan turbofan has the overall dimensions and main gas generator flow at sea-level static operation indicated in the figure. Stress limitations set the revolutions per minute of this assembly at 8500 rev/min. Since the original was designed, improvements in compressor blading allow operation at a maximum inlet relative Mach number of 1.2 and a rotor pressure coefficient of 0.3.

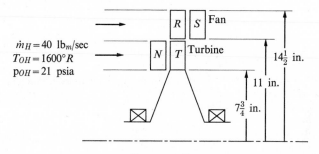

PROBLEM 8

For economic reasons it is not feasible to change the radial dimensions of this element, but it is desired to see what performance advantages can be gained by modifying blade shapes.

Make a preliminary design for this machine, to the extent of specifying hub, tip, and mean velocity triangles for the compressor, and the mean radius velocity triangle for the turbine. Sketch rough blade shapes (including stator and nozzle) for each velocity triangle. Some or all of the following design assumptions may be used:

(a) Constant work from hub to tip of fan ($u \Delta c_\theta = $ const). (b) No inlet guide vanes on fan. (c) Pressure coefficient applies at mean radius of fan *rotor*. (d) Axial outlet for fan and turbine. (e) Constant axial velocity in fan and in turbine. (f) Isentropic diffuser at fan inlet. [*Note:* compressor inlet velocity, density, and area determine fan flow rate and hence power and turbine power.] (g) Ambient conditions: $p_{\text{atm}} = $ 14.7 psia, $T_{\text{atm}} = 60°F$, engine not moving. (h) $R = 53.35$ ft·lb$_f$/lb$_m$·°R and $c_p = $ 0.24 Btu/lb$_m$·°R may be used for both hot and cold gas streams.

Check the pressure coefficient at the hub and tip of both rotor and stator.

If the turbine efficiency is assumed 0.90, what is the hot gas exhaust velocity for an isentropic hot exhaust nozzle? If the compressor efficiency is 0.85, what is the cold gas exhaust velocity for an isentropic cold gas nozzle?

PART 3

Rockets

Rocket vehicles in flight. Left, the Atlas, whose engines use liquid propellants. (Photograph courtesy U.S. Air Force.) Right, the Minuteman, which has solid-propellant engines. (Photograph courtesy Boeing Aircraft Company.)

10

Performance of
Rocket Vehicles

10-1 INTRODUCTION

Rockets are distinguished from air-breathing propulsion engines by the fact that they carry all of their propellant with them. They develop thrust by imparting energy and momentum to the propellant as it is expelled from the engine. The necessary energy source in a rocket may be chemical, nuclear, or solar. Momentum may be given to the fluid by pressure, or by electrostatic or electromagnetic forces. There is thus a variety of possible rocket engines, each with its own characteristic performance. The purpose of this chapter is to discuss in general the performance of rocket vehicles in order to show which characteristics of a rocket engine are most desirable for a given mission. The rocket most suitable for a long space journey, for example, is quite different from the one most suitable for launching from the earth's surface.

At present, chemical rockets are overwhelmingly the most important, and their characteristics and limitations are reasonably well known. The future performance of nuclear and electrical rockets is not easy to predict. Nevertheless, elementary considerations discussed in the following pages show that, for space missions, these rockets may be much superior.

In this chapter we shall discuss the thrust and acceleration of rocket engines and vehicles. We shall consider the effects of exhaust velocity and the proportionate distribution of vehicle mass between propellant, payload, engine, and structure, for both single and multistage vehicles. A brief discussion of space-flight mechanics and missions is included at the end of the chapter in order to show how requirements are established for propulsion systems to be used in space missions.

319

10–2   STATIC PERFORMANCE

Thrust

It is useful to begin by examining the performance of a rocket under static tests. An application of the momentum equation developed in Chapter 2 will show how the thrust developed depends on the propellant flow rate, the exhaust velocity and pressure, and the ambient conditions.

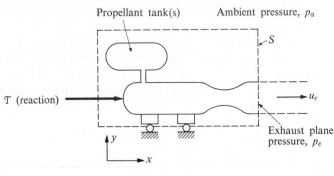

FIGURE 10–1

Consider the thrust of a stationary rocket indicated schematically in Fig. 10–1. For simplicity the flow may be assumed one-dimensional, with a steady exit velocity u_e and propellant flow rate $\dot{m}$. Consider a stationary control surface S which intersects the jet perpendicularly through the exit plane of the nozzle. Positive thrust T acts in the direction opposite to u_e. The *reaction* to the thrust is shown in Fig. 10–1 as it acts on the control volume. If the expelled fluid can be considered a continuum, it is necessary to consider the pressures just inside the exit plane of the nozzle, p_e, and in the environment, p_a. The cross-sectional area of the jet is A_e, the exit area of the nozzle. The momentum equation for any such control volume is

$$\sum F_x = \frac{d}{dt} \int_{cv} \rho u_x \, d\mathcal{V} + \int_{cs} u_x \, d\dot{m}, \tag{10–1}$$

where F_x = force component in the x-direction,

ρ = density of fluid,

u_x = velocity component of fluid in the x-direction,

$\mathcal{V}$ = volume, and

$\dot{m}$ = mass rate of flow (positive for outflow),

and where the subscripts cv and cs denote the control volume and control surface, respectively.

Since u_x is zero within the propellant tank or tanks (the control volume includes the tanks), and the flow is steady within the thrust chamber, the time-derivative term is zero. Also the momentum-flux term can be written

$$\int_{cs} u_x \, d\dot{m} = \dot{m} u_e. \tag{10-2}$$

Considering the pressure on the control surface to be uniformly p_a, except in the plane of the jet, the force summation may be written

$$\sum F_x = T + A_e p_a - A_e p_e. \tag{10-3}$$

Thrust is actually a result of pressure or stress distribution over interior and exterior surfaces, as shown typically for a chemical rocket in Fig. 10–2. However, the questions of how these forces depend on geometry and propellant behavior are deferred to later discussions of individual rocket types. For the present, the momentum equation permits calculation of the overall thrust in terms of conditions in the exhaust plane. Combining Eqs. (10–1), (10–2), and (10–3), we obtain

$$T = \dot{m} u_e + (p_e - p_a) A_e. \tag{10-4}$$

If the pressure in the exhaust plane is the same as the ambient pressure, the thrust is given by $T = \dot{m} u_e$.

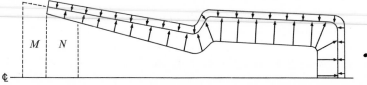

FIGURE 10–2

The condition $p_a = p_e$ is called correct or optimum expansion because it corresponds to maximum thrust for given chamber conditions. This can be shown quite simply. Suppose that a length of diverging nozzle passage (M in Fig. 10–2) is added to a nozzle undergoing correct expansion. Due to further expansion the pressure inside M will be less than the ambient pressure, so that the total pressure force on it will oppose the thrust. Thus the fluid should not be expanded below ambient pressure. On the other hand, expanding the fluid to a pressure above ambient is equivalent to removing length N from the correct nozzle, and it can be seen that this also reduces the thrust. Consequently the thrust of the rocket is maximized, for given chamber conditions and throat area, when the fluid expands to ambient pressure. As will be shown in Section 10–3, it is convenient to define an equivalent exhaust velocity, u_{eq}, such that

$$u_{eq} = u_e + \left(\frac{p_e - p_a}{\dot{m}}\right) A_e. \tag{10-5}$$

With this definition we can write the thrust equation as

$$T = \dot{m} u_{eq}. \tag{10-6}$$

Specific impulse

The impulse per unit mass of propellant will be shown to be an important performance variable. If the velocity u_{eq} is constant, Eq. (10–6) shows that the total impulse I imparted to the vehicle during acceleration is

$$I = \int \mathcal{T}\, dt = \mathfrak{M}_p u_{\mathrm{eq}},$$

where $\mathfrak{M}_p$ is the total mass of expelled propellant. The impulse per unit mass of propellant is therefore

$$\frac{I}{\mathfrak{M}_p} = \frac{\mathcal{T}}{\dot{m}} = u_{\mathrm{eq}}.$$

The term specific impulse, I_{sp}, is usually defined by

I_{sp} = amount of thrust produced by burning 1 lb of fuel for 1 sec

$$I_{\mathrm{sp}} = \frac{I}{\mathfrak{M}_p g_e} = \frac{u_{\mathrm{eq}}}{g_e}, \quad = \frac{\mathcal{T}}{\dot{m} g_e} \tag{10–7}$$

where g_e is the acceleration due to gravity at the earth's surface. The presence of g_e in the definition is arbitrary, but it does have the advantage that in all common systems of units the specific impulse is expressed in seconds.

Since the mass of the propellant is often a large part of the total mass, it would seem desirable to have as large a specific impulse as possible. This conclusion holds directly for chemical rockets. For electrical rockets, high specific impulse implies massive power-generating equipment; hence maximum specific impulse does not generally mean best vehicle performance. These factors will be discussed in more detail subsequently.

10–3 VEHICLE ACCELERATION

Since a large fraction of the total mass of a rocket before firing may be propellant, the mass of the vehicle varies a great deal during flight. This must be taken into account in order to determine the velocity the vehicle attains during the consumption of its propellant. Consider an accelerating rocket as indicated in Fig. 10–3. At some time t its mass is instantaneously $\mathfrak{M}$ and its velocity u. During a short time interval dt, it exhausts a mass increment dm with an exhaust velocity u_e relative to the vehicle as the vehicle velocity changes to $u + du$.

The change of momentum (in the u-direction) of the vehicle during the interval dt is

$$\mathfrak{M}(u + du) - \mathfrak{M}u = \mathfrak{M}\,du.$$

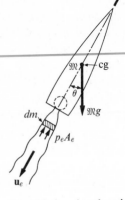

Fig. 10–3. Acceleration of rocket vehicle in a gravity field.

The change of momentum of the mass dm is

$$dm(u - u_e) - dmu = -dmu_e.$$

Consider the forces acting on a system composed of the masses $\mathfrak{M}$ and dm. Unless the exhaust pressure is equal to the atmospheric pressure, there will be a net pressure force $(p_e - p_a)A_e$ acting on the system in the positive **u**-direction. In general, a drag force $\mathfrak{D}$ will act in the negative **u**-direction, along with a gravitational force $(\mathfrak{M} + dm)g \cos \theta$, where g is the local gravitational acceleration and θ is the angle between **u** and **g** as in Fig. 10–3. The net force on the system in the **u**-direction, neglecting dm relative to $\mathfrak{M}$, is

$$\sum F = (p_e - p_a)A_e - \mathfrak{D} - \mathfrak{M}g \cos \theta.$$

The resultant impulse $(\sum F)\,dt$ must equal the momentum change of the system during this interval. Therefore

$$\mathfrak{M}\,du - dmu_e = [(p_e - p_a)A_e - \mathfrak{D} - \mathfrak{M}g \cos \theta]\,dt. \qquad (10\text{–}8)$$

Since

$$dm = + \dot{m}\,dt = -\frac{d\mathfrak{M}}{dt}\,dt,$$

where $\dot{m}$ is the propellant flow rate, Eq. (10–8) becomes

$$\mathfrak{M}\,du = [(p_e - p_a)A_e + \dot{m}u_e - \mathfrak{D} - \mathfrak{M}g \cos \theta]\,dt$$

or, using the definition of u_{eq} in Eq. (10–5), we obtain

$$du = -u_{\text{eq}}\frac{d\mathfrak{M}}{\mathfrak{M}} - \frac{\mathfrak{D}}{\mathfrak{M}}\,dt - g \cos \theta\,dt. \qquad (10\text{–}9)$$

In the absence of drag and gravity, integration of Eq. (10–9), with the assumption of constant equivalent velocity u_{eq}, leads to

$$\Delta u = -u_{\text{eq}}\ln\frac{\mathfrak{M}}{\mathfrak{M}_0} = + u_{\text{eq}}\ln\frac{\mathfrak{M}_0}{\mathfrak{M}}, \qquad (10\text{–}10)$$

where Δu is the change in vehicle velocity and $\mathfrak{M}_0$ is its initial mass. Thus the total change Δu in velocity during the burning period is

$$\Delta u = u_{\text{eq}}\ln\mathfrak{R}, \qquad (10\text{–}11)$$

where the mass ratio $\mathfrak{R} = \mathfrak{M}_0/\mathfrak{M}_b$, and $\mathfrak{M}_b = $ burnout mass, or mass at the end of the thrust period. It will be shown that this term can characterize the propulsion requirements for any mission, including those with gravity and drag effects.

Gravity

The variation in gravitational attraction with distance from the earth's surface can be deduced from Newton's law of gravitation:

$$g = g_e \left(\frac{R_e}{R_e + h} \right)^2, \tag{10-12}$$

where g = the local acceleration due to gravity,
   g_e = the acceleration due to gravity at the surface of the earth,
   R_e = the radius of the earth, and
   h = distance from the surface of the earth.

The thrust period of chemical rockets usually ends when the distance traveled by the vehicle is a small fraction of the earth's radius and the gravitational acceleration has not altered very much. At an altitude of 100 miles, for example, $g = 0.95g_e$. Assuming constant equivalent exhaust velocity, zero drag, and constant gravitational acceleration, we may integrate Eq. (10-9) with the result that

$$\Delta u = u_{eq} \ln \Re - g \, \overline{\cos \theta} t_b, \tag{10-13}$$

where t_b is the burning period and $\overline{\cos \theta}$ is the integrated average value of $\cos \theta$. This approach is convenient only for relatively short thrust periods. When gravity is considered, it is clear that the absolute thrust level is important as well as the exhaust velocity and mass ratio. When a vehicle is launched from the earth's surface, the thrust of the vehicle should be from one and one-half to two times the initial weight in order that the vehicle may leave the ground with reasonable acceleration.

Drag

The resistance of the atmosphere to the passage of a rocket through it can be estimated from empirical drag-coefficient information. It is conventional to express the drag on any body passing through a real fluid in the form

$$\mathfrak{D} = C_{\mathfrak{D}} \tfrac{1}{2} \rho u^2 A_f, \tag{10-14}$$

where $\mathfrak{D}$ = retarding force due to viscous and pressure forces,
   ρ = local air density,
   u = vehicle velocity,
   A_f = frontal cross-sectional area of the vehicle, and
   $C_{\mathfrak{D}}$ = drag coefficient.

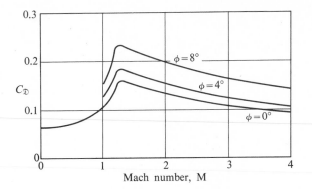

FIG. 10–4. Typical drag coefficient.

The coefficient $C_{\mathfrak{D}}$ depends on the vehicle shape, speed, and inclination, ϕ, to the flight direction. A typical variation is suggested by Fig. 10–4.

Atmospheric density may vary considerably during the thrust period of a rocket. Density varies with altitude approximately as

$$\rho(h) = 0.075 \exp\left(-7.4 \times 10^{-6} h^{1.15}\right),$$

where ρ = the atmospheric density in lb/ft^3, and

    h = distance above the earth in ft.

The atmospheric density is reduced to 1% of its sea-level value at an altitude of just over 100,000 ft. With adequate data on drag coefficient and density variations it is possible to calculate accurately the actual performance of a given rocket vehicle rising against gravitational forces through the atmosphere.

For guidance purposes, of course, accurate trajectory calculations are required. Equation (10–13) is then inadequate, since it requires foreknowledge of the trajectory. Figure 10–5 indicates an approximate calculation procedure which accounts for the effects of gravitational and drag forces on the path of a typical maneuver, a gravity turn. In a gravity turn the vehicle is assumed to begin at some angle θ_1 to the local gravity vector, with all subsequent turning being due only to gravity. The thrust and drag are always parallel to the vehicle velocity (i.e., no aerodynamic lift). Figure 10–5(a) illustrates the changing vehicle velocity components parallel and normal to the earth's surface (u_x and u_y, respectively). The calculation proceeds stepwise through small time and velocity increments from the initial condition. During each time interval δt, $\mathfrak{D}/\mathfrak{M}$, g, and θ are assumed constant. The vehicle velocity change δu in any time increment is the vector sum of three terms: δu_T due to thrust, parallel to u at the start of δt; $\delta u_{\mathfrak{D}}$ due to drag, parallel and opposite to δu_T; and δu_g due to gravity, parallel to g. Each of these terms can be calculated from the velocity at the beginning of δt. Of course, the accuracy of this procedure depends on the size of the time intervals δt compared to t_b.

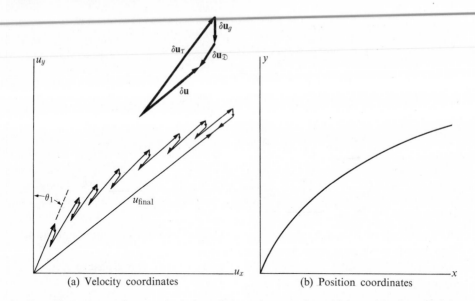

(a) Velocity coordinates   (b) Position coordinates

FIG. 10-5. Approximate calculation of the trajectory of a vehicle undergoing a gravity turn from launch at angle θ_1.

Having obtained the solution of $\mathbf{u}$ as a function of time, the trajectory, Fig. 10-5(b), is easily plotted from the velocity diagram. That is, $\delta y = \mathbf{u}_y \, \delta t$, $\delta x = u_x \, \delta t$, for each δt. This figure is intended only to indicate the nature of the complete problem and to point out that for our purposes we are interested simply in the summation of the magnitudes of the $\delta \mathbf{u}_T$ terms. Clearly, this is

$$ \sum |\delta \mathbf{u}_T| = |\mathbf{u}_{\mathrm{eq}}| \ln \mathfrak{R}. $$

Thus, having properly calculated the actual trajectory, the simple integration of Eq. (10-9), as given by Eq. (10-11), adequately describes the propulsion requirements for a given mission even in the presence of gravity and drag.

Single-stage sounding rocket

As a simple example, consider the height to which a single-stage rocket will rise if drag is neglected and the effective exhaust velocity is assumed constant.

For vertical flight, the altitude attained at burnout, h_b, is

$$ h_b = \int_b^{t_b} u \, dt, $$

where u is given by

$$ u = -u_e \ln \frac{\mathfrak{M}}{\mathfrak{M}_0} - g_e t. $$

vehicle mass at time t

velocity at time t

If the rate of fuel consumption is constant, the mass varies with time as

$$\mathfrak{M}(t) = \mathfrak{M}_0 - (\mathfrak{M}_0 - \mathfrak{M}_b)t/t_b.$$

Then

$$u = -u_e \ln\left[1 - \left(1 - \frac{1}{\mathfrak{R}}\right)\frac{t}{t_b}\right] - g_e t, \tag{10–15}$$

and

$$h_b = -u_e t_b \frac{\ln \mathfrak{R}}{\mathfrak{R} - 1} + u_e t_b - \frac{1}{2} g_e t_b^2. \tag{10–16}$$

Equating the kinetic energy of the mass at burnout, $\mathfrak{M}_b$, with its change of potential energy between that point and the maximum height, $h_{\max}$, we obtain

$$\mathfrak{M}_b \frac{u_b^2}{2} = \mathfrak{M}_b g_e (h_{\max} - h_b),$$

and thus

$$h_{\max} = h_b + \frac{u_b^2}{2g_e}. \tag{10–17}$$

Finally,

$$h_{\max} = \frac{u_e^2 (\ln \mathfrak{R})^2}{2g_e} - u_e t_b \left(\frac{\mathfrak{R}}{\mathfrak{R} - 1} \ln \mathfrak{R} - 1\right). \tag{10–18}$$

Burning time

The above result points out the desirability of reducing the burning time as much as possible while accelerating against a gravity field. Physically, short burning times reduce the energy consumed in simply lifting the propellant. Very short burning times, however, imply not only very high accelerations, which may impose severe stresses on the structure and instruments, but also exceedingly high propellant flow rates. The size of the machinery necessary to handle large flows is a limiting factor. In addition, atmospheric drag imposes a penalty if the vehicle is accelerated too quickly within the earth's atmosphere. Burning times for existing high-thrust rockets are usually in the range of 30 to 200 seconds.

In the absence of gravitation or drag, burning time has no influence on stage velocity increment, as may be seen from Eq. (10–11). In this case, the velocity of the vehicle is a function only of the fraction of propellant expended and not of the time consumed in acceleration.

10–4 CHEMICAL ROCKETS

Chemical rockets are essentially energy-limited devices, whereas electrical rockets are mainly power-limited. Their performance characteristics are quite different, and we will deal with them separately. We begin in this section with the single-stage chemical rocket and then discuss multistaging.

Single-stage rockets

Descriptions of the performance of both single and multistage rockets can be greatly facilitated if we first consider certain definitions. The total mass of a rocket stage may be considered the sum of several parts. Most important among these, although often the smallest, is the mass of the *payload*, $\mathfrak{M}_\ell$. The mass necessary to give the payload the desired motion is usually divided into the *propellant mass*, $\mathfrak{M}_p$, and the *structural mass*, $\mathfrak{M}_s$. Unless otherwise stated, the structural mass includes all mass other than payload and propellant. In particular this includes the engine and the guidance and control equipment, as well as tankage and supporting structures. The *initial mass* $\mathfrak{M}_0$ is the sum of all these quantities:

$$\mathfrak{M}_0 = \mathfrak{M}_\ell + \mathfrak{M}_p + \mathfrak{M}_s. \tag{10–19}$$

If the rocket consumes all its propellant during firing, the burnout mass consists simply of structure and payload:

$$\mathfrak{M}_b = \mathfrak{M}_\ell + \mathfrak{M}_s. \tag{10–20}$$

It has been found useful to arrange these variables in certain ratios. The first of these is the previously introduced mass ratio $\mathfrak{R}$, given by

$$\mathfrak{R} = \frac{\mathfrak{M}_0}{\mathfrak{M}_b} \simeq \frac{\mathfrak{M}_0}{\mathfrak{M}_\ell + \mathfrak{M}_s}, \tag{10–21}$$

the last expression applying when Eq. (10–20) is valid. Another is the payload ratio, λ, defined by

$$\lambda = \frac{\mathfrak{M}_\ell}{\mathfrak{M}_0 - \mathfrak{M}_\ell} = \frac{\mathfrak{M}_\ell}{\mathfrak{M}_p + \mathfrak{M}_s}. \tag{10–22}$$

Large payload ratios are, of course, desirable. To the scientist interested in performing certain observations requiring rocket transportation of instrument payloads, the payload ratio invariably seems too small. Obviously, the larger the payload a given rocket carries, the lower its maximum attainable velocity.

The structural coefficient, ϵ, is defined as

$$\epsilon = \frac{\mathfrak{M}_s}{\mathfrak{M}_p + \mathfrak{M}_s} \approx \frac{\mathfrak{M}_b - \mathfrak{M}_\ell}{\mathfrak{M}_0 - \mathfrak{M}_\ell}. \tag{10–23}$$

The last expression applies exactly when the burnout mass includes no propellant. For a chemical rocket, the structural coefficient ϵ is a measure of the vehicle designer's skill in designing a very light tank and support structure. Combining these definitions, the mass ratio $\mathfrak{R}$ may be expressed as

$$\mathfrak{R} = \frac{1 + \lambda}{\epsilon + \lambda}. \tag{10–24}$$

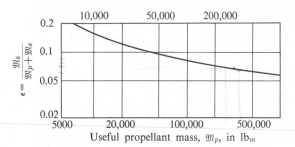

FIG. 10–6. Typical variation of the structural coefficient with the mass of the propellant (adapted from Sandorff [1]).

High values of $\mathfrak{R}$ for a given payload required very careful structural design. It is at present possible to design rockets with total structural masses only 6 or 7% of their total initial mass.

For a given mass of propellant carried and a given mass ratio, every decrease in the structural mass permits an increase of equal magnitude in the payload. Thus it is advantageous to reduce ϵ as much as possible, consistent with the strength requirements of the vehicle. The values of ϵ for the V-2 and Viking rockets were 0.27 and 0.18. In the past few years, however, advances in design have resulted in considerably lighter structures. Figure 10–6 indicates, for present materials and methods of construction, the way in which ϵ varies with the absolute mass of the propellant.

Multistage rockets

For many important missions with chemical rockets, the propellant mass is much larger than the payload. The mass of the propellant tanks and support structure may, in itself, be larger than the payload. Unless portions of this structure and tankage are discarded as they become empty, much energy is consumed in their acceleration and, therefore, less is available for acceleration of the payload. This is one of the reasons for designing rocket vehicles which can discard tank sections as they become empty. In addition, an engine large enough to accelerate the initial mass of the vehicle may produce excessive acceleration stresses when the propellant is nearly consumed. Since it is difficult to operate a given engine at reduced thrust, multistage rocket vehicles are often employed. A multistage rocket is a series of individual vehicles or stages each with its own structure, tanks, and engines. The stages are so connected that each operates in turn, accelerating the remaining stages and the payload before being detached from them. In this way excess structure and tankage are discarded, and the engines of each stage can be properly matched to the remaining vehicle mass. Stages are numbered in the order of firing, as illustrated in Fig. 10–7 for a three-stage chemical rocket carrying payload $\mathfrak{M}_\ell$.

The analysis of multistage rockets is similar to that for single-stage rockets, since the payload for any particular stage is simply the mass of all subsequent stages. The payload for the first stage of Fig. 10–7 is simply the sum of the masses of stages 2 and 3 (including $\mathfrak{M}_{\mathfrak{L}}$).

The definitions above may be extended to apply to the generalized ith stage of a rocket consisting of a total of n stages. The nomenclature used is as follows:

i = any of the stages $1 \leqq i \leqq n$,

$\mathfrak{M}_{0i}$ = the total initial mass of the ith stage prior to firing, including its effective payload,

$\mathfrak{M}_{bi}$ = the total mass of the ith stage after burnout, including its effective payload,

$\mathfrak{M}_{\mathfrak{L}}$ = the payload of the last stage to fire, and

$\mathfrak{M}_{si}$ = the structural mass of the ith stage alone, including the mass of its engine, controls, and instruments.

Since, as we mentioned, the payload of any stage is simply the mass of all subsequent stages, the payload ratio λ_i of the ith stage is

P/L wt not included in moi

$$\lambda_i = \frac{\mathfrak{M}_{0(i+1)}}{\mathfrak{M}_{0i} - \mathfrak{M}_{0(i+1)}}. \qquad (10\text{--}25)$$

The structural coefficient of the ith stage becomes

$$\epsilon_i = \frac{\mathfrak{M}_{si}}{\mathfrak{M}_{0i} - \mathfrak{M}_{0(i+1)}}, \qquad (10\text{--}26)$$

or, if the ith stage contains no propellant at burnout,

$$\epsilon_i = \frac{\mathfrak{M}_{bi} - \mathfrak{M}_{0(i+1)}}{\mathfrak{M}_{0i} - \mathfrak{M}_{0(i+1)}}. \qquad (10\text{--}27)$$

The mass ratio $\mathfrak{R}_i$ of the ith stage may be written

$$\mathfrak{R}_i = \mathfrak{M}_{0i}/\mathfrak{M}_{bi}, \qquad (10\text{--}28)$$

which, with the aid of the definitions, becomes

$$\mathfrak{R}_i = \frac{1 + \lambda_i}{\epsilon_i + \lambda_i}. \qquad (10\text{--}29)$$

Third stage

Second stage

First stage

$\mathfrak{M}_{\mathfrak{L}}$

Fig. 10–7. Typical multistage rocket.

Example

1000

14000

As an illustration of the gains to be made by multistaging, consider first a single-stage rocket whose exhaust velocity is 10,000 fps. Suppose we must accelerate a 1000-pound payload with a total initial mass of 15,000 pounds and a total structural mass of 2000 pounds (including engine, controls, and instruments). Using the foregoing definition with this data, we find that the structural coefficient of

2000

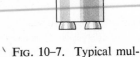

$$\epsilon = \frac{M_{s_1}}{M_{0_1} - M_{0_2}} = \frac{2}{15 - 1} = \frac{1}{7} = .143$$

$$\lambda = \frac{\mathfrak{M}_{02}}{\mathfrak{M}_{01} - \mathfrak{M}_{02}} = \frac{1}{15-1} = \frac{1}{14} = .0714$$

the rocket is $\epsilon = 0.143$, and the payload factor is $\lambda = 0.0714$. The velocity to which the payload could be accelerated in a zero gravity field can be obtained from

$$u = u_e \ln \mathfrak{R} = u_e \ln \frac{1 + \lambda}{\epsilon + \lambda},$$

$$= 10 \cdot 10^3 \, \ln \frac{1.0714}{.2144}$$

$$= 16,088 \text{ fps}$$

which yields a value of u of 16,300 fps.

Consider next a two-stage rocket of the same initial mass, total structural mass, payload, and exhaust velocity as the first rocket. If the payload ratios of the two stages are assumed to be equal,

$$\lambda_1 = \lambda_2 = \lambda = \frac{\mathfrak{M}_{02}}{\mathfrak{M}_{01} - \mathfrak{M}_{02}} = \frac{\mathfrak{M}_\mathfrak{L}}{\mathfrak{M}_{02} - \mathfrak{M}_\mathfrak{L}}.$$

With $\mathfrak{M}_\mathfrak{L} = 1000 \text{ lb}_m$ and $\mathfrak{M}_{01} = 15,000 \text{ lb}_m$, this requirement is satisfied by $\mathfrak{M}_{02} = 3875 \text{ lb}_m$ and $\lambda_1 = \lambda_2 = 0.348$. Assuming the structural coefficients of the two stages to be the same, we obtain

$$\epsilon_1 = \epsilon_2 = \epsilon = \frac{\mathfrak{M}_{s1}}{\mathfrak{M}_{01} - \mathfrak{M}_{02}} = \frac{\mathfrak{M}_{s2}}{\mathfrak{M}_{02} - \mathfrak{M}_\mathfrak{L}}.$$

Also

$$\mathfrak{M}_{s1} + \mathfrak{M}_{s2} = 2000 \text{ lb}_m.$$

Solving these two equations, we find that

$$\mathfrak{M}_{s1} = 1589 \text{ lb}, \qquad \mathfrak{M}_{s2} = 411 \text{ lb}, \qquad \epsilon_1 = \epsilon_2 = 0.143.$$

Since the mass ratio $\mathfrak{R} = (1 + \lambda)/(\epsilon + \lambda)$, and since the exhaust velocity of each stage is the same, the final velocity attained by the vehicle is

$$u = u_1 + u_2$$

$$u = 2u_{eq} \ln \left(\frac{1 + \lambda}{\epsilon + \lambda} \right) \qquad \text{or} \qquad u = 20,200 \text{ fps}.$$

The performance of the two rockets is summarized in Table 10–1, which shows that the two-stage rocket is clearly superior to the single-stage rocket in terminal velocity of the payload.

Returning to the general n-stage rocket, the stage velocity increment in the absence of gravity and drag is $\Delta u_i = u_{ei} \ln \mathfrak{R}_i$, and the payload terminal velocity is simply the sum of the n velocity increments:

$$u_n = \sum_{i=1}^{n} u_{ei} \ln \mathfrak{R}_i. \tag{10–30}$$

If the exhaust velocities are all the same, this reduces to

$$u_n = u_e \ln \left(\prod_{1}^{n} \mathfrak{R}_i \right), \tag{10–31}$$

Table 10-1   Performance of Single-Stage and Two-Stage Rockets*

	Single stage	Two stage	
		Stage 1	Stage 2
Total initial mass $\mathfrak{M}_0$, lb (including payload)	15,000	15,000	3,875
Payload, lb	1,000	3,875	1,000
Structural mass, lb	2,000	1,589	411
Structural coefficient, ϵ *const*	0.143	0.143 ⟷ 0.143	
Payload ratio, λ *const*	0.0714	0.348 ⟷ 0.348	
Mass ratio, $\mathfrak{R}$ *const*	5	2.72 ⟷ 2.72	
Velocity increment, u, fps	16,300	10,100 ⟷ 10,100	
Terminal payload velocity, fps	16,300		20,200

* Same payload, exhaust velocity and total structural mass.   Exhaust velocity 10,000 fps.   Zero gravity and drag.

where $\prod_1^n \mathfrak{R}_i$ is the product of the n mass ratios. If the exhaust velocities, structural coefficients, and payload ratios are all identical, this reduces further to

$$u_n = n u_e \ln \mathfrak{R},$$

or

$$u_n = n u_e \ln \frac{1 + \lambda}{\epsilon + \lambda}. \tag{10-32}$$

The stages are said to be similar if they have the same values of λ and ϵ. This special case is quite useful for approximate calculations, especially if the exhaust velocities we are considering are reasonably uniform and the structural coefficient does not vary markedly from stage to stage. It can be shown (see Appendix 1) that under these conditions it is desirable to have identical stage payload ratios.

The overall mass ratio is related to the stage payload ratios as follows. From Eq. (10-25), we have

$$\frac{\mathfrak{M}_{01}}{\mathfrak{M}_{02}} = \frac{1 + \lambda_1}{\lambda_1},$$

$$\frac{\mathfrak{M}_{02}}{\mathfrak{M}_{03}} = \frac{1 + \lambda_2}{\lambda_2},$$

$$\vdots \qquad \qquad \vdots$$

$$\frac{\mathfrak{M}_{0(n-1)}}{\mathfrak{M}_{0n}} = \frac{1 + \lambda_{n-1}}{\lambda_{n-1}},$$

$$\frac{\mathfrak{M}_{0n}}{\mathfrak{M}_{\mathfrak{L}}} = \frac{1 + \lambda_n}{\lambda_n}.$$

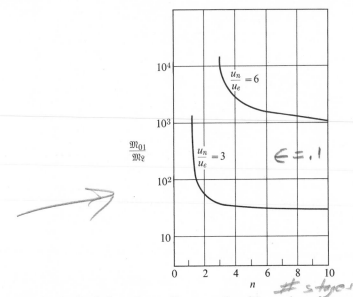

FIG. 10–8. Variation of overall mass ratio with number of stages for fixed terminal velocity ratios; similar stages and structural coefficient $\epsilon = 0.10$.

Multiplying these ratios together, we obtain as a result:

$$\frac{\mathfrak{M}_{01}}{\mathfrak{M}_{\mathfrak{L}}} = \prod_{i=1}^{n} \left(\frac{1 + \lambda_i}{\lambda_i}\right). \tag{10–33}$$

If the payload ratios are identical, the overall mass ratio is given by

$$\frac{\mathfrak{M}_{01}}{\mathfrak{M}_{\mathfrak{L}}} = \left(\frac{1 + \lambda}{\lambda}\right)^{n}. \tag{10–34}$$

The variation of payload terminal velocity with overall mass ratio can be determined from Eqs. (10–32) and (10–34) for the special case of the rocket with similar stages. The result is

$$\frac{u_n}{u_e} = n \ln \left\{ \frac{(\mathfrak{M}_{01}/\mathfrak{M}_{\mathfrak{L}})^{1/n}}{\epsilon[(\mathfrak{M}_{01}/\mathfrak{M}_{\mathfrak{L}})^{1/n} - 1] + 1} \right\}. \tag{10–35}$$

Gravitational forces may be taken into account approximately by replacing the term on the left side by

$$\frac{u_n + g_e t_b \,\overline{\cos \theta}}{u_e}.$$

Figure 10–8 illustrates Eq. (10–35) for a value of the structural coefficient $\epsilon = 0.10$. For a ratio of terminal and exhaust velocities of 3 (which corresponds

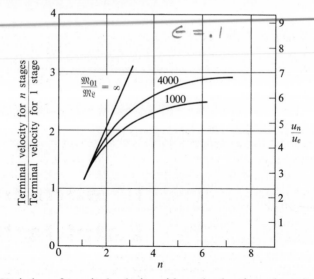

FIG. 10-9. Variation of terminal velocity with payload ratio and number of similar stages; $\epsilon = 0.10$.

approximately to earth escape using high-energy propellants), 2 or 3 stages could be used; but a single-stage rocket would not be able to do the job even if the payload mass were negligible.

The ratio of payload terminal velocities with several similar stages to that for one stage, and with the same overall mass ratio, is shown in Fig. 10-9, again for the assumption that $\epsilon = 0.10$. Here it is evident that increasing the number of stages yields diminishing returns for a given u_n/u_e. It is of interest to determine the terminal velocity obtained with an infinite number of similar stages for a given overall mass ratio $\mathfrak{M}_\mathfrak{L}/\mathfrak{M}_{01}$. Using Eq. (10-35), we find that this limit is given by

$$\lim_{n \to \infty} u_n = u_e \lim_{n \to \infty} \left\{ -n \ln \left[\epsilon + \frac{1 - \epsilon}{(\mathfrak{M}_{01}/\mathfrak{M}_\mathfrak{L})^{1/n}} \right] \right\}$$

or

$$\lim_{n \to \infty} u_n = u_e(1 - \epsilon) \ln \left(\frac{\mathfrak{M}_{01}}{\mathfrak{M}_\mathfrak{L}} \right). \qquad (10\text{-}36)$$

For the example summarized in Table 10-1, we obtain

$$\frac{\mathfrak{M}_{01}}{\mathfrak{M}_\mathfrak{L}} = 15, \qquad \epsilon = 0.143, \qquad u_e = 10{,}000 \text{ ft/sec},$$

and Eq. (10-36) yields

$$\lim_{n \to \infty} u_n = 23{,}200 \text{ ft/sec},$$

which compares with 16,300 ft/sec for one stage and 20,000 ft/sec for two.

It is also interesting to determine the limiting value of the terminal velocity for a massless structure. From Eq. (10–35),

$$\lim u_n = u_e \ln \frac{\mathfrak{M}_{01}}{\mathfrak{M}_{\mathfrak{L}}}, \qquad \epsilon \to 0.$$

As one might expect, with the limiting case of a massless structure there is no advantage in multistaging.

As it is in most engineering problems, economy is of importance in the design of rocket vehicles. Optimization analyses of the various vehicles have as their goal the design of the most economical system for the proposed mission. However, the approximations which must be made so that optimization can be done mathematically are usually severe, and thus the results of such analyses should be viewed as general guides rather than specific answers.

The analyses discussed here have to do with minimizing the initial mass of the vehicle for a given mission. It is implied that the significant cost considerations are closely related to the mass of equipment and propellant of which the vehicle is composed. This assumption ignores many important factors which help to determine the best design. The choice of propellant, for example, may be a major consideration. It often depends on factors such as availability, storability, and ease of handling. Another factor is the time necessary to design and develop a new rocket vehicle. Often it is not feasible to design and build a completely new vehicle for a particular application, and the best vehicle under some circumstances may consist of a combination of existing components or stages—a vehicle which is not necessarily optimum in the narrow sense already defined.

An important question arises in the design of multistage chemical rocket vehicles: For a given number of stages, what should their relative sizes be? The answer to this question employs the techniques of the calculus of variations. Appendix 1 shows that if the structural coefficients ϵ and the exhaust velocities u_e are uniform through the stages, the stage payload ratios λ should also be uniform in order to minimize the overall mass ratio for a given terminal velocity. If the structural coefficient or the exhaust velocity (or both) are variable, methods are available (Appendix 1) for determining the best variation in payload ratio through the stages.

10–5 ELECTRICAL ROCKET VEHICLES   *Skip*

Electrical and electromagnetic rockets differ fundamentally from chemical rockets with respect to their performance limitations. Chemical rockets are essentially *energy-limited*, since the quantity of energy (per unit mass of propellant) which can be released during combustion is limited by the fundamental chemical behavior of propellant materials. If, on the other hand, a separate energy source (e.g., nuclear or solar) is used, much higher propellant energy is possible. Further, if the temperature limitations of solid walls could be made unimportant by direct electrostatic or electromagnetic propellant acceleration without necessarily raising the fluid and solid temperatures, there would be no necessary upper limit to the energy which could be added to the propellant. However, the rate of conversion of nuclear or solar energy to electrical energy and thence to propellant kinetic energy is limited by the mass of conversion equipment required. Since this mass is likely to be a large portion of the total mass of the vehicle, the electrical rocket is essentially *power-limited*.

Electrostatic and electromagnetic rockets convert electrical energy directly to propellant kinetic energy without necessarily raising the temperature of the working fluid. For this reason the specific impulse is not limited by the temperature limitations of wall materials, and it is possible to achieve very high exhaust velocities, although at the cost of high power consumption.

Suppose a fraction η of the generated electrical power is transformed to propellant kinetic energy. For a given machine with constant η, or for different machines with similar η, the exhaust power per unit thrust is

$$\frac{\mathcal{P}_e}{T} \propto \frac{\dot{m}u_e^2/2}{\dot{m}u_e} = \frac{u_e}{2},$$

whereas the mass flow rate per unit thrust is $\dot{m}/T = 1/u_e$. From this it can be seen that high exhaust velocity (high specific impulse) requires high power, and hence large capacity for power conversion per unit thrust. On the other hand, low exhaust velocity requires high mass flow rate per unit thrust and hence large propellant storage capacity for a given thrust duration. Between these extremes lies an optimum exhaust velocity. The optimum exhaust velocity is that which gives a maximum payload ratio for the mission under consideration. An approximate method of establishing this optimum is given in the following paragraphs. It is assumed that the vehicle must accelerate through a velocity increment Δu during a period t_b of constant thrust. The way in which the variables Δu and t_b depend on the mission will be discussed subsequently. Variables of importance for the rocket itself are the *specific mass* α of the power plant and the efficiency of the thrust chamber η. The specific mass is here defined as the mass of power plant and structure per unit of electrical power output. The structural mass is included in the definition for convenience. It is likely to be a small fraction of the power-plant mass:

$$\alpha = \frac{\mathfrak{M}_{\mathcal{P}} + \mathfrak{M}_s}{\mathcal{P}}, \tag{10-37}$$

where $\mathcal{P}$ = electrical power output,

   $\mathfrak{M}_{\mathcal{P}}$ = mass of power plant, and

   $\mathfrak{M}_s$ = mass of structure.

The thrust chamber efficiency is the fraction of electrical power that is converted to exhaust kinetic energy,

$$\eta = \frac{\dot{m}u_e^2}{2\mathcal{P}}. \tag{10-38}$$

The velocity increment is given by Eq. (10–10),

$$\Delta u = u_e \ln \frac{\mathfrak{M}_0}{\mathfrak{M}_b},$$

or

$$\frac{\mathfrak{M}_b}{\mathfrak{M}_0} = e^{-\Delta u/u_e}. \tag{10–39}$$

If all the propellant is consumed, the mass of the vehicle at the end of the thrust period is $\mathfrak{M}_b = \mathfrak{M}_\wp + \mathfrak{M}_s + \mathfrak{M}_\wp$. The burnout mass ratio, when we use Eq. (10–37), is

$$\frac{\mathfrak{M}_b}{\mathfrak{M}_0} = \frac{\alpha\mathcal{P}}{\mathfrak{M}_0} + \frac{\mathfrak{M}_\wp}{\mathfrak{M}_0}. \tag{10–40}$$

Using Eq. (10–38), we obtain

$$\frac{\alpha\mathcal{P}}{\mathfrak{M}_0} = \frac{\alpha\dot{m}u_e^2}{2\eta\mathfrak{M}_0},$$

and if the propellant flow rate is constant,

$$\frac{\alpha\mathcal{P}}{\mathfrak{M}_0} = \frac{\alpha u_e^2\mathfrak{M}_p/t_b}{2\eta\mathfrak{M}_0} \quad \text{or} \quad \frac{\alpha\mathcal{P}}{\mathfrak{M}_0} = \frac{\alpha u_e^2}{2\eta t_b}\left(1 - \frac{\mathfrak{M}_b}{\mathfrak{M}_0}\right).$$

Combining this expression with Eq. (10–40) yields

$$\frac{\mathfrak{M}_b}{\mathfrak{M}_0} = \frac{(\alpha u_e^2/2\eta t_b) + (\mathfrak{M}_\wp/\mathfrak{M}_0)}{1 + (\alpha u_e^2/2\eta t_b)}.$$

Substituting this expression in Eq. (10–39), we find that the payload-to-initial-mass ratio is given by

$$\frac{\mathfrak{M}_\wp}{\mathfrak{M}_0} = e^{-\Delta u/u_e}\left(1 + \frac{\alpha u_e^2}{2\eta t_b}\right) - \frac{\alpha u_e^2}{2\eta t_b}. \tag{10–41}$$

Figure 10–10 illustrates the variation of this ratio with specific impulse for a one-month transfer between earth and moon orbits. During this period the thrust is assumed constant through a velocity increment $\Delta u \approx 24{,}000$ ft/sec (7350 m/sec).

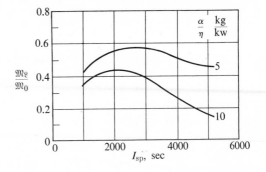

FIG. 10–10. Variation of payload ratio with specific impulse according to Eq. (10–41) for a one-month earth–moon transfer.

The specific mass ratio α of large nuclear turbo-generators for space-vehicle power plants is expected to fall within the range 1 to 10 kg/kw and the thrust chamber efficiency between 0.7 and 1.0.

The following section discusses briefly the way in which equivalent values of Δu and t_b are found for space missions.

10–6 SPACE MISSIONS

In order to demonstrate the relationship between engine performance and vehicle trajectory for a space mission, it is necessary to introduce the subject of orbital and ballistic mechanics. To start the discussion, we shall assume that the vehicle is accelerated impulsively at certain points and that it travels between these points without thrust under the influence of gravitational force alone.

High thrust

For our purposes it is satisfactory to simplify the general trajectory problem by considering the gravity force between the vehicle and only one other body. Even for the case of interplanetary transfer, this simplification will suffice for most calculations. That is, escape from or capture by a planet may be considered an interaction between the vehicle and that particular planet alone, while the transfer process is considered an interaction between the vehicle and the sun alone. The transition from one operation to another has only minor influence on propulsion requirements.

Since the vehicle will always be much smaller than the other body under consideration, the latter is uninfluenced by the vehicle motion and its center can serve as a stationary coordinate origin. Under these conditions it is easy to see that an unpowered vehicle must travel in a stationary plane which passes through the center of the influencing planet. At any time the instantaneous velocity of the vehicle and the center of mass of the planet together determine a plane which the vehicle cannot leave because the only force (gravitational) acting on the vehicle is in the plane.

Consider the motion of a vehicle of mass $\mathfrak{M}$ moving past a large body of mass M' as shown in Fig. 10–11. The minimum separation between the two is $r_{\min}$. The gravitational force acts in the negative r-direction, and is given by Newton's law of gravitation:

$$F_r = -\frac{GM'\mathfrak{M}}{r^2}, \qquad (10\text{–}42)$$

where G = universal gravitational constant = 6.670×10^{-11} newton-m^2/kg^2. Equating the gravitational force to the radial inertial forces, we have

$$-\frac{GM'\mathfrak{M}}{r^2} = \mathfrak{M}(\ddot{r} - r\dot{\theta}^2), \qquad (10\text{–}43)$$

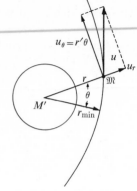

FIG. 10–11. Vehicle coordinates and velocity components.

where $\ddot{r} = d^2r/dt^2$ and $\dot{\theta} = d\theta/dt$. Since the only force acting on $\mathfrak{M}$ is radial, its angular momentum h must remain constant:

$$h = \mathfrak{M}r^2\dot{\theta} = \text{constant}. \tag{10-44}$$

Eliminating $\dot{\theta}$ from Eq. (10–43) by using Eq. (10–44), we obtain

$$\mathfrak{M}\ddot{r} - \frac{h^2}{\mathfrak{M}r^3} = -\frac{GM'\mathfrak{M}}{r^2}. \tag{10-45}$$

We may express the velocity and acceleration as follows:

$$\dot{r} = \frac{dr}{d\theta}\frac{d\theta}{dt} = \dot{\theta}\frac{dr}{d\theta} = \frac{h}{\mathfrak{M}r^2}\frac{dr}{d\theta}, \qquad \ddot{r} = \frac{d\dot{r}}{dt} = \frac{d\dot{r}}{d\theta}\frac{d\theta}{dt} = \frac{h}{\mathfrak{M}r^2}\frac{d\dot{r}}{d\theta}.$$

Using also the substitution [2] $Z = 1/r - GM'\mathfrak{M}^2/h^2$, we can show that Eq. (10–45) can be transformed to $d^2Z/d\theta^2 = -Z$, which has the solution

$$Z = A\cos\theta + B\sin\theta \qquad \text{or} \qquad \frac{1}{r} = \frac{GM'\mathfrak{M}^2}{h^2} + A\cos\theta + B\sin\theta.$$

Using the boundary conditions at $r = r_{\min}$,

$$\frac{d}{d\theta}\left(\frac{1}{r}\right) = 0, \qquad \theta = 0,$$

we find that

$$\frac{1}{r} = \frac{GM'\mathfrak{M}^2}{h^2} + \left(\frac{1}{r_{\min}} - \frac{GM'\mathfrak{M}^2}{h^2}\right)\cos\theta. \tag{10-46}$$

This may be compared with the equation of a conic (ellipse, parabola, or hyperbola) in polar coordinates:

$$\frac{1}{r} = \frac{1}{a(1 - \epsilon^2)} + \frac{\epsilon\cos\theta}{a(1 - \epsilon^2)} \tag{10-47}$$

where, referring to Fig. 10–12, a = semimajor axis, b = semiminor axis, ϵ = eccentricity = c/a, and $b^2/a^2 = 1 - \epsilon^2$.

Hence the vehicle travels in a hyperbola, a parabola, or an ellipse, with the planet as focus. By comparing terms in Eqs. (10–46) and (10–47), we can show that

$$\epsilon = \frac{h^2}{GM'\mathfrak{M}^2 r_{\min}} - 1 \tag{10-48}$$

and, referring to Fig. 10–12 for the ellipse, we have

$$\frac{a}{r_{\min}} = \frac{1}{1 - \epsilon}. \tag{10-49}$$

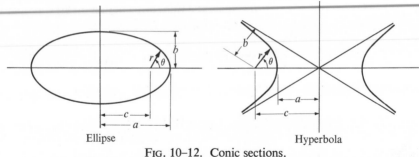

Ellipse   Hyperbola
FIG. 10–12. Conic sections.

If $\epsilon > 1$, the trajectory is hyperbolic; if $\epsilon = 1$, it is parabolic; and if $\epsilon < 1$, it is elliptic.

The conic is completely specified by its shape and size. The size, expressed by a, will now be shown to be a function of *orbital energy*. This orbital energy, defined as the sum of the kinetic and potential energies of the vehicle with respect to the influencing planet, must remain constant. The potential energy is defined as

$$\text{PE} = \int_{\infty}^{r} \frac{GM'\mathfrak{M}}{r^2}\, dr = -\frac{GM'\mathfrak{M}}{r}. \tag{10–50}$$

At infinity the potential energy is zero and for all finite r it is negative. The orbital energy E is then

$$E = \text{KE} + \text{PE} = \frac{\mathfrak{M}u^2}{2} - \frac{GM'\mathfrak{M}}{r} = \frac{\mathfrak{M}}{2}[\dot{r}^2 + (r\dot{\theta})^2] - \frac{GM'\mathfrak{M}}{r}. \tag{10–51}$$

When we evaluate at $r = r_{\min}$, $\dot{r} = 0$, and use Eq. (10–44), this becomes

$$E = \frac{1}{2}\frac{h^2}{\mathfrak{M}r_{\min}^2} - \frac{GM'\mathfrak{M}}{r_{\min}}, \tag{10–52}$$

which is constant for an unpowered vehicle.

After we combine Eqs. (10–52), (10–49), and (10–48), we obtain

$$E = -\frac{G\mathfrak{M}M'}{2a}. \tag{10–53}$$

Equations (10–48), (10–52), and (10–53) show that a and ϵ are functions only of E and h. Thus the orbit or trajectory of a given mass $\mathfrak{M}$ about a given mass M' is completely determined by the orbital energy E and the angular momentum h. Conversely, these quantities are *properties* of fixed orbits, as are, for example, the orbits of the various planets about the sun. Table 10–2 contains data on the planets and their orbits.

Since kinetic energy is positive while potential energy is negative, the total orbital energy E may be either positive or negative, and three distinctly different behaviors are observed corresponding to positive, zero, or negative orbital energy.

Table 10–2   Some Properties of Planets and Their Orbits Around the Sun*

Body	Relative mass†	Relative mean diameter†	Orbit radius, km $\times 10^{-6}$	Eccentricity	Period, years
Sun	332,488.	109.15	——	——	——
Mercury	0.0543	0.38	57.9	0.2056	0.241
Venus	0.8136	0.967	108.1	0.0068	0.615
Earth	1.0000	1.000	149.5	0.0167	1.000
Mars	0.1069	0.523	227.8	0.0934	1.881
Jupiter	318.35	10.97	777.8	0.0484	11.862
Saturn	95.3	9.03	1426.1	0.0557	29.458
Uranus	14.58	3.72	2869.1	0.0472	84.015
Neptune	17.26	3.38	4495.6	0.0086	164.788
Pluto	<0.1?	0.45	5898.9	0.2485	247.697
Moon	0.0123	0.273	0.3844	0.0549	27.322 (days)

* From Smithsonian Physical Tables [3].
† Relative to earth: Earth mass = 5.975×10^{24} kg; earth mean diameter = 12,742.46 km.

$E > 0$

In this case, the kinetic energy is greater than the rise in potential energy between the initial radius and infinity. Thus the vehicle possesses sufficient energy to coast away from the influencing planet, arriving at infinity with some excess kinetic energy. With $E > 0$, Eq. (10–52) shows that $h^2/GM'\mathfrak{M}^2 r_{\min} > 2$, so that, from Eq. (10–48), $\epsilon > 1$. Thus the vehicle trajectory will be hyperbolic.

$E = 0$

In this case the vehicle possesses kinetic energy just sufficient to coast to infinite radius with zero velocity. This special case is of importance in that it determines the minimum kinetic energy necessary to escape the planet's influence. The *escape velocity*, $u_{\rm esc}$, is then, setting $E = 0$ in Eq. (10–51):

$$\frac{\mathfrak{M}u_{\rm esc}^2}{2} - \frac{GM'\mathfrak{M}}{r} = 0,$$

or

$$u_{\rm esc} = \sqrt{2GM'/r}. \tag{10–54}$$

Escape velocity may also be found directly by equating kinetic energy with work done against gravity from r to infinity,

$$\frac{\mathfrak{M}u_{\rm esc}^2}{2} = \int_r^\infty F_r\, dr = \int_r^\infty \left(+ \frac{GM'\mathfrak{M}}{r^2} \right) dr,$$

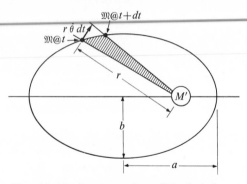

FIG. 10–13. Interpretation of Kepler's law.

with the same result. Again using Eqs. (10–52) and (10–48), we find that $\epsilon = 1$ for $E = 0$, and the escape trajectory is parabolic.

$E < 0$

In this case the vehicle possesses kinetic energy which is insufficient to allow it to escape the planet. With Eqs. (10–52) and (10–48), it may be seen that $\epsilon < 1$; thus the orbit is elliptic. For the special case of a circular orbit, $\epsilon = 0$, $u_{\text{circ}} = r\dot{\theta}$, $r = $ constant and, using Eqs. (10–44) and (10–48), we have

$$\epsilon = 0 = \frac{(\mathfrak{M}ru_{\text{circ}})^2}{GM'\mathfrak{M}^2 r} - 1,$$

or

$$u_{\text{circ}} = \sqrt{GM'/r}. \tag{10–55}$$

This result may also be found by equating centrifugal and gravitational forces for the circular case. Note, by comparing Eqs. (10–54) and (10–55), that the velocity for escape from circular orbit is simply a constant times the orbital velocity at that radius:

$$u_{\text{esc}} = \sqrt{2}\, u_{\text{circ}}. \tag{10–56}$$

The period of an elliptical orbit is given by Kepler's law, which expresses the conservation of angular momentum. From Eq. (10–44), the conservation of angular momentum requires

$$r^2\dot{\theta} = \frac{h}{\mathfrak{M}} = \text{constant}.$$

Referring to Fig. 10–13, we can see that the area swept out by the radius vector during time dt is $dA = \frac{1}{2}r^2\dot{\theta}\, dt$. Thus

$$\frac{dA}{dt} = \frac{1}{2}r^2\dot{\theta} = \frac{1}{2}\frac{h}{\mathfrak{M}}.$$

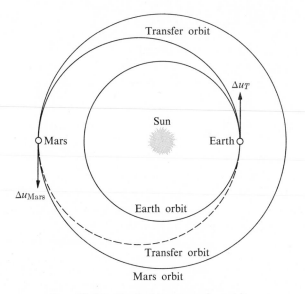

Fig. 10–14. Earth–Mars transfer orbit.

For a complete period τ, the vehicle sweeps out the entire ellipse area πab. Since dA/dt is constant, $(dA/dt)\tau = \pi ab$, and the period is $\tau = 2\pi ab/(h/\mathfrak{M})$ or, using Eqs. (10–48) and (10–49),

$$\tau = \frac{2\pi a^{3/2}}{\sqrt{GM'}}. \tag{10–57}$$

Interplanetary trajectories have been studied by Hohmann [4] and Oberth [5]. Hohmann treated earth escape and interorbital transfer as two separate problems (see Fig. 10–14). The first step in an earth-to-Mars trip, for example, would be an escape from the earth's influence achieved by attaining escape velocity in any direction from the surface of the earth. After escape, the vehicle finds itself free of the earth, but in an orbit about the sun, an orbit which is essentially identical to that of the earth about the sun. (The vehicle has zero velocity relative to the earth.) A second velocity increment Δu_T is then added, which places the vehicle in an elliptic "transfer" orbit about the sun. This orbit at its perigee (nearest approach to sun) is tangent to the earth's orbit, and at its apogee (greatest distance from sun) is tangent to the orbit of Mars. A third velocity increment, Δu_{Mars}, is necessary at this point to bring the vehicle to that velocity corresponding to the Mars orbit. The total velocity increment which the vehicle needs to arrive in the Mars orbit is then

$$\Delta u_{\text{Hohmann}} = u_{\text{esc}} + \Delta u_T + \Delta u_{\text{Mars}} \tag{10–58}$$

and the change in orbital energy per unit mass (with respect to the earth) needed in order to put the vehicle into the transfer orbit is

$$E_{\text{Hohmann}} = E_{\text{esc}} + \frac{(\Delta u_T)^2}{2} = \frac{(\Delta u_T)^2}{2}, \qquad (10\text{–}59)$$

since the escape energy is zero.

Escape from earth while carrying enough propellant to produce Δu_T involves unnecessary work done against gravity, work which can be avoided if the initial velocity is properly directed and of sufficient magnitude. Oberth [5] treated this problem, and arrived at a smaller total velocity increment. His maneuver requires a larger launch velocity u_{Oberth} at the earth's surface, which places the vehicle in the same transfer orbit as Hohmann's; i.e., the Hohmann ellipse. The orbital energy is therefore the same. With respect to the earth, this energy is given by

$$E_{\text{Oberth}} = \frac{u_{\text{Oberth}}^2}{2} - \frac{GM_e'}{R_e},$$

where R_e is the earth's radius and M_e' is its mass.

Equating these two energies and using Eq. (10–54) for GM_e'/R_e, we obtain

$$\frac{u_{\text{Oberth}}^2}{2} - \frac{u_{\text{esc}}^2}{2} = \frac{(\Delta u_T)^2}{2}$$

or

$$u_{\text{Oberth}} = \sqrt{u_{\text{esc}}^2 + (\Delta u_T)^2}. \qquad (10\text{–}60)$$

Thus the Oberth velocity is less than the direct sum of escape and transfer velocities necessary for the Hohmann transfer. The total Oberth velocity increment is then

$$\Delta u_{\text{Oberth}} = u_{\text{Oberth}} + \Delta u_{\text{Mars}}. \qquad (10\text{–}61)$$

This establishes the minimum propulsive requirement to arrive in the orbit of Mars. The transit time does not appear in this calculation since, for this case, it is a dependent variable determined by the characteristics of the orbits of the earth and Mars. The effects of atmospheric drag during the escape maneuver have been neglected.

Actually, one might be interested in more than simply reaching the orbit of a given target planet. One might want to achieve an orbit around the target planet itself, or at least pass within a certain observational distance of it. These goals require correct coordination of time as well as space and, generally, additional powered flight within the influence of the target planet.

In addition, the transfer time on a minimum-energy transfer might be excessive for some missions, especially those involving manned vehicles. Faster trajectories requiring greater Δu_{total} but fewer supplies might be advantageous. These faster trajectories, which involve nontangential transfer and initial orbit intersections, are wasteful of propellant, since the resultant velocity is not the summation of colinear orbital and incremental velocities.

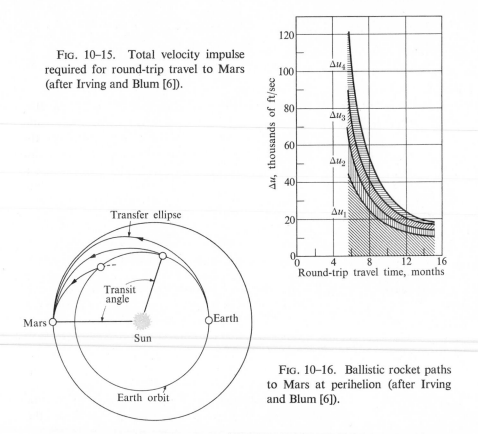

FIG. 10–15. Total velocity impulse required for round-trip travel to Mars (after Irving and Blum [6]).

FIG. 10–16. Ballistic rocket paths to Mars at perihelion (after Irving and Blum [6]).

Irving and Blum [6] have presented a study of the propulsive requirements of a round-trip mission to Mars as a function of total travel time. The results are shown in Fig. 10–15.

At least four separate velocity increments are necessary to perform the following tasks:

Δu_1: escape from the earth toward Mars with excess (over Hohmann transfer ellipse) velocity which determines one-way transit time. If the trajectory were a Hohmann ellipse, this combined escape and transfer maneuver would correspond to the Oberth velocity.

Δu_2: increment applied in gravitational field of Mars to effect "capture" in an orbit about Mars.

Δu_3: the opposite of Δu_2, to effect escape from Mars and return to a transfer ellipse back toward the earth.

Δu_4: increment applied to decelerate the vehicle to a safe re-entry velocity after it has been captured by the earth's gravitational field.

Typical transfer orbits are illustrated in Fig. 10–16.

For this study the orbit of the earth was assumed a circle of 92,833,000 miles radius, while the orbit of Mars was assumed a coplanar ellipse of eccentricity $\epsilon = 0.093366$ and 128,250,000 miles distant from the sun at perihelion (closest approach to sun). All transfer paths were assumed tangent to the Mars orbit at perihelion, although this is not necessarily desirable, since the appropriate relative positions of the earth and Mars for these missions do not occur frequently. Figure 10–17 presents the one-way transit time as a function of transit angle (Fig. 10–16), with the minimum-energy or Hohmann-ellipse trajectory occupying approximately 7.8 months.

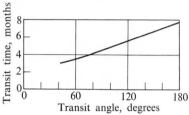

Fɪɢ. 10–17.   One-way earth-to-Mars transit time (after Irving and Blum [6]).

All trajectories arrive in the Mars orbit with less velocity than Mars has at that point. Thus the vehicle approaches Mars in a rearward direction, having energy greater than zero with respect to Mars. The shorter-transit trajectories have lower velocities relative to the sun at this point, and thus higher energy relative to Mars, requiring greater Δu_2 to effect capture by Mars. The assumption that Δu_3 is the mirror image of Δu_2 is artificial, in that it may require inordinate waiting until the vehicle, Mars, and the earth are all in favorable position for this maneuver, but the resultant figures are valid for comparisons.

Low thrust

Electrical rockets of very low thrust-to-mass ratio are capable of only very low accelerations, perhaps 10^{-5} to 10^{-3} times the acceleration of gravity at the earth's surface. Thus they are incapable of performing the initial launch function. However, once launched into orbit by conventional chemical rockets, these devices are intended to accelerate continuously over long periods of time until the required velocity increment is achieved. Since the electrical rocket is a power-limited device, it is advantageous to operate it at constant power, using variation of thrust direction, and perhaps magnitude, to achieve the desired motion. The escape trajectory then becomes a gradual outward spiral until escape velocity is reached. The transfer trajectory also becomes a spiral about the sun, and the capture maneuver a spiral around the target planet. In the absence of high-thrust chemical rockets, the approach to the target planet is quite critical in that it must not necessitate maneuvering beyond the very limited capability of the low-thrust device.

Tsien [7] considered the case of takeoff from a satellite orbit using a constant radial or tangential acceleration low-thrust vehicle. Since torque is applied to the vehicle by the circumferential thrust component, its angular momentum is no longer constant. If the thrust T_θ is applied in the circumferential direction, the tangential equation of motion may be written

$$T_\theta = \mathfrak{M}(r\ddot{\theta} + 2\dot{r}\dot{\theta})$$

or

$$\frac{T_\theta r}{\mathfrak{M}} = \frac{d}{dt}(r^2\dot{\theta}). \tag{10-62}$$

In the absence of a radial thrust component, Eq. (10–43) still applies:

$$-\frac{GM'\mathfrak{M}}{r^2} = \mathfrak{M}(\ddot{r} - r\dot{\theta}^2). \tag{10-43}$$

When we introduce the dimensionless variables,

$$\rho = \frac{r}{r_0}, \quad \tau = \sqrt{\frac{GM'}{r_0^3}}\, t, \quad \text{and} \quad \nu = \frac{T_0 r_0^2}{GM'\mathfrak{M}} = \frac{T_\theta}{\mathfrak{M}g},$$

where r_0 = initial orbit radius and g = acceleration of gravity in initial orbit, these two equations of motion can be written

$$\frac{d^2\rho}{d\tau^2} = \rho\left(\frac{d\theta}{d\tau}\right)^2 - \frac{1}{\rho^2}, \tag{10-63}$$

$$\frac{d}{d\tau}\left(\rho^2 \frac{d\theta}{d\tau}\right) = \nu\rho. \tag{10-64}$$

At $t = 0$, when thrust is initiated with the vehicle already in orbit, $\ddot{r} = \dot{r} = 0$ and $r = r_0$. Using Eq. (10–43), we can write these conditions (when $t = 0$) as

$$\dot{\theta}^2 = \frac{GM'}{r_0^3}, \quad \dot{r} = 0, \quad r = r_0 \tag{10-65}$$

or, in terms of the dimensionless variables (when $\tau = 0$) as

$$\frac{d\theta}{d\tau} = 1, \quad \frac{d\rho}{d\tau} = 0, \quad \rho = 1. \tag{10-66}$$

For $d\theta/d\tau = 1$, $\rho = 1$, Eq. (10–63) requires that

$$\frac{d^2\rho}{d\tau^2} = 0. \tag{10-67}$$

Combining Eqs. (10–63) and (10–64) yields a third-order differential equation for ρ,

$$\frac{d}{d\tau}\left(\rho^3 \frac{d^2\rho}{d\tau^2} + \rho\right)^{1/2} = \nu\rho, \tag{10-68}$$

with initial conditions given by (10–66) and (10–67). This equation cannot be

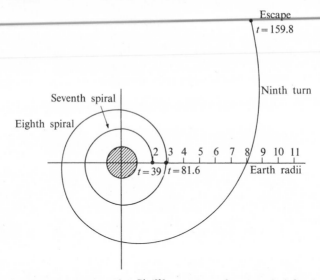

FIG. 10–18. Escape maneuver for $T_e/\mathfrak{M} = 0.005g$ from earth (after Irving [8]).

solved directly, but Tsien [7] has given an approximate solution, while Irving [8] has given a more accurate one.

Figure 10–18 illustrates the solution of Eq. (10–68) for a constant $T_\theta/\mathfrak{M}$ escape maneuver from an initial 200-mile circular orbit about the earth. Time is measured in multiples of the time required to travel one radian in the initial orbit, about 14.5 minutes in this case. For any solution of this type, the propulsion requirement for escape, expressed as the equivalent free space Δu (that is, $u_e \ln \mathfrak{R}$), can be calculated, since in this case Δu is simply $T/\mathfrak{M}$ times the propulsion time.

The relationship between Δu and $T/\mathfrak{M}$ is shown in Fig. 10–19 for several solutions of different $T/\mathfrak{M}$. Both circumferential and tangential (parallel to vehicle velocity) thrust solutions are shown; Δu and $T/\mathfrak{M}$ have been nondimensionalized with respect to the vehicle velocity and gravitational acceleration, respectively, in the initial orbit. It is apparent that for lower $T/\mathfrak{M}$ higher velocity increments are required

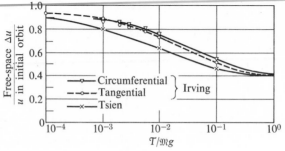

FIG. 10–19. Velocity increment for vehicle escape as a function of vehicle thrust-to-mass ratio (after Irving [8]).

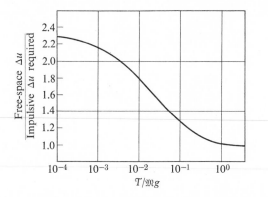

FIG. 10–20.   Increase of Δu for escape associated with low-thrust engine (after San-dorff [9]).

Sandorff [9] presents data for the tangential thrust case, as in Fig. 10–20, which indicate the "penalty" associated with escape using less than impulsive thrust. Low-thrust vehicles are advantageous in spite of this penalty because the extremely low propellant consumption more than compensates for the higher velocity increment required.

As an example of an interplanetary mission, Stuhlinger [10] has presented an account of a round trip to Mars by a low-thrust space ship of the following type:

Total initial mass	730 tons	Exhaust velocity	84 km/sec
Total propellant mass	365 tons	Initial acceleration	$0.67 \times 10^{-4}g$
Propellant consumption rate	5.8 gm/sec	Payload	150 tons

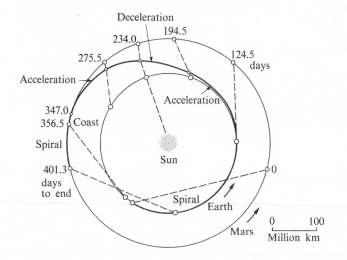

FIG. 10–21.   Low-thrust earth escape and transfer to Mars orbit (after Stuhlinger [10]).

From this one can calculate the total velocity increment required for this mission, assuming the difference between initial and burnout mass to be the total propellant mass. Actually, additional mass might be discarded or added in the course of the mission, but the propulsion requirements are known in any case in terms of the total propellant, the rate of consumption of propellant, and the exhaust velocity.

Assuming coplanar circular orbits for the earth and Mars, the outward trajectory would appear as in Fig. 10–21. The earth-escape spiral consumes 124 days. The deceleration and acceleration periods are found necessary to reach Mars at the correct time and with the correct velocity for capture. The energy relative to Mars should be nearly zero. After an exploration period of 472 days, the earth and Mars are favorably positioned for a return trip of 311 days.

References

1. SANDORFF, P. E., "Structural Considerations in the Design of Space Boosters." *Journal of the American Rocket Society*, November 1960

2. CONSTANT, F. WOODBRIDGE, *Theoretical Physics.*   Reading, Mass.: Addison-Wesley, 1954

3. FORSYTHE, W. E., *Smithsonian Physical Tables*, ninth revised edition. Washington, D. C.: Smithsonian Institution, 1954

4. HOHMANN, W., *Die Erreichbarkeit der Himmelskörper.*   Munich and Berlin: R. Oldenbourg, 1925

5. OBERTH, H., *Wege zur Raumschiffahrt.* Munich and Berlin: R. Oldenbourg, 1929

6. IRVING, J. H. and E. K. BLUM, "Comparative Performance of Ballistic and Low-Thrust Vehicles for Flight to Mars," *Vistas in Astronautics*, Volume II. New York: Pergamon Press, 1959

7. TSIEN, H. S., "Takeoff from Satellite Orbits," *Journal of the American Rocket Society*, **23**, July–August, 1953

8. IRVING, J. H., "Low-Thrust Flight: Variable Exhaust Velocity in Gravitational Fields," Chapter 10 in *Space Technology*, edited by Howard Siefert. New York: John Wiley & Sons, 1959

9. SANDORFF, P. E., "Orbital and Ballistic Flight." Cambridge, Mass.: Massachusetts Institute of Technology, Department of Aeronautics and Astronautics, 1960

10. STUHLINGER, E., "Flight Path of an Ion-Propelled Spaceship." *Jet Propulsion*, April 1957

Problems

1. Find the ratio of the velocities of two vehicles, one powered by a liquid chemical rocket and the other powered by a solid chemical rocket, when they are used for acceleration of a 10,000-lb_m payload in a zero gravity field. Both vehicles have a total initial mass of 510,000 lb_m. The liquid-propellant rocket has 60% greater specific impulse and 30% greater mass of empty vehicle (without propellant and payload), and the solid-propellant rocket has a structural coefficient of $\epsilon = 0.080$.

2. A vehicle which utilizes a liquid chemical rocket is to be used for escape of a payload of 100,000 lb_m from the earth. The vehicle is to have two stages, each with $\epsilon = 0.06$ and $I_{sp} = 400$ sec. Estimate the total mass of the vehicle. It may be assumed as a first approximation that $t_b \overline{\cos \theta} = 200$ sec. To assure acceptable vehicle performance at launch, what restriction must be placed on the burning period?

3. An astronaut is about to explore an asteroid of average density $\rho = 5000$ kg/m³. He is worried that he may accidentally jump from its surface and float off into space. How big (i.e., what diameter) must it be before he can neglect this possibility? The astronaut knows that his mass m, including space suit, is 91 kg (200 lb_m). He also knows that on the earth he can raise his center of gravity 0.6 m by jumping with his space suit on. (This is only about two feet, but try it yourself sometime, even without a space suit.) It seems reasonable to assume that the maximum energy output of his legs will be the same on the asteroid as on the earth. For simplicity assume $M \gg m$, such that M does not move significantly during the action, and then check this assumption.

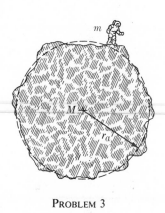

PROBLEM 3

4. (a) Calculate the minimum total velocity increment and transit time which would be required for a two-impulse transfer from the orbit of the earth into the orbit of Mars about the sun.

(b) Calculate the same quantities for a two-impulse transfer from the surface of the earth into the oribit of Mars.

5. The Mars trip described in Fig. 10–21 required the expulsion of 365 tons of propellant at a velocity of 84 km/sec for an initial vehicle mass of 730 tons. From this data, calculate the equivalent free space Δu of this mission, assuming that no other mass is expelled. For the stated propulsion time, compare the actual specific impulse with the optimum specific impulse obtained from Eq. (10–41), assuming $\alpha/\eta = 5$ kg/kw.

6. Show how to find the optimum specific impulse for an electrical propulsion system to be used for a particular mission. It may be assumed that the thrust is constant. Consider the mass of the electrical power generation equipment to be proportional to the power produced, and all other structural mass to be proportional to the sum of the payload and propellant masses. That is,

$$\mathfrak{M}_\mathscr{P} = \alpha \mathscr{P}, \qquad \mathfrak{M}_s = \beta(\mathfrak{M}_\mathscr{L} + \mathfrak{M}_p).$$

7. The "optimum" specific impulse for an earth-to-moon transfer of an electrically propelled rocket is about 2000 sec for

$$\frac{\alpha}{\eta} = 16 \frac{\text{lb}_m}{\text{hp}}, \qquad \Delta u = 24{,}000 \text{ ft/sec},$$

where α and η are defined by Eqs. (10–37) and (10–38), respectively. Ion rockets are not efficient at these low specific impulses. What payload ratio can be carried by an ion rocket of 10,000-sec specific impulse on a moon trip of three months duration, if the parameters α/η and Δu are unchanged?

8. In the year 1992, weekly bus service is to be established between Boston and the moon. To avoid the need for large expensive boosters for moon escape, the following plan has been suggested.

   Since there is no moon atmosphere, there is no drag loss associated with high velocity at low altitude. Therefore, it is proposed that a long smooth track be built on the surface of the moon. On this track a rocket-powered sled is to accelerate the bus to escape velocity, after which the bus leaves, carrying with it a few passengers, a member of the carmen's union, and a rocket engine powerful enough to effect guidance and landing in Boston.

   (a) What does a scheme of this type gain over conventional escape trajectories (neglecting friction of sled)? (b) Would there be any incentive for designing a very light sled for this purpose? (c) Is there any reason to multistage the sled? (d) What should the track configuration be at the point where escape velocity is reached? (e) If there is absolutely no limit on track length, even to the extent of circling the moon, is there any advantage in making the track very long? or very short?

11

Chemical Rockets:
An Introduction

11-1 CHARACTERISTICS OF CHEMICAL ROCKETS

Chemical rockets are characterized by the fact that the energy required to accelerate the propellant comes from the propellant itself. Thus the attainable kinetic energy per unit mass of propellant is limited primarily by the energy released in chemical reaction; the attainment of high exhaust velocity (shown to be desirable in Chapter 10) requires the use of high-energy propellant combinations. It will be shown in this chapter that combustion products with low molecular weight are also desirable for this reason. Currently, propellants with the best combinations of high energy content and low molecular weight seem capable of producing specific impulses in the range of 400 to 450 seconds or exhaust velocities of 13,000 to 14,500 ft/sec. As mentioned in Chapter 10, higher velocities would be desirable if they were available.

Chemical rockets may employ liquid or solid propellants or, in some schemes, various combinations and modifications thereof. Attention here will be limited to conventional liquid or solid propellant rockets, as illustrated in Fig. 11-1. Liquid rockets may utilize one (monopropellant), two (bipropellant), or more propellants. Bipropellants consist of a combination of a fuel (such as kerosene, alcohol, or hydrogen) and an oxidizer (such as oxygen, nitric acid, or fluorine). The liquids are stored separately in tanks and fed as needed into the combustion chamber, where they react and then expand through the exhaust nozzle.

In contrast, solid propellants are an intimate mixture containing all the material necessary for reaction. The entire block of solid propellant, called a *grain*, is stored within the combustion chamber itself. Combustion proceeds from the surface of the propellant.

Since liquid- and solid-propellant rockets have many common features, they will be examined simultaneously in the following chapters. When necessary,

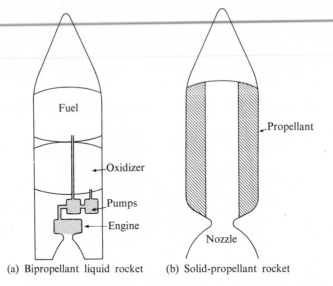

Fuel

Oxidizer

Pumps

Engine

Propellant

Nozzle

(a) Bipropellant liquid rocket   (b) Solid-propellant rocket

Fig. 11–1. Schematic diagrams of liquid- and solid-propellant rockets.

particular features will be discussed separately. We shall begin with the discussion of a simplified analytical model and the definition of common performance indices based on this model. Questions will arise from this simple analysis regarding the choice of actual operating conditions and deviations from the idealized performance. To answer these, more realistic treatments of the rocket components will be presented in turn.

11–2 ANALYSIS OF AN IDEAL ROCKET

Chemical rockets, whether powered by liquid or solid propellants, consist in varying complexity of a propellant supply and feed system, a combustion chamber, and an exhaust nozzle. To simplify our analysis of the thrust chamber (that is, the combustion chamber and nozzle), we assume the following:

(1) The working fluid is a perfect gas of constant composition.

(2) The chemical reaction is equivalent to a constant-pressure heating process.

(3) The expansion process is steady, one-dimensional, and isentropic.

With these assumptions the thrust chamber may be described schematically and on a T-s diagram as in Fig. 11–2. Propellant enters at state ①. A quantity of heat Q_R (the "heating value" of the propellant combination per unit mass) is added at constant pressure, and the propellant reaches the nozzle's inlet at state ②. The propellant expands isentropically from state ② through the throat, where the Mach number is unity, to the exhaust where the velocity is u_e. Application of the

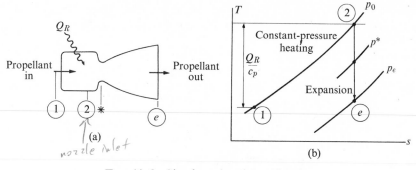

Fɪɢ. 11–2. Simple rocket thrust chamber.

energy equation, Eq. (2–8), to the heating process gives

$$\dot{Q} = \dot{m}(h_{02} - h_{01}) = \dot{m}c_p(T_{02} - T_{01})$$

or

$$T_{02} = T_{01} + \frac{Q_R}{c_p},\tag{11-1}$$

where $\dot{Q}$ = equivalent heating rate, and $\dot{m}$ = propellant mass flow rate.

Assuming adiabatic nozzle expansion, the energy equation requires constant stagnation enthalpy in the nozzle:

$$h_{02} = h_{0e} \quad \text{or} \quad \frac{u_e^2}{2} = h_{02} - h_e = c_p(T_{02} - T_e),$$

and if the expansion is isentropic, we use Eq. (3–6) to obtain

$$u_e = \sqrt{2c_p T_{02}\left[1 - \left(\frac{p_e}{p_{02}}\right)^{(\gamma-1)/\gamma}\right]}.\tag{11-2}$$

Expressing c_p in terms of γ, the universal gas constant $\overline{R}$, and the propellant molecular weight $\overline{M}$, we have

$$u_e = \sqrt{\frac{2\gamma\overline{R}}{(\gamma - 1)\overline{M}} T_{02}\left[1 - \left(\frac{p_e}{p_{02}}\right)^{(\gamma-1)/\gamma}\right]}.\tag{11-3}$$

Thus, at least for a given T_{02}, it is advantageous to employ propellants with a low molecular weight in order to achieve high u_e. In terms of the "heating value" Q_R per unit mass of propellant, substitution of Eq. (11–1) in (11–2) yields

$$u_e = \sqrt{2c_p\left(T_{01} + \frac{Q_R}{c_p}\right)\left[1 - \left(\frac{p_e}{p_{02}}\right)^{(\gamma-1)/\gamma}\right]} \approx \sqrt{2Q_R\left[1 - \left(\frac{p_e}{p_{02}}\right)^{(\gamma-1)/\gamma}\right]}.$$

$$\tag{11-4}$$

The last expression assumes $T_{01} \ll Q_R/c_p$, as is generally the case. Expressing Q_R in terms of the heating value per *mole* of propellant, $\bar{Q}_R$, we obtain

$$u_e \approx \sqrt{2 \frac{\bar{Q}_R}{\bar{M}} \left[1 - \left(\frac{p_e}{p_{02}} \right)^{(\gamma-1)/\gamma} \right]}. \qquad (11\text{–}5)$$

Calculation of $\bar{Q}_R$ according to the methods of Section 2–7 will be considered in Section 12–1. For this simple model we shall assume that $\bar{Q}_R$ is known, perhaps in terms of the "complete" chemical reaction assumption. The desirability of propellant mixtures with large $\bar{Q}_R/\bar{M}$ may be seen from Eq. (11–5). Both $\bar{Q}_R$ and $\bar{M}$ will depend on the fuel–oxidant ratio. In many cases fuel-rich mixtures are burned, as it is found that the resultant reduction in $\bar{M}$ more than offsets the accompanying reduction of $\bar{Q}_R$.

The propellant mass flow rate $\dot{m}$ is given in terms of the properties of the combustion-chamber fluid and the nozzle throat area by Eq. (3–14):

$$\dot{m} = \frac{A^* p_{02}}{\sqrt{RT_{02}}} \sqrt{\gamma \left(\frac{2}{\gamma+1} \right)^{(\gamma+1)/(\gamma-1)}}, \qquad (11\text{–}6)$$

where A^* = nozzle throat area and $R = \bar{R}/\bar{M}$.

For isentropic flow the results of Section 3–3 may be used to calculate propellant properties throughout the exhaust nozzle. In particular, the exhaust pressure p_e can be found in terms of the exhaust area ratio A_e/A^*. The thrust of the simple chemical rocket engine may, as shown in Chapter 10, be written

$$T = \dot{m} u_e + (p_e - p_a) A_e,$$

where p_a = ambient pressure.

Substituting Eqs. (11–3) and (11–6) for u_e and $\dot{m}$, we have

$$T = \left[\frac{A^* p_0}{\sqrt{(\bar{R}/\bar{M}) T_0}} \sqrt{\gamma \left(\frac{2}{\gamma+1} \right)^{(\gamma+1)/(\gamma-1)}} \right]$$

$$\times \left\{ \sqrt{\frac{2\gamma \bar{R}}{(\gamma-1)\bar{M}} T_0 \left[1 - \left(\frac{p_e}{p_0} \right)^{(\gamma-1)/\gamma} \right]} \right\} + (p_e - p_a) A_e,$$

or

$$\frac{T}{A^* p_0} = \sqrt{\frac{2\gamma^2}{\gamma-1} \left(\frac{2}{\gamma+1} \right)^{(\gamma+1)/(\gamma-1)} \left[1 - \left(\frac{p_e}{p_0} \right)^{(\gamma-1)/\gamma} \right]} + \left(\frac{p_e}{p_0} - \frac{p_a}{p_0} \right) \frac{A_e}{A^*}, \qquad (11\text{–}7)$$

where $(p_0, T_0 \equiv p_{02}, T_{02})$ are the stagnation properties of the combustion chamber. It should not come as a surprise that thrust can be expressed entirely in terms of geometry and pressure (for a given γ) since, after all, thrust is just the net pressure force on the thrust chamber.

The ideal rocket, aside from serving as a simple analytical model, provides a basis for the comparison and evaluation of real rockets. In order to describe separately the performance of each component of the thrust chamber, two coefficients are defined. For the combustion chamber the *characteristic velocity*, c^*, is defined as

$$c^* \equiv \frac{p_0 A^*}{\dot{m}}. \qquad actual \qquad (11\text{-}8)$$

Utilizing Eq. (11-6), we obtain

$$c^* = \sqrt{\frac{1}{\gamma}\left(\frac{\gamma+1}{2}\right)^{(\gamma+1)/(\gamma-1)} \frac{RT_0}{M}} \qquad theoretical \qquad (11\text{-}9)$$

for the ideal rocket. Thus c^* is primarily a function of combustion chamber properties. For the nozzle, the *thrust coefficient*, C_T, is defined as

$$C_T \equiv \frac{T}{p_0 A^*}. \qquad actual \qquad (11\text{-}10)$$

Using Eq. (11-7), we have

$$C_T = \sqrt{\frac{2\gamma^2}{\gamma-1}\left(\frac{2}{\gamma+1}\right)^{(\gamma+1)/(\gamma-1)}\left[1-\left(\frac{p_e}{p_0}\right)^{(\gamma-1)/\gamma}\right]} + \frac{p_e - p_a}{p_0}\frac{A_e}{A^*} \qquad Theory$$

$$(11\text{-}11)$$

for the ideal rocket. It can be seen that C_T is a function of nozzle geometry only, since the variable $(p_e - p_a)/p_0$ can be regarded as a measure of how well the nozzle geometry is suited to the actual pressure ratio. Combining Eqs. (11-8) and (11-10), we find that the thrust of a rocket is given by

$$T = \dot{m}c^* C_T,$$

in which c^* and C_T characterize the performances of combustion chamber and nozzle, respectively. For a real rocket, comparison of c^* and C_T, calculated from Eqs. (11-8) and (11-10) with ideal values from (11-9) and (11-11), indicates how well each component of the real thrust chamber is performing.

If the nozzle does not have a diverging portion, then the pressure ratio p_e/p_0 is given by Eq. (3-11) with $M = 1$:

$$p_e/p_0 = \left(\frac{2}{\gamma+1}\right)^{\gamma/(\gamma-1)},$$

and the thrust coefficient reduces to

$$C_{T\,\text{conv}} = \sqrt{\gamma^2\left(\frac{2}{\gamma+1}\right)^{2\gamma/(\gamma-1)}} + \left(\frac{2}{\gamma+1}\right)^{\gamma/(\gamma-1)} - \frac{p_a}{p_0}, \qquad (11\text{-}12)$$

in which $C_{T\,\text{conv}}$ is the thrust coefficient for a converging nozzle. The ratio of thrusts,

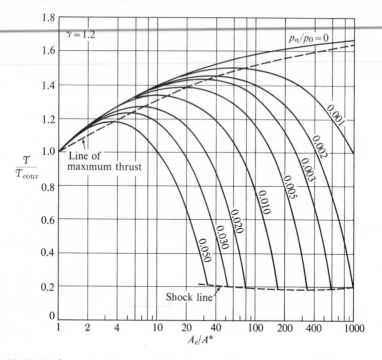

Fig. 11–3. Performance characteristics of a one-dimensional isentropic rocket nozzle; $\gamma = 1.2$. (After Malina [5].)

with and without a diverging section, is given by

$$\frac{T}{T_{\text{conv}}} = \frac{C_T}{C_{T\text{conv}}},$$

and may be evaluated by means of Eqs. (11–11) and (11–12) above as a function of γ, p_e/p_0 and p_a/p_0. For the special case $\gamma = 1.2$, Fig. 11–3 shows how the diverging portion of the nozzle contributes to the rocket thrust at various area and pressure ratios. It may be seen that for each value of the ambient pressure ratio p_a/p_0 there is an optimum area ratio for the nozzle; it is the one for which $p_e = p_a$ (see discussion of Fig. 10–2).

It may be seen from Fig. 11–3 that the thrust of a rocket will actually be reduced by adding a diverging section of too large an area ratio. However, for reasonable pressure ratios ($p_a/p_0 > 0.02$), this would not happen unless A_e/A^* were greater than about 30. For very small pressure ratios (such as $p_a/p_0 = 0.001$), it will be seen that very large area ratios are associated with maximum thrust.

Shown in Fig. 11–3 is a shock line which shows the pressure ratios and area ratios for which a shock enters the nozzle. It may be derived as follows: Suppose that a normal shock sits in the exit plane of the nozzle and that the conditions just ahead of it are p_e, M_e, etc.; the pressure just behind is p_a. Then, according to the formula quoted in Chapter 3 for normal shocks,

$$\frac{p_a}{p_e} = \frac{2\gamma}{\gamma + 1} \, M_e^2 - \frac{\gamma - 1}{\gamma + 1}.$$

For the isentropic nozzle the pressure ratio p_0/p_e is related to M_e by

$$\frac{p_0}{p_e} = \left(1 + \frac{\gamma - 1}{2} \, M_e^2\right)^{\gamma/(\gamma-1)}$$

so that

$$\frac{p_a}{p_0} = \left(\frac{2\gamma}{\gamma + 1} \, M_e^2 - \frac{\gamma - 1}{\gamma + 1}\right) \bigg/ \left(1 + \frac{\gamma - 1}{2} \, M_e^2\right)^{\gamma/(\gamma-1)}.$$

Further, the area ratio is related to M_e by [applying Eq. (3–15)]

$$\frac{A_e}{A^*} = \frac{1}{M_e} \left[\frac{2}{\gamma + 1} \left(1 + \frac{\gamma - 1}{2} \, M_e^2\right)\right]^{(\gamma+1)/2(\gamma-1)}.$$

These last two relations, then, show that for a given γ there is only one pressure ratio p_a/p_0 for which a normal shock will locate in the plane of a nozzle of a given area ratio A_e/A^*.

The above method for determining the ratios A_e/A^* and p_e/p_0 for which shocks will enter a nozzle is only a first approximation, since it assumes that the flow in the exit plane is exactly one-dimensional. A more realistic description of shock formation outside and inside nozzles is given in Chapter 13.

Real rockets are not accurately described by this simple model. The difference between actual and ideal performance can be reconciled, to some extent, in terms of empirical coefficients. However, the greatest complication in the simple analysis arises in the description of the combustion process. Complete combustion in the presence of excess oxidant may be defined as the absence of further possible exothermic reaction. However, in the more common case of excess fuel, when the fuel consists of more than one combustible element, the term "complete combustion" is not so easy to define. Further, it is inadequate simply to specify the actual T_0, since it is necessary to know propellant composition ($\overline{M}$ and γ) as well. Thus any realistic calculation of rocket performance requires detailed calculation of both the temperature and the composition of combustion products.

Nozzle performance, assuming constant (and known) composition flow, is more easily related to the simple model. The ideal nozzle flow can be refined slightly to account, in the case of conical nozzles, for the effect of nonaxial exhaust velocities. Consider the flow out of the conical nozzle of Fig. 11–4, and assume that the streamlines in the expanding part of the nozzle are straight lines which all intersect at the point O. A control surface is indicated, passing through the spherical segment (of radius R) over which exhaust properties are constant. The reaction to the thrust is again shown as the vector T. Applying the momentum equation, Eq.

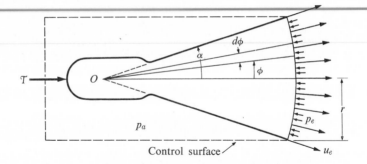

Fig. 11–4. Spherically symmetric nozzle flow.

(2–11), in the axial x-direction gives, for steady flow,

$$\sum F_x = T + (p_a - p_e)A_e = \int_{cs} \rho(\mathbf{u} \cdot \mathbf{n})u_x \, dA,$$

where A_e is the plane exhaust area πr^2. No mass crosses the control surface except over the spherical exhaust segment, and there it is convenient to let

$$dA = 2\pi R \sin \phi R \, d\phi.$$

Further, in this case,

$$\mathbf{u} \cdot \mathbf{n} = u_e \quad \text{and} \quad u_x = u_e \cos \phi.$$

Thus

$$T = \int_0^\alpha \rho(u_e)u_e \cos \phi 2\pi R^2 \sin \phi \, d\phi + (p_e - p_a)A_e.$$

Integrating, we obtain

$$T = 2\pi R^2 \rho u_e^2 \frac{1 - \cos^2 \alpha}{2} + (p_e - p_a)A_e.$$

Since the area of the spherical exhaust segment A_e' is $2\pi R^2(1 - \cos \alpha)$, the propellant flow rate is

$$\dot{m} = \rho u_e \times 2\pi R^2(1 - \cos \alpha),$$

and the thrust equation becomes

$$T = \dot{m}u_e \frac{1 + \cos \alpha}{2} + (p_e - p_a)A_e$$

or, since the spherical area segment $A_e' = [2/(1 + \cos \alpha)]A_e$, $A_e = \pi r^2$

$$T = \frac{1 + \cos \alpha}{2}[\dot{m}u_e + (p_e - p_a)A_e']. \qquad (11\text{–}13)$$

Since, for reasonable nozzles, both the difference between A_e and A_e' and the contribution of the pressure term to the thrust are small, it is adequate to replace

Table 11–1*

Performance parameter	Ordinary range	High range
T_0	2000–3000°K	3000–5000°K
c^*	4000–5500 ft/sec	5000–8000 ft/sec
C_T	1.3–1.5	1.5–1.6
I_{sp}	200–270 sec	270–400 sec
$\overline{M}$	20–25	8–20
$\overline{\gamma}$	1.15–1.25	1.15–1.20

* Courtesy Altman, *et al.* [1].

A_e' in Eq. (11–13) by the plane area A_e. Thus the thrust of an ideal conical rocket of half angle α is reduced by a factor λ,

$$\lambda = \frac{1 + \cos \alpha}{2}, \qquad (11\text{–}14)$$

over that of an axial-outlet rocket with the same exhaust conditions.

The performance of a real nozzle is further reduced by frictional effects. This may be accounted for in terms of an empirical *discharge coefficient*, C_d, such that

$$C_{T\text{actual}} = C_d \lambda C_{T\text{ideal}}. \qquad (11\text{–}15)$$

Typical values for C_d lie between 0.97 and 0.99 [1].

A comparison of Eq. (11–14) with experimental data will be discussed in Chapter 13.

Altman *et al.* [1] present typical values for these various parameters for ordinary and high-performance chemical rockets, as shown in Table 11–1. Note the combination of high temperature and low molecular weight of the high-performance group.

It is appropriate, after this description of a simplified rocket, to ask certain questions about the operation and analysis of real rockets. Foremost among the questions on operation is: how is the combustion pressure chosen? Equation (11–7) indicates that high p_0 is desirable, in that for a given thrust, rocket size (characterized by A^*) may be decreased as the combustion chamber pressure p_0 is increased. Offsetting this advantage are certain difficulties associated with high p_0. Thrust-chamber stresses increase with p_0, and so do heat-transfer rates. In addition, for pump-fed liquid rockets, pump power and size increase with p_0. In some cases, combustion behavior and products may be significantly influenced by combustion pressure. Thus many considerations enter into the choice of combustion pressure, most of which are sufficiently complex to prohibit simple analytical treatment. After considering the various rocket components separately we shall again approach the choice of chamber pressure, but it should be evident that no simple analytic answer will emerge.

The analytical simplifications of the ideal rocket should be compared with more realistic models. For example, is the actual combustion process accompanied by significant pressure losses? How nearly does the actual nozzle flow approach one-dimensional isentropic flow? It is known, of course, that real nozzles are not entirely frictionless nor adiabatic. Under certain conditions, nozzle flow patterns may even include shocks. Does the propellant composition remain constant throughout the nozzle, or does it shift in accordance with the rapidly changing propellant temperature? And finally, for a given area ratio, how does the shape of the nozzle affect performance? Questions such as these will be taken up in the following chapters.

References

1. ALTMAN, D., J. M. CARTER, S. S. PENNER, and M. SUMMERFIELD, *Liquid-Propellant Rockets,* Princeton, N. J.: Princeton Aeronautical Paperbacks, Princeton University Press, 1960

2. HUGGETT, C., C. E. BARTLEY, and M. M. MILLS, *Solid-Propellant Rockets.* Princeton, N. J.: Princeton Aeronautical Paperbacks, Princeton University Press, 1960

3. SUTTON, G. P., *Rocket Propulsion Elements.* New York: John Wiley & Sons, 1956

4. SEIFERT, H. S. (editor), *Space Technology.* New York: John Wiley & Sons, 1959

5. MALINA, FRANK J., "Characteristics of the Rocket Motor Unit Based on the Theory of Perfect Gases." *J. Franklin Inst.*, **203**, 4, 1940

Problems

1. The rocket nozzle shown in the figure operates in steady flow. Estimate the axial stresses at the throat if the stagnation conditions are 1000 psia and 6000°R. The molecular weight is 29 and the specific heat ratio $\gamma = 1.4$.

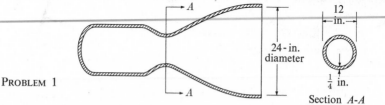

PROBLEM 1

Section *A-A*

2. A rocket is to be designed to produce 1,000,000 lb of thrust at sea level. The pressure in the combustion chamber is 500 psia and the temperature is 5000°R. If the working fluid is assumed to be a perfect gas with the properties of air at room temperature, determine the following: (a) specific impulse, (b) mass flow, (c) throat diameter, (d) exit diameter.

For the same nozzle, find (e) thrust at 100,000 ft altitude, (f) thrust at sea level if chamber pressure were increased to 1000 psia, (g) thrust with hydrogen at the same inlet stagnation conditions, and (h) thrust with stagnation temperature increased to 6200°R.

3. A rocket thrust chamber contains a propellant of molecular weight 25 and specific heat ratio 1.2. The chamber pressure is 400 psia and the combustion temperature is 4900°R. The throat area is 0.1 ft².
Determine the thrust for the following conditions:

	Ambient pressure, psia	Exhaust area, ft²
1	0	0.1
2	0	1.38
3	14.7	0.1
4	14.7	1.38

Compute the axial force exerted by the products of combustion on the expanding nozzle for cases 2 and 4, and compare with the corresponding values of the thrust.

4. Using a propellant of molecular weight 15 and flame temperature 6000°R, determine the rocket throat and exhaust areas required for a thrust of 100,000 lb and an ideal specific impulse of 300 sec. The ambient pressure is 15 psia and the specific heat ratio of the propellant is 1.4.
How much thrust would this rocket develop if the ambient pressure were changed to 6 psia? How much thrust would be developed by a rocket designed to expand to 6 psia if it had the same stagnation conditions, throat area, and propellant?

5. A toy rocket (see figure) containing water and compressed air weighs 0.21 lb when empty and can carry 1 lb of water. The initial air pressure is 50 psia.
(a) What is the minimum nozzle diameter (assuming reversible flow in nozzle) that will permit vertical takeoff? Sketch a suitable nozzle shape. (b) What is the specific impulse under these conditions? (c) Why is the water added? (d) Show how exhaust velocity depends on time during a static "firing" for given values of

PROBLEM 5

$\mathcal{v}$ internal volume of rocket,

p_0 initial pressure,

T_0 initial temperature,

m_w initial mass of water,

A_e exhaust area,

p_e exhaust pressure.

6. Consider the performance of a solid-propellant rocket under conditions which change with time. The stagnation temperature is constant at 5000°R. The stagnation pressure changes slowly with time, according to

$$p_0 = 1000 - 3t, \quad 0 < t < 100,$$

for p_0 in psia and t in seconds. The nozzle has an area ratio of 5 and a throat area of 1 ft^2. It exhausts into a vacuum. The fluid may be considered to have $\gamma = 1.4$ and molecular weight $\overline{M} = 20$. How do the following vary with time? (1) Exhaust velocity. (2) Mass flow. (3) Thrust.

7. A new high-energy liquid propellant whose chemical reactions are not entirely known is to be fired in a static-test thrust chamber. From the following measurements, show how to estimate the thrust and specific impulse which this propellant would develop in any other rocket nozzle.

Thrust	$\mathcal{T}$
Pressure in chamber near injection nozzles	p_1
Pressure in nozzle near exit plane	p_2
Ambient pressure	p_a
Rate of mass flow of liquid propellant into the chamber	$\dot{m}$
Exit area	A_e
Throat area	A_t

What factors might contribute error in such an estimate?

12

Chemical Rockets: Propellants and Combustion

12-1 INTRODUCTION

In Chapter 11 we showed that the specific impulse of a rocket propellant depends quite significantly on the composition and temperature of the combustion products. Another important incentive for studying combustion in rockets is that the intensity of combustion (energy transformed per unit time per unit volume) directly determines the size, and therefore the mass, of the combustion chamber. The performance of the rocket vehicle depends to some extent on the mass of the engine as well as its specific impulse.

The products of combustion in a well-designed rocket combustion chamber are very nearly in thermodynamic equilibrium. That is, at typical combustion-chamber temperatures and pressures, the time required for mixing of fuel and oxidant and for chemical reactions is small compared to the average residence time (several milliseconds) of the propellant within the combustion chamber. Therefore the equilibrium thermodynamic methods of Chapter 2 may be employed to calculate the state of the propellant as it enters the exhaust nozzle. Deviations from equilibrium results may be accounted for by using an empirical combustion efficiency. Sutton [2] defines combustion efficiency as the ratio of the actual enthalpy change per unit propellant mass to the enthalpy change accompanying the attainment of equilibrium. Measured efficiencies vary from 0.94 to 0.99.

12-2 EQUILIBRIUM COMPOSITION

At typical rocket combustion temperatures (e.g., 5000°F or more), the propellant composition at equilibrium could differ considerably from the composition resulting from complete combustion. For example, consider a stoichiometric mixture of hydrogen and oxygen, initially at a temperature of 300°K and maintained at a pressure of 20 atm. If it is assumed that combustion proceeds to completion according to $H_2 + \frac{1}{2}O_2 \rightarrow H_2O$, then the resultant chamber temperature can be easily calculated. According to Eq. (2–33), the net heat of reaction Q_{net} is

$$Q_{net} = (H_{P2} - H_{Pf}) - (H_{R1} - H_{Rf}) + H_{RPf} = 0, \qquad (2\text{–}33)$$

in which H_P and H_R are the enthalpies of given masses of products and reactants, respectively. The symbol H_{RPf} is the change of enthalpy during a transformation from reactants to products at a reference temperature T_f. From Eq. (2–32) and Table 2–1,

$$H_{RPf} = n_{H_2O}Q_{fH_2O} - n_{O_2}Q_{fO_2} - n_{H_2}Q_{fH_2}$$
$$= 1 \times (-57.8) = -57.8 \text{ kcal/gm-mole } H_2O \text{ at } T_f \, 300°K.$$

Since the reactants enter the combustion chamber at the reference temperature T_f, $H_{R1} - H_{Rf} = 0$, and for adiabatic combustion we find, from Eq. (2–33), that

$$H_{P2} - H_{Pf} = 57.8 \text{ kcal/gm-mole of } H_2O.$$

Using Fig. 2–5, H_{Pf} at 300°K is zero, so $H_{P2} = 57.8$, and (again from Fig. 2–5), $T_{P2} \approx 5000°K$. This very high temperature would actually be accompanied by considerable product dissociation, as may be seen by examination of the various equilibrium constants of Fig. 2–12. For example, for the reaction $H_2O \rightarrow H_2 + O$, the equilibrium constant, K_p, is about 19.5 at $T_0 = 5000°K$. In terms of mole fractions, the equilibrium constant is

$$\frac{\chi_{H_2}\chi_O}{\chi_{H_2O}} = K_n = 20^{-1} \times 19.5 \approx 1.$$

Thus the mole fractions of H_2 and O would be appreciable compared to that of H_2O at this temperature. In fact, as is shown by the discussion at the end of Chapter 2 of the equilibrium of hydrogen and oxygen, the actual adiabatic combustion temperature would be approximately 3500°K and the product composition quite different from the product resulting from complete combustion.

To demonstrate the striking effect of dissociation on rocket performance, Fig. 12–1 shows the results of calculations of specific impulse as a function of fuel–oxidant ratio for the two extremes:

(1) Complete combustion; i.e., either all the fuel or all the oxidant consumed.

(2) Equilibrium; i.e., combustion product composition determined by thermodynamic equilibrium in the combustion chamber.

For both of these cases it has been assumed that there is no chemical reaction in the nozzle.

For complete combustion the variation of specific impulse with fuel–oxidant ratio can be estimated easily. With excess hydrogen the reaction is

$$(1 + n)H_2 + \frac{1}{2}O_2 \rightarrow H_2O + nH_2,$$

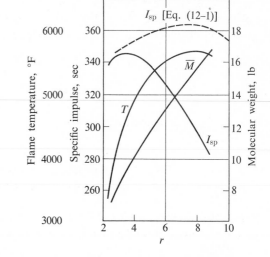

FIG. 12-1. Performance of liquid-oxygen–liquid-hydrogen rocket as a function of mixture ratio; r = oxidizer flow rate/fuel flow rate with Eq. (12–1) added for comparison. Dashed line = no dissociation; solid line = equilibrium dissociation. (Adapted from Sutton [2].)

where n can have any positive value. From this and Eq. (11–4), it may be shown that

$$I_{sp} = \frac{u_e}{g_e} = \frac{1}{g_e}\sqrt{2\left(c_p T_{01} + \frac{\bar{Q}_R}{18 + 2n}\right)\left[1 - \left(\frac{p_e}{p_0}\right)^{(\gamma-1)/\gamma}\right]} \qquad (12\text{–}1a)$$

in which $\bar{Q}_R$ is the heat of reaction per mole of H_2O. For excess oxygen reacting according to $H_2 + (\frac{1}{2} + m)O_2 \rightarrow H_2O + mO_2$, where m can have any positive value, the specific impulse is given by

$$I_{sp} = \frac{1}{g_e}\sqrt{2\left(c_p T_{01} + \frac{\bar{Q}_R}{18 + 32m}\right)\left[1 - \left(\frac{p_e}{p_0}\right)^{(\gamma-1)/\gamma}\right]}. \qquad (12\text{–}1b)$$

The specific impulse determined by Eqs. (12–1) is shown in Fig. 12–1 as a function of the ratio r of the oxidizer–fuel mass. The stoichiometric value of r is 8, so that Eq. (12–1a) applies when $r < 8$ and Eq. (12–1b) when $r > 8$; the two equations are identical when $r = 8$. Figure 12–1 shows that if the combustion reaction goes to completion, the best value of the fuel–air ratio would be stoichiometric.

A very different result is obtained if the much more realistic equilibrium assumption [(2) above] is employed. The full lines of Fig. 12–1 show the results of specific-impulse calculations which follow from this assumption. The product composition, specific heat ratio, and molecular weight have been determined by methods discussed in Chapter 2. The specific impulse has been calculated from Eq. (11–3), using a nominal value of the nozzle pressure ratio.

It may be seen from Fig. 12–1 that, due to dissociation, the oxidizer–fuel ratio must be much less than stoichiometric for maximum specific impulse. The difference between the results of the "complete combustion" calculation [Eqs. (12–1)] and the more realistic equilibrium calculation is largely due to the reduced release of energy if the reactants remain partially dissociated. This effect becomes greater as the mixture approaches the stoichiometric ratio and the combustion temperature increases. In general, although dissociation usually results in slightly lower molecular weight $\overline{M}$, the most important effect is that of reduced energy release and hence reduction in T_0 and I_{sp}.

12–3   LIQUID PROPELLANTS

Digital computers are generally used to calculate equilibrium composition for interesting combinations of rocket propellants. Figure 12–2 shows typical results for several liquid propellants as a function of oxidizer–fuel ratio r. Table 12–1 gives the equilibrium gas composition in the combustion chamber of several bipropellant combinations at mixture ratios for maximum specific impulse. The negligible dissociation of the H_2-O_2 combination accounts for the agreement between the simple and exact calculations of Fig. 12–1 at low r.

The effect of increasing chamber pressure is generally to reduce dissociation; consider the following dissociation reaction:

$$\alpha A + \beta B \rightarrow \mu M + \nu N.$$

The equilibrium concentration in terms of mole fractions must satisfy (see Chapter 2)

$$K_n = \frac{\chi_M^\mu \chi_N^\nu}{\chi_A^\alpha \chi_B^\beta} = p_{mix}^{(\alpha+\beta)-(\mu+\nu)} K_p,$$

Table 12–1   Equilibrium Gas Composition—Bipropellants*

Propellant†		
Gasoline-oxygen, 300 psia	Hydrogen-oxygen, 300 psia	RFNA-aniline, 300 psia
CO 0.322	H_2O 0.312	CO_2 0.159
CO_2 0.137	H_2 0.687	CO 0.251
H_2 0.105	H 0.001	H_2O 0.283
H_2O 0.330		H_2 0.063
H 0.028		N_2 0.192
OH 0.061		H 0.011
O_2 0.008		OH 0.032
O 0.006		

* Courtesy Altman, *et al.* [1].
† Products of reaction are given in mole fractions.

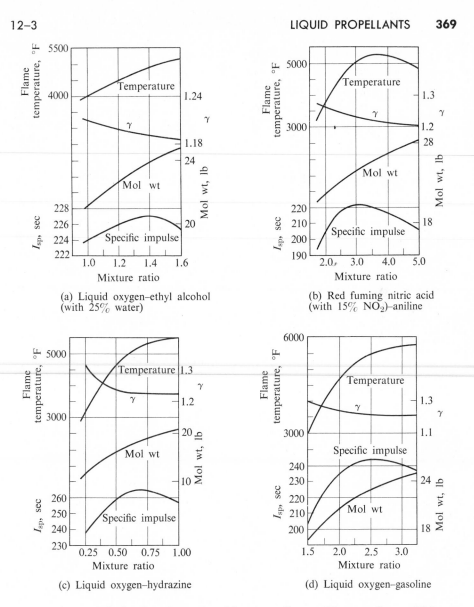

FIG. 12–2. Calculated performance of four propellants. (Courtesy Sutton [2].)

in which p_{mix} is the pressure of the mixture. For a given temperature (and thus a given K_p), it can be seen that high pressure reduces the relative concentration of M and N if $(\mu + \nu)$ is greater than $(\alpha + \beta)$, as is the case with the dissociation of H_2O, for example:

$$H_2O \rightarrow H_2 + \tfrac{1}{2}O_2.$$

Increasing the mixture pressure at constant temperature reduces the dissociation of the H_2O.

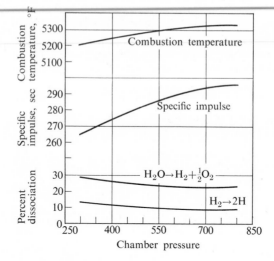

FIG. 12–3. Theoretical effect of pressure on the combustion of liquid oxygen–hydrazine propellant. (Courtesy Sutton [5].)

This effect is illustrated in Fig. 12–3 for an oxygen–hydrazine propellant combination of near-optimum mixture ratio (see Fig. 12–2). As shown by Eq. (11–3), the specific impulse may be written

$$I_{sp} \propto \sqrt{c_p T_{O_2} \left[1 - \left(\frac{p_e}{p_0} \right)^{(\gamma-1)/\gamma} \right]}.$$

The tendency toward recombination of H_2 and O_2 raises the mixture temperature so that both T_{02} and the term $[1 - (p_e/p_0)^{(\gamma-1)/\gamma}]$ increase with pressure. The effect of pressure on mixture specific heat is not so obvious, but Fig. 12–3 shows that the specific impulse increases appreciably with pressure.

Table 12–2 Equilibrium Gas Composition—Monopropellants*

Theoretical decomposition products, in mole fractions			Performance characteristics		
Ethylene oxide, C_2H_4O	Hydrazine, N_2H_4	Nitro-methane, CH_3NO_2	Ethylene oxide, C_2H_4O	Hydrazine, N_2H_4	Nitro-methane, CH_3NO_2
CO 0.481	H_2 0.656	CO_2 0.057	I_{sp} 160	170	218
CH_4 0.404	N_2 0.331	CO 0.277	T_0 1860	1125	3950
H_2 0.077	NH_3 0.013	H_2O 0.277	γ 1.17	1.37	1.25
C_2H_2 0.038		H_2 0.223	$\overline{M}$ 44	32.1	61
		N_2 0.166			

* Courtesy Altman, et al. [1].

Table 12-3 Liquid Rocket Propellants[1]

Fuel	Oxidizer	Best oxidizer–fuel mixture ratio (by weight)	Theoretical combustion temperature, °F	Ratio of specific heats	Average molecular weight of combustion products	Bulk density, 80°F propellant combination temperature, gm/cm³	Specific impulse, 400 psia, sea-level expansion, sec	Storability[2] Fuel	Storability[2] Oxidizer
Ammonia	Oxygen	1.3	4940	1.23	19	0.88	255	F–P	F–P
Ammonia	RFNA (22% NO$_2$)	2.15	4220	1.24	21	1.12	230	F–P	G
Diborane	Fluorine	5	7880	1.3	21	1.07	270	G	F–P
Diethylenetriamine	Oxygen	1.5	6500	1.24	21	1.06	245	G	F–P
Diethylenetriamine and hydrazine	Oxygen	2.5	6000	1.22	22		245	G	F–P
Hydrazine (anhydrous)	Chlorine trifluoride	2.4	6000	1.33	23	1.46	255	G	F–P
Hydrazine (anhydrous)	Fluorine	2	7740	1.33	19	1.3	290	G	F–P
Hydrazine (anhydrous)	H$_2$O$_2$ (90%)	1.5	4170	1.25	18	1.2	245	G	G
Hydrazine (anhydrous)	H$_2$O$_2$ (99.6%)	1.7	4690	1.22	19	1.24	255	G	F–P
Hydrazine (anhydrous)	Nitrogen tetroxide	1.2	5000	1.26	19	1.2	250	G	F–P
Hydrazine (anhydrous)	Oxygen	0.75	5370	1.25	18	1.06	265	G	F–P
Hydrazine (anhydrous)	RFNA (15% NO$_2$)	1.3	4980	1.25	20	1.26	247	G	G
Hydrogen (max. I$_{sp}$)	Fluorine	9.42[3]	8100	1.31	10	0.46	390	F–P	F–P
Hydrogen (max. I$_{sp}$)	Fluorine	3.8	4600	1.30	7.8	0.27	360	F–P	F–P
	Oxygen	8	5870	1.22	16	0.43	360	F–P	F–P
	Oxygen	3.5	4500	1.26	9	0.26	348	F–P	F–P
JP-4	H$_2$O$_2$ (99.6%)	6.5	4830	1.2	22	1.28	238	G	F–P
JP-4	Oxygen	2.3	5770	1.24	22	0.98	247	G	F–P
Kerosene	Oxygen	2.2–2.5	5200	1.24	22	0.99	240	G	F–P
Unsymmetrical dimethyl hydrazine	Oxygen	1.4	5650	1.24	20	0.96	249	G	F–P
Unsymmetrical dimethyl hydrazine	RFNA (22% NO$_2$)	2.6	5200	1.23	22	1.23	241	G	G
Unsymmetrical dimethyl hydrazine	WFNA	2.7	5100	1.23	22	1.22	240	G	G

[1]Courtesy *Space/Aeronautics* [17].   [2]G = good; F = fair; P = poor. "Good" storability means liquid can be stored in ordinary vessels or tanks over long periods and at many temperatures without decomposition or change of state.   [3]Mixture ratio yielding the highest loading density, or mass ratio.

Monopropellants are single-component chemicals which decompose and release energy in the presence of a suitable catalyst or at suitable temperatures. Table 12–2 illustrates typical equilibrium gas compositions in the combustion chamber for several monopropellants at 300 psia, and performance for equilibrium expansion to 14.7 psia [1].

Table 12–3 indicates typical performance and important characteristics of a variety of liquid propellant combinations. Many of the fuels are dangerous—for example, diborane, hydrazine, and hydrogen. The dangerous oxidizers are RFNA (red fuming nitric acid), WFNA (white fuming nitric acid), fluorine, and chlorine trifluoride. In general, the fuels have good coolant qualities, and the oxidizers do not.

12–4 SOLID PROPELLANTS

Solid propellants contain all the elements necessary for their combustion. Two types of solid propellant are distinguished, according to the distribution of fuel and oxidant. If fuel and oxidant are contained within the same molecule, the propellant is called *homogeneous*. Double-base propellants, combinations of nitroglycerin–nitrocellulose $[C_3H_5(NO_3)_3\text{-}C_6H_7O_2(NO_3)_3]$ with small quantities of additives, are the most common example of homogeneous solid propellants. Table 12–4 gives the composition of a typical double-base propellant, along with the functions of the various ingredients. The potassium salt promotes smooth burning at low temperatures, while the diethyl pthalate serves as an auxiliary plasticizer to improve mechanical properties. Ethyl centralite is added to counteract the autocatalytic decomposition of the major constituents. Such decomposition, if unchecked, could lead to self-ignition of the propellant. The carbon black is added to the nearly transparent propellant to prevent the transmission of radiant energy, which might cause internal ignition around internal voids or impurities. The candelilla wax is added to propellants formed by an extrusion process. In some cases metallic powders such as aluminum have been added, along with additional oxidizer of the composite type, to achieve higher performance.

Heterogeneous mixtures of oxidizing crystals in an organic plastic-like fuel binder are called *composite* propellants. Oxidizers usually consist of ground crystals

Table 12–4   Typical Double-Base Propellant (JPN)*

Material	Weight %	Purpose
Nitrocellulose (13.25%N)	51.40	Polymer
Nitroglycerin	42.93	Explosive plasticizer
Diethyl pthalate	3.20	Nonexplosive plasticizer
Ethyl centralite	1.00	Stabilizer
Potassium sulfate	1.20	Flash suppressor
Carbon black (added)	0.20	Opacifying agent
Candelilla wax (added)	0.07	Die lubricant

* Courtesy Geckler and Klager [15].

of potassium, lithium, or ammonium perchlorate or nitrate. The weight percent
of available oxygen and the density of these substances are given in Table 12–5.
A great variety of fuels or binders is available, including asphalt, rubber, and
synthetic resins and elastomers. Table 12–6 gives approximate compositions of
several composite solid propellants.

Table 12–5 Available O_2 in Solid-Propellant Oxidizers*

Oxidizer	Weight % O_2 available	Density, lb/in^3
Potassium perchlorate, $KClO_4$	46.0	0.090
Ammonium perchlorate, NH_4ClO_4	34.0	0.070
Lithium perchlorate, $LiClO_4$	60.0	0.087
Potassium nitrate, KNO_3	39.5	0.076
Ammonium nitrate, NH_4NO_3	20.0	0.061

* Courtesy Geckler and Klager [15].

Table 12–6 Typical Formulations of Composite Propellants*

Vinylpolyester, composition, %			Polysulfide, composition, %	
NH_4NO_3	—	72.50	NH_4ClO_4	71.00
NH_4ClO_4	64.60	—	Binder	26.00
$KClO_4$	11.40	—	Catalysts	2.00
Binder	23.60	25.50	Additives	1.00
Catalysts	0.25	—		100.00
Additives	0.15	2.00		
	100.00	100.00		

Polyurethane, composition, %			Polybutadiene copolymer, composition, %	
$KClO_4$	—	12.00	NH_4ClO_4	68.00
NH_4ClO_4	62.00	70.00	Aluminum	16.00
Binder	21.40	17.50	Binder	16.00
Aluminum	15.50	—	Additives	—
Additives	1.10	.50		100.00
	100.00	100.00		

Rubber, composition, %		
NH_4NO_3	—	86.50
NH_4ClO_4	90.00	—
Binder	7.00	11.30
Plasticizer	2.90	—
Catalyst	0.10	2.20
	100.00	100.00

* Courtesy Geckler and Klager [15].

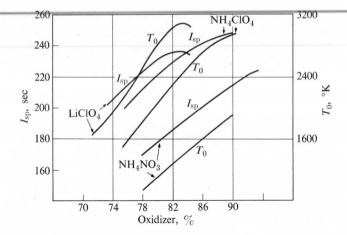

F<small>IG</small>. 12–4. Performance of several solid propellants as a function of oxidizer content. (Courtesy Geckler and Klager [15].)

The relatively high oxidizer content of the composite propellants is due primarily to the low available oxygen content of their oxidizers, and also to the fact that typical combustion temperatures are relatively low. Figure 12–4 indicates the variation of performance with oxidizer content for several hydrocarbon binder systems. At such temperatures dissociation is not important and maximum

Table 12–7*   Characteristics of Some Solid Propellants†

Property	Extruded JPN ballistite	Asphalt-base thermoplastic composite	NDRC molded composite
Specific impulse, sec	220	186	170
Exhaust velocity, ft/sec	7100	6000	5500
Density, lb_m/ft^3	101.5	110	100 to 115
Specific impulse/unit volume, %	100	96	80 to 90
Burning rate, in/sec	1.4	1.6	0.25 to 1.32
Burning-rate exponent	0.73	0.76	0.40
Area ratio**	185	175	200 to 1000
Low-pressure combustion limit, psia	500	1000	No lower limit
Flame temperature, °F	5300	3000	3000
Exhaust jet	Smokeless	Smoky	Smoky

* Courtesy Huggett, *et al*. [4].

† All pressure-dependent properties assume good exhaust nozzle design, 2000 psia combustion pressure, and 14.7 psia external pressure. The combustion process is temperature-sensitive, and 70°F has been assumed.

** Ratio of burning surface area to exhaust-nozzle throat area.

performance occurs with near-stoichiometric mixture ratios. High oxidizer content gives the added advantage of increased propellant density, since the oxidizers are generally more dense than the fuel. This, of course, permits smaller combustion chambers. A disadvantage of high oxidizer content in a propellant is that it generally results in poorer castability and less desirable mechanical properties. The selection of suitable additives and the choice of partially oxidized fuels can offset this disadvantage [5]. Generally the mechanical characteristics of the propellant depend on how much plastic binder and fuel are present.

Table 12–7 gives typical performance parameters for several solid propellants. Burning rate, burning rate exponent, area ratio, and low-pressure combustion limit are all important parameters which will be discussed in the following section.

12–5 COMBUSTION CHAMBERS

The function of the combustion chamber is to generate high-temperature and high-pressure combustion products. For a liquid-propellant rocket, the combustion chamber must have an appropriate array of propellant injectors, and a volume in which the propellant constituents can vaporize, mix, and burn, attaining near-equilibrium composition before entering the nozzle. For a solid-propellant rocket, the combustion chamber is a high-pressure tank containing the solid propellant and sufficient void space to permit stable combustion. An ignition system is required unless, as for certain liquid propellants, the chemicals ignite on contact (hypergolic propellant combination). The combustion chamber may be cooled or uncooled.

An appreciation of the intensity of rocket combustion may be gained by comparison with other combustion processes. Typical rates of energy release for steam power plants are about 10 Btu/sec/ft^3, and for jet engines, 10,000 Btu/sec/ft^3. In rocket chambers, rates of the order of 100,000 Btu/sec/ft^3 are achieved. Thus it is understandable that serious combustion problems have been encountered in the development of chemical rockets. In the following paragraphs, we shall discuss combustion chambers for liquid and solid propellants separately, after which we shall consider combustion instabilities qualitatively.

Liquid-propellant combustion chambers

Liquid propellants must be injected into the combustion chamber, atomized, and sufficiently mixed (except for monopropellants) before they can react. The *injector* is a series of small nozzles through which the propellant or propellants are introduced as a fine spray into the combustion chamber. These nozzles should be arranged to promote adequate mixing of the propellants and uniform properties across the combustion chamber. For certain hypergolic propellant combinations, such as nitric acid with unsymmetrical dimethyl hydrazine (UDMH), where ignition occurs as a result of liquid mixing [1], adequate mixing requires an unlike-stream-impinging injector, as illustrated in Fig. 12–5. Non-hypergolic propellant

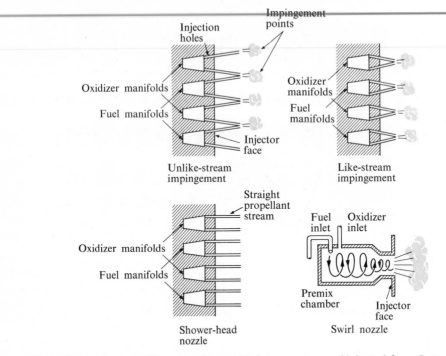

Fig. 12–5. Schematic diagrams of several injector systems. (Adapted from Sutton [2].)

combinations require rapid vaporization to promote adequate mixing of the gaseous phase. Like-stream or unlike-stream impingement, shower-head, and swirl-nozzle injectors have been successful. Monopropellants, while not requiring mixing, do require rapid vaporization and adequate distribution. Swirl injectors [1] have been successful in this case.

Injector configuration can substantially affect local wall temperatures throughout the rocket. Often it is possible, by controlling the injection pattern, to create a layer of relatively cool gas near the walls. Poor injection can lead to local hot spots and even to actual burnout of the wall material.

Figure 12–6 shows the injector configuration of the Rolls-Royce RZ.2 engine installed in the thrust chamber.

The mass flow rate of propellant and the pressure drop across the injector are related by an equation of the form

$$\dot{m} = C_d \rho A_i \sqrt{2(\Delta p / \rho)},\qquad(12\text{–}2)$$

where ρ = propellant density (liquid),

C_d = discharge coefficient,

A_i = injector nozzle area, and

Δp = pressure drop across injector nozzle.

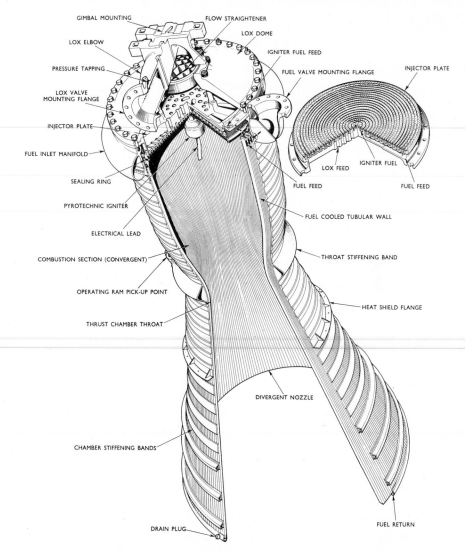

FIG. 12–6.   Rolls-Royce RZ.2 rocket engine thrust chamber.   Propellant is liquid oxygen and kerosene.   Sea-level thrust, 137,000 pounds.   (Courtesy Rolls-Royce.)

For injector nozzles which are essentially short tubes with rounded inlets, discharge coefficients from 0.97 to 0.99 are attainable.   For sharp-cornered tubes or orifices, the discharge coefficient drops to 0.6 to 0.8, primarily because of contraction of the jet cross section.   If Δp is measured across the entire inlet manifold system, rather than just across the inlet nozzle itself, C_d will, of course, be lower than these values.   Typical pressure drops for injectors are several hundred psi. A large drop in pressure promotes good atomization and tends to suppress those combustion instabilities associated with pressure oscillations in the chamber coupled with pressure oscillations in the propellant supply system. On the other hand, high pressure drop also produces high propellant inlet velocity, which we shall show to be undesirable. Further, high pressure drop requires increased power in the propellant pump or, for pressure-fed propellants, high-pressure propellant tanks. Thus the "best" injector pressure drop is not well defined, except by saying that it would probably be desirable to have the minimum Δp consistent with good atomization and combustion stability.

The size of the combustion chamber should be sufficient to permit the attainment of near-equilibrium product composition. The average traverse time Δt of a fluid element in a chamber is given by

$$\Delta t = \frac{\rho_0 \mathcal{V}_0}{\dot{m}}, \tag{12–3}$$

in which $\mathcal{V}_0$ is the combustion chamber volume and ρ_0 is the average gas density in the chamber. Chamber size is often described by the *characteristic length* L^*, which is defined by

$$L^* \equiv \frac{\mathcal{V}_0}{A^*}, \tag{12–4}$$

in which A^* is the exhaust nozzle throat area. Since, from Eq. (3–14),

$$\frac{\dot{m}}{A^*} = \frac{p_0}{\sqrt{RT_0}} \sqrt{\gamma} \left(\frac{2}{\gamma + 1}\right)^{(\gamma+1)/2(\gamma-1)},$$

the transit time Δt can be related to propellant properties and to L^* by

$$\Delta t = \frac{L^*}{\sqrt{\gamma RT_0}} \left(\frac{\gamma + 1}{2}\right)^{(\gamma+1)/2(\gamma-1)}. \tag{12–5}$$

It is obviously desirable to attain minimum chamber size or L^* for a given propellant, since this permits thinner chamber walls and lighter thrust chambers in general. In addition, the total chamber-wall heat-transfer rate decreases as size is decreased. The minimum acceptable chamber volume (i.e., minimum L^*) depends directly on the time Δt required for propellant injection, mixing, vaporization, and chemical reaction. Of these processes, the slowest (for typical propellants) is vaporization of the liquid drops, and this fact is the basis of simplified but enlightening one-dimensional analyses [9, 11] of liquid-propellant combustion chamber behavior. In order to illustrate what may be learned from these analyses, let us now consider their approximations and typical results.

Following Spalding [11], we consider the idealized cylindrical combustion chamber of Fig. 12–7, making the following assumptions:

(1) Drops of uniform size and velocity are injected into the chamber at plane 1. No distinction is made between drops of fuel and oxidant.

(2) The flow in the chamber is one-dimensional; all properties of the gas-liquid mixture are uniform at any plane of constant x.

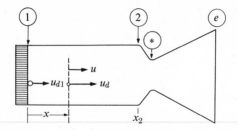

FIG. 12–7. Idealized liquid propellant combustor.

(3) Chemical reactions proceed infinitely fast; thus the composition of the gas is everywhere in local equilibrium.

(4) The time required for combustion is equal to the time required for vaporization of the drops.

(5) The pressure and temperature of the chamber are uniform.

In order to determine the behavior of the drops injected into this idealized chamber, we must formulate equations for (1) the rate of drop evaporation, and (2) the frictional interaction (drag) between the drops and the gas flow. In addition, we must write equations which express the conservation of mass, momentum, and energy for the gas-liquid mixture as a whole. Spalding has shown how differential equations which express these physical requirements can be formulated and integrated, using the following dimensionless variables:

Drop size $\quad\quad \rho = \dfrac{r}{r_0}$,

Gas velocity $\quad\quad \omega = \dfrac{\rho u A_c}{\dot{m}}$,

Drop velocity $\quad\quad \chi = \dfrac{\rho u_d A_c}{\dot{m}}$,

Axial distance $\quad\quad \xi = \dfrac{\chi}{r_0^2}\,\dfrac{k}{c_p \rho_l}\,\dfrac{\rho A_c}{\dot{m}}\ln(1 + B)$,

Drag $\quad\quad \mathcal{S} = \dfrac{9}{2}\,\dfrac{c_p \mu}{kB}$,

in which r = local drop radius, r_0 = initial drop radius, u = gas velocity, u_d = drop velocity, ρ = gas density, ρ_l = liquid density, $\dot{m}$ = total mass flow through the chamber, A_c = chamber cross-sectional area, k = thermal conductivity, and μ = viscosity. The variable B is defined by

$$ B = \dfrac{c_p(T_0 - T_f)}{h_{fg}}, $$

in which c_p = specific heat of gas, T_0 = gas stagnation temperature, T_f = saturation temperature of liquid, h_{fg} = heat of vaporization of the liquid.

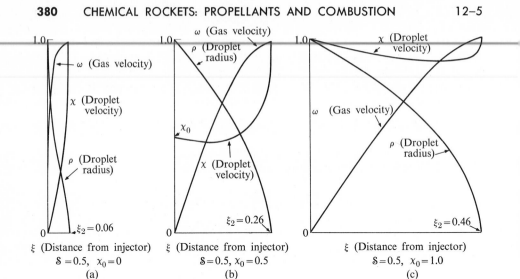

FIG. 12–8. Effect of initial drop size on dimensionless combustor performance variables. Variation of χ, ρ, and ω (defined on page 379) as a function of ξ for $\mathcal{S} = 0.5$. (Courtesy Spalding [14].)

Representative results of Spalding's calculations are shown in Fig. 12–8 for the typical value of $\mathcal{S} = 0.5$, showing the effect of variation of initial injection velocity χ_0. Since gas is formed only by liquid evaporation, the gas velocity u is zero at entrance to the chamber. Hence the drops are initially decelerated by the slower-moving gas and then accelerated as the gas velocity increases. Of particular interest is the value of $\xi = \xi_2$, at which the drop vanishes ($\rho \rightarrow 0$), since this determines the chamber length theoretically required for complete combustion. Spalding obtains for ξ_2,

$$\xi_2 = \frac{\chi_0 + 3\mathcal{S}/10}{2 + \mathcal{S}}, \tag{12–6}$$

and for the characteristic length of the chamber,

$$L^* = \xi_2 \frac{r_0^2((2/\gamma + 1)\{1 + [(\gamma - 1)/2]\mathrm{M}_2^2\})^{1/2(\gamma+1)}\sqrt{\gamma R T_0}}{(k/c_p \rho_l) \ln (1 + B)}, \tag{12–7}$$

in which M_2 is the Mach number of the flow after the drops have disappeared. Equations (12–6) and (12–7) and Fig. 12–8 show that minimum chamber size depends strongly on propellant inlet velocity χ_0 and on inlet drop size r_0.

As an example, Spalding has estimated L^* for the V2 rocket (liquid oxygen–alcohol) based on the following estimated data:

$$a = \sqrt{\gamma R T_0} \approx 3000 \text{ ft/sec}, \qquad r_0 \approx 0.005 \text{ cm},$$

$$(k/c_p \rho_l) \ln (1 + B) \approx 2 \times 10^{-3} \text{ cm}^2/\text{sec}, \qquad \chi_0 \approx 0.2, \qquad \mathcal{S} \approx 0.5,$$

with the result that $L^*_{min} = 63$ in. The actual value was about 113 in. [2], which suggests that the actual chamber may have been larger than necessary. Altman, *et al.* [1], give the following empirically determined values of minimum L^* at 300 psia chamber pressure: LOX–hydrogen, 10 in., nitric acid–aniline, 40 in., nitric acid–gasoline, 50 in., LOX–alcohol, 50 in., and nitromethane, 300 in.

This oxygen–alcohol value is in better agreement with Spalding's theory. However, the agreement may be misleading since these values are obtained with injectors of such small initial drop sizes that the chemical reaction rates may not be much faster than vaporization rates.

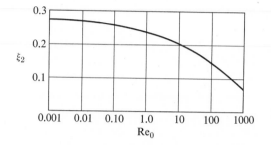

FIG. 12–9. Variation of ξ_2 with Reynolds number for $X_i = 0.5$, $S = 1.0$. (Adapted from Spalding [14].)

The real value of these one-dimensional analyses lies in the qualitative indication of the importance of various variables within the designer's influence (drop size and injection velocity, for example) and in the significance of the propellant variables. The effect of inlet drop size on the required chamber length variable ξ_2 is shown in Fig. 12–9. The dimensionless drop size variable Re_0 can be considered a Reynolds number, and is defined by

$$\mathrm{Re}_0 = \frac{2\rho u_2 r_0}{\mu} = \frac{2r_0(\dot{m}/A_c)}{\mu}.$$

A typical experimental value of Re_0 is about 100, so that the results given in Fig. 12–8, which were obtained with the assumption $\mathrm{Re}_0 < 10$, are conservative estimates of ξ_2.

The elementary rocket analysis given in Chapter 11 assumes constant-pressure combustion. Within actual combustion chambers there may be an appreciable drop in both static and stagnation pressure between the injector end and the nozzle end of the chamber. This pressure drop is due primarily to the equivalent "heat" addition, and it increases with increasing Mach number within the chamber. An accurate calculation of combustion pressure drop would be exceedingly difficult even for a geometrically simple combustion chamber, because of the very complex fluid dynamic and chemical phenomena involved. However, an approximate estimate of the pressure drop and its dependence on Mach number can be very easily obtained for a constant-area combustion chamber, using the relationships developed in Chapter 3. Returning again to the combustion chamber of Fig. 12–7, Eq. (3–16) becomes

$$\frac{p_{02}}{p_{01}} = \frac{1 + \gamma M_1^2}{1 + \gamma M_2^2} \left\{ \frac{1 + [(\gamma - 1)/2]M_2^2}{1 + [(\gamma - 1)/2]M_1^2} \right\}^{\gamma/(\gamma-1)}. \qquad (3\text{–}16)$$

Also the static pressure ratio p_2/p_1 is

$$\frac{p_2}{p_1} = \frac{1 + \gamma M_1^2}{1 + \gamma M_2^2}.$$

At the injector the Mach number will be very low. Thus, assuming $M_1 \approx 0$, these equations become

$$\frac{p_{02}}{p_{01}} = \frac{\{1 + [(\gamma - 1)/2]M_2^2\}^{\gamma/(\gamma-1)}}{1 + \gamma M_2^2} \quad \text{and} \quad \frac{p_2}{p_1} = \frac{1}{1 + \gamma M_2^2}. \qquad (12\text{–}8)$$

Assuming that the flow in the exhaust nozzle is isentropic, M_2 is determined by the area contraction ratio A_c/A^*, in accordance with Eq. (3–15):

$$\frac{A_c}{A^*} = \frac{1}{M_2} \left[\frac{2}{\gamma + 1} \left(1 + \frac{\gamma - 1}{2} M_2^2 \right) \right]^{(\gamma+1)/2(\gamma-1)}. \qquad (3\text{–}15)$$

Figure 12–10 indicates the Mach numbers and pressure ratios derived from this equation and Eq. (12–8) as a function of the contraction ratio A_c/A^* for a propellant of $\gamma = 1.2$. Typical combustion chamber cross sections are also indicated in Fig. 12–10. Although straight combustion chambers ($A_c/A^* = 1$) have been used, the contraction ratio is more typically between 1.5 and 2.0.

The walls of a liquid-propellant combustion chamber require intense cooling to maintain their temperature well below that of the combustion chamber. The wall structure is usually an extension of the regeneratively cooled nozzle structure discussed in Chapter 14.

Solid-propellant combustion chambers

The combustion chamber of a solid-propellant rocket is essentially a high-pressure tank containing the entire solid mass of propellant, or *grain* as it is called. Combustion proceeds from the surface of the grain. The rate of generation of gaseous propellant is, of course, equal to the rate of consumption of solid material. Hence it is given by

$$\dot{m}_g = \rho_p A_b r, \qquad (12\text{–}9)$$

where $\dot{m}_g$ = gas generation rate, mass per unit time,

ρ_p = solid propellant density,

A_b = area of burning surface, and

r = surface recession rate, length per unit time.

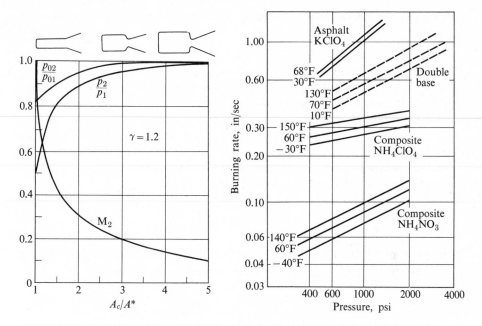

FIG. 12–10. Variation of static and stagnation pressure ratios and Mach number with contraction ratio for $\gamma = 1.2$. A_c = chamber cross-sectional area.

FIG. 12–11. Relation of burning rate to pressure and initial temperature for typical propellants. (Courtesy Geckler and Klager [15].)

The surface recession rate r, called the *burning rate*, is an empirically determined function of the propellant composition and certain conditions within the combustion chamber. These conditions include propellant initial temperature, combustion pressure, and the velocity of the gaseous combustion products over the surface of the solid. Figure 12–11 illustrates the burning rates of several propellants as a function of pressure and initial temperature in the absence of appreciable gas velocity. These curves can be approximated by

$$r = ap_0^n, \tag{12–10}$$

where a and n are constants for any one curve and p_0 is the pressure within the combustion chamber. The pressure exponent n, associated with the slope of the curves in Fig. 12–11, is almost independent of propellant temperature. The coefficient a is a function of initial propellant temperature T_p, but not a function of pressure. Often a can be expressed [4] as

$$a = \frac{A}{T_1 - T_p}, \tag{12–11}$$

where T_1 and A are empirical constants. Other equations, for example [4]

$$r = c + bp_0^n,$$

have been used to describe burning rates of certain propellants. A correction for the so-called "erosive burning" associated with high gas velocities over the solid surface will be discussed subsequently.

Both the rate of mass flow through the nozzle $\dot{m}_n$ and the gas generation rate $\dot{m}_g$ are strongly dependent on the combustion chamber pressure p_0, which in turn depends on the mass $\mathfrak{M}_s$ of gas within the chamber.

The gas generation rate $\dot{m}_g$ is given by Eqs. (12–9) and (12–10). The rate of increase of gas "stored" within the combustion chamber is

$$\frac{d\mathfrak{M}_s}{dt} = \frac{d}{dt}(\rho_0 \mathcal{V}_0),$$

where ρ_0 is the instantaneous gas density in the chamber and $\mathcal{V}_0$ is the instantaneous gas volume. Considering the recession of the burning surface, $d\mathcal{V}_0/dt = rA_b$, so that

$$\frac{d\mathfrak{M}_s}{dt} = \rho_0 r A_b + \mathcal{V}_0 \frac{d\rho_0}{dt}.$$

The rate at which gas flows through the nozzle is given by Eq. (3–14):

$$\dot{m}_n = \frac{p_0}{\sqrt{RT_0}} \sqrt{\gamma} \left(\frac{2}{\gamma+1}\right)^{(\gamma+1)/2(\gamma-1)} A^*. \tag{3–14}$$

Conservation of mass requires the balance

$$\dot{m}_g = \frac{d\mathfrak{M}_s}{dt} + \dot{m}_n \quad \text{or} \quad \rho_p A_b r = \rho_0 A_b r + \mathcal{V}_0 \frac{d\rho_0}{dt} + \dot{m}_n.$$

Using Eqs. (12–10) and (3–14), we obtain

$$\rho_p A_b a p_0^n = \rho_0 A_b a p_0^n + \mathcal{V}_0 \frac{d\rho_0}{dt} + \frac{p_0}{\sqrt{RT_0}} \sqrt{\gamma} \left(\frac{2}{\gamma+1}\right)^{(\gamma+1)/2(\gamma-1)} A^*. \tag{12–12}$$

The density derivative may be expressed as

$$\frac{d\rho_0}{dt} = \frac{1}{RT_0} \frac{dp_0}{dt},$$

since T_0 is practically independent of changes in combustion pressure for typical solid propellants. Equation (12–12) may then be written

$$\frac{\mathcal{V}_0}{RT_0} \frac{dp_0}{dt} = A_b a p_0^n (\rho_p - \rho_0) - \sqrt{\frac{\gamma}{RT_0} \left(\frac{2}{\gamma+1}\right)^{(\gamma+1)/(\gamma-1)}} A^* p_0. \tag{12–13}$$

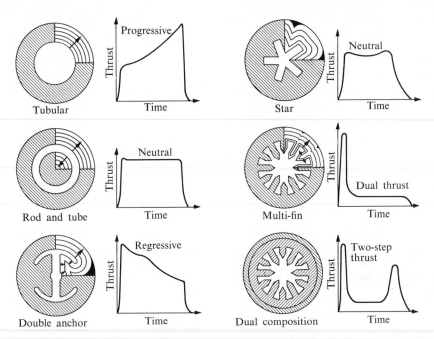

FIG. 12–12. Internal-burning charge designs with their thrust-time programs. (Courtesy Shafer [16].)

If the combustion pressure remains constant, the left-hand side of this equation vanishes, and therefore p_0 may be determined from

$$p_0 = \left[\frac{A_b}{A^*} \frac{a(\rho_p - \rho_0)}{\sqrt{\dfrac{\gamma}{RT_0} \left(\dfrac{2}{\gamma+1} \right)^{(\gamma+1)/(\gamma-1)}}} \right]^{1/(1-n)}. \tag{12–14}$$

Generally ρ_0 may be neglected in Eq. (12–14), since it is much less than ρ_p.

For constant p_0 (and hence constant thrust), it is clear that the burning area A_b must remain constant. Even if the burning area is not constant, however, the left-hand term of Eq. (12–13) may be negligible compared to the other terms, so that the instantaneous stagnation pressure is given by Eq. (12–14), in which A_b is the instantaneous burning area. The variation of A_b with time depends on the burning rate and the initial geometry of the propellant grain. Geometric variations can provide increasing thrust (*progressive* burning), constant thrust (*neutral* burning), or decreasing thrust (*regressive* burning) with time. Figure 12–12 indicates several internal-burning sections, along with their subsequent partially burned shapes and their resultant thrust histories. High-thrust rockets, requiring large A_b, are usually of the internal-burning type. Lower thrust may be provided by solid cylindrical grains which burn only on the end surface.

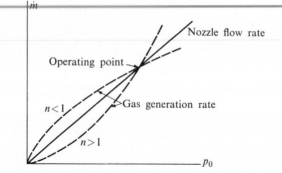

FIG. 12–13. Stability argument for solid-propellant combustion.

The burning-rate exponent n must be less than unity for stable combustion, as may be demonstrated by the following simple argument. Neglecting the relatively small gas storage terms, nozzle flow rate and gas generation rate must be equal. Equation (3–14) shows that the nozzle gas flow rate is directly proportional to p_0. It has also been shown that the gas generation rate is proportional to p_0^n. Figure 12–13 indicates typical curves of nozzle flow rate and gas generation rate versus pressure. Two gas generation rate curves, one for $n < 1$ and one for $n > 1$, are shown, both having the same operating point with the given nozzle. Consider a small decrease from the operating-point pressure. It can be seen that for $n < 1$ the resultant gas generation rate will then exceed the nozzle flow rate, and thus the chamber pressure will return toward the original value. For $n > 1$, any small decrease in chamber pressure means that nozzle flow rate will exceed gas generation rate and the pressure will drop even further. Thus for $n > 1$, small disturbances are amplified and the combustion cannot be stable. Hence for stable combustion n must be less than unity. Typical values of burning-rate exponent n for solid propellants are given in Table 12–7.

The variation of burning rate r with temperature, indicated in Fig. 12–11 and accounted for by an equation of the form of Eq. (12–10), can seriously alter the performance of a given rocket under different firing-temperature conditions. For example, Fig. 12–14 indicates the time variation of chamber pressure for three different initial temperatures. The thrust histories would be similar since, from Eq. (11–7), thrust is proportional to p_0. However, since at constant T_0 the mass flow rate will also be proportional to p_0, the thrust *duration* will be *inversely* proportional to p_0. Hence the total impulse, $\int T \, dt$, will be practically constant, as will the area under the above p_0-t curves, regardless of propellant temperature. However, excessive variation of burning rate can have serious consequences. If it is too high, excessive vehicle acceleration or chamber failure can occur. If it is too low, insufficient thrust may prevent satisfactory performance, especially for operation within a gravity field. Efforts have been made to reduce this temperature-sensitive behavior, and some of the composite propellants are particularly good in this respect.

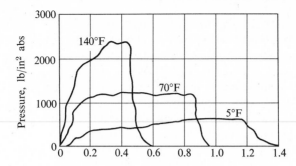

FIG. 12–14. The variation of chamber pressure with time for a 9-lb cruciform, semi-restricted-burning, ballistite-grain, 3.25-in. rocket motor. (Courtesy Huggett, *et al*. [4].)

The sensitivity of burning rate to propellant temperature at constant pressure, Π_r, is defined as

$$\Pi_r = \frac{1}{r}\left(\frac{\partial r}{\partial T_p}\right)_{p_0}. \tag{12–15}$$

Since the temperature variation of r is contained in a, Eq. (12–11) gives

$$\Pi_r = \frac{1}{r}\left(\frac{\partial r}{\partial T_p}\right)_{p_0} = \frac{1}{a}\frac{da}{dT_p} = -\frac{1}{T_1 - T_p}. \tag{12–16}$$

Thus as T_p approaches the empirical constant T_1, the burning rate becomes very temperature-sensitive. Propellants should have high values of T_1 relative to their expected operational temperatures.

The variation of equilibrium pressure with initial temperature, shown in Fig. 12–14, is also of importance. From Eq. (12–14) it can be seen that if n approaches unity the pressure can become an extremely sensitive function of initial temperature, through the factor $a^{1/(1-n)}$. The dependence of the chamber pressure p_0 on the propellant temperature T_p may be expressed as

$$\Pi_p = \frac{1}{p_0}\left(\frac{\partial p_0}{\partial T_p}\right)_{A_b/A^*,T_0}. \tag{12–17}$$

Differentiation of Eq. (12–14) yields

$$\Pi_p = \frac{1}{1-n}\frac{1}{a}\frac{da}{dT_p} = \frac{1}{1-n}\Pi_r.$$

Thus the sensitivity of chamber pressure to grain temperature is greater than the sensitivity of burning rate to grain temperature by the factor $1/(1-n)$. Hence, in addition to high T_1, a desirable propellant should have low n. As mentioned before, n must be less than one for stability.

The surface-combustion process is a complex interaction that requires heat transfer from the combustion products to the solid material to bring about vaporization and perhaps decomposition of the solids. As in more conventional heat-transfer processes, high fluid velocity over the surface can substantially increase the heat-transfer rate. Although the phenomena occurring in solid-propellant combustion are complicated by the outflow of mass from the solid surface, it is observed that burning rate increases with increased combustion-product velocity. This effect is called *erosive burning*, although it seems that the increased heat transfer is much more important than any actual erosion of the solid material. The increased burning rate can be accounted for by a variety of empirical corrections of the form [4]

$$r = r_0(1 + k_1 u), \qquad r = r_0(1 + k_2 \rho u),$$

or

$$r = r_0, \quad u < u_0, \qquad r = r_0[1 + k_3(u - u_0)], \quad u > u_0,$$

in which r is the burning rate at gas velocity u and r_0, u_0, k_1, k_2, k_3 are empirical constants. The velocity u_0 may be thought of as a threshold value below which no appreciable erosive effect occurs.

Within the void space of an internal-burning grain the gas velocity varies appreciably, and it is actually zero only at the head end of the grain. As a result the burning rate can vary significantly along the grain length, thus complicating the grain design problem. It is desirable to have the grain burning completed instantaneously along its entire length, since nonuniform burn-through may reduce the chamber pressure enough to extinguish combustion before all the propellant is consumed. Even if combustion did not cease, the prematurely exposed chamber wall could fail due to overheating.

Nonuniform burning rates may be compensated for by simultaneous axial variations of the flow area and the thickness and shape of the grain. In one design the effect of erosion is made uniform by increasing the port (gas flow) area in the flow direction, thus tending to reduce the variations in gas velocity. The propellant web thickness (burning-path length) may be maintained constant by varying the shape of the void cross section or by varying the outside diameter of the grain. In another design the web thickness is increased near the nozzle end of the chamber to compensate for the greater burning rate there. This may result in decreased port area and further increased erosive burning effects. Rather long and thin grains have been constructed in this manner.

The relatively continuous variation of burning characteristics with variation of combustion-chamber conditions is subject to certain limitations. It has been observed that for very low chamber pressures combustion becomes very unsteady or ceases altogether. The pressure at which this occurs is called the *combustion limit* of a particular propellant. There is also an upper *pressure limit*, above which combustion again becomes erratic and unpredictable. For most propellants this is above 5000 psi [4]. Variation of propellant mechanical properties with temperature places limits on the temperatures at which solid-propellant rockets may be stored and fired. For example, at high temperature the propellants may tend to flow or "slump," especially if stored for a considerable time. At low tempera-

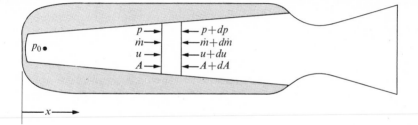

FIG. 12–15. Incremental control volume for the calculation of pressure drop within a solid-propellant combustion chamber.

ture the propellant may become brittle and crack, exposing a much larger burning surface than intended.

As in the liquid-propellant combustion chamber, there is a pressure drop along the axis of the combustion chamber, a drop which is necessary to accelerate the increasing mass flow of gas out of the void space. The axial variation of pressure may offset to some extent the uneven erosive burning due to axial variations in gas velocity. Figure 12–15 illustrates an incremental control volume within a typical internal-burning propellant grain. In addition to the changes in pressure, velocity, and area, there is an incremental increase in mass flow rate $d\dot{m}$ over the length of the control volume. Neglecting the effect of wall friction on the gas flow, the momentum equation [Eq. (2–4)] may be used for this control volume to obtain

$$-A\,dp = \dot{m}\,du + u\,d\dot{m} = d(\dot{m}u). \tag{12–18}$$

Integrating from $x = 0$, $p = p_1$ to any position x where the static pressure is p yields

$$p_1 - p = \int_{x=0}^{x} \frac{d(\dot{m}u)}{A}, \tag{12–19}$$

and for the special case, $A \approx$ constant,

$$p_1 - p = \frac{(\dot{m}u)_x}{A}. \tag{12–20}$$

Since $u = \dot{m}/\rho A$, where ρ is the gas density, Eq. (12–20) becomes

$$p_1 - p = \frac{1}{\rho}\left(\frac{\dot{m}}{A}\right)^2. \tag{12–21}$$

If the port area and shape do not vary with x, and if the compensating effects of pressure variation and erosive burning have negligible net effect on local gas generation rate, then

$$\dot{m} = cr\rho_p x,$$

where c is the port circumference and ρ_p the propellant (solid) density. Using

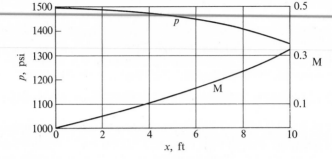

FIG. 12–16. Variation of pressure according to Eq. (12–22) for a cylindrical port one foot in diameter; typical propellant properties.

this approximation and the perfect gas law in combination with Eq. (12–21) gives us

$$\frac{p}{p_1} = \frac{1}{2}\left[1 + \sqrt{1 - 4RT\left(\frac{cr\rho_p x}{p_1 A}\right)^2}\right].$$
(12–22)

If, as a further approximation, it is assumed that the axial variation in temperature is negligible, Eq. (12–22) may be used to estimate the axial variation in pressure.

Figure 12–16 illustrates this distribution of pressure and Mach number along a cylindrical port one foot in diameter, for typical propellant properties. As burning proceeds and the port area increases, axial variations of both pressure and gas velocity decrease.

Early solid-propellant grains were often molded or extruded prior to being inserted into the combustion chamber. The grain was then held in place within the combustion chamber by suitable, and sometimes relatively complex, support structures. There were several disadvantages to this practice. The support structure, which had to withstand combustion temperatures, represented undesirable inert mass. In most such configurations, a part if not all of the chamber walls were directly exposed to combustion temperatures. For a given combustion pressure, this necessitated relatively heavy chamber walls. In many cases the reduced grain thickness near the end of combustion resulted in mechanical failure of the grain, with subsequent erratic or destructive burning.

The modern tendency has been to use internal-burning case-bonded grains, as illustrated in Fig. 12–12. In this configuration the grain is bonded directly to the chamber wall, or perhaps to a thin layer of inert material (inhibitor) which is itself bonded to the wall. The primary advantage of this design is that the propellant itself serves as an insulator, enabling the chamber walls to remain relatively cool throughout their operation (provided the propellant does not prematurely burn through locally). Much lighter combustion chambers are thus possible, since no internal structures are necessary and thinner walls are permissible. A much higher fraction of the total chamber volume (greater than 80% as opposed to about 65% for separately formed grains [4]) may be filled with propellant, and the grain remains adequately supported right up to burnout. Very large grains

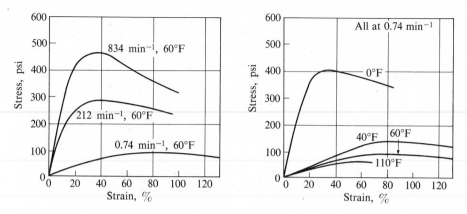

FIG. 12–17. Stress-strain properties for a polyurethane-type propellant at various strain rates and temperatures. (Courtesy Geckler and Klager [15].)

may be constructed by casting directly in the combustion chamber; although extremely large grains, contemplated for the launching of space vehicles, may require internal support elements [18].

Grain design must satisfy mechanical requirements as well as the desired thrust program. The mechanical properties of some propellants are extremely complex, as may be seen in Fig. 12–17, which illustrates stress-strain curves for a typical composite propellant. It can be seen that the stress-strain relationship is highly dependent on both the temperature and strain rate of the propellant. Low strain rates are characteristic of thermal strains during curing and storage, while high rates represent ignition conditions. Note the relatively high strains encountered. Stress in these figures is based on initial cross section, and at such strains the true stress would be substantially higher.

The stress loading of a propellant grain can be quite complex. The axial pressure distribution, viscous shear from the gas flow, and vehicle acceleration all tend to move the propellant toward the nozzle. Opposing these forces is the support system or wall shear force for the case-bonded grain. The resultant bulging of the propellant near the nozzle may reduce port area, thus accentuating pressure and viscous forces. In fact, for a given configuration and propellant there exists a critical Young's modulus for the propellant, below which bulging and pressure rise continue until the chamber bursts. Reference [4] treats this problem in detail.

While the solid-propellant combustion chamber is, in principle, a simple pressure vessel, the actual design of a suitable chamber is not an easy task. In the cylindrical wall, the calculation of stress is quite simple. The principal stresses σ, assuming the combustion-chamber wall is a thin-walled tube, are:

$$\text{Tangential} \quad \sigma = p_0 \left(\frac{r}{\tau}\right), \qquad \text{Axial} \quad \sigma = p_0 \left(\frac{r}{2\tau}\right),$$

$$\text{Radial} \quad \sigma = 0,$$

where p_0 = chamber pressure, τ = wall thickness, and r = chamber radius.

The necessity of providing ports for the nozzle, an igniter, perhaps a blowout port for thrust termination, and external structural members for the attachment of payloads, etc., considerably complicates the stress pattern. Stress concentrations around these objects, coupled with the need to operate at extremely high stress levels (to obtain light chambers) mean that very careful design is necessary. If detailed stress distributions are difficult to calculate (as is usually the case), experimental stress measurements must be made on models of contemplated designs.

Stress levels above 200,000 psi (tangential stress in thinnest sidewall section) have been used in steel chambers, while experimental values above 300,000 psi have been achieved. Other metals, for example titanium, may be used to advantage at lower stress levels since their densities are lower than steel. Titanium has been used at stress levels corresponding to well over 300,000 psi in an equivalent weight of steel.

Such stresses are possible, of course, only in the relatively cool wall of an internal-burning case-bonded rocket. In fact, even glass-fiber-reinforced plastic chambers are feasible. In some cases the weight of such a chamber is less than that of a conventional steel one, owing to the much lower density of the glass fiber structure.

12–6   COMBUSTION INSTABILITIES

Both liquid- and solid-propellant rocket engines are sometimes subject to combustion instabilities in the form of large pressure oscillations within the chamber. Such instabilities can cause engine failure either through excess pressure, increased wall heat transfer, or a combination of the two. Instabilities are especially prevalent in the larger rockets. When instability is encountered it is usually necessary to take corrective measures on the full-scale engine itself, since similar behavior cannot often be observed on smaller-scale models. Hence experimental study is quite difficult and expensive and once a "fix" has been found, experimentation often ceases. For these reasons, and also because of the extreme complexity of the problem, rocket combustion instabilities cannot be said to be fully understood.

Instabilities occurring within liquid-propellant combustion chambers may be divided into at least two distinct categories. Low-frequency (less than 100 cycles per second) oscillations, often called *chugging*, result from a coupling between the combustion process and the feed system. Qualitatively it may be explained as follows. Suppose the chamber pressure were to rise instantaneously above the steady-state value. This would be followed by a decrease in rate of delivery of the propellant. This in turn would bring about a decrease in chamber pressure and an increase in pump delivery pressure, followed by an increase in rate of delivery of the propellant, an increased chamber pressure, and so on. A self-sustaining oscillation is therefore possible, depending on the dynamic behavior of the system components. Chugging is relatively well understood [19] and easily corrected. Usually a greater drop in injector pressure, tending to make chamber pressure oscillations small relative to pump-delivery pressure, will eliminate chugging in a given rocket. This, of course, requires higher pump pressure and power so that an arbitrarily large injector pressure drop is undesirable.

High-frequency oscillations (of the order of thousands of cycles per second), often called *screaming*, are not so well understood. They are usually associated with one or more of the acoustic vibration modes within the chamber: longitudinal, radial, transverse, or tangential. Ordinarily such vibrational motion of the propellant would be damped by viscous effects as well as carried out through the nozzle. However, if the combustion process is pressure-sensitive, a mechanism of nonuniform energy release is available which can maintain the oscillations. Such energy release must occur in the proper place and at the proper time relative to the pressure oscillations if it is to feed energy to the oscillating propellants. Hence changes in combustion pattern, as produced for example by small modifications of the injector, can mean the difference between stable and unstable operation. With the possibility of so many vibrational modes it is not at all simple to determine which corrective action is called for. Screaming, unlike chugging, can occur entirely within the combustion chamber and with essentially steady inlet conditions.

A third or intermediate frequency range has been described [1] in which pressure oscillations within the chamber interact with the injector-spray pattern but not with the propellant-delivery rate or system.

Solid-propellant rockets are also subject to nonuniform pattern of energy release since, as we have seen, the combustion rate is influenced by propellant pressure and velocity. It is possible that the energy release and the propellant velocity and/or pressure pattern which causes nonuniformity can interact to produce sustained oscillations. Such oscillations produce high rates of erosive burning which may change the combustion geometry to a stable one rapidly enough to avoid damage, or they may lead to engine failure. It has been found that placing an irregular rod of nonburning material within the burning volume, or drilling radial holes at intervals along the grain, can reduce instabilities [5]. The rod or the combustion gases issuing from radial holes tend to break up the wave patterns within the burning gases. Reference [4] contains an extensive discussion of the instabilities of solid-propellant combustion.

References

1. ALTMAN, D., J. M. CARTER, S. S. PENNER, and M. SUMMERFIELD, "Liquid Propellant Rockets." Princeton, N. J.: Princeton Aeronautical Paperbacks, Princeton University Press, 1960

2. SUTTON, G. P., *Rocket Propulsion Elements*. New York: John Wiley & Sons, 1956

3. SEIFERT, H. S., "Twenty-Five Years of Rocket Development." *Jet Propulsion*, **25**, 11, November 1955, pages 594–603

4. HUGGETT, C., C. E. BARTLEY, and M. M. MILLS, *Solid Propellant Rockets*. Princeton, N. J.: Princeton Aeronautical Paperbacks, Princeton University Press, 1960

5. SEIFERT, H. S. (editor), *Space Technology*. New York: John Wiley & Sons, 1959

6. PRIEM, R. J., "Propellant Vaporization as a Criterion for Rocket Engine Design: Calculations of Chamber Length to Vaporize a Single Heptane Drop." *NACA* TN 3985, 1957

7. PRIEM, R. J., "Propellant Vaporization as a Criterion for Rocket Engine Design: Calculations Using Various Log-Probability Distributions of Heptane Drops." *NACA* TB 4098, 1957

8. PRIEM, R. J., "Propellant Vaporization as a Criterion for Rocket Engine Design: Calculation of Chamber Length to Vaporize Various Propellants." *NACA TN* 3883, 1958

9. WILLIAMS, F. A. (*see* S. S. Penner), *Chemical Rocket Propulsion and Combustion Research*. New York: Gordon & Breach, 1962, Chapter 3

10. HEIDMAN, M. F., and R. J. PRIEM, "Propellant Vaporization as a Criterion for Rocket Engine Design: Relation between Percentage of Propellant Vaporized and Engine Performance," *NACA* RM *E57 E*03, 1957

11. SPALDING, D. B., "Combustion in Liquid-Fuel Rocket Motors," *Aero Quarterly*, **10**, 1, 1, January 1959

12. SPALDING, D. B., *Some Fundamentals of Combustion*. London: Butterworth's Scientific Publications, 1955

13. HUNSAKER, J. C., and B. G. RIGHTMIRE, *Engineering Applications of Fluid Mechanics*. New York: McGraw-Hill Book Company, 1947

14. SPALDING, D. B., "A One-Dimensional Theory of Liquid-Fuel Rocket Combustion." *Aeronautical Research Council*, C.P. 445

15. GECKLER, R. D., and K. KLAGER, "Solid Propellant Rocket Engines," Chapter 19 of *Handbook of Astronautical Engineering*, edited by H. Koelle. New York: McGraw-Hill, 1961

16. SHAFER, JOHN I., "Solid Rocket Propulsion." Chapter 16 of *Space Technology* (editor, H. S. Seifert). New York: John Wiley & Sons, 1939

17. R and D Technical Handbook, *Space/Aeronautics*. New York: Conover-Mast, 1961–1962, pages C–8, C–9

18. BOYD, A. B., W. M. BURKES, and J. E. MEDFORD, "Grain Design and Development Problems for Very Large Rocket Motors." Presented at ARS Solid Propellant Rocket Conference, Princeton, N. J., January 28–29, 1960; and published in *Solid Propellant Rocket Research* (editor, Martin Summerfield). New York: Academic Press, 1960

19. TSIEN, H. S., "Servo Stabilization of Combustion in Rocket Motors," *Journal of the American Rocket Society*, **25**, 2, March–April, 1953

Problems*

1. A certain solid propellant burns at a rate shown in the figure. It is being used in a rocket combustion chamber in which the stagnation pressure is at 500 psia. The initial grain temperature is 500°R. Suddenly a large crack develops in the propellant grain, which increases the burning area by 10%. What will happen to the chamber pressure?

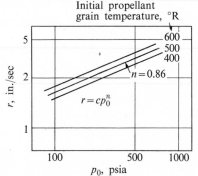

PROBLEM 1

2. The throat liner material of a solid-propellant rocket motor erodes at a constant rate $dD^*/dt = k$. This may significantly alter the rocket thrust with time. Show how the grain burning area must vary with time if the thrust is to remain constant despite this erosion. Assume that the burning rate is proportional to the nth power of the pressure at a given grain temperature and that the chamber and grain temperature remain constant. The nozzle pressure ratio $p_e/p_0 \ll 1$.

3. How long would it take for the thrust of a rocket to diminish to 10% of its steady value if the fuel and oxidant flows into the chamber were suddenly stopped? Consider, for example, the following conditions:

Initial combustion chamber pressure	$p_0 = 600$ psia
Initial combustion chamber temperature	$T_0 = 6000°R$
Combustion chamber volume	$\mathcal{V} = 5$ ft^3
Throat area	$A^* = 1$ ft^2
Molecular weight of propellant	$\bar{M} = 10$
Ratio of specific heats	$\gamma = 1.4$
Ambient pressure	$p_a = 0$

4. Shown in the figure is a much-idealized rocket combustion chamber (1–2) whose diameter is the same as the nozzle throat diameter. The combustion process is assumed

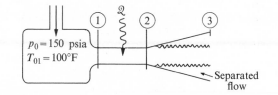

PROBLEM 4

*See Chapter 2 for other problems on chemical combustion.

equivalent to a heat addition rate $\dot{Q}$ which heats the fluid to a stagnation temperature T_{02} of 4000°F. What exhaust pressure will just cause choking, assuming that the subsonic flow in the nozzle does not follow the walls? What maximum back pressure will permit full nozzle flow if $A_3/A_2 = 5$? ($\gamma = 1.4$.)

5. Hydrogen and oxygen at 77°F and 300 psia are introduced into a combustion space in the proportion, by mass, of 1:2. Find the flame temperature, assuming that the equilibrium constant is

$$K_p = \frac{(p_{H_2})(p_{O_2})^{1/2}}{p_{H_2O}} = 0.05$$

where p is the partial pressure in atmospheres. Assume that only H_2, O_2, and H_2O are present, and use the following average values of specific heat between the flame temperature and 77°F:

$$c_{pH_2O} = 11.0 \text{ Btu/lb-mole·°F},$$
$$c_{pO_2} = 8.6 \text{ Btu/lb-mole·°F},$$
$$c_{pH_2} = 7.8 \text{ Btu/lb-mole·°F}.$$

The heat of formation of H_2O is $-104{,}000$ Btu/lb$_m$; H_2O at 77°F. What is the specific impulse which could be obtained on expansion to 14.7 psia, assuming frozen equilibrium?

6. In the process of computing an equilibrium combustion temperature, one assumes the answer and calculates the composition based on equilibrium constants dependent on that assumption. Suppose you have gone that far and have obtained the following reaction for the products of gasoline and oxygen:

$$C_8H_{18} + 12.5O_2 \rightarrow 6.3CO_2 + 6.5H_2O + 1.7CO + 0.5O_2$$
$$+ 0.9H_2 + 0.7O + 2.5OH + 0.7H.$$

Based on the same assumed temperature, the specific heats of the various substances are

Compound	c_p, Btu/lb-mole·°F	Heat of formation, Btu/lb-mole
C_8H_{18}	—	$-105{,}000$
CO_2	14.0	$-169{,}300$
H_2O	11.5	$-104{,}000$
CO	8.6	$-47{,}550$
O_2	8.9	0
H_2	8.0	0
O	5.0	$+105{,}400$
OH	5.0	$+18{,}000$
H	5.0	$+92{,}910$

Calculate the adiabatic flame temperature of the above reaction. Assume that reactants enter at standard temperature and pressure ($H_{R1} - H_{Rf} = 0$).

13

Chemical Rockets:
Expansion in Nozzles

13-1 ROCKET EXHAUST NOZZLES

If the flow in a rocket nozzle is assumed one-dimensional and isentropic (as in Chapter 11), then the only important geometric variable is the area ratio. However, real nozzle flows are never truly one-dimensional, and the shape of the nozzle walls can be quite important. The design of an actual nozzle requires the specification of the entire nozzle shape, and generally takes into account variations in velocity and pressure on surfaces normal to the streamlines. In addition, the influence of friction, heat transfer, composition change, or shocks must be considered. However, let us first consider the determination of suitable nozzle shapes without these latter complications.

Nozzle shape

The exact shape of the subsonic or converging portion of the nozzle is not, within limits, a matter of great importance. In a region of favorable pressure gradient such as this, almost any reasonably smooth contour provides good subsonic flow. On the other hand, the shape of the supersonic or diverging portion of the nozzle is important since, even in the absence of boundary layer effects, improper shaping can result in shock formation and substantial performance loss.

Simple conical divergent sections can provide shockfree performance (when expanding to the appropriate pressure) and they are often used. It has been found empirically that divergence half-angles between 12 and 18 degrees provide best performance [2]. Smaller angles lead to excessive nozzle lengths (for a given area ratio), and large friction losses. Larger angles lead to radial flow losses and, possibly, to flow separation from the walls. Approximate experimental verification of the thrust correction due to radial flow of Eq. (11–14) is shown in Fig. 13–1, which compares actual to ideal (axial exhaust velocity) thrusts for a series of

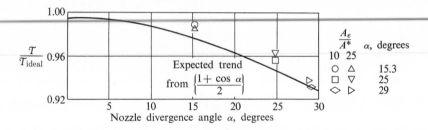

Fig. 13–1. Radial-flow thrust loss of conical nozzles [3].

conical nozzles. In this figure the expected performance has been reduced arbitrarily about $\frac{1}{2}\%$ to account for boundary layer losses.

The radial flow loss can be eliminated if the nozzle is suitably contoured to give a uniformly axial exhaust velocity. The design of a suitable nozzle contour to produce such a flow utilizes an analytical procedure called the *method of characteristics*. The many applications of this method in supersonic flow calculations warrant a brief discussion of its essential features here. A thorough discussion with more extensive background material may be found in Reference [4], Chapters 15 and 17.

The method of characteristics depends on the fact that, in supersonic flow, the influence of a small pressure disturbance is limited to a specific region. In subsonic flow the region of influence is unlimited and there is no comparable calculation procedure. The region of influence is illustrated in Fig. 13–2, which shows a small disturbance originating at a point A in a field of uniform parallel supersonic flow. If the pressure disturbance is small, it will propagate *relative to the fluid* as a spherical pressure wave at the local velocity of sound a. The center of the spherical sound wave is moving downstream with velocity u. Thus the zone of influence is limited to a cone of half-angle α, which is given by

$$\alpha = \sin^{-1}\left(\frac{at}{ut}\right) = \sin^{-1}\left(\frac{1}{M}\right)$$

or

$$\alpha = \tan^{-1}\frac{at}{\sqrt{(ut)^2 - (at)^2}} = \tan^{-1}\frac{1}{\sqrt{M^2 - 1}}, \qquad (13\text{–}1)$$

where M is the Mach number. The angle α is called the *Mach angle* and the limit of influence is called the *Mach line*. If the flow upstream of a given Mach line is uniform it will be straight, and all properties of the flow immediately downstream of it will be uniform. In a sense, changes in fluid properties may be thought of as propagating along Mach lines.

Suppose that uniform parallel supersonic flow enters the divergent portion of a duct, as in Fig. 13–3. Due to the wall curvature there are established pressure gradients which turn the streamlines so that the fluid follows the wall. Since we are ignoring friction in this discussion, the fluid may be assumed to slide freely over

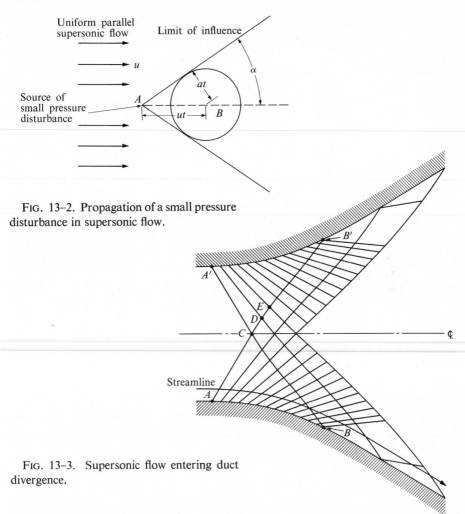

FIG. 13–2. Propagation of a small pressure disturbance in supersonic flow.

FIG. 13–3. Supersonic flow entering duct divergence.

the wall surface. At A an incremental turning $d\theta$ causes a small pressure disturbance dp which propagates into the fluid along the Mach line A–C at an angle determined by the local Mach number. Similarly at A' the wall's initial influence propagates along A'–C. Continuous wall curvature beyond A or A' may be considered to generate an infinite number of successive Mach lines, a few of which are illustrated. Those Mach lines emanating from the lower duct wall are called left-running Mach lines, and those from the upper wall right-running Mach lines. The designations refer to the direction in which the Mach lines appear to propagate downstream to an observer looking downstream.

In a region such as ABC or $A'B'C$ where Mach lines of only one type are present, the fluid properties are constant just upstream or downstream of each Mach line. Thus specification of fluid properties along any streamline (such as the wall

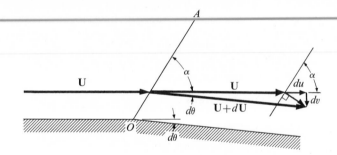

FIG. 13–4. Expansion at an infinitesimal corner.

streamline) is sufficient to describe the entire region. Suppose, as in Fig. 13–4, the wall undergoes an infinitesimal turning $d\theta$. The flow upstream has a Mach number M; thus the disturbance emanating from the infinitesimal corner propagates into the flow field on line OA at an angle

$$\alpha = \tan^{-1}\frac{1}{\sqrt{M^2 - 1}}$$

to the upstream flow direction. The fluid on any streamline crossing the Mach line must have its direction changed so that it will continue to travel parallel to the wall. Thus it must turn through angle $d\theta$. Further, the vectorial change $d\mathbf{U}$ in its velocity must be normal to the Mach line, since there can be no change in the momentum component parallel to the Mach line (the fluid is not acted on by any forces in that direction). These considerations fix the magnitude of $d\mathbf{U}$ and its components du and dv. It may easily be verified that

$$dv = U\, d\theta, \tag{13-2}$$

$$du = dU, \tag{13-3}$$

$$\tan\alpha = \frac{du}{dv}. \tag{13-4}$$

Combining Eqs. (13–1), (13–2), (13–3), and (13–4), we obtain

$$\frac{dU}{U} = \frac{d\theta}{\sqrt{M^2 - 1}}. \tag{13-5}$$

Combining the definitions of Mach number and stagnation temperature,

$$U^2 = M^2\gamma RT \quad \text{and} \quad \frac{T_0}{T} = 1 + \frac{\gamma - 1}{2}M^2,$$

yields

$$U^2 = \frac{\gamma RT_0 M^2}{1 + [(\gamma - 1)/2]M^2}. \tag{13-6}$$

From this it may be shown that since the stagnation temperature is constant along a streamline,

$$\frac{dU}{U} = \frac{dM^2}{2M^2\{1 + [(\gamma - 1)/2]M^2\}}, \qquad (13\text{--}7)$$

and therefore Eq. (13–5) may be transformed to

$$dM^2 = \frac{2M^2\{1 + [(\gamma - 1)/2]M^2\}\, d\theta}{\sqrt{M^2 - 1}}. \qquad (13\text{--}8)$$

Thus the change in Mach number is easily related to the change in direction of the streamline, so long as the flow may be considered isentropic. Changes in all fluid properties may be determined from the change in Mach number evaluated from Eq. (13–8). In particular Eq. (13–7) gives the change in fluid velocity. Changes in temperature, density, and pressure may be obtained from

$$\frac{dT}{T} = -\frac{\gamma - 1}{2}\frac{dM^2}{1 + [(\gamma - 1)/2]M^2}, \qquad (13\text{--}9)$$

$$\frac{d\rho}{\rho} = -\frac{dM^2}{2\{1 + [(\gamma - 1)/2]M^2\}}, \qquad (13\text{--}10)$$

$$\frac{dp}{p} = -\frac{\gamma}{2}\frac{dM^2}{1 + [(\gamma - 1)/2]M^2}. \qquad (13\text{--}11)$$

Equations (13–7) through (13–11) may be used to evaluate, by a numerical process, the conditions in the flow field ABC or $A'CB'$ in Fig. 13–3 by replacing the wall curvature by a finite number of straight line segments.

In the region downstream of BCB' in Fig. 13–3 the Mach lines cross each other, and conditions downstream of a given Mach line are no longer uniform.

Consider the flow in the neighborhood of a general Mach line intersection as indicated in Fig. 13–5. The Mach number and flow direction immediately upstream of OA_2 are not the same as they are upstream of OA_1. Therefore the

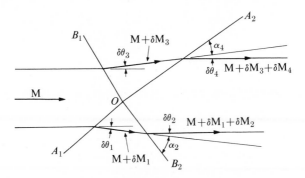

FIG. 13–5. Intersection of Mach lines.

Mach lines change direction upon intersecting. If we know the flow in regions B_1OA_2 and A_1OB_2, we may determine the flow in region A_2OB_2 as follows.

The deflection of the streamlines crossing OA_2 and OB_2 is fixed by Eq. 13-8, and by the requirement that the streamlines in region A_2OB_2 must all be parallel; they must therefore have a uniform Mach number.

Referring to Fig. 13-5, we can see that the streamlines in A_2OB_2 will be parallel if

$$\delta\theta_1 - \delta\theta_2 = \delta\theta_3 - \delta\theta_4, \tag{i}$$

and the Mach number will be uniform if

$$\delta M_1 + \delta M_2 = \delta M_3 + \delta M_4. \tag{ii}$$

A finite-difference approximation of Eq. (13-8) yields

$$\delta\theta_2 = m_1 \, \delta M_2 \tag{iii}$$

and

$$\delta\theta_4 = m_3 \, \delta M_4, \tag{iv}$$

in which

$$m_1 = \frac{\sqrt{(M + \delta M_1)^2 - 1}}{(M + \delta M_1)\{1 + [(\gamma - 1)/2](M + \delta M_1)^2\}}$$

and

$$m_3 = \frac{\sqrt{(M + \delta M_3)^2 - 1}}{(M + \delta M_3)\{1 + [(\gamma - 1)/2](M + \delta M_3)^2\}}.$$

Equations (i) through (iv) can easily be solved for $\delta\theta_2$, $\delta\theta_4$, δM_2, and δM_4. For example,

$$\delta M_4 = \frac{(\delta\theta_3 - \delta\theta_1) - m_1(\delta M_1 - \delta M_3)}{m_3 - m_1}.$$

Thus so long as M, $\delta\theta_1$, $\delta\theta_3$, δM_1, δM_3 in Fig. 13-5 are known, the flow immediately downstream of O may be solved directly. If $\delta\theta_3 = \delta\theta_1$, we can see that $\delta M_1 = \delta M_3$ and $m_1 = m_3$ so that the formula for δM_4 becomes indeterminate. However, in this case the flow in A_2OB_2 is parallel to M so that $\delta\theta_4$ will be known and δM_4 may be calculated directly from Eq. (iv).

Now, referring to Fig. 13-3, we see how we can determine the flow downstream of $B'CB$. First we can calculate the flow just downstream of C by the above method, since the flow in regions $A'CB'$ and ACB has already been determined. Next we may obtain the flow just downstream of D in a similar way, and so on to E and all the other points of Mach-line intersection.

For a subsonic–supersonic nozzle, the starting boundary for the calculation merits some consideration. In the vicinity of the throat, contours of constant Mach number may be obtained, either by experiment or by calculation of the subsonic flow. Figure 13-6 shows contours for $M = 0.9$ and $M = 1.0$, and it

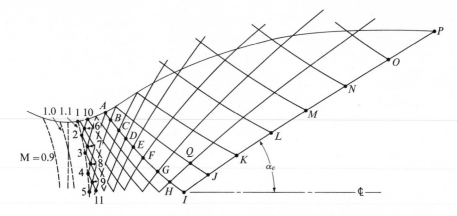

Fig. 13–6. Supersonic flow in a converging-diverging nozzle.

may be noted that the flow is not exactly one-dimensional. It is not possible to start the finite-difference calculation process right on the $M = 1$ line because here the Mach angles are 90°. However, it will make negligible difference to the calculation of the flow field as a whole if the location of the $M = 1.1$ line is assumed and the calculation is started from there. First a number of starting points (for example, 1–5) are selected and the characteristic lines are drawn from them, giving points of intersection (6–11). The flow immediately downstream of points (2–5) may be obtained by the method just outlined. Along the nozzle center line and immediately adjacent to the wall the flow direction is known; hence the flow downstream of points (6–11) may be determined in the same way. In this manner the whole flow field can be determined.

The purpose of the preceding discussion is to illustrate simply the way in which an isentropic, two-dimensional, supersonic flow can be calculated, using a finite-difference method. As mentioned earlier, the method in general use for solving problems of this type is called the method of characteristics [4]. It is adaptable to axisymmetric flows and is very convenient to use as a graphical computation process. However, it is considerably more complicated to derive than the procedure described above. The method of characteristics also entails the construction of a mesh of Mach lines or "characteristics," as illustrated in Fig. 13–6.

It is of interest to consider the design of a so-called "perfect" nozzle, i.e., one giving uniform axial exhaust velocity and uniform exhaust pressure. Figure 13–6 may be used to illustrate the design of this nozzle by the method of characteristics. Suppose again that the location of the $M = 1.1$ contour is assumed and, in addition, a small portion of the shape of the supersonic part of the nozzle; that is, 1–10–A in Fig. 13–6. Then the method of characteristics may be used to solve the flow field. At some point along the axis the design exhaust Mach number will be reached. For illustration, suppose this occurs at I (although it is unlikely that it would occur exactly on a mesh point). Downstream of this point, and hence

Minimum-surface nozzle
$A/A^* = 25$ $D^* = 0.750$ in.
Sharp corner at throat $\gamma = 1.4$

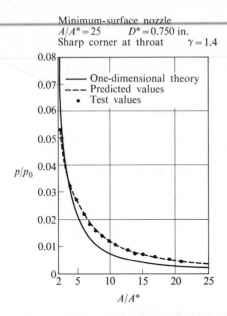

FIG. 13–7. Comparison of actual and predicted pressure distributions. (Courtesy Ahlberg, *et al*. [5].)

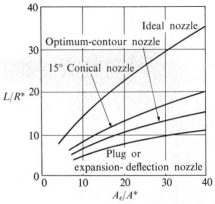

FIG. 13–8. Comparison of lengths of various types of nozzles for $\gamma = 1.23$. (Courtesy Rao [6].)

downstream of a Mach cone extending from the point, fluid properties remain constant and the velocity is parallel to the axis. Hence the characteristic or Mach line *I–P*, a straight line, may be drawn at the exit Mach angle α_e, and at points along this line, such as *J*, *K*, and *O*, the fluid properties are known. The solution from line 1–5 has extended downstream to line *A–I*, utilizing the arbitrary wall contour 1–10–*A*. Downstream of *A* the wall contour is no longer arbitrary since streamlines in this region are determined by the known properties at *A, B, . . . , I* and *I, J, . . . , P*. That is, from these known points the solution may be filled in between *A–I* and *I–P*. For example, *Q* is found from the points *H* and *J*, and so on. Once the mesh is obtained, streamlines may be drawn throughout the region. In particular, that streamline *A–P* passing through *A* is the required portion of the wall contour 1–*A–P*.

With a sufficiently fine mesh, the method of characteristics can provide good accuracy. For example, Fig. 13–7 shows predicted and measured pressures along a particular nozzle. For comparison, the simple one-dimensional prediction of pressure is also shown.

The length of a perfect nozzle is usually large for a given area ratio. The length is somewhat dependent on the assumed wall contour 1–10–A, and the minimum length for a given area ratio corresponds to a sharp corner at which the points 1 and A converge. Although this configuration is perfectly suitable aerodynamically (some wind tunnels requiring uniform test section properties employ a sharp corner at the throat), it is not desirable in rocket applications where sharp corners can create heat-transfer difficulties [1]. Figure 13–8 compares the lengths of perfect (or "ideal") nozzles and three other types. The perfect nozzle throat contour was assumed to be a circular arc of radius 0.4 times the throat radius. The so-called optimum-contour, plug, and expansion-deflected (E–D) nozzles will be discussed in the following.

Examination of Fig. 13–6 indicates that near the exhaust end of the nozzle a large portion of the nozzle wall is nearly parallel to the axis. A consideration of the pressure forces on this portion suggests that it might be removed with a saving in weight that would more than compensate for the loss of thrust. In fact, when wall friction is considered, removal of part of the nozzle can result in an *increased* thrust, as we shall see.

The desirability of improving on the "perfect" nozzle has led to various optimization procedures which attempt, within prescribed restrictions, to determine a nozzle shape which provides maximum thrust for a given propellant flow. One method [7] starts with given throat flow characteristics, nozzle length, and ambient pressure, and provides a nozzle contour yielding maximum thrust. It uses variational calculus techniques to find the desired exhaust properties (along a characteristic line analogous to *I–P* of Fig. 13–6, but not a straight line, since the exhaust properties are not uniform) and then employs the method of characteristics to determine a contour which provides the desired flow. Since the method of characteristics (which assumes reversible flow) is employed in the solution, this optimization cannot include the effects of wall friction. The optimum-contour nozzle of Fig. 13–8 represents the minimum-length nozzle found by this solution which has the same vacuum thrust coefficient* as the 15° conical nozzle.

A second method seeks maximum thrust from a family of perfect nozzles (i.e., nozzles designed for parallel constant-pressure exhaust flow) which have been cut off or truncated at various lengths. Figure 13–9 shows the results of calculations [5] on a family of perfect nozzles designed for various area ratios $(A_e/A^*)_p$ and then truncated at various length ratios L/R^*. The thrust has been stated in terms of the vacuum thrust coefficient. In the calculations on which Fig. 13–9 is based, frictional losses were estimated by the use of an empirical skin friction coefficient defined by

$$\tau_0 = C_f(\rho u^2/2), \tag{13–12}$$

---

* The vacuum thrust T_v of a nozzle is the thrust corresponding to zero ambient pressure. In accordance with Eq. (11–10), the vacuum thrust coefficient C_{T_v} is defined by $C_{T_v} = T_v/p_0 A^*$. For finite back pressure p_a, the thrust coefficient C_T is related to C_{T_v} by

$$C_T = C_{T_v} - A_e p_a/A^* p_0.$$

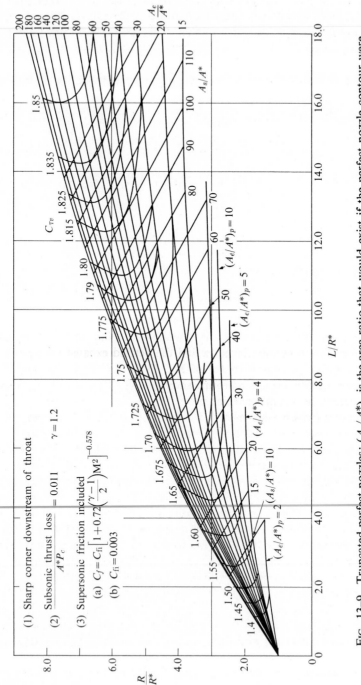

FIG. 13–9. Truncated perfect nozzles: $(A_e/A^*)_p$ is the area ratio that would exist if the perfect nozzle contour were not truncated; (A_s/A^*) is the surface area of the nozzle divided by the throat area; C_{Tv} is the vacuum thrust coefficient. (Courtesy Ahlberg, *et al.* [5].)

in which τ_0 is the wall shear stress and u is the local velocity. Various boundary layer calculations and empirical data lead to values for C_f which vary with local Mach number in a manner similar to that given by the equation in Fig. 13–9, where C_{fi} designates the skin friction for small Mach number or incompressible flow.

For a given thrust coefficient, three kinds of optimum truncated perfect nozzles may be selected from Fig. 13–9:

(1) The nozzle with minimum surface area corresponds to the point at which the appropriate C_{Tv} curve is tangent to a surface area ratio (A_s/A^*) curve. The local values of $(A_e/A^*)_p$ and (L/R^*) indicate its design conditions and truncated length, respectively.

(2) The nozzle of minimum length is identified by the point of vertical slope on the appropriate C_{Tv} curve.

(3) The nozzle of minimum exit diameter is found from the point where the C_{Tv} curve has zero slope.

The question as to which of these three optima is most desirable depends on the rocket vehicle design conditions.

The effects of friction and heat transfer

Propellant flow in rocket nozzles, like other fluid flows, is altered by viscous effects near the fluid boundaries. However, since the pressure gradient within a nozzle is generally favorable, the boundary layer is usually quite thin. Apart from its important effects on heat transfer (to be discussed shortly), the boundary layer may be considered to affect nozzle performance in three ways.

First, the presence of the boundary layer slightly alters the free-stream characteristics. By definition, the boundary layer *displacement thickness* is that thickness of free-stream flow which is apparently lost due to the velocity defect within the boundary layer. Hence the free stream is in effect displaced from the wall by this thickness. At the throat the result is a slight reduction of A^* and hence of mass-flow rate. At the exit, the result is a slight reduction in A_e/A^*, with negligible influence on average exhaust properties. For nozzles whose exit flow pattern is very important—as in wind tunnels, for example—it may be desirable to calculate the displacement thickness and enlarge the nozzle contour by this amount to ensure truly uniform test section flow. However, in rocket applications, where thrust rather than flow pattern is important, this is an unnecessary refinement.

The experimental nozzle of Fig. 13–7 was relatively small ($D^* = 0.75$ in.), so that the Reynolds number would have been relatively low and viscous effects could conceivably have been significant. Nevertheless the experimental pressure distribution agrees very well with that predicted by the method of characteristics, *without* modifying the contour to account for the boundary layer. For reasonably large rocket nozzles, then, the effect of the boundary layer on the free-stream characteristics is negligible.

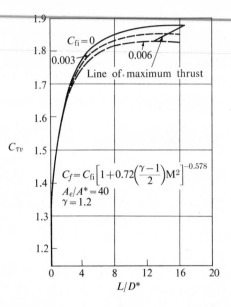

FIG. 13-10. Effect of friction on the thrust of the truncated perfect nozzles of Fig. 13-9. (Courtesy Ahlberg, *et al.* [5].)

Second, the boundary layer affects the rocket thrust directly through the shear stress or skin friction on the nozzle wall, since thrust is simply the integrated stress over the thrust chamber surface. For typical rocket propellants, viscous stresses are very small relative to pressure stresses, so that viscous contributions to thrust are usually negligible. An exception occurs near the exhaust of a perfect nozzle. Examination of the pressure forces in this region reveals that although they may be large, they are almost radially directed, with very little axial component. In contrast, the skin friction is very nearly axially directed and, since it detracts from thrust, a net negative thrust may exist on this portion of the nozzle. Figure 13-10 indicates that, depending on the magnitude of the skin friction, the vacuum thrust coefficient of a truncated perfect nozzle may be greater than that of the untruncated perfect nozzle. As expected, for no friction ($C_{\mathrm{fi}} = 0$), maximum thrust occurs with the parallel exhaust velocity of the perfect nozzle. However, Fig. 13-9 shows that, for $C_{\mathrm{fi}} = 0.006$, it would be desirable to remove about one-quarter of the length of this particular perfect nozzle, since this would save weight [5] as well as slightly increasing the thrust.

The third influence of the boundary layer on nozzle performance has to do with shock–boundary layer interactions. Under certain ambient pressure conditions, shocks can occur even within well-designed nozzles. The position and strength of the shock or shock system depend on boundary layer behavior as well as on free-stream and ambient conditions. We shall discuss this phenomenon when we consider the effects of back pressure on nozzle performance.

Direct calculation of boundary layer growth on the nozzle walls is quite difficult, since little is known regarding boundary layer behavior in regions of high favorable pressure gradient and strong wall curvature, as in rocket nozzles. However, there exist approximate methods of calculation, such as those used in Fig. 13–10, which predict friction factors and, more importantly, heat-transfer coefficients which will be dealt with in Chapter 14. One author [8] assumes a turbulent boundary layer throughout the nozzle, although there is some evidence that the strong favorable pressure gradient may actually cause a "reverse transition" from turbulent to laminar behavior. Reference [8] employs the momentum integral method of Chapter 4, along with an analogous energy integral equation, assuming both the velocity and temperature defect are described by $\frac{1}{7}$th power law profiles. That is,

$$\frac{u}{U} = \left(\frac{y}{\delta}\right)^{1/7} \quad \text{and} \quad \frac{t_0 - T_w}{T_0 - T_w} = \left(\frac{y}{\Delta}\right)^{1/7},$$

where u, t_0 = velocity and stagnation temperature within boundary layer,

$\quad\quad U, T_0$ = local free-stream velocity and stagnation temperature,

$\quad\quad\quad T_w$ = local wall temperature,

$\quad\quad\quad\quad y$ = distance from wall, and

$\quad\quad \delta, \Delta$ = thicknesses of velocity and thermal boundary layers, respectively.

In the absence of better information it is assumed that the local skin friction coefficient is the same as that for a flat-plate boundary layer (i.e., one with no axial pressure gradient) of the same thickness. An expression similar to Eq. (4–19) is chosen:

$$C_{\text{fi}} = 0.0228 \left(\frac{U\delta}{\nu}\right)^{-1/4},$$

where the subscript i denotes that this expression describes the behavior of an incompressible boundary layer. The variation of temperature and hence of viscosity ν through the boundary layer in compressible, nonadiabatic flow introduces a Mach number dependence for the compressible C_f similar to that assumed in Fig. 13–10.

Figure 13–11 illustrates typical results of this method of calculating the boundary layer in nozzles. Two cases assuming different initial thicknesses are shown. Note that in either case δ at the throat is quite small, and that downstream of the throat the assumed initial thickness is of no importance. If δ is assumed zero at the throat the boundary layer grows linearly with distance along the wall. This is in agreement with Reference [1], which suggests that displacement thickness δ^* in the supersonic portion of the nozzle grows approximately as $\delta^* = 0.004x^*$, where x^* is the distance from the throat.

The major importance of the boundary layer development is its influence on the heat transfer to the nozzle walls; this heat transfer is important primarily in its effects on the nozzle structure, as discussed in Chapter 14. The effect of con-

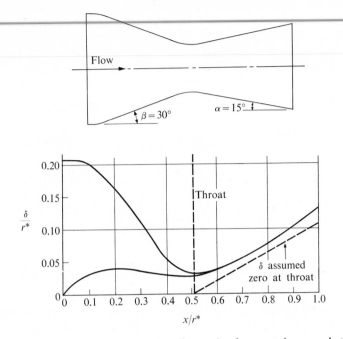

Fig. 13–11.   Typical calculated boundary layer development in a rocket nozzle; r^* = nozzle throat radius.   (Courtesy Bartz [8].)

vective heat transfer on the flow is very slight in practical nozzles, since its influence extends only so far as the thermal boundary layer thickness, and this is of the order of the velocity boundary layer thickness (for typical gases). Radiant heat transfer is also of little importance to propellant behavior, since it is generally smaller than the convective heat transfer.

The effect of back pressure on nozzle flow

Thrust chambers which are called on to operate within the atmosphere may be subject to wide variations in ambient or "back" pressure. Launch vehicles, especially, which operate from sea level to high altitudes, experience large variations in back pressure. Under certain conditions the performance of a thrust chamber designed for one back pressure can be significantly reduced at other back pressures.

So long as the ambient pressure is *less* than the nozzle exhaust pressure, the flow within the nozzle remains undisturbed. That is, the influence of a pressure *decrease* downstream of a supersonic nozzle cannot propagate upstream into the nozzle, since an expansion wave travels only at the local speed of sound. Thus the pressure adjustment occurs beyond the exit plane as a series of expansion waves whose initial influence propagates at the exit Mach angle α_e. Such an *underexpanded nozzle* (so called because the flow has not expanded to ambient pressure within

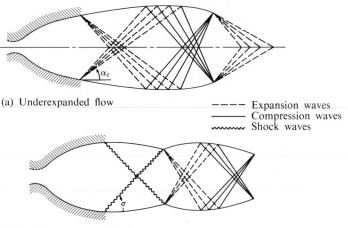

(a) Underexpanded flow

- - - - Expansion waves
———— Compression waves
〰〰〰 Shock waves

(b) Slightly overexpanded flow

Fig. 13–12. Exhaust jet behavior at various back pressures.

the nozzle) is illustrated in Fig. 13–12(a). The expansion waves reflect from the jet boundary as compression waves which may actually converge to form a shock. The pattern of alternate expansion and compression is repeated downstream and the associated temperature pattern may, because of varying gas incandescence, render the pressure pattern visible. Note that conditions within the nozzle are entirely undisturbed.

In contrast, an ambient pressure *greater* than the intended exhaust pressure can disturb the flow within the nozzle. A pressure *rise* occurs as a sudden increase or shock which propagates relative to the fluid at supersonic velocity. A slightly overexpanded ($p_e < p_a$) nozzle might appear as in Fig. 13–12(b), where the required pressure adjustment occurs through an oblique shock. The shock angle σ is determined by the required pressure ratio, p_a/p_e, and the exhaust Mach number. As p_a increases the shock angle also increases, since stronger shocks propagate at higher speeds. For sufficiently high p_a/p_e the shock pattern is significantly affected by shock–boundary layer interaction. Since a portion of the boundary layer is subsonic, the influence of the shock-induced pressure rise can influence boundary layer behavior slightly upstream of the wall–shock intersection point. If the required pressure rise is large enough, the boundary layer will separate slightly upstream of the shock. The resultant boundary disturbance causes the shock itself to move upstream. If the pressure rise is still high enough to cause separation, the shock continues to move upstream until an equilibrium position is found, at which the required pressure rise is just small enough to avoid separation upstream of the shock.

Figure 13–13 illustrates a typical wall pressure distribution with shocks inside the nozzle. Note that, if the separated jet does not reattach, the downstream pressure is essentially p_a regardless of shock position, while the upstream pressure

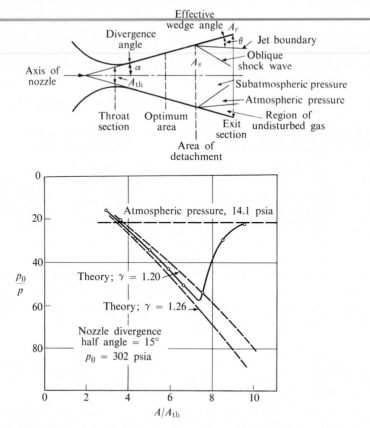

FIG. 13–13. Effect of shock–boundary layer interaction on the pressure distribution within a nozzle. (Courtesy Altman, *et al.* [1].)

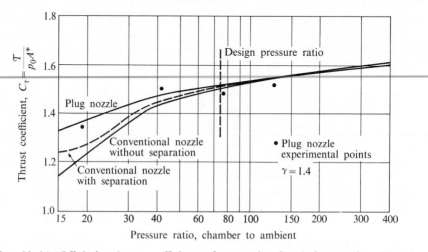

FIG. 13–14. Off-design thrust coefficients of conventional and plug nozzles. (Courtesy Berman [11].)

p_s increases as the shock moves upstream. In some contoured nozzles jet reattach-
ment may occur, especially near the exit, so that the downstream pressure may
also vary somewhat with shock position [3, 9]. By assuming typical boundary
layer characteristics and neglecting friction throughout the rather sudden pressure
rise, it is possible to predict what pressure rise will just produce separation, and
hence to locate the shock position. Reference [10] concludes that a Mach number
ratio of 0.76 across an oblique shock should separate a typical turbulent boundary
layer. Alternatively, Reference [1] suggests a shock–pressure ratio of $p_s/p_a \approx 0.36$.
These figures agree quite well with the observed behavior in conical nozzles, but
they do not adequately describe separation within contoured nozzles [3, 9]. Presumably the wall curvature in such a nozzle produces a variation of boundary
layer characteristics along the wall which is not adequately described by such a
simple separation criterion.

It can be seen that back pressure only affects the flow within the nozzle if it is
sufficiently above the design exhaust pressure. However, a nozzle designed for low-
altitude (high p_a) operation will not produce as much thrust at high altitude
(low p_a) as a nozzle specifically designed for the higher altitude. As a result, rocket
engines for launch vehicles are often designed for an intermediate pressure, with
overexpanded sea-level operation and underexpanded high-altitude operation.
Figure 13–14 indicates the performance loss of conventional (i.e., internal flow)
nozzles associated with off-design back pressures, both with and without separation.

Plug and expansion-deflection nozzles

In an attempt to offset the thrust loss associated with overexpansion, and for
several other reasons, nozzle shapes other than conventional internal-flow configurations have been developed. Plug and expansion-deflection nozzles are
shown schematically in Fig. 13–15. Flow from the combustion chamber passes
through an annular throat in both configurations. For the plug nozzle the flow
expands around a corner at the throat to ambient pressure on one side, and along
a central body or plug on the other. Expansion-deflection nozzles achieve the
same end in an inverse geometry; that is, with the solid boundary surrounding
the free jet boundary.

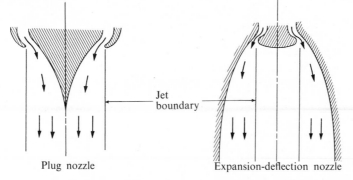

FIG. 13–15. Schematic cross sections of plug and expansion-deflection nozzles.

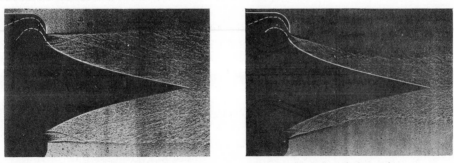

(a) At design p_0/p_e (b) Below design p_0/p_e

FIG. 13–16. Shadowgraphs of flow around plug nozzle at design back pressure and above design back pressure. (a) $p_0/p_a = 75$; no external flow. (b) $p_0/p_a = 20$; no external flow. (Courtesy Berman [11].)

In both configurations the presence of a free jet boundary to the expansion process lends a kind of self-adjustment to back-pressure variations. Figure 13–16 indicates the variation of exhaust pattern for a plug nozzle flow with increasing back pressure. The flow expands around the corner to the actual ambient pressure, and thus overexpansion is essentially avoided. Figure 13–14 illustrates how the performance of a plug nozzle exceeds that of a conventional nozzle at high back pressures.

References [11] and [12] indicate that the design-point performance of plug nozzles can be as good as that of conventional nozzles. For plug shapes designed by the method of characteristics to give axial exhaust velocity (as in Fig. 13–16) the thrust was measured to be 98% of the ideal thrust (isentropic flow, axial exhaust velocity, $p_e = p_a$) at a design pressure ratio of 16. However, when the back pressure was increased to obtain a pressure ratio of only 1.5, the thrust was observed to decrease only to 96% of the ideal. The complicated contoured shape may be replaced by a mechanically simpler conical shape without serious performance penalty. With cones of half-angles 30° to 40°, the performance is reduced by about 1% over that of the contoured plug, while the insensitivity to variations in back pressure is retained [12].

Further advantages of these configurations are as follows: Both types are substantially shorter than comparable internal-flow nozzles. This may be seen in Fig. 13–8, where the length required of a plug nozzle in order to produce 99.5% of the vacuum thrust coefficient of an ideal nozzle of the same area ratio is shown. The annular combustion chamber of the plug-nozzle design might be built up of relatively small and easily developed segments. The propellant flow from these segments might be separately controlled in such a way as to produce nonaxial thrust, thus avoiding the necessity of gimbaling the entire engine to achieve thrust vector variations. The combustion chamber of an expansion-deflection nozzle design would be very similar to that for a conventional engine. The main difficulties associated with these configurations arise in the process of adequately supporting and cooling the central plug. Thus far expansion-deflection nozzles have not displaced conventional nozzles, but for very high-thrust engines their size advantage is impressive.

The effect of propellant property variations on nozzle flow

Except for preliminary calculations, it is usually not satisfactory to assume that the specific heats of expanding propellants are constant. One can calculate the isentropic flow of a gas of variable specific heat, using suitable tables of properties, as discussed in Chapter 2; and the method of characteristics may be modified to treat fluids of variable γ [14]. In addition to such ordinary property variations, however, the propellant may exhibit much more important variations of chemical composition. We have seen that the *equilibrium* composition of the combustion products is dependent on pressure and temperature and that, at common combustion-chamber conditions, these products may include large quantities of dissociated material. The propellant, when it expands in the exhaust nozzle, experiences large temperature and pressure drops which, if equilibrium composition is maintained, are accompanied by a decrease in the fraction of dissociated material. This (exothermic) chemical reaction during expansion acts as an apparent heat source in the flow, in addition to causing changes in c_p and γ.

Figure 13–17 illustrates the division of total enthalpy for equilibrium combustion and expansion of a stoichiometric mixture of hydrogen and oxygen. "Sensible energy," made up of microscopic molecular (and electronic) motion, is what we have called static enthalpy; the dissociation energy is that additional sensible

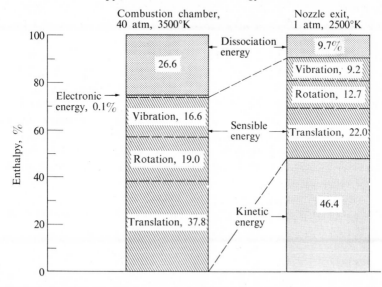

FIG. 13–17.   Typical equilibrium energy distribution in rocket gases.   (Courtesy Olson [15].)

~~energy which would be available if the composition were the same at high tempera-~~ ture as at low temperature. Note that in the expansion process some of the dissociation energy is converted to sensible energy, and thence to useful kinetic energy. Since conversion from dissociation to sensible energy occurs within the nozzle where the pressure is lower than in the combustion chamber, the subsequent conversion to kinetic energy is not as productive as if the dissociation had never occurred.

In order for the propellant to maintain equilibrium composition at the local temperature, as assumed in this figure, it is necessary that the various recombination processes occur in a time which is short relative to the expansion time. Since the expansion process is very rapid, this condition is not always met and a loss in performance, due to unregained dissociation energy, results. In the limit, if reaction rates are very slow, no recombination occurs at all and the propellant is said to be "frozen," usually at the combustion-chamber composition. The difference between "equilibrium" and "frozen" flow performance can be appreciable for some propellants.

It is, of course, desirable to know in advance how any propellant will behave since, in addition to the influence of propellant behavior on specific impulse, variations of composition can substantially alter the required nozzle geometry. Reference [14] contains examples with calculations showing contour variations and as much as 20% variation in required exit area.

The maintenance of equilibrium or near-equilibrium composition as the propellant expands requires that the speed of the important reactions be high relative to the speed at which changes take place in the state of the propellant. Suppose that the general reaction

$$\alpha A + \beta B \rightarrow \mu M + \nu N$$

describes an important recombination occurring during expansion. In general it is observed that the rate of change of the concentration, C, of any species is described by an equation of the form

$$-\frac{dC_A}{dt} = -\frac{\beta}{\alpha}\frac{dC_B}{dt} = \frac{\mu}{\alpha}\frac{dC_M}{dt} = \frac{\nu}{\alpha}\frac{dC_N}{dt} = kC_A^{\alpha}C_B^{\beta}$$

where k is the temperature-dependent reaction rate constant, *provided* that the reaction occurs directly as written; that is, without the formation of intermediate species. At the same time this recombination is proceeding, the temperature is dropping so that k will drop rapidly. The temperature change of a small "particle" of fluid is given by

$$\frac{dT}{dt} = \frac{dT}{dx}\frac{dx}{dt} = u\frac{dT}{dx},$$

where u and dT/dx may be calculated throughout the nozzle, assuming either frozen or equilibrium flow. In principle, the temperature changes due to reduction of pressure of the expanding fluid and also to changes in its composition may be

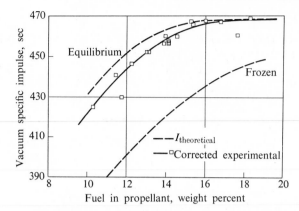

FIG. 13–18.   "Corrected" specific impulse of hydrogen–oxygen ·rocket.   (Courtesy Olson [15].)

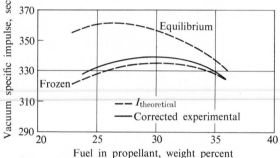

FIG. 13–19. "Corrected" specific impulse of JP4-oxygen rocket. (Courtesy Olson [15].)

found by simultaneously satisfying the continuity, momentum, and energy equations if the appropriate rate constants are known.

Unfortunately, accurate calculations of the actual state of the expanding propellant are not often possible. Actual propellant reactions are exceedingly complex in that they seldom, if ever, proceed directly from reactants to final products. In many cases the reaction paths themselves are not known, and in other cases where the reactions are known there are insufficient rate data.   Reference [1] contains an analytical procedure due to Penner which indicates when equilibrium conditions might be expected for specific reactions for which sufficient data are available. Fortunately, in many cases one need only treat a few reactions to account for most of the dissociation energy.

The performance of a given propellant combination can, of course, be determined experimentally.   Reference [15] reports experimental data for several propellants, comparing actual (corrected) specific impulse with theoretical frozen and equilibrium performance for various fuel–oxidizer ratios, as shown in Figs. 13–18 and 13–19. The data are "corrected" to eliminate performance losses due to poor combustion and exhaust divergence. The remaining losses are thus due to lack of recombination and to nozzle boundary layer losses (presumably small). From these data it can be concluded that H_2-O_2 propellant remains essentially at equilibrium, while JP4-O_2 remains frozen at combustion-chamber composition. Note also that the optimum fuel–oxidizer ratio in both cases depends significantly on the extent to which equilibrium is maintained.

The effect of solid or liquid particles in the flow

Solid propellants often contain metals or metallic compounds as constituents. Many of the oxides of these metals are refractory materials and they may appear as solid or liquid particles in the exhaust. Measurements have shown that the average particle diameter is about two or three microns [15]. A complete analysis of fluid-particle flow systems would require calculation of the fluid-particle forces (drag) and heat transfer and, for very small particles, consideration of the thermal (random) motion of the particles themselves. However, the size of the particles is such that greatly simplified analyses yield useful results. Reference [1] states that the thermal motion of particles larger than about $10^{-3}\,\mu$ in diameter may be neglected, and for diameters greater than about $10^{-1}\,\mu$ it is reasonable to neglect particle acceleration, at least downstream of the throat. In any case, it seems that fluid-particle heat transfer has little influence on performance, as will be shown.

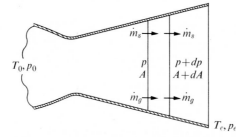

FIG. 13–20. Incremental control volume for the development of two-phase flow equations.

Let us consider a kind of "frozen-flow" model, in which the particles are formed within the combustion chamber, there being no subsequent formation or growth of particles within the nozzle. Under such conditions the mass flow rates of solid (or liquid) and gaseous materials, $\dot{m}_s$ and $\dot{m}_g$, respectively, are constant throughout the nozzle, as shown in Fig. 13–20. When we apply the general momentum relation, Eq. (2–4), to the incremental control volume of Fig. 13–20, the result is

$$-A\,dp = \dot{m}_s(u_s + du_s) + \dot{m}_g(u_g + du_g) - \dot{m}_s u_s - \dot{m}_g u_g, \qquad (13\text{–}13)$$

where u_s = particle velocity and u_g = gas velocity. From continuity, we see that

$$\dot{m}_s = \rho_s A u_s, \qquad \dot{m}_g = \rho_g A u_g, \qquad (13\text{–}14)$$

where ρ_s = solid density (mass of solid material per unit volume of mixture) and ρ_g = gas density.

Combining Eqs. (13–13) and (13–14), we obtain

$$-dp = \rho_s u_s \, du_s + \rho_g u_g \, du_g. \tag{13–15}$$

The energy equation for steady adiabatic flow relates the particle and gaseous enthalpies and kinetic energies according to

$$\dot{m}_s(c_s \, dT_s + u_s \, du_s) + \dot{m}_g(c_p \, dT_g + u_g \, du_g) = 0, \tag{13–16}$$

where T_s = temperature of solid material,

T_g = temperature of gaseous material,

c_s = specific heat of solid, and

c_p = specific heat at constant pressure of gas.

When we define X as the fraction of total mass flow in solid (or liquid) form, then $1 - X$ is the gaseous fraction, and we can write Eq. (13–16) as

$$X c_s \, dT_s + (1 - X)c_p \, dT_g + X u_s \, du_s + (1 - X)u_g \, du_g = 0. \tag{13–17}$$

From Eq. (13–15), we obtain

$$u_g \, du_g = -\frac{dp}{\rho_g} - \frac{\rho_s}{\rho_g} u_s \, du_s,$$

and, from the definition of $X = \rho_s u_s / (\rho_g u_g + \rho_s u_s)$,

$$u_g \, du_g = -\frac{dp}{\rho_g} - \frac{X}{1 - X} u_g \, du_s.$$

Substituting this expression in Eq. (13–17), we have

$$\frac{X}{1 - X} c_s \, dT_s + c_p \, dT_g + \frac{X}{1 - X}(u_s - u_g) \, du_s = \frac{dp}{\rho_g}. \tag{13–18}$$

In general, the solution of this equation requires knowledge of heat-transfer and drag processes in order to express T_s and u_s in terms of overall flow properties. However, four limiting solutions can be obtained quite easily for the following extremes of behavior:

(1) The heat transfer may be very fast, $T_s \approx T_g$, or very slow, $T_s \approx$ constant.

(2) The particle drag may be very high, $u_s \approx u_g$, or very low, $u_s \approx$ constant.

For example, consider the case $T_s = T_g = T$, $u_s = u_g = u$. Then Eq. (13–18) reduces to

$$\left(\frac{X}{1 - X} c_s + c_p\right) dT = \frac{dp}{\rho_g}.$$

Assuming that the gas behaves as a perfect gas,

$$\rho_g = \frac{p}{RT} \quad \text{and} \quad \left\{ \frac{[X/(1 - X)]c_s + c_p}{R} \right\} \frac{dT}{T} = \frac{dp}{p}.$$

Integrating from stagnation (combustion-chamber) conditions, T_0 and p_0, to exhaust conditions, T_e and p_e, we obtain

$$\frac{T_e}{T_0} = \left(\frac{p_e}{p_0}\right)^n \quad \text{where} \quad n = \frac{R}{[X/(1 - X)]c_s + c_p}.$$

For these conditions Eq. (13–17) becomes

$$[Xc_s + (1 - X)c_p]\, dT + u\, du = 0,$$

which may be integrated to give the exhaust velocity in terms of T_0 and T_e. There results

$$u_e = \sqrt{2[Xc_s + (1 - X)c_p]T_0[1 - (p_e/p_0)^n]}$$

in which, as before,

$$n = \frac{R}{[X/(1 - X)]c_s + c_p}.$$

For a second example, suppose that T_s = constant and u_s = constant. In this case Eq. (13–18) becomes

$$c_p\, dT_g = \frac{dp}{\rho_g}.$$

Proceeding as above, one obtains the familiar relationship

$$\frac{T_{eg}}{T_{0g}} = \left(\frac{p_e}{p_0}\right)^{(\gamma - 1)/\gamma}$$

and, from Eq. (13–17), the exhaust *gas* velocity is

$$u_{eg} = \sqrt{2c_p T_0[1 - (p_e/p_0)^{(\gamma - 1)/\gamma}]}.$$

In this case the gas velocity is not affected by the presence of the solid particles. However, the specific impulse is reduced. If several constituents are present, the specific impulse may be written

$$I_{sp} = \frac{1}{g_e} \frac{\sum \dot{m}_i u_{ei}}{\sum \dot{m}_i},$$

in which i denotes a particular constituent. Considering the case $\dot{m}_s u_{es} \ll \dot{m}_g u_{eg}$, $T_s = T_0$,

$$I_{sp} \cong \frac{1}{g_e} (1 - X)u_{eg},$$

where u_{eg} is independent of the presence of the particles. Thus the effect of the particles is to reduce the specific impulse by the factor $1 - X$.

The other two combinations of T_s and u_s variations may be found by similar developments of Eqs. (13–17) and (13–18). When we know the gas exhaust velocity and the fraction of solid material, we can calculate the effect of particle behavior on specific impulse (for an assumed u_s if $u_s \neq u_g$). For example, consider the following typical conditions:

$$T_0 = 5000°R, \qquad \overline{M} = 25 \text{ (molecular weight of gas)}, \qquad \gamma = 1.2,$$

$$c_s = 0.5 \text{ Btu/lb}_m \cdot °R, \qquad p_0 = 800 \text{ psia}, \qquad p_e = 14.7 \text{ psia}, \qquad X = 0.10.$$

The effect of the particles on specific impulse under these limiting conditions is indicated in the following table. In those cases where the particles were not accelerated, their momentum flux was neglected in the calculation of thrust (that is, $\dot{m}_s u_s \ll \dot{m}_g u_g$). It can be seen that the mere presence of the particles brings about a significant loss (since the condensed material does not expand in the nozzle). The velocity lag of nonaccelerated particles accounts for a substantial additional loss of specific impulse. Beyond this, however, note that the presence or absence of fluid-particle heat transfer is of little importance. In practice it is observed that the thermal lag of the particles is greater than the velocity lag but, as expected, its influence on performance is small. The effect of the velocity lag can be substantial, however.

	No particles present	Accelerated particles		Nonaccelerated particles	
		Complete heat transfer	No heat transfer	Complete heat transfer	No heat transfer
I_{sp}	236	$u_s = u_g,$ $T_s = T_g$ 228	$u_s = u_g,$ $T_s = \text{const}$ 224	$u_s \simeq 0,$ $T_s = T_g$ 216	$u_s \simeq 0,$ $T_s = \text{const}$ 213

It is possible that condensation of certain components of the gas flow may take place during expansion. The effects of condensation within the nozzle are mainly the continuous decrease in gas-phase flow per unit area and the transfer of the latent heat of vaporization from the condensed material to the remaining gas flow. To predict the effects of condensation on the flow, it is necessary to be able to estimate the rate of formation of nuclei and their subsequent growth as a function of the degree of supersaturation of the component in question. The nucleation rate typically is a very strong exponential function of the supersaturation, so that as a component expands to a supersaturated state the rate is very low at first, then at some point suddenly expands to exceedingly high values, for example 10^{20} nuclei formed per cc per second. If the density of the vapor is not too low, these nuclei will grow very rapidly and the temperature of the flow will rise until the component approaches saturation and thereafter expands in equilibrium. The fact that the pressure and temperature adjustments in the stream due to condensation are commonly very rapid has led to usage of the term "condensation shock." Reliable prediction of the condensation of complex propellant mixtures has not yet been shown to be possible, though the subject is under continuing study [15] [16] [17].

References

1. ALTMAN, D., J. M. CARTER, S. S. PENNER, and M. SUMMERFIELD, "Liquid-Propellant Rockets." Princeton, N. J.: Princeton Aeronautical Paperbacks, Princeton University Press, 1960

2. SUTTON, G. P., *Rocket Propulsion Elements.* New York: John Wiley & Sons, 1956

3. CAMPBELL, C. E., and J. M. FARLEY, "Performance of Several Conical Convergent-Divergent Rocket-Type Exhaust Nozzles." *NASA TN D*467, September 1960

4. SHAPIRO, ASCHER H., *The Dynamics and Thermodynamics of Compressible Fluid Flow.* New York: The Ronald Press Company, 1953; Volumes I and II, Chapters 14, 15, 17, and 21

5. AHLBERG, J. H., S. HAMILTON, and E. N. NILSON, "Truncated Perfect Nozzles in Optimum Nozzle Design." *J. Amer. Rocket Soc.,* **31,** 5, May 1961; pages 614–620

6. RAO, G. V. R., "Recent Developments in Rocket Nozzle Configurations." *J. Amer. Rocket Soc.,* **31,** 11, November 1961; pages 1488–1494

7. RAO, G. V. R., "Exhaust Nozzle Contour for Optimum Thrust." *Jet Propulsion,* **28,** 6, June 1958; pages 377–382

8. BARTZ, D. R., "An Approximate Solution of Compressible Turbulent Boundary Layer Development and Convective Heat Transfer in Convergent-Divergent Nozzles." *Transactions of the ASME,* **77,** November 1955; pages 1235–1245

9. FARLEY, J. M., and C. E. CAMPBELL, "Performance of Several Method of Characteristics Exhaust Nozzles." *NASA TN D*293, 1960

10. ARENA, M., and E. SPIEGLER, "Separated Flow in Overexpanded Nozzles at Low Pressure Ratios." *The Bulletin of the Research Council of Israel,* Section C, Technology, **11C,** April 1962, pages 45–55

11. BERMAN, K., "The Plug Nozzle: A New Approach to Engine Design." *Astronautics,* **5,** 4, April 1960

12. KRULL, H. G., and W. T. BEALE, "Effect of Plug Design on Performance Characteristics of Convergent-Plug Exhaust Nozzles." *NACA RM E*54 *H*05, October 1954

13. KRULL, H. G., and W. T. BEALE, "Effect of Outer-Shell Design on Performance Characteristics of Convergent-Plug Exhaust Nozzles." *NACA RM E*54 *K*22, April 1955

14. GUENTERT, E. C., and H. E. NEUMANN, "Design of Axisymmetric Exhaust Nozzles by Method of Characteristics Incorporating a Variable Isentropic Exponent." *NASA TR R*33, 1959

15. OLSON, W. T., "Recombination and Condensation Processes in High Area Ratio Nozzles." *J. Amer. Rocket Soc.,* **32,** 5, May 1962, pages 672–680

16. COURTNEY, W. G., "Recent Advances in Condensation and Evaporation." *J. Amer. Rocket Soc.,* **31,** 1961, page 751

17. Courtney, W. G., "Condensation in Nozzles." Ninth Symposium (International) on Combustion. New York: Academic Press, 1963, page 799

18. Wegener, P. P., and L. M. Mack, "Condensation in Supersonic and Hypersonic Wind Tunnels." *Advances in Applied Mechanics*, Volume 5 (editors, H. L. Dryden and Theodore von Karman). New York: Academic Press, 1958

Problems

1. A rocket is designed for isentropic expansion from 500 psia chamber pressure to 2 psia ambient pressure. At what pressure would a normal shock stand in the exit plane? Show qualitatively how thrust varies with ambient pressure from zero pressure to pressures greater than the values calculated above. Assume $\gamma = 1.4$.

2. Suppose that for a rocket nozzle of given throat area A_t and exit area A_e, the momentum and displacement thicknesses can be estimated at the throat and at the exit. Show how these values could be used to correct the one-dimensional flow estimation of propellant flow rate, exhaust velocity, and rocket thrust (for fixed p_0 and T_0).

3. To investigate the possible effects of combustion taking place in the nozzle of a chemical rocket, consider the following simplified case. Suppose that the combustion proceeds at such a rate that the fluid (static) temperature remains constant throughout the nozzle. For a pressure ratio of 100 and a given inlet stagnation temperature, how does the exit velocity compare with that for an isentropic expansion, if $\gamma = 1.2$? Would the occurrence be beneficial or detrimental to the performance of chemical rockets?

4. For a given stagnation pressure and a given nozzle designed for "correct" expansion (exhaust pressure = ambient) at 3 miles altitude, is there danger of shock-induced separation at launch? Show reasoning. It may be assumed that the ambient pressure at 3 miles altitude is 1,147 lb/ft² and that the stagnation pressure is between 300 and 1000 lb/ft². For simplicity, take $\gamma = 1.4$.

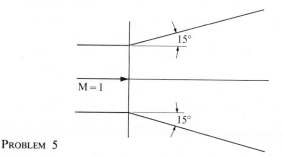

PROBLEM 5

5. In a step-by-step numerical procedure, show how to calculate the flow in a symmetrical two-dimensional nozzle which diverges at a half-angle of 15°, as shown in the figure. In particular, determine the pressure distributions along the walls and center line.

6. A mixture of very small solid particles and a perfect gas expands adiabatically through a given pressure ratio. The particles, which are some fraction μ of the total mass flow, are so small that they may be considered to have the same speed as the gas at all points in the flow. It may be assumed further that the relative volume of the solid material is negligible and that the specific heats of gas and solid are identical. (a) Assuming that during the expansion no heat transfer takes place between solid and gaseous phases, show that the presence of the solid particles reduces the exit velocity by the factor $\sqrt{1 - \mu}$. (b) Assuming that the solid particles always have the same temperature as the gas, show that the ideal exhaust velocity is given by

$$\frac{u}{u_{\mu=0}} = \sqrt{\frac{1 - (p_2/p_1)^{(1-\mu)(\gamma-1)/\gamma}}{1 - (p_2/p_1)^{(\gamma-1)/\gamma}}},$$

in which p_2/p_1 is the pressure ratio across this nozzle.

14

Chemical Rockets:
Thrust Chambers

14–1 INTRODUCTION

This chapter is concerned essentially with the thrust chamber as a whole. A major problem in its construction is the cooling of the surfaces exposed to the hot gases. In this chapter, therefore, the problems of heat transfer and cooling will be examined in some detail, for both solid- and liquid-propellant rockets. Also we shall consider the machinery necessary for pumping liquid propellants, and other mechanisms necessary for the proper functioning of the thrust chamber. The selection of chamber pressure will also be discussed in relation to combustion, heat transfer, and other factors.

14–2 ROCKET HEAT TRANSFER

Combustion temperatures of rocket propellants typically are higher than the melting points of common metals and alloys, and perhaps even of many refractory materials. Short of the actual melting point, of course, the strength of most materials declines rapidly at high temperature, as may be seen in Fig. 14–1. Since the thickness (and therefore the mass) of the thrust chamber wall depends strongly on the stresses which it can support, it is desirable to use highly stressed materials, and the wall should be cooled to a temperature considerably below its melting point, and therefore much below the propellant stagnation temperature. In order to design a suitable cooling system, it is of course necessary to understand nozzle heat-transfer phenomena.

We can gain a rough idea of the magnitude of the cooling problem by considering typical heat-transfer rates, which are of the order of several Btu/sec-in^2 near the throat of a liquid-propellant rocket, and even greater in a high-pressure solid-propellant rocket. This is an order of magnitude greater than typical rates in the combustion chamber of a high-performance aircraft engine. Three basic cooling methods are commonly used.

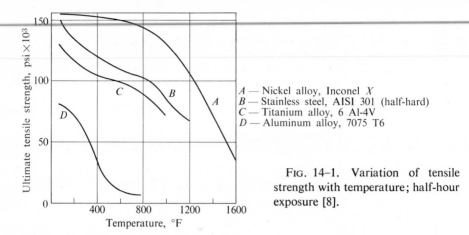

FIG. 14–1. Variation of tensile strength with temperature; half-hour exposure [8].

A — Nickel alloy, Inconel *X*
B — Stainless steel, AISI 301 (half-hard)
C — Titanium alloy, 6 Al-4V
D — Aluminum alloy, 7075 T6

(1) For liquid propellants it is common to use the fuel or oxidizer as a coolant, flowing in tubes or passages directly outside the chamber wall. Such a system is called *regenerative cooling*, since heat lost from the hot propellant is added to the incoming propellant.

(2) For solid propellants it is common to surround the nozzle walls with a mass of metal or other material which absorbs heat from the hot surface. The time during which such a heat sink can be effective is limited by its heat capacity, the allowable material temperature, and the heat-transfer rate.

(3) Additional cooling may be attained by the vaporization or sublimation of material from the inner surface of the chamber wall or from the wall itself. The injection of liquids or gases through porous walls is called *sweat cooling*, and the intentional loss of wall material is called *ablation cooling*.

In addition to these cooling techniques, it is possible to protect the chamber walls to some extent from the high-temperature gases. The use of an injection pattern which provides a layer of relatively cool gas near the walls, called *film cooling*, can substantially reduce heat transfer to the walls. The heat flux may also be reduced by inserting suitable refractory insulating materials or lining materials.

It is not as yet possible to predict accurately the heat-transfer rates in rocket nozzles and combustion chambers, though approximate methods are useful. The most important gas-to-wall heat-transfer mechanism is convection, although radiation may be important in some cases (see below). As shown in Chapter 4, convective heat transfer is directly related to the development of the boundary layer. Relatively little is known of boundary layer behavior under the extremes of wall curvature, pressure and temperature gradients, and heat fluxes prevalent in rocket nozzles. However, even if the boundary layer were well understood, accurate calculation would still be prevented by other important and largely unpredictable effects. For example, a seemingly unimportant alteration in injection pattern, which changes specific impulse by only a few percent, can change local heat-transfer rates by a factor of two [1]. The deposition of material on thrust chamber walls, either intentionally or unintentionally, is highly unpredictable, yet Sellers [9] indicates that carbon deposition may contribute a major share of the total resistance to heat transfer in an oxygen-RP1 thrust chamber.

Despite all these difficulties there exist methods, largely empirical, for the rough estimation of heat-transfer rates and wall temperatures. In the following sections we shall consider several of these methods as applied to simple regenerative and heat-sink cooling systems. From the foregoing discussion two features of such analyses should be evident: first, that elaborate boundary layer calculations are not really justified; and second, any real design must be based on direct experimental measurements. However, calculations can be very useful in evaluating the feasibility of proposed designs or design changes, and in designing test engines.

Regenerative cooling

In its simplest form regenerative cooling consists of the steady flow of heat from a hot gas through a solid wall to a cool liquid. Although the actual configuration may be a series of tubes or coolant channels, as in Fig. 14–2(a), the essential features of regenerative cooling within a rocket can be studied without geometric complications by considering a one-dimensional model, as in Fig. 14–2(b). For the moment we ignore the heat transfer through the walls which separate the coolant passages (Fig. 14–2a). The convective heat transfer at the solid–fluid interfaces may be described by a *film coefficient*, as discussed in Chapter 4. For the hot gas side, this is often written

$$\dot{\mathcal{Q}}_c = h_g(T_{\text{wa}} - T_{\text{wh}}), \tag{14–1}$$

where $\dot{\mathcal{Q}}_c$ = convective heat-transfer rate per unit area,

h_g = hot gas film coefficient [defined by Eq. (14–1)],

T_{wa} = adiabatic wall temperature,

T_{wh} = hot side wall temperature.

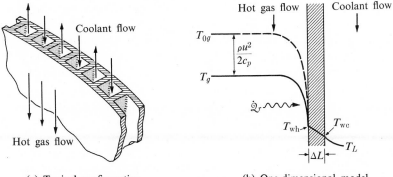

(a) Typical configuration   (b) One-dimensional model

Fɪɢ. 14–2.   Regenerative cooling definitions.

The *adiabatic wall temperature* (or recovery temperature) T_{wa} is, as Eq. (14–1) shows, that temperature which would be attained by the surface of an adiabatic or insulating wall ($\dot{Q}_c = 0$). One might expect that T_{wa} would be the stagnation temperature of the fluid, since the fluid adjacent to the wall has been brought to rest. However, especially for high-speed flow as encountered in rocket nozzles, the temperature rise accompanying stagnation is large enough that the viscous slowing-down process is not exactly adiabatic. That is, there is significant heat transfer *from* the low-speed (high static temperature T_g) fluid near the wall to the higher-speed (lower T_g) fluid further from the wall. Hence the stagnation temperature of the fluid at the wall is *less* than that of the free stream, as shown schematically in Fig. 14–3. On the other hand, to satisfy the steady-flow energy equation, there must be a region in which the stagnation temperature is slightly greater than that in the free stream. The magnitude of T_{wa} may be related to T_{0g} and T_g by the *recovery factor*, r, which is defined by

$$r = \frac{T_{wa} - T_g}{T_{0g} - T_g}.\tag{14–2}$$

For compressible turbulent boundary layers up to Mach numbers of 4, the recovery factor is about 0.91 for typical rocket propellants [1].

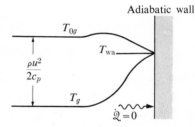

FIG. 14–3.   The adiabatic wall or recovery temperature.

Near the nozzle throat, where the maximum heat transfer rate occurs (as will be shown), the difference between T_{wa} and T_{0g} is rather small. For instance, if γ is 1.2,

$$\frac{T^*}{T_0} = \frac{2}{\gamma + 1} = 0.91,$$

and if $T_{0g} = 6000°R$, $T_g^* = 5460$, and for $r = 0.91$, $T_{wa} = 5950$. This difference of only about 1% is well within the uncertainty of heat-transfer calculations and, since the throat region presents the most severe cooling requirements, T_{wa} is often replaced by T_{0g} for rocket heat-transfer calculations. That is, Eq. (14–1) is often written simply as

$$\dot{Q}_c = h_g(T_{0g} - T_{wh})$$

with little loss of accuracy.

One-dimensional heat transfer through the solid wall is described by

$$\dot{Q}_w = -k_w \frac{dT_w}{dx} = \frac{k_w}{\Delta L}(T_{wh} - T_{wc}),$$ (14–3)

where k_w = thermal conductivity of the wall material, ΔL = wall thickness, and T_{wc} = cold-surface wall temperature.

The conductivity of the liquid-side film is described by an additional film coefficient h_L defined by

$$\dot{Q}_L = h_L(T_{wc} - T_L),$$ (14–4)

where T_L = liquid free-stream temperature ($T_L \cong T_{0L}$ for an incompressible fluid), and $\dot{Q}_L$ is the heat transfer from the wall to the liquid film.

For steady conditions the heat flow from the hot gas (including radiation heat transfer $\dot{Q}_r$) may be written

$$\dot{Q} = \dot{Q}_c + \dot{Q}_r = \dot{Q}_w = \dot{Q}_L.$$ (14–5)

The (usually small) radiant heat transfer rate $\dot{Q}_r$ to the hot wall surface will be discussed subsequently. Equations (14–1), (14–3), and (14–4) may be combined with Eq. (14–5) to give

$$\dot{Q} = \frac{T_{wa} - T_L + (\dot{Q}_r/h_g)}{(1/h_g) + (\Delta L/k_w) + (1/h_L)} \approx \frac{T_{0g} - T_L + (\dot{Q}_r/h_g)}{(1/h_g) + (\Delta L/k_w) + (1/h_L)}.$$ (14–6)

Of course, Eq. (14–1) may be used to express T_{wa} more accurately in terms of T_{0g}, T_g, and r, but the assumption $T_{0g} \approx T_{wa}$ is adequate near the throat. Having determined $\dot{Q}$, we may use the previous equations to calculate the local temperatures. The hot wall temperature is, of course, of great interest:

$$T_{wh} = T_{wa} - \frac{\dot{Q}_c}{h_g} = T_{wa} - \frac{\dot{Q} - \dot{Q}_r}{h_g} \approx T_{0g} - \frac{\dot{Q} - \dot{Q}_r}{h_g}.$$

This one-dimensional model may be modified to account for the actual geometry of the coolant system. The most common configuration consists of an array of rectangular (or truncated pie-shaped) thin-walled tubes welded or brazed together as in Fig. 14–4. The thin walls extend into the coolant passages much as cooling fins and, as such, they significantly decrease the coolant side heat-flow resistance. A simple modification to account for this would be to define "effective" wall and liquid side areas which are larger than the actual areas. Equation (14–5) would then be written

$$\dot{Q}A_g = (\dot{Q}_c + \dot{Q}_r)A_g = \dot{Q}_w A_w = \dot{Q}_L A_L$$

where A_g = gas-side area, A_w = effective wall area, and A_L = effective liquid-side area. Equation (14–6) would become

$$\dot{Q} = \frac{T_{wa} - T_L + (\dot{Q}_r/h_g)}{(1/h_g) + (A_g \Delta L/A_w k_w) + (A_g/A_L h_L)}.$$

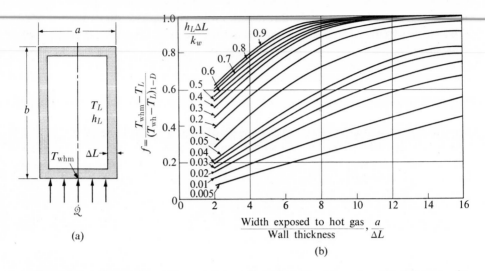

Fɪɢ. 14–4.   Comparison of one- and two-dimensional heat transfer. The influence of b is small so long as $b/a > 1$.   (Courtesy Sellers [10].)

A more detailed analysis would include the essentially two-dimensional heat-transfer pattern. That is, the gas-side wall temperature would not be uniform, being lower near the "cooling fins" or tube corners and reaching a maximum at the tube center plane. Sellers [10] presents an analysis for this situation and defines a fin factor f as

$$ f = \frac{T_{\mathrm{whm}} - T_L}{(T_{\mathrm{wh}} - T_L)_{\mathrm{one\text{-}dimensional}}} \qquad \text{at constant } \dot{\mathcal{Q}}, \qquad (14\text{–}7)$$

where

$$ T_{\mathrm{whm}} = \text{maximum (center plane) hot wall temperature,} $$

and

$$ (T_{\mathrm{wh}} - T_L)_{\mathrm{one\text{-}dimensional}} = \text{one-dimensional wall temperature drop} $$

as in Eq. (14–3).

Figure 14–4 shows typical results for rectangular tubes of constant wall thickness, assuming $\dot{\mathcal{Q}}$ distributed evenly over the hot wall surface. This assumption and the comparison, in Eq. (14–7), of one- and two-dimensional temperatures *for the same* $\dot{\mathcal{Q}}$ is reasonable for $T_{0g} \gg T_{\mathrm{wh}}$. Note that T_{whm} can be substantially less than the one-dimensional result, T_{wh}.

Heat sinks

The nozzles of solid-propellant rockets and some short-duration liquid-propellant rockets are cooled by heat sinks. In contrast to the above steady-state process, heat sink cooling is necessarily an unsteady process. In the absence of internal heat sources, conduction within an isotropic solid is described, in cartesian coordinates, by the following equation (see for examples References [7] and [8] of Chapter 4):

$$\frac{\partial^2 T}{\partial x^2} + \frac{\partial^2 T}{\partial y^2} + \frac{\partial^2 T}{\partial z^2} = \frac{1}{\alpha}\frac{\partial T}{\partial t},$$

where $\alpha = k_w/\rho_w c$, the thermal diffusivity of the solid, and k_w, ρ_w, and c are the thermal conductivity, density, and specific heat, respectively, of the solid.

For heat conduction into a wall which is thin relative to its radius of curvature one might, as in the previous regenerative cooling analysis, consider a one-dimensional model. In this case, the conduction equation reduces to

$$\frac{\partial T^2}{\partial x^2} = \frac{1}{\alpha}\frac{\partial T}{\partial t}. \tag{14–8}$$

A solution to this equation is

$$\theta = \frac{T_{0g} - T}{T_{0g} - T_i} = e^{-\alpha\beta^2 t}(c_1 \cos \beta x + c_2 \sin \beta x), \tag{14–9}$$

where T_{0g} = gas stagnation temperature,

T = wall temperature = $T(x, t)$,

T_i = initial wall temperature = $T(x, 0)$, and

β, c_1, c_2 = constants.

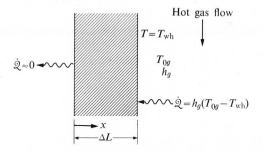

FIG. 14–5.   One-dimensional heat sink cooling model.

Several boundary conditions may be considered, but the most realistic simple case for the present application is illustrated in Fig. 14–5. Consider the case of heat transfer from a gas of constant stagnation temperature, T_{0g}, into an infinite plane wall of thickness ΔL. Assume that the hot wall heat transfer is described by a constant film coefficient h_g with negligible radiation heat transfer and, for simplicity, assume that there is negligible heat transfer from the cold side. Assuming $T_{wa} \approx T_{0g}$, we may then write the boundary conditions (noting that $\dot{\mathcal{Q}}$

is negative, that is, in the negative direction)

at $x = 0$: $\dfrac{\partial T}{\partial x} = 0$ or $\dfrac{\partial \theta}{\partial x} = 0$;

at $x = \Delta L$: $-k_w \dfrac{\partial T}{\partial x} = -h_g(T_{0g} - T)$ or $\dfrac{\partial \theta}{\partial x} = -\dfrac{h_g}{k_w}\theta.$ (14–10)

Differentiating Eq. (14–9) with respect to x yields

$$\frac{\partial \theta}{\partial x} = e^{-\alpha\beta^2 t}(-\beta c_1 \sin \beta x + \beta c_2 \cos \beta x).$$

Satisfaction of the boundary condition at $x = 0$ requires that $c_2 = 0$. The second boundary condition then requires that

$$-e^{-\alpha\beta^2 t}\beta c_1 \sin \beta \, \Delta L = -\frac{h_g}{k_w} e^{-\alpha\beta^2 t} c_1 \cos \beta \, \Delta L$$

or

$$(\beta \, \Delta L) \sin (\beta \, \Delta L) = \frac{h_g \, \Delta L}{k_w} \cos (\beta \, \Delta L).$$

This equation can be satisfied by a series of values of $\beta \, \Delta L$, so defining the nth of these as B_n we have

$$B_n \tan B_n = \frac{h_g \, \Delta L}{k_w}, \qquad n = 1, 2, 3, \ldots . \qquad (14\text{–}11)$$

The group $h_g \, \Delta L/k_w$ is a kind of Nusselt number, except that ΔL and k_w are here properties of the *solid* rather than the gas or gas-flow geometry. For each solution B_n there is a corresponding value c_n for c_1 which is, as yet, undetermined. A complete solution may be written as an infinite sum of the form

$$\theta(x, t) = \sum_{n=1}^{\infty} c_n \exp\left(-B_n^2 \frac{\alpha t}{\Delta L^2}\right) \cos B_n \frac{x}{\Delta L}. \qquad (14\text{–}12)$$

The remaining boundary condition, that $\theta(x, 0)$ must satisfy the initial wall temperature distribution θ_i, serves to determine the constants c_n. In a manner similar to that for determining the constants in an ordinary Fourier expansion, Schneider [11] shows that

$$c_n = 4\theta_i \sin \frac{B_n}{2B_n + \sin 2B_n}. \qquad (14\text{–}13)$$

The number of terms n actually included in the infinite series of Eq. (14–12) is, of course, dictated by the desired accuracy of the result. Reference [11] contains a table of the first five roots of Eq. (14–11) for various $h_g \, \Delta L/k_w$, along with an interesting sample calculation.

Example [11]

Suppose that a rocket nozzle is constructed of $\frac{1}{4}$-in.-thick high-temperature steel. (This wall thickness is very much smaller than the nozzle diameter.) Under the following conditions, how long may the rocket be fired before the wall temperature reaches 2000°F? The gas stagnation temperature is 4000°F and, initially, the wall temperature is uniformly 80°F. The heat-transfer coefficients are

$$h_g = 1500 \text{ Btu/hr·ft}^2\cdot°F, \qquad k_w = 18 \text{ Btu/hr·ft·}°F,$$

$$\alpha = \frac{k_w}{\rho_w c} = 0.25 \text{ ft}^2/\text{hr}, \quad \text{and} \quad \frac{h_g \, \Delta L}{k_w} = 1.74.$$

We shall assume that only the first two terms of the infinite series of Eq. (14–12) need to be calculated. For $h_g \, \Delta L/k_w = 1.74$, Eq. (14–11) yields $B_1 = 1.06$, $B_2 = 3.56$ (radians). Then from Eq. (14–13), for $\theta_i = 1$ (uniform initial temperature), $c_1 = 1.176$ and $c_2 = 0.206$. The final value θ_f of the temperature variable is given by

$$\theta_f = \frac{T_{0g} - T_f}{T_{0g} - T_i} = 0.51.$$

Since the allowable wall temperature is first reached on the hot side, where $x = \Delta L$, Eq. (14–12) becomes

$$\theta(\Delta L, t) = 0.51 = (1.176 e^{-1.123\tau} \cos 1.06) - (0.206 e^{-12.68\tau} \cos 3.56)$$

or

$$0.51 = 0.976 e^{-1.123\tau} + 0.188 e^{-12.63\tau}$$

where $\tau = \alpha t/(\Delta L)^2$. By trial and error, we find that the solution to this equation is $\tau = 0.155$. Hence the wall should reach 2000°F in

$$t = \frac{(\Delta L)^2}{\alpha} \tau = 0.97 \text{ sec.}$$

In this case the allowable firing duration is extremely limited. For practical rockets it is necessary to use higher-temperature materials and/or greater cooling effects, that is, greater solid conductivity and heat capacity.

In practice the required wall thickness is usually of the order of the nozzle-throat radius, so the above one-dimensional model is inadequate. It is more accurate, though still only an approximation of course, to consider radial heat flow within cylindrical walls, neglecting axial heat flow and axial variations in geometry and fluid properties. The appropriate equation, instead of Eq. (14–8), is

$$\frac{\partial^2 T}{\partial r^2} + \frac{1}{r} \frac{\partial T}{\partial r} = \frac{1}{\alpha} \frac{\partial T}{\partial t} \qquad \text{for} \quad a < r < b, \quad t \geq 0, \qquad (14\text{–}14)$$

where a and b are the inner and outer wall radii, respectively. The boundary conditions, replacing those of Eq. (14–10) and again neglecting cold-side heat transfer, are

$$\text{at } r = a: \quad -k_w \frac{\partial T}{\partial r} = h_g(T_{0g} - T);$$

$$\text{at } r = b: \quad \frac{\partial T}{\partial r} = 0.$$

(14–15)

The solution of Eq. (14–14) with these boundary conditions is, in a sense, quite similar to the foregoing solution for Eq. (14–8). It is, however, in the form of an infinite series of Bessel functions rather than trigonometric functions. Equation (14–11) is replaced by a rather complex transcendental equation in Bessel functions which must be solved for the desired number of roots. Medford [12] presents a summary of the solution method along with graphical relationships between the temperature variables

$$\frac{T_{0g} - T_a}{T_{0g} - T_i},$$

and

$$\frac{T_{0g} - T_b}{T_{0g} - T_i}$$

and the Nusselt number, $h_g a/k_w$, the time, $\alpha t/a^2$, and the ratio a/b. The inner wall temperature is T_a and the outer wall temperature is T_b. As before, T_i is the initial temperature.

Further complications arise in actual geometries, of course. Near the nozzle throat, where heat transfer is most severe, the wall curvature and axial variations may significantly alter the wall temperature distribution. In many cases the wall consists of a composite structure of varying thermal diffusivities, and in some cases the materials are highly anisotropic. Consideration of such factors leads to mathematically complex analyses, usually requiring computer solutions. In addition, certain non-analytic phenomena, such as surface erosion or chemical reaction, may be of great importance. As a result, the development of an actual nozzle requires considerable experimental work, just as in the case of a liquid-cooled nozzle.

Convective film coefficients

Among the terms in Eqs. (14–6) or (14–12) which we must know in order to calculate the local heat-transfer rate, the gas side film coefficient, h_g, is the most difficult to predict. While it is true that the calculation of fin effects or of transient heat conduction within a solid may be difficult, these difficulties are largely mathematical, arising from the application of well-verified equations to complicated geometries. On the other hand, the physical phenomena controlling h_g are not well understood.

There is considerable experimental data pertaining to heat transfer to a fluid flowing in long smooth tubes, as discussed in Chapter 4. Hence the liquid-side film coefficient in a regeneratively cooled rocket can be reasonably predicted by an equation of the form

$$\frac{h_L D}{k_b} = 0.023 \left(\frac{GD}{\mu_b}\right)^{0.8} \left(\frac{\mu c_p}{k}\right)^{0.33}_b,$$

or the equivalent form

$$\frac{h_L}{G c_p} = 0.023 \left(\frac{GD}{\mu_b}\right)^{-0.2} \left(\frac{\mu c_p}{k}\right)^{-0.67}_b,$$

where D = tube diameter, G = average mass flow per unit area = $4\dot{m}/\pi D^2 = \overline{\rho u}$, and subscript b denotes fluid properties evaluated at the bulk mean temperature.

Lacking a complete and detailed understanding of the flow processes, there is need of a correlation of this type relating h_g to the appropriate geometries and fluid properties within rocket nozzles. Although nozzles are not of constant diameter and the propellant flow is not fully developed, this relation nonetheless appears to be a reasonably good approximation of the actual nozzle heat transfer, if D is the local nozzle diameter [2], so that we may write

$$\frac{h_g}{G c_p} = 0.023 \left(\frac{GD}{\mu_b}\right)^{-0.2} \left(\frac{\mu c_p}{k}\right)^{-0.67}_b.$$

Although this equation may not always provide suitable accuracy as to magnitude, it does correctly predict the trends of the variation of h_g with nozzle diameter and pressure. Since G is $4\dot{m}/\pi D^2$, and $\dot{m}$ is constant for a given nozzle, it can be seen from the above that

$$h_g \propto \frac{1}{D^{1.8}}.$$

Thus, neglecting the rather small effects of the variation of Prandtl number ($\mu c_p/k$) and conductivity with temperature, it can be seen that the maximum h_g and hence the most severe cooling requirements for a given rocket and propellant occur at the throat, where D is a minimum. Experimentally, the maximum heat transfer rate is observed slightly upstream of the throat.

Second, if G is written as ρu, it may be seen that since the gas temperature is rather insensitive to pressure for most propellants, and μ, c_p, and k vary primarily with temperature, not pressure, then

$$h_g \propto \rho^{0.8} \propto p_0^{0.8}.$$

Thus cooling problems increase rather rapidly with increasing chamber pressure (although it is experimentally observed that the exponent is not precisely 0.8).

Empirically determined refinements of this equation are suggested in Reference [1]. In the combustion chamber (especially near the nozzle end) and at the throat the equation is adequate as it stands except, perhaps, for a refinement in the determination of fluid properties at the throat. In the supersonic portion of the nozzle an alternate form has been suggested:

$$\frac{h_g L}{k} = a \left(\frac{GL}{\mu}\right)^{0.8} \left(\frac{\mu c_p}{k}\right)^{0.33}, \tag{14-16}$$

where L is the distance downstream of the throat and a varies from 0.025 to 0.028.

Since the gas properties ρ, c_p, μ, and k vary with temperature, and since large temperature variations occur across high-velocity, high-heat-flux boundary layers, it is necessary to specify a temperature at which these properties should be evaluated. In these correlations a so-called film temperature T_f is used. In the combustion chamber or at the throat the film temperature may be defined as the arithmetic mean temperature

$$T_f = \frac{T_g + T_{\mathrm{wh}}}{2}.$$

Reference [1] suggests, for the supersonic portions of the flow and perhaps at the throat, an empirical formula,

$$T_f = T_{\mathrm{wh}} + 0.23(T_g - T_{\mathrm{wh}}) + 0.19(T_{\mathrm{wa}} - T_{\mathrm{wh}}).$$

Near the throat the difference between these two methods of specifying T_f is slight and the fluid properties are usually not precisely known anyway. Furthermore, the wall temperature, T_{wh}, may not be known with precision.

Bartz [13] has developed an equation which can be shown to be almost equivalent in form to Eq. (14–16). According to Bartz,

$$h_g = \left[\frac{0.026}{(D^*)^{0.2}} \left(\frac{\mu^{0.2} c_p}{\mathrm{Pr}^{0.6}}\right)_0 \left(\frac{p_0}{c^*}\right)^{0.8} \left(\frac{D^*}{r_c}\right)^{0.1}\right] \left(\frac{A^*}{A}\right)^{0.9} \sigma, \tag{14-17}$$

in which the subscript 0 signifies stagnation conditions and

$$\mathrm{Pr} = \frac{\mu c_p}{k}, \text{ the Prandtl number,}$$

$$c^* = \frac{p_0 A^*}{\dot{m}}, \text{ the characteristic velocity,}$$

$$D^* = \text{throat diameter,}$$

$$r_c = \text{throat radius of curvature in a plane which contains the nozzle axis,}$$

and

$$\sigma = \frac{1}{\left[\frac{1}{2}\frac{T_{\mathrm{wh}}}{T_{0g}}\left(1 + \frac{\gamma - 1}{2}M^2\right) + \frac{1}{2}\right]^{0.8-0.2\omega} \left(1 + \frac{\gamma - 1}{2}M^2\right)^{0.2\omega}},$$

in which M = local Mach number, and ω = exponent of the viscosity-temperature relation $\mu \alpha T^\omega$ ($\omega = 0.6$ for diatomic gases).

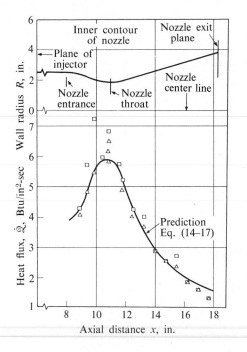

	Test conditions		
	p_0, psia	C^*, ft/sec	T_{wh}, °R
△	200	5421	860
□	199	5374	860

FIG. 14–6. Experimental heat flux compared with Eq. (14–17) for a nitrogen tetroxide (N_2O_4) and hydrazine (N_2H_4) rocket. (Courtesy California Institute of Technology Jet Propulsion Laboratory [14].)

An advantage of this unusual grouping of terms is the fact that the term in square brackets in Eq. (14–17) is a constant, since all fluid properties are evaluated at the stagnation temperature. There is reasonably good agreement between Eq. (14–17) and experiment, as indicated in Fig. 14–6.

It can be seen that the peak heat flux occurs slightly upstream of the throat. The deviation between predicted and measured heat flux in this region appears to be dependent on combustion pressure. It was observed that for variations of Reynolds number at a given nozzle position, due to variations of combustion pressure, the Nusselt number did not vary with $Re^{0.8}$ as predicted. Hence the complete validity of equations of this type is not certain, and they should not be relied on for the design of radically different nozzle types.

Radiative heat transfer

A combination of complex geometric considerations and radiation, absorption, and reflection mechanisms makes the calculation of radiative heat transfer from the hot propellant to the thrust-chamber walls an enormously complicated task. Radiation from a comparatively transparent hot gas may reach a particular point on the chamber wall from any point in the gas volume. On the other hand, in the typically nonuniform conditions within a combustion chamber, energy radiated from one location within the gas may be effectively absorbed by another cooler portion of the gas before reaching the walls. For example, a layer of cool gas near the wall (film cooling) may be as effective in stopping radiant heat transfer as in stopping convective heat transfer. Also gases, unlike solid radiators, do not radiate over a continuous spectrum of wavelengths but rather over discrete intervals or "bands." These bands are characteristic of the gas. Furthermore, the total energy radiated from a mixture of radiating gases, if some of the bands of the different gases possess common wavelengths, is not simply the sum of energies from the separate gases. The important radiating gases from common propellants, in approximate order of descending emissive powers, are CO_2, H_2O, CO, NO, OH, and HF [1]. These gases display considerable overlapping of bands.

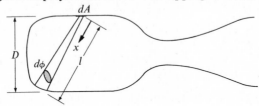

Fig. 14-7.   Radiative heat transfer to thrust chamber wall.

Consider the radiation to an element of area dA in the wall of a thrust chamber, as shown in Fig. 14-7. The energy reaching dA from within the solid angle $d\phi$ would be the integral over the "beam length" l of the energy emitted toward dA from the volume elements $x\,dx\,d\phi$ (assuming it is not absorbed in the intervening gas). The energy radiated per unit volume is proportional to the concentration (or partial pressure) of the radiating material, and represents a summation over the important radiating materials and the appropriate wavelengths (with corrections for overlapping band effects). The total energy reaching dA is the integral over the entire solid angle "visible" from dA. It is obvious that a calculation by this method which considered the actual nonuniform gas properties (if they were known) and the actual geometry would be exceedingly difficult. For simplicity the gas is usually assumed uniform throughout the volume. The actual geometry may be replaced by an "equivalent" hemisphere, with dA as the center, if an appropriate radius is chosen. The appropriate radius for an element on the wall of an infinite cylinder of diameter D is $0.9D$ [15], and this value is often used for calculations within combustion chambers [1]. Defining the emissivity ϵ_g of the volume of gas visible from dA to include all the geometric and radiative complexities, the total radiant energy per unit area *reaching* dA may be written

$$\dot{\mathcal{Q}}_r = \epsilon_g \sigma T_g^4, \tag{14-18}$$

where σ = *the Stefan-Boltzmann constant* = 0.1713×10^{-8} Btu/ft^2·hr·°R^4. The energy radiated *from dA* is also a complex function of the gas properties, since the gases absorb radiation as they radiate it, over discrete wavelength bands rather than over a complete spectrum. However, since $T_{wh} \ll T_g$, the wall is usually assumed to absorb all incident radiation (i.e., it absorbs as a blackbody) while reradiating negligible energy. Hence Eq. (14-18) represents the net radiant heat flux to the wall.

If the gases contain solid or liquid particles (as, for example, soot from excess fuel), an appreciable portion of the radiant energy may originate from the particles. The particles radiate over a continuous spectrum of wavelengths just as solids do. In fact, for a high concentration of particles or for large beam lengths, the radiating volume may behave much as a blackbody; that is, ϵ_g may approach unity [15, 1].

In the absence of complete calculations (for which data are often lacking), Sutton [2] suggests that for temperatures ranging from 3500 to 7000°F, radiation accounts for 5 to 25 percent of the total heat transfer. It is obvious that the importance of radiant heat transfer increases with chamber size, since the ratio of radiating volume to radiated surface is proportional to the chamber dimensions. In an analytical sense the beam length is proportional to the size of the chamber.

14–3 ROCKET CONSTRUCTION

The choice of thrust-chamber operating conditions usually requires consideration of many aspects of the vehicle and its function. Often there are conflicting requirements, necessitating compromise solutions which best suit overall performance. Hence, although it is impossible here to discuss system design in detail, it is appropriate to illustrate typical systems, and to indicate some of the more important considerations which lead to the choice of component designs. Liquid-propellant and solid-propellant systems will be treated separately.

Liquid-propellant rockets: regenerative cooling systems

The walls of large regeneratively cooled liquid-propellant rockets usually consist of an array of suitably shaped tubes, brazed together to form the thrust chamber as shown in Fig. 12–6. This particular engine consists of 312 nickel tubes of 0.012-in. wall thickness. The stresses due to chamber pressure are carried by external bands of steel. A double layer of bands is needed at the high-pressure end of the chamber, but bands that are relatively widely spaced suffice near the nozzle exit. For large chambers, this construction has replaced the much heavier double-walled construction of the V-2 engine. For very low-thrust rockets the double-wall construction is utilized, and for some intermediate units the thrust chambers are machined from solid material, the coolant passages being drilled by rather ingenious techniques.

Returning to Fig. 12–6, we can see that the fuel (kerosene in this case) flows radially from an inlet manifold into the top of alternate tubes, thence down the tubes, through a return manifold, and up through the intervening tubes. From these tubes it flows to the injector nozzles through radial holes and annular grooves in the stainless steel injector plate. Liquid oxygen travels directly through the injector plate from the "LOX dome" on the back side. This particular injector is of the like-on-like impinging type, with excess fuel injected near the chamber wall for film cooling. With minor modifications (for example, one-way coolant flow or a combination of one- and two-way flow), this construction is typical of present practice in large rocket thrust chambers.

The success of regenerative cooling systems is dependent on the fact that the total cooling load is a rather small fraction of the heat generated in the combustion chamber. Most propellants can absorb only a few percent of their heat of reaction before they suffer a temperature rise which renders them unsuitable as a coolant. For example, kerosene may decompose at temperatures above about 350°F, forming harmful insulating deposits of carbon on the tube walls. Hydrazine may explode if it gets too hot (as Table 12–2 shows, it is a reasonably good monopropellant). It is necessary with almost all coolants to keep the bulk temperature below the local boiling temperature. Although local surface boiling may be permissible, overall boiling of the fluid is usually accompanied by rapid burnout of the chamber wall, since most vapors are very poor coolants.

Bulk liquid temperature is dependent on the total heat transfer to the liquid up to the point in question. At any particular point the peak liquid temperature is T_{wc}, and this is dependent on the local heat-transfer rate as well as on the local bulk temperature. Total heat transfer may be found by integration of an equation such as (14–6) or—preferable in the case of a final design—by direct experimental measurement. Knowing the allowable coolant temperatures, we must check T_{wc} in at least two positions: at the throat where the rate is highest, and at the injector end of the chamber where the bulk temperature is highest. In general, calculations of this type, including actual geometric and heat flux variations, can be quite tedious. It is fortunate that since the coolant temperature is so very much less than the propellant temperature the exact distribution of coolant temperature has little influence on the heat-transfer rate. This feature enables us to avoid the need for iterative calculation procedures and also enables us to

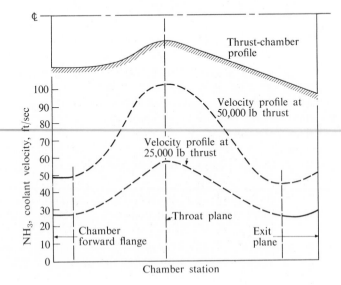

Fig. 14–8. Coolant velocities in the Thiokol S1 R99-RM-1 variable-thrust engine. (Courtesy Thiokol Chemical Corporation and USAF.)

measure the heat-transfer rates of experimental chambers which are not identical (insofar as cooling system is concerned) to the final design.

To maintain a low cold surface temperature T_{wc} at the throat, thus avoiding both excessive T_{wh} and local boiling or decomposition of the coolant, we need to increase the liquid-side film coefficient, h_L, in this region. This may be done by creating high local coolant velocities near the throat through proper distribution of coolant passage area. Figure 14–8, which is a plot of a typical coolant velocity distribution, shows a marked increase in the throat region.

It happens that gaseous hydrogen is a very good coolant, having both a high specific heat ($c_p = 3.42$ Btu/lb$_m$·°R at room temperature, as compared to 0.24 for air) and a high thermal conductivity [2.13×10^{-6} (Btu/in)/(sec·in^2·°R) at 32°F,

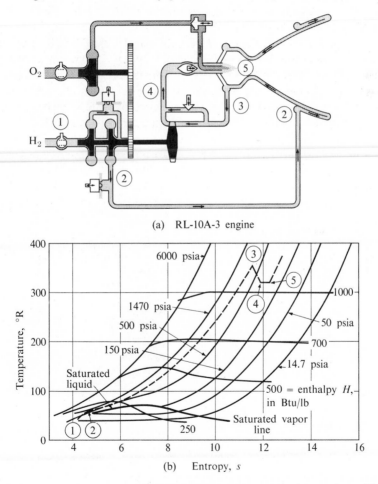

(a)   RL-10A-3 engine

(b)   Entropy, s

FIG. 14–9. Schematic of the RL-10A-3 hydrogen–oxygen rocket engine, and typical T-s diagram of the hydrogen flow. (Courtesy Pratt & Whitney Aircraft, Division of United Aircraft Corp.)

as compared to about 0.32×10^{-6} for air] relative to other gases. The possibility of using a low-temperature gaseous coolant has led to the unique system of the Pratt & Whitney Aircraft RL10 liquid oxygen–liquid hydrogen engine. As shown schematically in Fig. 14–9(a), these engines utilize gaseous hydrogen, formed in the coolant passages, to drive the turbopump. In this way no separate gas generator is needed and all the propellants are burned at high pressure in the combustion chamber with no loss of specific impulse. The system is capable of self-starting from the pressure available in the propellant tanks and from the residual heat in the thrust chamber.

The *T-s* diagram of Fig. 14–9(b) indicates a typical thermodynamic path of the hydrogen as it flows through the system. Note that the rather high liquid pressure, necessary to provide turbine pressure drop, is well above the critical pressure of hydrogen. Hence the hydrogen does not actually boil in the coolant passages. This high pressure requires a two-stage centrifugal pump, as shown in the schematic diagram.

Several other features of this system are of interest. The large nozzle area ratio (40:1) requires a "pass-and-a-half" tube design, in which hydrogen enters alternate tubes part way down the nozzle, flows toward the exit, through a return manifold, and back up to the combustion chamber in alternate tubes. Thus a portion of the nozzle contains twice as many tubes as the combustion chamber, as can be seen in Fig. 14–10. The gear train which drives the liquid-oxygen pump is cooled with gaseous hydrogen and run dry to avoid low-temperature problems which would accompany more conventional lubrication systems. The very low turbine-inlet temperature minimizes turbine stress problems and, at the same time, greatly reduces the problem of heat transfer between turbine and pump. Hence a relatively compact turbopump results, as may be seen in the photograph.

Fig. 14–10. The Pratt & Whitney Aircraft RL-10 hydrogen–oxygen rocket engine. (Courtesy Pratt & Whitney Aircraft, Division of United Aircraft Corp.)

Solid-propellant rockets: casings

A solid-propellant rocket casing is primarily a pressure vessel which must be provided with suitable ports for nozzles, igniters, etc. However, to achieve the desired light weight, design stresses are commonly much higher than those encountered in more conventional, earthbound pressure vessels. In addition, it is necessary either to protect the walls from the hot combustion products or to use materials which retain reasonable strength at very high temperatures. The modern practice of employing internal-burning case-bonded propellant grains serves to protect the chamber walls very well except at the ends, where additional insulation must be provided. As a result, reasonably cool wall temperatures are maintained and, with careful design and manufacture, very high stresses may be used.

High-strength steel alloys may be stressed to above 200,000 psi, and experimental models, at least, have withstood stresses above 300,000 psi. Obviously strength-to-weight ratio is more important than absolute strength. Hence, since the operating temperatures are reasonable, some of the lighter alloys and even glass-fiber structures are attractive. Test vessels made of titanium alloy have withstood stress levels as high as 260,000 psi before bursting, and this alloy is only about 65% as heavy as steel. Casings wound with glass filaments have been stressed to values several times the permissible limit of stress in an equal weight of steel. The third- and fourth-stage engines of the Minuteman are of glass filament construction. Very large casings may be built in segments, with each segment containing its own cast-in propellant [18]. Such segmented construction may be desirable for very large booster rockets since, if made in one piece, they would be too large to transport. Experimental engines have demonstrated successful construction and joining techniques on relatively small-scale engines.

Solid-propellant rockets: nozzles

The absence of a liquid cooling medium makes the design of a solid-propellant rocket nozzle a very difficult task. Since the wall heating rate varies considerably throughout a given nozzle (reaching a maximum near the throat), many nozzles are composite structures, as indicated in Fig. 14–11. In this way the best material is used only in the throat region, while other materials—which may be lighter, cheaper, or easier to form—are sufficient in other regions. Figure 14–11 [19] is typical of present concepts for rather high-performance solid-propellant rockets, while simpler structures would be adequate for operation at lower temperatures and pressures. This composite construction has been called a "modified hot structure" [20], since the lining material tolerates a temperature near that of the

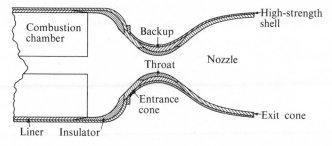

FIG. 14–11. Schematic diagram of an advanced solid-propellant nozzle. (Courtesy Kelble and Bernados [19].)

propellant but is cooled somewhat by a backup or heat-sink material. Note that the construction includes a high-strength outer shell which remains at relatively low temperature and carries most of the stress in the nozzle wall. Note also the insulating lining on the end wall of the combustion chamber.

A large number of materials have been used in nozzle construction. Graphite has been a popular material; although it is not particularly strong at low temperature, its strength increases with temperature up to about 4500°F, and at high temperature it is considered a strong material [20].

Depending on manufacturing processes, graphite is available in a wide variety of forms [21]. Single-crystal graphite is extremely anisotropic, but ordinary bulk graphite, consisting of many crystals of random orientation, can be isotropic. The strength of the material increases as the density becomes greater. Typical bulk densities are in the range from 0.05 to 2.1 gm/cm^3. Thus the throat insert in Fig. 14–11 might consist of high-density material to provide the necessary erosion resistance, while less-dense material might be suitable in other regions. Graphite is considered a good heat-sink material because of its relatively high specific heat and its high sublimation temperature.

Pyrolytic graphite, formed by a vapor-deposition process, consists largely of well-aligned crystals and possesses highly anistropic properties approaching those of single-crystal graphite. It is an extremely good conductor in one plane [234 (Btu/ft)/(hr·ft^2·°F) as compared with 227.5 for copper at room temperature] and an extremely good insulator in the direction normal to this plane [0.54 (Btu/ft)/ (hr·ft^2·°F) as compared to 17 for alumina] [22]. Pyrolytic graphite may be applied to a nozzle surface in such a way that it acts as an insulator in the radial direction, thus protecting underlying material. In this case the practically uncooled inner graphite surface attains the propellant temperature almost instantaneously. For higher-temperature propellants, nozzles may be constructed of pyrolytic graphite "washers" which conduct in the radial direction. In this case the graphite serves as a very effective heat sink, and propellant temperatures above the sublimation temperature can be tolerated for some time. In test nozzles, temperatures as high as 6551°F at pressures above 900 psi have been satisfactorily withstood for 44 seconds [22]. In any form, the rather high thermal expansion of graphite requires careful design to avoid failure due to the severe thermal shock which inevitably occurs upon ignition.

Tungsten, with a melting point of 6160°F, and various tungsten alloys have been used in solid-propellant rocket nozzles. High density and difficult fabrication problems have limited tungsten applications largely to throat inserts. Sintered tungsten is available and, with suitable organic or metallic filler material, it can withstand higher temperatures than pure tungsten [23, 24]. The filler material, for example a metal with a low melting point, is destroyed or vaporized so that it absorbs energy and provides a relatively cool gas layer along the surface as a result of a kind of ablation process.

A great many additional materials may be used, of course. The references cited contain considerable additional material and extensive bibliographies.

14-4 LIQUID-PROPELLANT PRESSURIZATION

Liquid propellants, stored in suitable tanks, must be carried to the combustion chamber and injected into it at relatively high pressure. For at least one of the propellants, the process may include passage through a regenerative cooling system. Two feed systems are in common usage: gas-pressurized systems and turbopumps.

Gas pressurization

The simplest feed system utilizes regulated high-pressure gas to pressurize the propellants within their storage tanks, as in Fig. 14-12. Inert gases such as helium or nitrogen, stored at relatively high pressure (5000 psia) to minimize gas-tank volume, are often used. The main advantage of this system is, of course, simplicity and attendant reliability. The weight of the pressurizing system itself can be substantially less than that of a comparable turbopump system. Offsetting this advantage, however, is the requirement that the pressurized propellant tanks must withstand internal pressure slightly greater than the combustion pressure. Thus, for large propellant tanks or high-combustion pressures, the overall weight exceeds that of a turbopump low-pressure tank system. As a result, compressed-gas feed systems are usually restricted to rather low thrust durations (of the order of 30 seconds). Making the right choice between compressed-gas and turbopump systems is not at all simple; we shall return to this question after discussing turbopump systems.

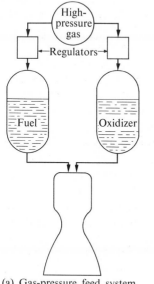

(a) Gas-pressure feed system

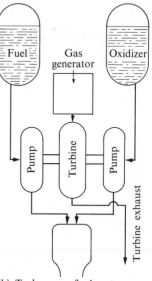

(b) Turbopump feed system

FIG. 14-12. Propellant feed-system schematics. Control valves, check valves, safety devices, fill and drain systems, etc., are not shown.

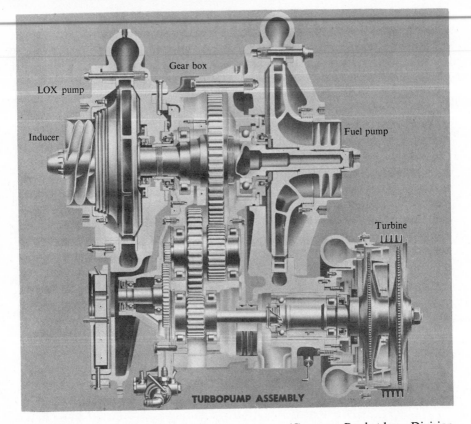

Fig. 14–13. Rocketdyne Mark III turbopump. (Courtesy Rocketdyne Division, North American Aircraft, Inc.)

Turbopumps

All the currently used liquid-propellant launch vehicles, and many of the upper stages as well, employ turbopump propellant-feed systems. The propellants are each pumped from a relatively low storage pressure up to the required pressure by a centrifugal pump. The pump is driven either directly or by means of a gear train, by a turbine which in turn is driven by hot high-pressure gases from a gas generator as in Fig. 14–12. The gas generator may utilize main rocket propellants, separate propellants, a liquid monopropellant, or even a solid-propellant charge. In one case (see Fig. 14–9) the turbine is driven by gaseous hydrogen which is vaporized in the chamber cooling system. Figure 14–13 is a cutaway view of a typical turbopump unit in which liquid oxygen and kerosene pumps, on a common shaft, are driven through gearing (for speed reduction) by a two-stage impulse turbine. From the cross section it may be seen that the pump blading is not radial. Also each pump has an axial inducer at its inlet. We shall now discuss the fluid dynamics of these turbomachines.

Centrifugal pumps

The dynamics of rotating machinery as developed in Chapter 9 are, for the most part, directly applicable to turbopump components. The power input $\mathcal{P}$ to a centrifugal pump, assuming no swirl in the incoming flow, is given by an expression identical to that for a centrifugal compressor:

$$\mathcal{P} = \dot{m} U_2 c_{\theta_2}, \qquad (14\text{-}19)$$

where $\dot{m}$ = fluid mass flow rate,

$\quad\quad\quad U_2$ = impeller tip speed, and

$\quad\quad\quad c_{\theta_2}$ = tangential component of absolute fluid velocity leaving impeller tip.

For incompressible flow the tangential component of the fluid leaving velocity, c_{θ_2}, is rather easily related to the exit fluid angle and the impeller geometry. Thus, referring to Fig. 14–14, we see that

$$c_{\theta_2} = U_2 + w_{\theta_2} = U_2 + w_{r_2} \tan \beta_2 \quad \text{or} \quad c_{\theta_2} = U_2 + \frac{\dot{m}}{2\pi r_2 b \rho} \tan \beta_2,$$

where w = relative velocity,

$\quad\quad\quad \beta$ = angle of relative velocity from radial, and

$\quad\quad\quad \rho$ = fluid density.

As in Chapter 9, the fluid angle β may be related to the blade angle in terms of an empirical slip factor. As a further refinement the radial flow area, $2\pi r_2 b$, might be reduced to account for the blockage of finite-thickness blades. However, neglecting these rather minor effects, we can write Eq. (14–19) as

$$\mathcal{P} = \dot{m} U_2^2 \left(1 + \frac{\dot{m}}{2\pi r_2 b \rho U_2} \tan \beta_2 \right)$$

or, using the steady flow energy equation (2–8), we can show that, since the flow is adiabatic,

$$\frac{\mathcal{P}}{\dot{m}} = \Delta h_0 = U_2^2 \left(1 + \frac{\dot{m}}{2\pi r_2 b \rho U_2} \tan \beta_2 \right), \qquad (14\text{-}20)$$

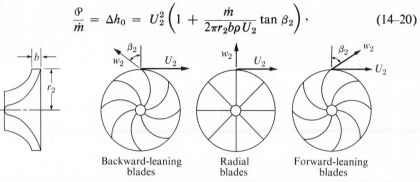

Backward-leaning blades   Radial blades   Forward-leaning blades

FIG. 14–14.   Centrifugal impellers.

~~in which Δh_0 is the increase in stagnation enthalpy of the fluid. Using the definition~~
of enthalpy, $h = e + pv$, we can write

$$T \, ds = dh - \frac{dp}{\rho},$$

in which $\rho = 1/v$, the density. Applied to the stagnation states of an incompressible fluid, this relationship becomes

$$T_0 \, ds_0 = dh_0 - \frac{dp_0}{\rho}.$$

Thus if the process under consideration is isentropic, $dh_0 = dp_0/\rho$, and Eq. (14–20) may be used to determine the ideal pressure rise through the impellers of Fig. 14–14:

$$\frac{\Delta p_0}{\rho} = U_2^2 \left(1 + \frac{\dot{m}}{2\pi r_2 b \rho U_2} \tan \beta_2 \right). \tag{14–21}$$

The pump efficiency η_p is conventionally defined as the ratio of ideal (isentropic) work to the actual work for the same pressure rise. Thus

$$\eta_p = \frac{\Delta h_{0s}}{\Delta h_0} = \frac{\Delta p_0}{\rho \, \Delta h_0} = \frac{\Delta p_0}{(\Delta p_0)_{\text{ideal}}},$$

in which $\Delta p_0/\rho$ is often called the pump head. Thus the actual pressure rise resulting from the work input is given by

$$\frac{\Delta p_0}{\rho} = \eta_p U_2^2 \left(1 + \frac{\dot{m}}{2\pi r_2 b \rho U_2} \tan \beta_2 \right).$$

Prediction of η_p is difficult; in practice η_p is either determined experimentally or is estimated from past experience rather than from calculation procedures.

Defining the dimensionless pressure and flow variables

$$\psi = \frac{\Delta p_0}{\rho U_2^2} \quad \text{and} \quad \phi = \frac{\dot{m}}{2\pi r_2 b \rho U_2},$$

we can write Eq. (14–21) (for an ideal pump) as

$$\psi = 1 + \phi \tan \beta_2. \tag{14–22}$$

Thus the ideal performance of the centrifugal impeller is as shown in Fig. 14–15. For a given impeller a single curve determines the pressure rise as a function of speed, flow rate, and fluid density.

In Eq. (14–22), frictional effects are entirely ignored and the adiabatic efficiency is implicitly assumed unity. Further, the flow was assumed to leave the impeller always at the blade angle β (relative to the rotor). Neither of these assumptions is satisfactory in reality, so that Fig. 14–15 cannot be regarded as an accurate prediction of real performance.

In order to show in a simple manner from experimental data how actual pump performance differs from the ideal, we resort to the technique of dimensional analysis. For the moment we ignore cavitation (which will be discussed subsequently) and assert that the pump pressure rise Δp_0 and efficiency η_p depend essentially on the following variables:

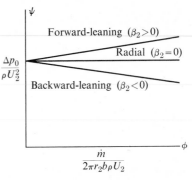

FIG. 14–15.   Ideal performance of centrifugal impeller, showing work input per unit mass at constant speed (U_2).

$$\Delta p_0 = f(\Omega, \dot{m}, \rho, \mu, \text{design}, D)$$

$$\eta_p = g(\Omega, \dot{m}, \rho, \mu, \text{design}, D)$$

in which η_p = pump efficiency

$$= \dot{m}\,\Delta p_0/\rho\mathcal{P},$$

Ω = rotational speed of the pump,

$\dot{m}$ = mass flow rate,

ρ = density,

μ = viscosity,

and "design" signifies a sufficient number of dimensionless geometrical ratios to completely specify the shape, with D indicating the size or scale of the pump; D can refer to any typical dimension of the pump, e.g., rotor tip diameter. By dimensional analysis it may be shown that the performance of the pump may be represented by

$$\frac{\Delta p_0}{\rho\Omega^2 D^2} = f\left(\frac{\dot{m}}{\rho\Omega D^3}, \frac{\rho\Omega D^2}{\mu}, \text{design}\right), \tag{14–23}$$

$$\eta_p = g\left(\frac{\dot{m}}{\rho\Omega D^3}, \frac{\rho\Omega D^2}{\mu}, \text{design}\right). \tag{14–24}$$

For a given pump design this represents a valuable improvement of the initial statement of physical dependence, since the number of independent variables has been reduced from four to two. For a given design it may easily be shown that

$$\psi \propto \frac{\Delta p_0}{\rho\Omega^2 D^2} \quad \text{and} \quad \phi \propto \frac{\dot{m}}{\rho\Omega D^3},$$

in which ψ and ϕ are the variables previously defined in the analysis of the idealized pump. Further, the viscosity variable may be identified as a Reynolds number, Re. With these variables, relationships (14–23) and (14–24) can be written as

$$\psi = f(\phi, \text{Re}, \text{design}), \tag{14–25}$$

$$\eta_p = g(\phi, \text{Re}, \text{design}). \tag{14–26}$$

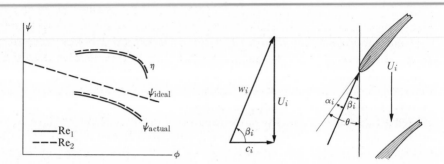

FIG. 14–16. Typical performance of a   FIG. 14–17. Typical pump inlet geometry.
backward-leaning pump; cavitation absent.

In the absence of cavitation, performance tests of a geometrically similar series of pumps (i.e., of a given design) will yield results shown typically in Fig. 14–16.

Note first of all that variations in Reynolds number appear to have little effect. This is not to say that friction is unimportant; it *is* important, and η_p is appreciably less than unity. Nevertheless, beyond a certain value for which the flow in the pump is turbulent, further increases in Re have little effect on the flow pattern or the efficiency. Rocket turbopumps will generally operate at high values of the Reynolds number. An approximate indication of the Reynolds number above which variations cease to have significant influence on performance is perhaps Re = $\rho w L/\mu = 10^5$, in which w and L are a typical relative velocity and passage width. For a given design, a precise limit can really only be established by tests. Figure 14–16, of course, refers to a series of pumps which are geometrically similar. If test data on a series of pumps which are not exactly geometrically similar are plotted, one could not expect their performance to be so nearly described by a single curve as in Fig. 14–16.

The actual pressure coefficient ψ is always less than ideal for two reasons:

(1) The fluid does not leave the impeller at the blade angle β, and is said to "slip" past the impeller. Consequently the impeller does not do so much work on the fluid as the idealized analysis implies.

(2) There are frictional losses in the rotor and stator, and these are especially magnified if the flow is partially stalled. The presence of separations leads to energy losses due to the mixing of slow- and fast-moving fluid.

The dimensionless flow variable ϕ has an important physical interpretation. For a given design it can be shown to be uniquely related to the flow angles at entrance to the rotor and stator. For example, at the rotor inlet the axial velocity component c_i may be written $c_i = \dot{m}/\rho A_i$. The rotor speed at some radius r_i is $U_i = \Omega r_i$; thus, referring to Fig. 14–17, we observe that the angle of the flow relative to the rotor at its radius r_i is

$$\tan \beta_i = \frac{c_i}{U_i} = \frac{\dot{m}}{\rho A_i \Omega r_i}.$$

For a given pump design we can see that $\tan \beta_i \propto \phi$. A similar argument shows that the flow incident to the stator is also uniquely dependent on ϕ, and therefore the incidence angle of the flow entering both rotor and stator blades depends only on ϕ. At large positive or negative incidence, pressure gradients near certain parts of the blade surface may be large enough to stall the boundary layer. Thus at sufficiently high or low values of ϕ (relative to the design value) the performance of the pump may drop off as indicated in Fig. 14-16.

Propellant pumps are usually of the backward-leaning type ($\beta_2 < 0$), for several reasons. As mentioned in Chapter 9, if the pressure rises with flow rate, the flow can become unstable. Further, although the work input of a radial-bladed impeller is higher (at a given speed), the backward-leaning blades are capable of producing the required pressure rise at speeds which produce tolerable stresses in the impeller. Generally the backward-leaning impellers have somewhat greater efficiency. Figure 14-18 indicates the pressure rise and efficiency characteristics of a typical rocket pump.

The data of Fig. 14-18 were compiled using water at a single shaft speed. These data may be used to predict the pump pressure rise with liquid oxygen, if it is assumed that cavitation is absent and Reynolds number changes are unimportant. Relationship (14-25) may be expressed, for a given pump, as

$$\frac{\Delta p_0}{\rho \Omega^2} = f\left(\frac{Q}{\Omega}\right).$$

in which Q is the volume flow rate. From this and the data of Fig. 14-18, we can predict the pump performance at general values of Q, Ω, and ρ, so long as the cavitation and Reynolds number variables remain insignificant.

It is often desirable to characterize a pump design in terms of a parameter which does not involve size, since size is usually a quantity to be determined. For example, the requirements of a given application may be specified by $\Delta p_0/\rho$,

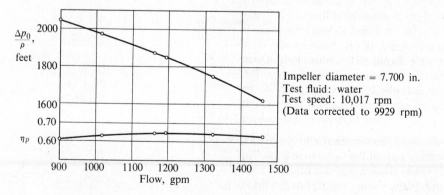

Impeller diameter = 7.700 in.
Test fluid: water
Test speed: 10,017 rpm
(Data corrected to 9929 rpm)

FIG. 14-18. Typical performance of the Rocketdyne Mark IV liquid oxygen pump of the sustainer engine (57,000-lb thrust) used on the Atlas MA-3. (Courtesy Rocketdyne Division, North American Aviation, Inc.)

Q, and Ω, with size to be determined. The common parameter for this purpose is *specific speed*, which may be obtained by eliminating D from the preceding groups in the following way:

$$\frac{(Q/\Omega D^3)^{1/2}}{[(\Delta p_0/\rho)/\Omega^2 D^2]^{3/4}} = \frac{\Omega Q^{1/2}}{(\Delta p_0/\rho)^{3/4}}.$$

As it stands this new group is of no particular value, since it may take on practically any value, depending on the operating point of the pump. However, when evaluated at the *maximum efficiency point* it becomes a single number characteristic of the pump shape. This number is usually called the specific speed N_s, defined by

$$N_s = \left(\frac{\Omega Q^{1/2}}{(\Delta p_0/\rho)^{3/4}}\right)_{\eta p \max}.$$

As derived, N_s is dimensionless, but in the literature the separate variables are commonly expressed as rpm, gallons per minute, and "feet" (ft-lb$_f$/lb$_m$), so that N_s has the peculiar units (rpm) (gpm)$^{1/2}$(ft)$^{-3/4}$.

The value of N_s is that it provides a rough means of selecting the "best" pump (in the sense of best efficiency) for a given application. The specific speed N_s is a characteristic of pump shape which has been evaluated (experimentally) for a complete range of pumps, from axial-flow through mixed-flow to centrifugal-flow pumps. If Ω, Q, and $\Delta p_0/\rho$ are specified, the designer will calculate $\Omega Q^{1/2}/(\Delta p_0/\rho)^{3/4}$ and pick a pump type having a specific speed near this value. A well-designed pump of this type will then operate at or near its peak efficiency which, presumably, will be better than that of other pump types which would be operating under less favorable conditions.

Cavitation

Up to this point the phenomenon of cavitation has been ignored, though this is a matter of great importance for rocket pumps. Cavitation is the spontaneous formation of bubbles of vapor in the liquid as the static pressure falls below the vapor pressure. It can be seriously detrimental to pump performance in two ways:

(1) The vapor bubbles may grow into a sufficiently large region in the impeller passages that the pressure rise across the pump falls toward zero. It may be seen from Eq. (14–22) and the definition of ψ that the pressure rise Δp_0 is directly proportional to the density of the fluid within the impeller.

(2) Even when cavitation is confined to so small a region of the impeller passage that the pressure rise is virtually unaffected, the blading of the impeller can be seriously eroded if the period of operation is sufficiently lengthy. Generally the point of minimum pressure in an impeller passage (as in the flow about an airfoil) is located on a solid surface. Thus cavitation begins near the surface, and the rapid growth and collapse of a vast number of minute bubbles immediately adjacent to the surface can lead to serious erosion, depending on the kind of material and

the time duration. The mechanism of cavitation damage is far from fully under-stood.

Damage caused to the surface by cavitation is generally insignificant in rocket pumps because their operating life is of the order of minutes. However, the effect of cavitation on pressure rise is most important and exercises a controlling influ-ence on pump design. We must therefore take account of it as well as viscosity in any satisfactorily general account of rocket pump behavior.

With highly purified liquids it has been shown possible under laboratory condi-tions to reduce the pressure of the liquid far below the vapor pressure without vapor formation. It is even possible for the liquid to exist under tension without any vapor formation. In general the presence of dust particles or other impurities to be expected in untreated fluids is sufficient to promote nearly immediate produc-tion of vapor bubbles when the static pressure falls below the vapor pressure. Minute cracks in the surface of the impeller blade are also thought to play a role in nucleating vapor bubbles. General experience with pumps suggests that the assumption that cavitation begins when the local pressure reaches the vapor pressure is approximately valid. Thus the difference $(p_{01} - p_v)$ between the vapor pressure and the inlet stagnation pressure can be regarded as a significant per-formance variable.

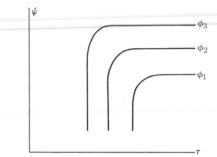

FIG. 14–19. Typical effect of cavitation on turbopump per-formance; given design of pump.

If the dimensional analysis of pump performance is repeated, this time recog-nizing $(p_{01} - p_v)$ as a significant variable, the following more general result may be obtained:

$$\psi = f(\phi, \text{Re}, \tau, \text{design}), \qquad \eta = g(\phi, \text{Re}, \tau, \text{design}),$$

in which

$$\tau = \frac{p_{01} - p_v}{\rho U_2^2/2}.$$

For a given design, τ is proportional to $(p_{01} - p_v)/\rho \Omega^2 D^2$. It is only one of many possible variables which could be used to express the influence of cavitation on performance.

Cavitation can drastically alter the performance as indicated in Fig. 14–19, and its effect depends very much on the pump geometry as well as on the value of τ. The effect of cavitation generally leads to a large drop in pressure coefficient, below a value of τ which depends on the flow coefficient. The critical value of τ depends strongly on blade geometry as well as on flow coefficient. Figure 14–20 shows experimental data on the effects described by Fig. 14–19. The variable $(p_{01} - p_v)/\rho$ is called the *net positive suction head*, NPSH, of the pump. As this variable is reduced, other variables (speed, pressure rise, flow rate) being held constant, cavitation and performance deterioration can be expected.

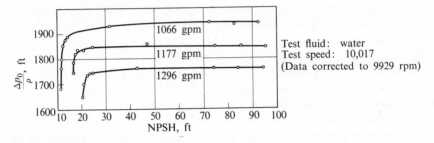

FIG. 14–20.   Cavitation performance of the pump of Fig. 14–18. (Courtesy Rocketdyne Division of North American Aviation, Inc.)

Considering Fig. 14–17, we see that the difference between the static pressure p_1 just outside the blade row and the minimum pressure $p_{\min}$ on the blade may be written (for high Reynolds numbers)

$$\frac{p_1 - p_{\min}}{\rho w_i^2/2} = f(\alpha_i, \text{ blade geometry}),$$

in which α_i is the incidence angle of the fluid approaching the blade and w_i is the relative velocity of the fluid at inlet. The blade geometry is specified by the blade shape and orientation. For given blade geometry, α_i depends uniquely on ϕ. In addition, since c_i/U and w_i/U depend only on ϕ, and also since

$$\frac{p_{01}}{\rho} = \frac{p_1}{\rho} + \frac{c_i^2}{2},$$

the above relationship may be written

$$\frac{p_{01} - p_{\min}}{\rho U^2/2} = f(\phi, \text{ blade geometry}).$$

If it is assumed that head breakdown begins to occur when the minimum static pressure equals the vapor pressure, the minimum tolerable value of τ is given by

$$\tau_{\min} = \frac{p_{01} - p_v}{\rho U^2/2} = f(\phi, \text{ blade geometry}). \tag{14–27}$$

Rocket pumps are required to operate with low values of $p_{01} - p_v$ (to avoid heavy propellant tanks) and high values of U (in order to produce high pressure rise). Thus it is necessary to develop a special inlet blade geometry which makes it possible to operate at low values of τ without performance breakdown due to cavitation. In order to meet this requirement a separate impeller called an inducer (see Fig. 14–13) is generally affixed coaxially to the pump rotor. Usually it consists of helical passages bounded by thin blades whose inlet direction is only a few degrees from the tangential. The tip blade angle θ in Fig. 14–17 may be as low as 5 or 6 degrees and only two or three blades may be used. The fluid enters the inducer which gently raises its static pressure until the flow can enter the main pump without cavitating. For inducers of this type, Stripling and Acosta [6] have shown, on the basis of a theoretical analysis, that the relationship (14–27) may be represented quite acceptably by a correlation of this form:

$$\frac{p_{01} - p_v}{\rho U_t^2/2} = f\left(\frac{\sin \theta}{1 + \cos \theta}\phi\right), \tag{14–28}$$

in which U_t is the blade speed at the outer radius of the inlet and θ is the tip blade angle.

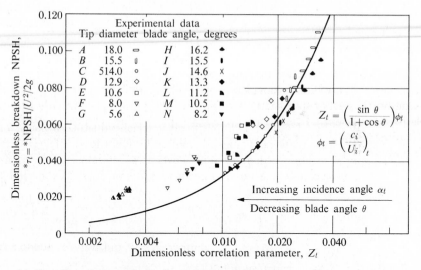

FIG. 14–21. Correlation of the breakdown of cavitation performance for several helical inducers; mean data curve. (Courtesy Stripling [6].)

Data on the cavitation breakdown of performance of several turbopumps with helical inducers varying widely in design correlate fairly well in this way as shown by Fig. 14–21. The scatter in the data is not unreasonable, since, in the derivation of the above correlation parameters, no account is taken of leading-edge blade shapes and of flow conditions other than at the tip of the inducer inlet. To be able to pump a fluid of low NPSH, it is desirable to have a very small value of θ.

The suction specific speed S is a variable which characterizes the cavitation performance of a pump or inducer design. Like the specific speed N_s, it does not contain the size variable, and is defined by

$$S \equiv \left\{ \frac{\Omega Q^{1/2}}{[(p_{01} - p_v)/\rho]^{3/4}} \right\}_{\text{cavitation}} . \qquad (14\text{–}29)$$

The cavitation "point" may not be distinct, and it may be necessary to define cavitation as that operating condition for which head rise has dropped by an arbitrary amount, say 2%. The value of S at cavitation will depend on ϕ (like the critical value of τ in Fig. 14–19), so that if a single value of S is to characterize the design, ϕ must be fixed; e.g., at the optimum efficiency point. If we want a pump to provide a given flow rate at a given speed, knowledge of S for pumps of the type considered will permit us to estimate the minimum tank pressure which must be provided to avoid cavitation. It is clearly desirable to design pumps having high values of S, since the higher the allowable shaft speed Ω, the smaller the pump for a given overall pressure rise. Sutton [2] states that "poor" pumps have suction specific speeds near 5000 (rpm) (gpm)$^{1/2}$ (ft)$^{-3/4}$, while the best pump designs without cavitation have values near 10,000 to 15,000. Pumps with local cavitation may operate with values exceeding 30,000.

Turbines

The driving turbines used in turbopumps are usually of the impulse type. In most cases they are directly coupled to the pump, and owing to the possibility of cavitation the turbopump shaft usually runs at speeds for which the turbine is relatively inefficient. In order to minimize the required number of stages, it is, of course, desirable to produce high power per stage. As discussed in Chapter 9, the power output per stage of an impulse turbine is greater than that of a reaction

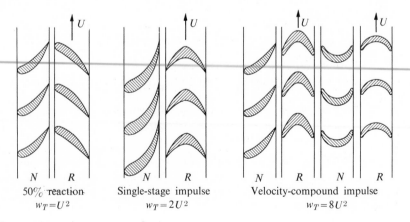

FIG. 14–22. Blade cross sections and work outputs of three typical turbines, for constant axial velocity and axial outlet velocity.

turbine for the same blade speed (given axial exhaust velocities). Figure 14–22 indicates three turbine types. For each, the work output per unit mass flow is stated for cases in which the axial velocity is held constant through the stage and the outlet velocity has no swirl component. Thus, at a given rpm, the single-stage impulse turbine can deliver twice the power of a 50% reaction turbine of the same size.

The velocity-compound impulse turbine is, in a sense, a single impulse stage since all the pressure drop occurs in the first nozzle. The fluid leaves the first rotor with swirl in the direction opposite to the rotation. This is redirected in the second "nozzle" (without pressure drop) to enter the last rotor. If the fluid leaves the last rotor without swirl it is easily shown that the power output of the two rotors is four times that of a single-impulse rotor, and thus about twice that of an ordinary two-stage impulse turbine. For this reason the velocity compound impulse turbine is commonly used in turbopumps for those cases in which a single-impulse stage is inadequate. For lightness, two rows of rotor blades are often mounted on a single disc which branches near the blade roots since, at typical speeds, the stress problem is not severe. The velocity-compound turbine is, however, significantly less efficient than a single-impulse stage (which in turn is less efficient than the 50%-reaction stage) owing largely to the very high fluid velocities encountered in the first nozzle and rotor blades, and to the requirement that the fluid travel through three successive blade rows without the beneficial effect of pressure drop on the boundary layer.

Designating the turbine inlet as state (1) and the outlet as (2), the turbine power is given by

$$\mathcal{P}_T = \dot{m}_T(h_{01} - h_{02}),$$

where $\dot{m}_T$ is the turbine mass flow rate. Since the useful turbine output is entirely shaft power, it is appropriate to describe its performance in terms of total-to-static turbine efficiency, η_{ts}, as defined in Eq. (9–15):

$$\eta_{ts} = \frac{h_{01} - h_{02}}{h_{01} - h_{2s}},$$

where h_{2s} is that enthalpy attained by *isentropic* expansion to the outlet *static* pressure, p_2. Assuming the working fluid a perfect gas with constant c_p, we may write this as

$$\eta_{ts} = \frac{1 - (T_{02}/T_{01})}{1 - (p_2/p_{01})^{(\gamma-1)/\gamma}},$$

and the turbine power as

$$\mathcal{P}_T = \dot{m}_T \eta_{ts} c_p T_{01} \left[1 - \left(\frac{p_2}{p_{01}}\right)^{(\gamma-1)/\gamma} \right]. \tag{14–30}$$

Because designers aim at light-weight turbines, they must tolerate turbine efficiencies that are quite low; even less than 50% in some cases.

Turbines must supply enough power for the pumps, plus power to make up for losses incurred in the power transmission system. If the system includes a gear train, the losses can be appreciable. Thus, defining η_m, the mechanical efficiency of the power transmission system, as the ratio of power transmitted to power received,

$$\mathcal{P}_T = \frac{1}{\eta_m}\left\{\left[\frac{\dot{m}(\Delta p_0/\rho)}{\eta_p}\right]_{\text{oxidant}} + \left[\frac{\dot{m}(\Delta p_0/\rho)}{\eta_p}\right]_{\text{fuel}}\right\}. \qquad (14\text{–}31)$$

From this, and using Eq. (14–30), we can determine the turbine mass flow for given inlet conditions, exhaust pressure, and turbine efficiency. Requirements vary widely since T_{01}, p_{01}, and η_{ts} are all subject to wide variations, but typical turbopumps consume something like several percent of the thrust chamber flow. This turbine fluid, exhausted without providing appreciable thrust, detracts from the overall thrust per unit mass flow or specific impulse of the rocket engine. In at least one engine (which utilizes liquid oxygen and hydrogen), the turbine is powered by the entire mass flow of hydrogen which has been vaporized in the regenerative cooling process. The hydrogen is then injected into the combustion chamber, burned, and accelerated with no loss in specific impulse. This desirable procedure

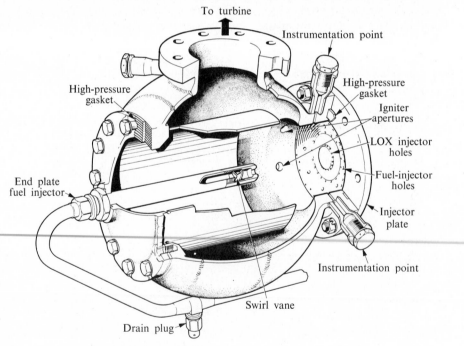

FIG. 14–23.   Gas generator for the Rolls-Royce RZ-2 rocket engine (Fig. 12–6) and for several Rocketdyne engines. About 90% of the kerosene is injected near the LOX injector, the remaining 10% being injected from the opposite end to provide cooling. (Courtesy of Rolls-Royce.)

seems limited to those engines utilizing hydrogen fuel, however; it has been discussed previously in the section on regenerative cooling systems.

Gas generators often utilize the main engine propellants. Figure 14–23 is a cutaway view of a typical combustor for such a gas generator. Liquid oxygen and kerosene are burned in a very rich fuel mixture (0.351:1 LOX/kerosene) in order to keep the temperature within allowable turbine limits (about 1200°F in this case). The exhaust of such a mixture burns on contact with the atmosphere and is often visible as a jagged, smoky flame during rocket launchings, as one can see in the frontispiece on page 317.

The generation of gas resulting from the decomposition of hydrogen peroxide is also commonly used, the combustion chamber being replaced by a catalytic decomposition chamber. While such systems are somewhat simpler and easier to start than those in which gas is generated from propellant, they require separate storage and a supply of working fluid.

The design of a turbopump system obviously entails many compromises between conflicting requirements. For example, the detrimental effect of low turbine blade speed on turbine efficiency can be overcome by using higher turbine speeds along with suitable gear drives. This reduces the fuel consumed by the turbopump unit, but increases the mass of the unit itself. Whether the benefits of reduced fuel consumption outweigh the disadvantage of increased turbopump mass is dependent on the time duration of the thrust, and on other design conditions.

The pump cavitation problem can be alleviated by increasing the pressurization of the propellant storage tanks. However, this increases the tank weight. Again a compromise is required between the fixed turbopump weight and a duration-dependent tank weight. Since so many complicated interactions are involved, the final choice of a design is usually made only after comparing a number of designs for the specific application.

Similarly, the choice between turbopump and gas pressurization systems is not simple. For high-combustion pressure systems which consume large quantities of propellant, the mass of a pressurized tank is prohibitively large. But if the quantity of the propellant is moderate or if the combustion pressure is reasonably low, only detail designs of alternate proposals can determine the best choice. The gas-pressurized Able-Star rocket of Aerojet-General Corporation provides an interesting example [16]. At first glance, its moderately high thrust (7890 lb) and long duration (300 sec) suggest a turbopump application. However, since it is to be used in space, where the back pressure is very low, high combustion pressure is not necessary to achieve high nozzle pressure ratio. Figure 12–3 indicates a performance loss due to reduction in combustion pressure, but this must be considered in relation to the performance benefit due to decreasing the tank mass or eliminating a turbopump. Even a turbopump system would require some pressurization to suppress cavitation. The simplicity, reliability, and ease-of-starting characteristics of pressurized systems are very attractive for space operation. All these considerations, and more, led to the choice of a helium-gas pressurization system and a combustion chamber pressure of 206 psia.

14-5 SELECTION OF COMBUSTION PRESSURE

Having considered some of the aspects of rocket operation, let us again take up the question of combustion pressure. It should be evident that the choice of combustion pressure influences overall vehicle performance through a great many, sometimes conflicting, effects. In particular, it is very likely that the best pressure for thrust-chamber performance is *not* the best pressure for overall performance.

Liquid-propellant engines

Suppose, for example, that a given rocket produces a given thrust at a given combustion pressure, and consider the effects of producing that same thrust at a higher pressure. The elementary analysis leading to Eq. (11–7) is adequate to indicate that higher p_0 would result in a smaller thrust chamber. However, higher pressure would also require thicker chamber walls and perhaps a greater nozzle area ratio. As a result, the thrust-chamber mass may increase or decrease with increased p_0. Chemical calculations such as those leading to Fig. 12–3 indicate that increased p_0 leads to increased specific impulse. In a conventional turbopump feed system higher pressure would require a higher ratio of turbine flow to rocket flow, thus offsetting to some degree the gain in performance. Furthermore, a heavier turbopump and gas generator would be required. For a pressure feed system the tank weight would increase rapidly with combustion pressure. Finally, as shown in Section 14–1, the increased pressure and accompanying increase in wall thickness would both intensify wall temperature problems.

The best choice between such conflicting effects is dependent on the performance and design of the various components. It is usually impossible to make a purely analytical choice, since many of the effects cannot be reduced to algebraic analysis. Hence we must make several rather detailed designs and calculate the performance of each in order to find the best design. Figure 14–24 indicates the results of such a procedure applied to a ballistic vehicle, assuming range to be the ultimate measure of performance.

Better component performance and lighter designs lead to higher optimum combustion pressures. Reference [1] states that for turbopump systems the best pressure is usually near 1000 psi. At the present time, however, pressures are limited by cooling problems to lower values. To take advantage of higher performance at higher pressures, designers are expending extensive effort on solutions to the cooling problem. One promising solution is to coat the interior chamber walls with an insulating ceramic material, thus decreasing the heat-transfer rate and the peak metal temperature. A difficulty with this solution is that the ceramic coating often cracks

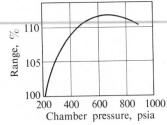

FIG. 14–24. Typical performance optimization study in terms of range increase for a V-2-type ballistic rocket. (After Sutton [2].)

or flakes off locally. The resultant surface roughness may cause increased transfer rate and rapid burnout.

As one might expect, gas-pressurized feed systems are much more sensitive to combustion pressure, and a lower optimum is observed. The best pressure usually lies between 250 and 400 psi [1].

Solid-propellant engines

The total impulse of a given solid-propellant mass is relatively independent of combustion pressure. Hence, assuming chamber size to be a function only of propellant mass or volume, and neglecting gravity for the moment, it would seem that the best combustion pressure for a solid propellant would be the lowest pressure permitting stable combustion, since this would result in the thinnest and hence lightest chamber wall. This simple conclusion is, however, not true, since the *total* chamber volume is a function of chamber pressure as well as propellant volume. For example, higher p_0 requires a smaller nozzle throat area, and a correspondingly smaller flow area through the propellant may be permissible. Further, since burning rate is a function of chamber pressure, a smaller burning area is required as p_0 is increased. Thus combustion chamber pressure, in addition to influencing nozzle size and mass, influences the entire combustion chamber geometry in a rather complex fashion which is, among other things, a function of the propellant properties and the geometric ingenuity of the designer. Limitations on thrust magnitude (for example, provision of acceptable acceleration in a gravity field), also place restrictions on the choice of combustion pressure. Reference [4] states that for any particular mission and set of design restrictions, an optimum pressure, usually between 500 and 1500 psi, can be found.

14–6 IGNITION

A complete rocket propulsion unit requires a large number of auxiliary systems. For example, start-up and shutdown processes require elaborate controls to avoid the buildup of explosive mixtures before ignition. Start-up of a turbopump system usually requires an external energy source until near steady state is reached. The feed system requires additional controls to maintain proper fuel–oxidizer ratio (which may vary throughout a firing in order to use up a maximum of the available propellant) and to establish and maintain the desired combustion pressure. This latter function is complicated by the variation of pump inlet pressure due to changing tank levels and vehicle acceleration.

Ignition of non-hypergolic propellants is usually achieved by a pyrotechnic igniter: that is, a small chemical charge which burns vigorously for a few seconds after it is ignited by an electrical filament. Such an ignition system is obviously a one-shot device, and if restart capability is desired more than one igniter must be provided. Alternatively, high-energy electrical spark igniters may be used to provide multiple starts. For large rockets it may be necessary to spark-ignite a small propellant flow in a pilot chamber and use the flame from this to ignite the main chamber. Igniters are often mounted on the injector plate as shown in Fig. 12–6, with electrical leads passing through the nozzle (to be blown out after ignition).

The solid-propellant rocket, like the liquid-propellant rocket, requires additional mechanisms in order to assure proper starting, stopping, and control during flight. Ignition is conventionally achieved by a pyrotechnic device, but the requirements for solid propellants may be more complicated than for liquid ones. First, it is necessary to achieve ignition over a rather large surface area to assure a uniform and proper recession of the burning face. At the propellant surface, ignition is the result of a combination of convective and radiative heat transfer and the impingement of hot particles from the igniter. However, the igniter must provide, in addition to sufficient surface heating, a chamber pressure high enough to sustain steady combustion (that is, greater than the combustion limit defined on page 388). On the other hand, too high a pressure can result either in immediate chamber rupture or in propellant cracking and subsequent chamber rupture. The initial pressure is usually provided by the rapid gas generation of the igniter, but for high-altitude starts it may be necessary to obstruct the nozzle with a plug which blows out upon ignition. Experimental ignition systems based on injection of a suitable hypergolic fluid have been tried successfully [25].

Thrust termination, which must be precise and repeatable, is not as easy with solid propellants as with liquid, since the propellant cannot be simply shut off. Usually, thrust is terminated by blowing off suitable ports or the nozzles themselves, thus rapidly reducing the chamber pressure and extinguishing the flame. This precludes restarting, of course, since the combustion chamber is essentially destroyed in the process. It may be possible to extinguish the flame through the injection of an inert gas or even water, and restartable solid-propellant rockets may soon become practical.

14–7 THRUST VECTORING

Thrust vectoring (altering the thrust direction to control vehicle motion) of a liquid-propellant rocket is achieved by pivoting the entire thrust chamber about a gimbal mounting on the thrust axis. Motion is provided by hydraulic actuators which often use rocket fuel at pump-outlet pressure to avoid the necessity of a separate fluid system. The turbopump is usually attached to and moves with the thrust chamber, thus avoiding flexible joints in high-pressure propellant lines.

A great many more components and subsystems go into an actual rocket engine, of course, but they are not our concern here. References [1], [2], and [5] treat some aspects of rocket-system design in more detail.

Thrust vectoring, to achieve in-flight control, is more difficult with solid propellants. It is, of course, impossible to pivot the thrust chamber since this includes the entire vehicle structure. Early rockets achieved control through movable vanes immersed in the exhaust flow. Such vanes present very difficult

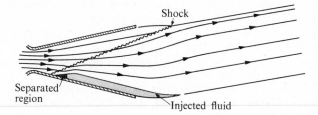

Fɪɢ. 14–25. The effect of the injection of secondary fluid on rocket nozzle flow. (Adapted from Stambler [26].)

cooling problems and they detract substantially from nozzle performance. Many present rockets achieve thrust vectoring by pivoting the nozzle relative to the combustion chamber, but this is rather difficult, since it requires a movable joint which must contain the high temperature–high pressure propellants and at the same time be reasonably easy to move. Considerable research is being done on the problem of achieving thrust vectoring through non-axisymmetric injection of a secondary fluid within the nozzle. The injection of secondary fluid disrupts the supersonic nozzle flow, causing shocks resulting in a nonaxial exhaust momentum flux, as shown in Fig. 14–25. Some systems of this type are already in operation [26], and they may be of increasing importance in the future, especially for very high thrust applications.

References

1. Aʟᴛᴍᴀɴ, D., J. M. Cᴀʀᴛᴇʀ, S. S. Pᴇɴɴᴇʀ, M. Sᴜᴍᴍᴇʀғɪᴇʟᴅ, "Liquid Propellant Rockets." Princeton, N. J.: Princeton Aeronautical Paperbacks, Princeton University Press, 1960

2. Sᴜᴛᴛoɴ, G. P., *Rocket Propulsion Elements.* New York: John Wiley & Sons, 1956

3. Sᴇɪғᴇʀᴛ, H. S., "Twenty-Five Years of Rocket Development." *Jet Propulsion*, **25,** 11, November 1955, pages 594–603

4. Hᴜɢɢᴇᴛᴛ, C., C. E. Bᴀʀᴛʟᴇʏ, and M. M. Mɪʟʟs, *Solid-Propellant Rockets.* Princeton, N. J.: Princeton Aeronautical Paperbacks, Princeton University Press, 1960

5. Sᴇɪғᴇʀᴛ, H. S. (editor), *Space Technology.* New York: John Wiley & Sons, 1959

6. Sᴛʀɪᴘʟɪɴɢ, L. B., "Cavitation in Turbopumps," Part 2. *Trans. of the ASME*, Series D, *J. Basic Eng.*, **84,** September 1962, page 339; *see also* Sᴛʀɪᴘʟɪɴɢ, L. B., and A. Aᴄoꜱᴛᴀ, Jʀ., "Cavitation in Turbopumps." Part I, *Trans. of the ASME*, Series D, *J. Basic Eng.*, **84,** September 1962, page 326

7. Joꜱᴛ, W., *Explosion and Combustion Processes in Gases*, translated by H. O. Croft. New York: McGraw-Hill, 1946, pages 242–247 and 255

8. "Strength of Metal Aircraft Elements." Military Handbook MIL-HDBK-5, March 1959, Armed Forces Supply Support Center, Washington 25, D.C.

9. Sᴇʟʟᴇʀs, J. P., Jʀ., "Effect of Carbon Deposition on Heat Transfer in a LOX/RP-1 Thrust Chamber." *J. Amer. Rocket Soc.*, **31,** 5, May 1961, pages 662–663

10. SELLERS, J. P., JR., "Effect of Two-Dimensional Heat Transfer and Wall Temperatures in a Tubular Thrust Chamber." *J. Amer. Rocket Soc.*, **31,** 3, March 1961, pages 445–446; and "Two-Dimensional Heat Conduction in a Tubular Thrust Chamber." *J. Amer. Rocket Soc.*, **32,** 7, July 1962, pages 1111–1112

11. SCHNEIDER, P. J., *Conduction Heat Transfer.* Reading, Mass.: Addison-Wesley, 1955; especially Chapters 5 and 10

12. MEDFORD, J. E., "Transient Radial Heat Transfer in Uncooled Rocket Nozzles." *Aerospace Engineering*, **21,** 10, October 1962, pages 15–21

13. BARTZ, D. R., "A Simple Equation for Rapid Estimation of Rocket Nozzle Convective Heat Transfer Coefficients." *Jet Propulsion*, **37,** 1, January 1957, pages 49–51

14. WITTE, A. B., and E. Y. HARPER, "Experimental Investigation and Empirical Correlation of Local Heat-Transfer Rates in Rocket-Engine Thrust Chambers." Technical Report No. 32–244, Jet Propulsion Laboratory, California Institute of Technology, Pasadena, Cal., March 19, 1962

15. ROHSENOW, W. M., and H. Y. CHOI, *Heat, Mass and Momentum Transfer.* Englewood Cliffs, N. J.: Prentice-Hall, 1961

16. STAMBLER, I., "Simplicity Boosts Able-Star." *Space/Aeronautics*, **36,** 2, August 1961, pages 59–64

17. COAR, J. R., "The RL10A-3 Engine." *Astronautics*, **7,** 2, February 1962

18. COHEN, W., "Solid Rockets for Space Vehicles." *Astronautics*, **6,** 8, August 1961, page 22

19. KELBLE, J. M., and J. E. BERNADOS, "High-Temperature Nonmetallic Materials." *Aerospace Engineering*, **22,** 1, January 1963, pages 56–75

20. UNGAR, E. W., "Applications of Materials to Solid-Rocket Nozzles." *Astronautics*, **6,** 4, April 1961, page 24

21. BUSHONG, R. M., "Graphite as an Aerospace Material." *Aerospace Engineering*, **22,** 1, January 1963, pages 40–45

22. KRAUS, G., "Uncooled Rocket Nozzles for Ultra-High Temperature Propellants." SAE 595 J, presented at SAE National Aerospace Engineering and Manufacturing Meeting, Los Angeles, Cal., October 8–12, 1962

23. COOPER, T. D., and O. O. SRP, "Refractory Metals and Their Protection." *Aerospace Engineering*, **22,** 1, January 1963, pages 46–55

24. MALOOF, S. R., "Tungsten-Base Composites." *Astronautics*, **6,** 4, April 1961, page 36

25. PRIAPI, J. J., "Advanced Ignition System for Solid-Propellant Rocket Motors." *J. Amer. Rocket Soc.*, **31,** 7, July 1961, page 1029

26. STAMBLER, I., "Secondary Injection: Leading Technique for Thrust Vector Control." *Space/Aeronautics*, **38,** 3, August 1962, pages 58–61

Problems

1. Consider the pumps necessary to supply a booster engine of 1.5 million pounds of thrust. The engine consumes liquid oxygen and gasoline in the ratio 2.5:1. The flame temperature is 5470°F, the chamber pressure 300 psia, and the specific impulse 242 sec. Densities of liquid oxygen and gasoline are 71.0 and 46.8 lb_m/ft^3, respectively. What total shaft power is necessary, assuming a pump efficiency of 0.65 and propellant tank pressure of 50 psia?

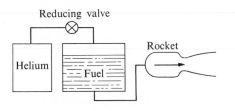

PROBLEM 2

2. A 100-ft³ fuel tank shown in the figure is pressurized by helium at 1000 psia. The helium enters the fuel tank from a storage tank through a pressure reducer which keeps the pressure in the fuel tank constant. If the pressure in the helium tank is 4000 psia when the fuel tank is full, determine the volume of the helium tank necessary to pump all the fuel into the rocket. Is the initial helium temperature of any consequence? Would the use of air instead of helium alter the performance?

3. For the LOX pump whose performance is shown in Fig. 14–18, what will the speed and power consumption be if the pump is to produce an outlet pressure of 1050 psia at a flow rate of 1100 gal/min, pumping LOX from a tank at 50 psia? If the data of Fig. 14–20 were for water at 60°F, what would the suction specific speed of this pump be, assuming the critical NPSH is that for which the head rise drops 2%? How constant is this for the range of flow rates given?

4. Show that, for constant axial velocity and axial outlet velocity, the work output per unit mass of fluid flowing through a velocity compound impulse turbine is $8U^2$, as indicated in Fig. 14–22.

5. A liquid oxygen ($\rho = 71$ lb_m/ft^3) pump is to run at 10,000 rev/min and pump 175 lb_m/sec against a pressure rise $p_0 = 800$ psi. From tests on geometrically similar pumps, its dimensionless suction specific speed is known:

$$S = \frac{\Omega Q^{1/2}}{[(p_{01} - p_v)/\rho]^{3/4}} = 10.9.$$

If the LOX is just saturated at atmospheric pressure (i.e., if its vapor pressure is 14.7 psia), what is the minimum allowable pressure (psia) in the LOX tank? Neglect pressure losses in lines from tank to pump and gravitational (hydrostatic) head of tank.

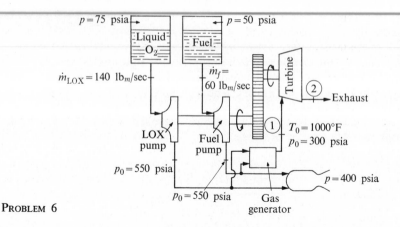

PROBLEM 6

6. Sketched in the figure is a turbopump feed system pumping a total propellant mass flow of 200 lb_m/sec. If the component efficiencies are as below, what fraction of the total propellant flow is diverted through the gas generator and turbine?

Fuel pump efficiency	$\eta_{Pf} = 0.70$
LOX pump efficiency	$\eta_{PL} = 0.65$
Turbine efficiency	$\eta_{tts} = 0.70$
Turbine discharge pressure	$p_2 = 20$ psia
Mechanical efficiency	$\eta_m = 0.90$

Fluid properties are:

$$c_p = 200 \text{ ft·lb}_f/\text{lb}_m\text{·°R} = \text{const for turbine fluid}$$
$$\gamma = 1.3 = \text{const for turbine fluid}$$
$$\rho_{LOX} = 71 \text{ lb}_m/\text{ft}^3$$
$$\rho_{fuel} = 50 \text{ lb}_m/\text{ft}^3$$

7. An experimental rocket thrust chamber has an outside wall temperature of 300°F at the throat, with a 6000°F chamber temperature. The local heat transfer rate is measured to be 4.3×10^6 Btu/hr·ft², of which 25% may be assumed due to radiation. If the wall is of stainless steel 0.1 in. thick with $k = 15$ Btu/hr·ft²·°F/ft, and if the coolant surface area is the same as the hot gas surface area, (a) what is the inner wall temperature? (b) It is expected that this temperature will cause failure in the actual application. The throat is to be lined with a ceramic of conductivity $k = 5$ Btu/hr·ft²·°F/ft to protect the metal. Assuming that the fraction of heat transfer by radiation is unchanged and that all gas properties are unchanged, what ceramic thickness is necessary to reduce the peak metal temperature to 2000°F while the coolant side remains at 300°F?

8. The thrust of a regeneratively cooled rocket engine is varied by controlling the propellant flow rate, which in turn controls the combustion pressure. Assuming that the combustion temperature is constant, and considering the rise in temperature of both wall and coolant, how would the relative difficulty of cooling the thrust chamber depend on the thrust level?

9. A liquid-propellant combustion chamber is 3 ft long and 1 ft in diameter. The temperature and pressure in the chamber are uniform at approximately 6000°R and 1000 psia,

and the diameter Reynolds number of the flow is of the order of 10^7. The chamber wall is type 301 stainless steel 0.100 in. thick, and is maintained at 200°R on the outside of the inner surface if radiation is one-third of the total heat flux.

Conductivity of stainless steel   15 Btu/hr·ft·°F,
Conductivity of chamber gases   0.1 Btu/hr·ft·°F.

Will the chamber wall be able to stand this temperature?

10. An early rocket design using a propellant at 6000°R is limited by wall cooling problems to a chamber pressure of 300 psia. Under these conditions a hot-side wall temperature, T_{wh}, of 800°R is observed at the throat. A new material will allow an increase of T_{wh} to 1000°R. What new p_0 will be allowed by this new material, if it is assumed that its resistance to heat transfer ($\Delta L/k_w$) is about the same as the old? Data from the original rocket (at the throat):

T_L = 500°R (coolant temperature)
T_{0g} = 6000°R (propellant temperature)
ΔL = 0.015 in. (wall thickness)
k_w = 2.8 × 10^{-4} Btu/sec·in^2·°F/in. (wall conductivity)
h_L = 0.10 Btu/sec·in^2·°F (coolant side film coefficient)
$\dot{Q}$ = 4.68 Btu/sec·in^2 (measured heat transfer rate at throat)

Comment on the seriousness of assuming that $\Delta L/k_w$ is about constant. Why would it not be constant?

11. An experimental rocket shown in the figure has a combustion chamber temperature of 6000°F, and the nozzle throat diameter is one foot. On the basis of reasonably accurate data on heat transfer in tubes, conductivity of the tube material, and the geometry of the coolant tube system, we can calculate that, at the throat,

$$\frac{A_g t}{A_w k_w} + \frac{A_g}{A_L h_L} = 5 \times 10^{-5} \frac{\text{hr·ft}^2·°\text{F}}{\text{Btu}},$$

where

t = wall thickness,
k_w = wall conductivity,
h_L = liquid or coolant side heat transfer coefficient,
A_g, A_w, A_L = area normal to heat transfer on gas side, in the wall, and on the coolant side.

In one experimental test, with water-cooled walls and water temperature T_L = 100°F, the total heat transfer rate, based on the gas side area, was measured as

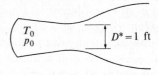

$$\frac{\dot{Q}}{A_g} = 1.7 \times 10^7 \frac{\text{Btu}}{\text{hr·ft}^2}.$$

PROBLEM 11

If radiant heat transfer is neglected, what will the heat transfer rate per unit area be if p_0 is doubled?

[*Note:* In both cases assume the recovery factor, r, to be one. That is, $T_{wa} = T_{0g}$. The viscosity and thermal conductivity of the gas may be assumed functions of temperature only.]

15

Nuclear Rockets

15-1 INTRODUCTION

In the preceding chapters we have seen that the specific impulse obtainable from a chemical rocket is limited to something in the neighborhood of 400 seconds, or a little more. We showed in Chapter 10 that much higher specific impulses would be desirable and that, in fact, substantial performance gains could be realized if the specific impulse could be increased by a factor of two. The nuclear rocket has been proposed in an attempt to gain such an increase in specific impulse by using nuclear reaction to replace chemical reaction as the energy source. Thermal energy is generated in a nuclear reactor and added to the propellant in a heat-transfer process, whereupon the propellant is expanded in a conventional exhaust nozzle.

The limitations on chemical rocket propellants, viewed from an atomic standpoint appropriate to this chapter, are twofold. First, the energy released is relatively low. The chemical identity and activity of an atom are determined by its electron structure, and energy released in chemical reactions can be related to changes in the energies of these electrons. Since the binding energy of an electron is of the order of several electron volts (see Chapter 5), the order of magnitude of the energy change or release accompanying chemical reaction must be a few electron volts per reacting pair of atoms. Second, since the chemical rocket requires that the propellant act as its own energy source, propellant choice is limited to those elements or compounds which undergo suitable reactions.

The energy released in a nuclear reaction, compared with that released in a chemical rocket, is very large. As we shall see, the binding energy of a nucleon (i.e., a proton or neutron) is of the order of several *million* electron volts; hence very much more energy is released per reaction. Further, since the propellant need not provide its own energy, a wider choice of propellant is available.

As we pointed out in Chapter 11, for high specific impulse the propellant should have low molecular weight and high stagnation temperature. From Eq. (11-3) we have

$$u_e \propto \sqrt{\frac{T_0}{M}}.$$ (15-1)

Thus hydrogen, with the lowest molecular weight, would be a good propellant, and would provide the highest exhaust velocity for a given stagnation temperature. It might be thought that in nuclear rockets the abundant energy available would mean that indefinitely high stagnation temperatures T_0 could be employed. This is definitely not so, however, since (at least in the solid-core reactor, which is the only one experimentally tested to date) heat is transferred from a solid reactor to the propellant. Thus the structural components within a nuclear rocket, unlike those in chemical rockets, *must be hotter than the propellant*, and T_0 cannot exceed the limiting temperature of the structure or reactor material. It is presently hoped that reactors may be run at temperatures as high as 5500°R for short durations. This is considerably below the temperatures attained in some chemical rockets, of course, but the use of hydrogen as the propellant more than offsets this temperature disadvantage of nuclear rockets. Thus, as far as specific impulse is concerned, the increased performance of a nuclear rocket is entirely due to the use of a propellant with low molecular weight. The high nuclear energy release is naturally still of importance, since it reduces the necessary fuel mass to almost negligible proportions. However, the reactor greatly increases the mass of the thrust chamber.

Methods have been proposed for transferring nuclear energy directly from the nuclear reaction products to the propellant without necessarily raising the temperature of the surrounding solid materials. Such concepts, which are still of uncertain feasibility, are discussed very briefly at the end of the chapter.

In the following section we shall assume that the reader is familiar with elementary concepts of atomic configuration: that atoms consist of a nucleus of protons and neutrons surrounded by a number of orbiting electrons, that particular chemical species often consist of a number of isotopes of different atomic mass, and that certain elements are naturally radioactive. It would not be possible to give here a comprehensive review of the subject of nuclear reactions.* Nevertheless nuclear reactions will be discussed to some extent in Section 15–3, in relation to reactors which could be used for nuclear propulsion.

15–2 NUCLEAR REACTIONS AS ENERGY SOURCES

A chemical reaction entails a rearrangement of the electron structures of atoms. In contrast, a nuclear reaction entails a rearrangement of the nucleus of an atom, sometimes bringing about a change of chemical identity and sometimes creating more than one product atom. In both chemical and nuclear reactions the energy released may be related to the change in mass of the material undergoing reaction. From the special theory of relativity, the energy release ΔE accompanying a decrease of mass Δm is

$$\Delta E = c^2 \, \Delta m. \qquad (15\text{--}2)$$

If Δm is in kilograms and the velocity of light c is in meters per second ($c = 2.998 \times 10^8$ m/sec), then ΔE has the units of newton-meters or joules. For

---

* See, for example, References [1] and [4].

typical chemical reactions this mass change is too small to be observed, but for very much more energetic nuclear reactions it is measurable. Furthermore, for our purposes, this mass–energy interchange provides a convenient basis for the discussion of nuclear energy sources.

It is observed that, except for the most abundant isotope of oxygen, which *by definition* has a mass of 16 amu (atomic mass units, equal to 1.66×10^{-24} gm), atomic masses are *not* whole numbers. This implies that the neutrons, protons, and electrons which add up to a mass of 16 amu in this particular configuration do not have the same masses in other configurations. In particular the mass of a free proton is about 1.00759 amu, that of a free neutron 1.00898 amu, and that of an electron, 0.000549 amu [1]. Thus for the case of oxygen, consisting of eight protons, eight neutrons, and eight electrons, the total of the free particle masses is 16.13695 amu and there is a *mass defect* of 0.13695 amu when these particles exist together as an oxygen atom. This mass defect is proportional, according to Eq. (15–2), to the amount of energy which would be *released* if an oxygen atom were formed by the assembly of individual particles. Conversely, this same amount of energy would be required to completely disassemble an oxygen atom. According to this latter interpretation, the energy increment associated with the mass defect is called *binding energy*. Since one amu equals 1.66×10^{-24} gm, and since we are given the definition of an electron volt (Chapter 5), Eq. (15–2) indicates that a loss of one amu corresponds to the release of 931×10^6 eV (931 Mev) of energy. Thus in the case of oxygen the binding energy is about 127.5 Mev.

As another example, consider that isotope of uranium which has an atomic number (number of protons) of 92, and a mass number (number of protons plus neutrons) of 235. This isotope, which is designated $_{92}U^{235}$ in common notation, is of great interest as a fuel for nuclear reactors. The mass of 92 free protons and electrons and 143 free neutrons (that is, $235 - 92$) is 237.03293 amu, while the mass of the isotope is about 235.124 amu [1]. This leaves a mass defect of 1.909 amu or a binding energy of about 1777 Mev.

It is of greater interest to examine and compare the average binding energy *per nucleon* (proton or neutron) in the various atoms. Thus, for $_8O^{16}$, it is $127.5/16 = 7.97$ Mev, while for $_{92}U^{235}$ it is only $1777/235 = 7.56$ Mev. A similar treatment of the other elements leads to the curve which we see in Fig. 15–1, which shows binding energy per nucleon for the various nuclei. This variation of binding energy per nucleon with mass number accounts for the enormous energy release accompanying nuclear fission. In the fission of $_{92}U^{235}$, for example, two product nuclei are formed, having mass numbers in a range near the peak of the binding energy curve. Thus the product nucleons have an average binding energy, according to the values from Fig. 15–1, of about $(8.4 - 7.5)$ or 0.9 Mev per nucleon greater than those in the parent nucleus. Since binding energy is that energy which corresponds to the formation of a nucleus from its constituent nucleons, it is clear that the net energy release accompanying this reaction is equal to the number of nucleons involved times the *change* in their average binding energy. Assuming that fission occurs as the result of the capture of a neutron by the $_{92}U^{235}$, there are 236 nucleons involved and the energy release is 236×0.9 or about 200 Mev! This is an enormous energy release compared with the several eV of a chemical reaction. In passing, it is interesting to note the tremendous potential of a *fusion* reaction, in which helium ($_2\text{He}^4$) is formed from hydrogen ($_1\text{H}^1$). Figure 15–1 indicates the great difference between the binding energies of these elements.

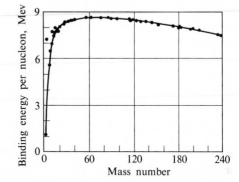

FIG. 15–1.   Binding energy per nucleon, as a function of mass number [1].

The controlled production of a desired nuclear reaction is quite different from that of a chemical reaction. Some nuclear reactions, responsible for so-called natural radioactivity, occur spontaneously in certain naturally present nuclides. These nuclides may eject particles such as α- or β-particles, and neutrinos, or they may emit γ-rays.

An α-particle is a helium nucleus, consisting of two protons and two neutrons. Thus the ejection of an α-particle causes in the parent nucleus a decrease in atomic number of two and a decrease in mass number of four. A β-particle is an electron (β^-) or positron (β^+) which is ejected from a nucleus. A β^- may be considered to originate from the disintegration of a neutron into an electron, a proton, and perhaps a neutrino. The net result is an increase of one in atomic number with no change in mass number. Similarly, a β^+ emission may be considered to originate from the disintegration of a proton into a positron, a neutron, and a neutrino. In this case the atomic number of the resultant nucleus is one less than that of the original. A neutrino is a particle of very little mass (when at rest) as compared even with an electron, and having no charge. Its existence was first proposed on theoretical grounds to account for the conservation of energy during β-decay, but it has since been experimentally detected. Gamma rays are a form of electromagnetic radiation originating in atomic nuclei. They permit the loss of energy from an "excited" nucleus without changing either its mass number or atomic number. For example, β-emission often leaves the nuclide with excess energy which is lost almost immediately through γ-emission.

Naturally occurring radioactive elements are all members of one of three "chains." Each chain begins with an element for which it is named ($_{92}\text{U}^{238}$, $_{92}\text{U}^{235}$, $_{90}\text{Th}^{232}$), and proceeds through a series of α- and β-emissions to a stable isotope of lead [1]. The energy released during the various disintegrations can be related to the mass changes of the material, as previously discussed. Indeed, the study of the particle and γ-ray energies is a source of information on nuclide masses and energy levels. Particle energies are of the order of several Mev. Hence radioactive decay would be a very good energy source relative to a chemical reaction, but it is not nearly so energetic as a fission process. Radioisotope energy sources have been used as energy supplies for small power needs in space. Artificially radioactive fuels are usually employed, but the decay process is identical to that for natural radioactivity. The word "artificial" refers to a fuel which is not a naturally occurring isotope, but rather one which is man-made as a result of other nuclear reactions.

Artificial nuclear disintegrations may be produced by bombarding nuclei with various projectile particles; α- and β-particles, protons, deuterons ($_1H^2$), and neutrons all serve as suitable projectiles. A successful "hit" may be considered to result in the capture of the bombarding particle by the target nuclei to form a "compound nucleus" which, usually, breaks down in a relatively short time. (For very high energy bombarding particles—greater than 50 Mev—this compound-nucleus concept is not valid.) To cite an example of the variety of possible reactions and to demonstrate the energy interactions, let us consider the bombardment of aluminum $_{13}Al^{27}$ with neutrons. The compound nucleus, upon capture of the neutron, would have an atomic number of 13 and a mass number of 28, and would be in an "excited" state. That is,

$$_{13}Al^{27} + _0n^1 \rightarrow [_{13}Al^{28}],$$

and the mass defect may be found from the particle masses. The target plus projectile mass is $26.99012 + 1.00898$ or 27.99910 amu, while the compound nucleus has a mass of 27.99081 amu [1]. This mass defect appears as part of the excitation energy of the compound nucleus. The total excitation energy includes, in addition, a fraction of the initial kinetic energy of the neutron. Some of the neutron energy goes into kinetic energy associated with translation of the compound nucleus, of course, since momentum must be conserved. Thus, assuming that the target nucleus is initially at rest, one can easily show that the fraction of projectile energy which is contributed to excitation energy is $1 - (m_x/m_{cn})$, where m_x is the projectile mass and m_{cn} is the compound nucleus mass. In this example, then, the excitation energy of the compound nucleus is

$$(27.9910 - 27.9908)\,931 + E_n(1 - \tfrac{1}{28}) = 7.72 + 0.964E_n \text{ Mev},$$

where E_n is the neutron energy.

The compound nucleus may lose this energy by a variety of processes, which depend on the magnitude of the total and on how it happens to be distributed within the nucleus. In this case, several possibilities [1] are:

$$_{13}Al^{27} + _0n^1 \rightarrow [_{13}Al^{28}] \rightarrow _{11}Na^{24} + _2He^4,$$
$$_{13}Al^{27} + _0n^1 \rightarrow [_{13}Al^{28}] \rightarrow _{12}Mg^{27} + _1H^1,$$
$$_{13}Al^{27} + _0n^1 \rightarrow [_{13}Al^{28}] \rightarrow _{13}Al^{26} + _0n^1 + _0n^1,$$
$$_{13}Al^{27} + _0n^1 \rightarrow [_{13}Al^{28}] \rightarrow _{13}Al^{28} + \gamma.$$

Examination of the various masses indicates that in the first three cases there is a net mass *gain* from reactants to products which, of course, requires a net energy *input*. This energy must come from the kinetic energy of the neutron, and these reactions are not possible without sufficiently energetic neutrons. Such a reaction would be called *endoergic*, but there are many other reactions with lighter target nuclei which are *exoergic* (i.e., they give off energy). The last reaction, in which the [$_{13}Al^{28}$] loses energy through γ-emission, results in a mass decrease, and therefore it may be initiated by slow neutrons. This process, called *radiative capture*, is very common. The resultant nuclide is often radioactive, and this or similar reactions may be used to form artificial radioisotopes. For example, $_{13}Al^{28}$ in its ground state is a β^- emitter with a half-life* of 2.3 minutes [1] according to

$$_{13}Al^{28} \rightarrow {}_{14}Si^{28} + {}_{-1}e^0 + \gamma,$$

where the electron, $_{-1}e^0$, is, of course, the β^- particle. As is often the case, the excited product nucleus emits a γ-ray almost immediately.

The chance for any of the above reactions to occur, say for a given neutron energy, may be expressed in terms of a reaction cross section. As we mentioned in Chapter 5, the cross section is merely a proportionality constant such that the number of reactions of a particular type is equal to the number of target nuclei times the neutron flux times the cross section. The cross section for a particular reaction is likely to be energy-dependent, and in fact it may show a regular pattern of maxima called *resonances*, but these will not be discussed further here.

Fission is a nuclear reaction which is of great interest as an energy source since, as indicated, the energy release per nuclide is very large. Fission is distinguished from the nuclear-disintegration processes described above by the fact that the parent nucleus breaks into two (or rarely, three) relatively large product nuclides, along with several neutrons, γ-rays, and perhaps an occasional light nuclide. Of the naturally occurring elements, uranium is the most important fissionable material. Natural uranium consists of two isotopes, $_{92}U^{238}$ and a small amount (0.72%) of $_{92}U^{235}$. Both isotopes are radioactive, emitting α-particles in decay processes having half-lives of 4.50×10^9 and 7.10×10^8 years, respectively. With such long half-lives, they would not ordinarily be considered strongly radioactive materials. However, upon bombardment by suitable neutrons (or α- and β-particles, deuterons, and γ-rays, but these reactions are not of practical importance), the fission reaction is observed. The term "suitable neutrons" refers

---

* The rate of disintegration dN/dt, where N is the number of nuclides present, is proportional to N. This leads to the well-known exponential decay: that is,

$$\frac{dN}{dt} = -\lambda N \quad \text{or} \quad \frac{N}{N_0} = e^{-\lambda t},$$

where N_0 is the number at $t = 0$ and λ is the disintegration constant. The half-life τ is that time for which

$$\frac{N}{N_0} = 0.5 \quad \text{or} \quad \tau = \frac{-\ln 0.5}{\lambda} = \frac{0.693}{\lambda}.$$

to the fact that $_{92}U^{238}$ requires fast $(E_n > 1 \text{ Mev})$ neutrons for fission, while $_{92}U^{235}$ undergoes fission with fast or slow neutrons. In fact, "thermal" neutrons (neutrons which have achieved thermal equilibrium with the surrounding material through repeated collisions), having energies of only a small fraction of one Mev, have a relatively good chance of causing fission of $_{92}U^{235}$. Because of this favorable fission behavior, and because thermal neutron reactors are relatively well understood, and also for other reasons such as availability and physical properties, $_{92}U^{235}$ is a very popular reactor fuel. In fact, the early experimental nuclear rockets have used graphite-moderated, $_{92}U^{235}$-fueled reactors [2]. For this reason, let us illustrate the production of energy by fission processes for the case of capture of a thermal neutron by this material.

Upon capture of a thermal neutron by $_{92}U^{235}$, the excited compound nucleus $[_{92}U^{236}]$ is formed. The excitation energy in this case would be just equal to the binding energy of the captured neutron, since its initial kinetic (thermal) energy is negligible. As in the previous example for $[_{13}Al^{28}]$, the compound nucleus may react in a variety of ways. The reaction of interest is, of course, fission. However, the excited nucleus may also decay to the ground state of $_{92}U^{236}$ by the emission of γ-rays. This is a radiative capture reaction. Finally, the nucleus may eject a neutron to return to $_{92}U^{235}$, in a process called *scattering* (since, in effect, a neutron is just momentarily stopped and perhaps deflected). The probability of each of these reactions can be expressed in terms of a reaction cross section. The cross sections of $_{92}U^{235}$ for thermal neutrons [1] are: fission 580 b, radiative capture, 107 b, scattering 9.0 b. The symbol "b" stands for barns, a unit of area equal to 10^{-24} cm^2. For our purposes only the relative cross sections are of importance, and it can be seen that, of the collisions resulting in neutron loss, about 80% $[580/(580 + 107 + 9)]$ result in fission.

The products or "fission fragments" include a wide range of nuclides. However, 97% of the fissions are such that the two fission fragments fall into either a "light" group having mass numbers near 95 or a "heavy" group with mass numbers near 139 [1]. Most fission fragments contain too many neutrons, since the heavy nuclides such as uranium generally have a much higher neutron–proton ratio than the stable lighter nuclides do. Thus the fission fragments are highly radioactive. Most emit β-particles, in effect replacing a neutron with a proton, while some emit neutrons directly. This β^- emission may, of course, be accompanied or followed by neutrino and γ-emission, and the total fission energy release includes that released in these various product-decay processes.

A very important feature of the fission process is the emission of neutrons. Most (more than 99% [1]) are emitted during fission or very shortly thereafter, while some originate in the decay of the fission fragments. These neutrons, if captured by additional fissionable nuclides, can cause further fissions. Thus, under the proper circumstances, a self-sustaining *chain reaction* can occur in which the fissioning material releases enough neutrons to cause continued fission of remaining material at the same rate. This of course requires the emission of more than one neutron per fission since, as we have mentioned, only a fraction

Table 15–1   Energy Release in the Fission of U^{235} by Thermal Neutrons[1]

Kinetic energy of fission fragments	167 Mev
Kinetic energy of fission neutrons	5
Prompt γ-rays	5
β-decay energy	5
γ-decay energy	5
Neutrino energy	11
Total fission energy	198 Mev

(about 80% for $_{92}U^{235}$) of the neutrons captured by fissionable material result in fission. Furthermore, not all neutrons emitted will be captured by fissionable material. On the average, fission of $_{92}U^{235}$ is accompanied by the release of about 2.5 neutrons [1].

Most of the fission energy appears in the form of kinetic energy of the fission fragments. Such fragments, being relatively massive and usually very highly ionized, lose their kinetic energy very rapidly through collisions with surrounding material. In effect, then, a solid-core reactor, such as uranium in graphite, may be considered a "heat source." Additional energy is carried by neutrons, β-particles, neutrinos, and γ-rays, all of which may appear during fission, or later, during fragment decay. A typical distribution of energy [1] would appear as in Table 15–1, in which the term "prompt" denotes immediate release, in contrast to gradual decay. The γ-rays and neutrinos, both being rather penetrating, may in fact carry their share of the fission energy outside of the region of interest.

15–3 NUCLEAR REACTORS

A nuclear reactor is an assembly of a suitable fissionable material in a structure which contains other materials necessary to produce the desired neutron properties, to provide adequate control, and, for steady operation, to provide cooling. The attainment of the desired chain reaction depends on the proper utilization of the neutrons present. In general, a fission neutron may suffer one of three fates:

(1) It may be captured in a process causing fission resulting in the emission of additional neutrons.

(2) It may be captured in a nonfission process (such as radiative capture) either by the fissionable material or by other materials.

(3) It may escape the nuclear reactor.

For steady operation those excess neutrons generated by (1) must exactly balance those lost by processes (2) and (3). The first two processes depend on the nuclear properties of the fissionable and other material within the reactor and on how this material is distributed. The third process is dependent on the geometry of the reactor. For a given shape and material distribution pattern, it is dependent on the *size* of the reactor: larger reactors have lower surface-to-volume ratios and hence a lower tendency for escape of neutrons. Thus for a given shape, fuel, etc., there is a *critical size* below which the neutron loss rate will not permit a steady chain reaction. The critical size does, of course, depend strongly on the constituents and design of the reactor.

In order to have a relatively high probability of fission it is desirable to utilize a fuel having a large fission cross section. For low-energy neutrons the fission cross section of $_{92}U^{235}$ increases as the neutron velocity decreases. Hence it is desirable to use thermal neutrons. (Fast-neutron reactors have been built, and although their core at least can be lighter than that of a thermal neutron reactor, they have not as yet been applied to nuclear rockets.) Prompt neutrons from the thermal fission of $_{92}U^{235}$ have energies ranging from less than 0.05 Mev to greater than 17 Mev, with an average value of about 2.0 Mev [1]. These neutrons are slowed to thermal velocities (about 0.025 eV at room temperature [1]), through multiple collisions (elastic or inelastic collisions, usually elastic collisions with light nuclei) with an appropriate material called a *moderator*. Considerations of elementary mechanics show that for high energy loss per collision via elastic collisions, the moderator nuclei should have a low mass number. In order to avoid neutron loss by nonfission capture, it is also necessary that the moderator have a low neutron-capture cross section. Of the light nuclei, carbon (graphite) and beryllium are particularly good moderator materials, with beryllium being slightly better in the sense that it is required in a thickness only about half the required thickness of carbon [3]. On the other hand, the high-temperature characteristics of carbon are also attractive. Hydrogen is a reasonably good moderator, although deuterium ($_2H^2$ or D), having a much lower capture cross section, is better. Since chemical state has no apparent influence on nuclear properties, any of these materials may be used as compounds; for example, BeO, H_2O, D_2O ("heavy water"), and some of the hydrocarbon materials.

To avoid excessive neutron loss by nonfission capture, one must consider the nuclear properties of all materials used in or near the reactor core, and choose those materials having small capture cross sections if possible. This additional consideration, over and above ordinary strength and temperature requirements, serves to intensify design difficulties.

Neutron escape is a function of reactor geometry: those portions nearer the edges show a greater loss rate than those near the center. The falloff in thermal neutron density near the edges may be counteracted by a nonuniform fuel loading in such a way that high fission rates are maintained near the edges. Alternatively, the reactor core may be surrounded by a "reflector," which tends to stop leakage. This solution is more economical of fuel, of course. A reflector is really just a thermalizing material which slows escaping neutrons. A region of high neutron density is built up, from which some neutrons diffuse back to the reactor, while some continue to diffuse on outward and are eventually lost. There is no reflection of neutrons in the usual sense of the word. Clearly, most good moderator materials would also be good reflector materials. Hydrogen is an exception since, being very light, it exhibits "forward scattering" of neutrons. That is, the neutrons,

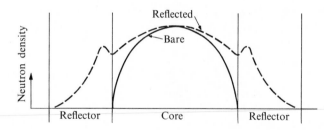

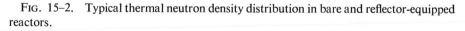

FIG. 15–2. Typical thermal neutron density distribution in bare and reflector-equipped reactors.

being of about the same mass as hydrogen, maintain after collision a marked preference for travel in their original direction.

The calculation of thermal neutron density is a rather complicated task, dependent on the particular reactor geometry and the distribution of fissionable material within the core. Figure 15–2 shows a typical result for the radial variation of thermal neutron density in bare and reflector-equipped cylindrical reactors [4]. With a reflector, the neutron density in the core is much more uniform. (The boundaries shown are the so-called "extrapolated boundaries," that is, the positions at which the neutron density appears to go to zero. The physical boundary of the core would be of the order of 2 cm within the extrapolated boundary, and the neutron density is not zero there [4].) The neutron density peak within the reflector is of interest. Thermal neutrons are formed, in both the core and the reflector, from the slowing of fast neutrons. The fast-neutron flux just within the reflector is about the same as that within the reactor. However, the neutron absorption or capture rate within the reflector should be less than that within the core; hence the same fast-neutron flux will create a greater thermal-neutron density.

A reactor may be controlled by regulating either of the neutron loss processes (2) or (3) described on page 475. Varying amounts of a material which is especially neutron-absorbent may be placed within the reactor to control it by the nonfission neutron-capture process. Movable "control rods" containing boron are an example of this technique. Alternatively, geometric variations which affect neutron escape may be employed. Movable reflectors are an example of this technique.

To provide an adequate response rate to changes in power demand, a reactor should have some excess reactivity. That is, it should be somewhat above critical size for the particular geometry and fuel distribution employed. During steady-state operation this excess reactivity is offset by the necessary neutron loss to the control system. In addition, since the amount of fissionable material gradually decreases with time, the reactor must initially have excess reactivity if it is to have a long life. Besides burnup of fuel, the reactor undergoes internal changes or "poisoning" effects which gradually increase the rate of nonfission neutron capture. For example, some of the fission products may have large neutron-capture cross sections. This also indicates that a reactor should have excess reactivity if it is to have long life. Actually, the life and total burnup of a rocket reactor are both

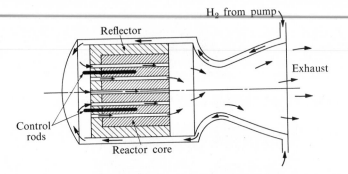

FIG. 15–3. Schematic of solid-core nuclear-heated hydrogen rocket.

rather small compared to earthbound power-generation reactors, so these effects are relatively unimportant. On the other hand, a rocket reactor would be called on to provide a very rapid power buildup.

The engineering problems connected with the construction of nuclear reactors for rocket vehicles are indeed formidable. Figure 15–3 is a schematic diagram of a possible solid-core nuclear rocket. The propellant, most likely liquid hydrogen, is fed through the nozzle in a conventional regenerative-cooling arrangement, thence through perforations in the reactor. The reactor is a cylindrical structure having about a third of its cross section occupied by heat-transfer channels. It is surrounded by reflector material on the sides, and perhaps on one or both ends. Control rods are shown schematically.

As mentioned before, the reactor should raise the propellant temperature to a very high point. Also the reactor size and mass should be as small as possible, so that high power density (heat release rate per unit volume) is desirable. Therefore the core material will be subject not only to high temperatures but also to high temperature gradients, so that the heat may flow from within the core material to the coolant channel surface. As we shall see in Section 15–4, the core must also support large axial-pressure forces stemming from the substantial pressure drop through the coolant passage.

The high-temperature strength of graphite, along with its desirable moderator characteristics, makes this an attractive core material. The experimental nuclear rocket reactors to date have used $_{92}U^{235}$-fueled, graphite-moderated reactors, as in Fig. 15–4. Unfortunately, hot hydrogen has an embrittling effect on carbon, so that it may be necessary to cover the coolant channel walls with a protective material for applications involving high temperature or long duration. This is, of course, undesirable from the heat-transfer standpoint, since a protective coating necessitates even greater temperatures within the core. Hydrogen, being a good moderator, also has the effect—when introduced at high density as in the liquid state—of substantially altering the reactivity of the reactor. This and the severe thermal shock encountered during a rapid power buildup pose difficult transient behavior problems. Rapid shutdown also presents severe problems, since the

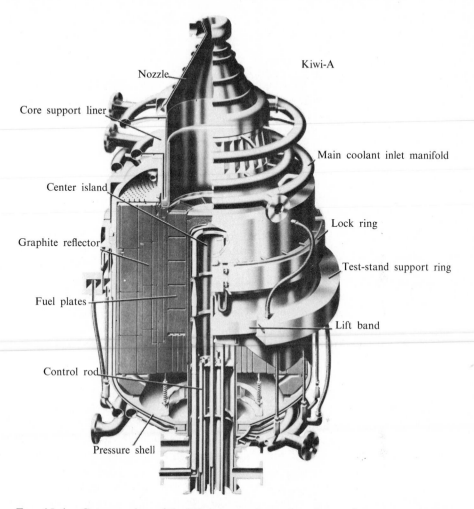

FIG. 15–4. Cutaway view of the Kiwi-A experimental nuclear-rocket reactor. (Courtesy Los Alamos Scientific Laboratory, University of California.)

highly radioactive fission products will continue to release energy after the chain fission reaction has ceased. Propellant used to provide aftercooling of the reactor is, of course, largely wasted so far as the overall performance of the vehicle is concerned.

The Kiwi-A experimental reactor [2, 5, 6] shown in Fig. 15–4 was designed to explore the problems of high-temperature hydrogen-cooled reactors with high power density. It was not intended to be the prototype for flying rockets. The fuel, in the form of uranium oxide, was loaded in many graphite plates, each about $\frac{1}{4}$ inch thick, and spaced so as to provide 0.050-inch-thick passages for coolant (H_2) flow. Each plate was uniformly loaded with fuel, although the loading of the various plates was varied according to the desired heat transfer and fission distribution. An internal island contained circulating heavy water coolant-moderator, along with control rods containing cadmium. The core was surrounded on the sides and one end by a graphite reflector. The propellant entered as gaseous hydrogen at the top of the reflector, flowed through the reflector to a plenum at the base, thence through the core and out the nozzle. In this case the nozzle was water-cooled [2, 5]. This reactor and several subsequent experimental reactors have successfully demonstrated the feasibility of solid-core hydrogen-cooled nuclear rockets. There remain, however, substantial problems before such rockets can be used in flight.

15–4   THE SOLID-CORE NUCLEAR ROCKET

Heat transfer to the rocket propellant

In the propellant heat-transfer channels shown in Fig. 15–3, two effects are of great importance. First, the heat transfer increases the fluid stagnation temperature and Mach number and decreases the stagnation pressure. This effect was analyzed in Chapter 3, with the results given by Eqs. (3–20) to (3–23), and presented graphically in Fig. 3–4 for the case $\gamma = 1.4$.

The second effect is the result of wall friction, which tends to lower the stagnation pressure and increase the local Mach number. This effect was also analyzed in Chapter 3, with the results given by Eqs. (3–25) to (3–28), and presented in Fig. 3–5 for $\gamma = 1.4$.

In the actual heat-transfer problem, both these effects are important. In fact it was shown in Chapter 4, Eq. (4–33), that the local rate of heat transfer from wall to fluid is directly proportional to the local wall shear stress. The reason for this is that the mechanism for transporting both thermal energy and momentum through the (turbulent) boundary layer is the same; it is the diffusion associated with turbulent fluctuations. Since heat transfer and friction are so intimately linked, neither of the solutions discussed in Chapter 3 (i.e., frictionless heat transfer and adiabatic friction) are applicable here, and it is necessary to consider both effects simultaneously. Considering the heat-transfer channel as a whole, one sees that the effects are not simply additive and that it is necessary to examine their interaction. First it is assumed that the axial power density is uniform, as might be the case (approximately) for a reactor having reflectors at both ends.

(1) *Uniform total axial power density*

If we neglect heat conduction axially along the reactor, the case of uniform total axial power density means constant local heat flux along the heat-transfer channels. Equation (4–31) provides a means for estimating the heat transfer coefficient h, so long as this flow is turbulent and fully developed. Since the Reynolds numbers will be high, the flow will certainly be turbulent. Also, since the channels will have large ratios of length to hydraulic diameter, the flow may be considered fully developed over a large part of its length. In view of these considerations, it may be quite acceptable to assume that the heat-transfer coefficient h_f is constant along the length of tube. This approximation allows the main features of the problem to be dealt with quite easily. Equation (4–33) shows that constant heat transfer coefficient h is equivalent to constant skin-friction coefficient C_f. The local heat-transfer rate per unit area is

$$\dot{\mathcal{Q}} = h_f \Delta T, \tag{15-3}$$

where ΔT is the difference between wall temperature and fluid stagnation temperature. If $\dot{\mathcal{Q}}$ and h_f are uniform along the tube, then ΔT will be also. Considering the control volume indicated in Fig. 15–5 and applying the energy equation for steady one-dimensional flow, Eq. (3–3), we obtain

$$c_p \, dT_0 = dh + u \, du = dq,$$

in which q is the heat transfer per unit mass of fluid. This equation may also be written as

$$c_p \, dT_0 = \frac{\dot{\mathcal{Q}} \, dA_s}{\dot{m}},$$

in which $\dot{\mathcal{Q}}$ is the heat transfer per unit area of wall and A_s is the wall surface area. Applying the definition of hydraulic diameter D_H, $dA_s/dx = 4A/D_H$, where A is the cross-sectional area of the passage. After substituting Eq. (15–3), we can write the energy equation as

$$\frac{dT_0}{dx} = \frac{h_f A \, \Delta T}{\dot{m} c_p} \frac{4}{D_H}. \tag{15-4}$$

For the case of uniform heat flux, the increase of stagnation temperature of the fluid is

$$T_{02} - T_{01} = \frac{h_f A \, \Delta T}{\dot{m} c_p} \left(\frac{4L}{D_H} \right),$$

in which L is the length of the channel and subscripts 1 and 2 denote its entrance and exit, respectively.

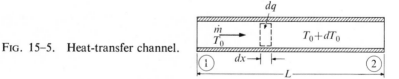

FIG. 15–5.   Heat-transfer channel.

The maximum wall surface temperature T_{sm} occurs at ② and is given by

$$T_{sm} - T_{01} = \Delta T \left[1 + \frac{h_f A}{\dot{m} c_p} \left(\frac{4L}{D_H} \right) \right].$$

Thus

$$\frac{T_{sm} - T_{01}}{T_{02} - T_{01}} = 1 + \frac{\dot{m} c_p}{h_f A} \left(\frac{D_H}{4L} \right).$$

Using the heat transfer–skin friction analogy, Eq. (4–33), we have

$$\frac{h_f}{\rho c_p \overline{U}} \mathrm{Pr}^{0.67} = \frac{C_f}{2},$$ (4–33)

with $\mathrm{Pr} \approx 1$. Then, noting that $\dot{m} = \rho \overline{U} A$, we obtain

$$\frac{T_{\mathrm{sm}} - T_{01}}{T_{02} - T_{01}} = 1 + \frac{D_H}{2 C_f L}.$$ (15–5)

The pressure drop in the heat-transfer channel may be found from an approximate momentum balance. Applying Eq. (3–2), we have

$$- A\, dp - \tau_0\, dA_s = \dot{m}\, du.$$

This equation may be approximately integrated in the form

$$(p_1 - p_2)A - \bar{\tau}_0 A_s = \rho_2 u_2^2 A - \rho_1 u_1^2 A,$$ (15–6)

in which $\bar{\tau}_0$ is the average wall shear stress and τ_0 is related to the skin friction coefficient C_f by

$$\tau_0 = C_f(\tfrac{1}{2}\rho u^2).$$

For an assumed uniform C_f the following approximation [7] may be adopted for simplicity:

$$\bar{\tau}_0 = C_f \frac{1}{2}\left(\frac{\rho_1 u_1^2}{2} + \frac{\rho_2 u_2^2}{2}\right).$$ (15–7)

Substituting Eq. (15–7) into (15–6) and using the definition of hydraulic diameter, we have

$$p_1 - p_2 = \rho_2 u_2^2\left(1 + \frac{C_f L}{D_H}\right) - \rho_1 u_1^2\left(1 - \frac{C_f L}{D_H}\right).$$

Then, using the definition of Mach number, $M^2 = \rho u^2 / \gamma p$, and ignoring variations in γ along the channel, we obtain

$$\frac{p_2}{p_1} = \frac{1 + \gamma M_1^2[1 - (C_f L / D_H)]}{1 + \gamma M_2^2[1 + (C_f L / D_H)]}.$$ (15–8)

As the hydrogen enters the heat-transfer channel, it may still be in the liquid state. If so it will vaporize very quickly, but the Mach number of the gas phase near the entrance will be very much less than unity. Thus Eq. (15–8) is approximately

$$\frac{p_2}{p_1} \simeq \frac{1}{1 + \gamma M_2^2[1 + (C_f L / D_H)]}.$$ (15–9)

Using Eqs. (3–11) and (15–8), we obtain the stagnation pressure ratio

$$\frac{p_{02}}{p_{01}} = \frac{1 + \gamma M_1^2[1 - (C_f L/D_H)]}{1 + \gamma M_2^2[1 + (C_f L/D_H)]} \left\{\frac{1 + [(\gamma - 1)/2]M_2^2}{1 + [(\gamma - 1)/2]M_1^2}\right\}^{\gamma/(\gamma-1)} \tag{15-10}$$

or, since $M_1 \ll 1$,

$$\frac{p_{02}}{p_{01}} \approx \frac{\{1 + [(\gamma - 1)/2]M_2^2\}^{\gamma/(\gamma-1)}}{1 + \gamma M_2^2[1 + (C_f L/D_H)]}. \tag{15-11}$$

It now remains to find the exit Mach number M_2. Using the continuity equation, Eq. (3–1), we have $\rho_2 u_2 = \rho_1 u_1$. For a perfect gas, $p = \rho R T$, and therefore,

$$\frac{p_2}{p_1} = \frac{u_1}{u_2}\frac{T_2}{T_1}.$$

Using the definitions of Mach number and stagnation temperature for a perfect gas, we can write this statement as

$$\frac{p_2}{p_1} = \frac{M_1}{M_2} \sqrt{\frac{T_{02}}{T_{01}}} \left\{\frac{1 + [(\gamma - 1)/2]M_1^2}{1 + [(\gamma - 1)/2]M_2^2}\right\}. \tag{15-12}$$

Combining Eqs. (15–8) and (15–12), we obtain the result

$$\frac{T_{02}}{T_{01}} = \frac{M_2^2\{1 + [(\gamma - 1)/2]M_2^2\}\{1 + \gamma M_1^2[1 - (C_f L/D_H)]\}^2}{M_1^2\{1 + [(\gamma - 1)/2]M_1^2\}\{1 + \gamma M_2^2[1 + (C_f L/D_H)]\}^2}, \tag{15-13}$$

which gives M_2 in terms of the temperature ratio and M_1.

The maximum fluid temperature T_{02} is limited by Eq. (15–5) for any given value of the maximum wall temperature T_{sm}. Therefore Eqs. (15–5) and (15–13) fix the exit Mach number of the fluid for given inlet conditions and average friction factor C_f. Equation (15–10) or (15–11) fixes the stagnation pressure loss in the heat-transfer channel for any value of the exit Mach number.

(2) *Cosine axial power density* [7]

For homogeneous reactors having bare ends, the axial power distribution may be approximated by a cosine function [4]. For this case, and again neglecting axial conduction, the wall heat transfer must also be a cosine function of the axial distance. Referring to Fig. 15–6, we have

$$\dot{\mathcal{Q}} = \dot{\mathcal{Q}}_{max} \cos\frac{\pi x}{L}.$$

If the heat-transfer coefficient h_f is again assumed constant, the difference between the wall temperature and the local fluid stagnation temperature must then also be a cosine function of the axial distance:

$$\Delta T = \Delta T_{max} \cos\frac{\pi x}{L}.$$

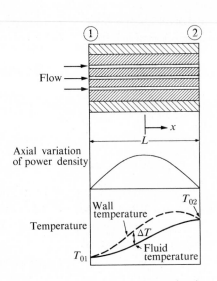

FIG. 15–6.   Cosine axial power density.

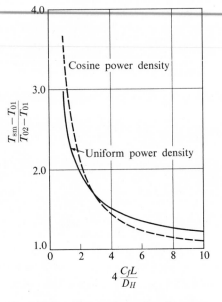

FIG. 15–7.   Comparison of wall temperatures for uniform and cosine power distributions.   (Courtesy Stenning [7].)

Equation (15–4) is still applicable, so therefore

$$\frac{dT_0}{dx} = \frac{hA}{\dot{m}c_p}\frac{4}{D_H}\Delta T_{\max}\cos\frac{\pi x}{L},$$

or, if we use Eq. (4–33) with unity Prandtl number,

$$\frac{dT_0}{dx} = \frac{2C_f}{D_H}\Delta T_{\max}\cos\frac{\pi x}{L}.$$

Integrating from $x = -L/2$ to x, we obtain

$$T_0 - T_{01} = \frac{2C_f L}{\pi D_H}\left(1 + \sin\frac{\pi x}{L}\right)\Delta T_{\max} \tag{15–14}$$

and

$$T_{02} - T_{01} = \frac{4}{\pi}\frac{C_f L}{D_H}\Delta T_{\max}. \tag{15–15}$$

The local wall temperature, $T_s = T_0 + \Delta T$, is then given by

$$T_s - T_{01} = \Delta T_{\max}\left[\cos\frac{\pi x}{L} + \frac{2C_f L}{\pi D_H}\left(1 + \sin\frac{\pi x}{L}\right)\right]. \tag{15–16}$$

Figure 15–6 shows qualitatively the variation in fluid and wall temperatures along the heat-transfer passage. The wall temperature has a maximum at a position (x_m) which may be obtained from maximizing T_s in Eq. (15–16). The result of this maximization is

$$\tan \frac{\pi x_m}{L} = \frac{2C_f L}{\pi D_H}. \tag{15–17}$$

Substituting Eq. (15–17) into Eq. (15–16) and using Eq. (15–15), we may write the result as

$$\frac{T_{sm} - T_{01}}{T_{02} - T_{01}} = \frac{1}{2}\left[1 + \sqrt{\left(\frac{\pi D_H}{2C_f L}\right)^2 + 1}\right]. \tag{15–18}$$

Equation (15–18) is compared with Eq. (15–5) for the uniform power density case in Fig. 15–7. For the cosine power density, Eqs. (15–10) and (15–12) relate the stagnation pressure ratio and the exit Mach number M_2 to the stagnation temperature ratio T_{02}/T_{01}. The stagnation temperature ratio is determined by Eq. (15–18) for any given value of the maximum wall surface temperature T_{sm} and the skin friction factor $C_f L/D_H$.

Rocket performance

As we showed in Chapter 11, the ideal specific impulse of a rocket is

$$I_{sp} = \frac{1}{g_e}\sqrt{\frac{2\gamma}{\gamma - 1}\frac{\overline{R}T_{02}}{\overline{M}}\left[1 - \left(\frac{p_e}{p_{02}}\right)^{(\gamma-1)/\gamma}\right]}, \tag{15–19}$$

in which (p_{02}, T_{02}) are the stagnation pressure and temperature of the propellant, $\overline{M}$ is the molecular weight, p_e is the nozzle exit pressure, and g_e is the gravitational acceleration at the earth's surface. If it is assumed that the loss in stagnation pressure at the exit of the reactor heat-transfer channels is negligible, this expression may be used to estimate the specific impulse of the nuclear-heated rocket. The temperature T_{02} is obtained from Eq. (15–5) if the axial power density is uniform, or Eq. (15–18) for the cosine distribution. Figure 15–7 shows that the difference between these two results is not really large, considering the possible uncertainty in C_f and other variables.

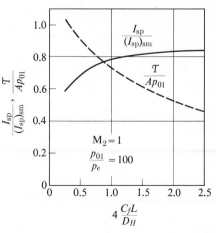

FIG. 15–8. Specific impulse and thrust as functions of $C_f L/D_H$. (Courtesy Stenning [7].)

The maximum possible specific impulse for a given wall temperature limitation T_{sm} and for an infinitely large coolant channel (that is, one with no stagnation pressure loss) may be defined by

$$(I_{sp})_{sm} = \frac{1}{g_e} \sqrt{\frac{2\gamma}{\gamma - 1} \frac{\overline{R}T_{sm}}{M} \left(1 - \frac{p_e}{p_{01}}\right)^{(\gamma-1)/\gamma}}, \qquad (15\text{–}20)$$

in which p_{01} is the reactor *inlet* stagnation pressure. The ratio of the impulses given by Eqs. (15–19) and (15–20) may be computed from Eq. (15–18) (for the cosine power distribution). The result is shown in Fig. 15–8 for $\gamma = 1.4$.

Using Eq. (3–14), we may express the mass flow as

$$\dot{m} = \frac{A p_{02}}{\sqrt{\overline{R}T_{02}/M}} \frac{\sqrt{\gamma}\, M_2}{\{1 + [(\gamma - 1)/2]M_2^2\}^{(\gamma+1)/2(\gamma-1)}}.$$

Using Eq. (15–11), we can write this statement as

$$\dot{m} = \frac{A p_{01}}{\sqrt{\overline{R}T_{02}/M}} \frac{\sqrt{\gamma}\, M_2\{1 + [(\gamma - 1)/2]M_2^2\}^{1/2}}{1 + \gamma M_2^2[1 + (C_f L/D_H)]}. \qquad (15\text{–}21)$$

If A is the total cross-sectional area of the reactor heat-transfer channels, then the thrust of the rocket is $T = \dot{m}g_e I_{sp}$. Then, using Eq. (15–21), we obtain

$$\frac{T}{A p_{01}} = \sqrt{\frac{2\gamma^2}{\gamma - 1}}\, M_2 \frac{\{1 + [(\gamma - 1)/2]M_2^2\}^{1/2}}{1 + \gamma M_2^2[1 + (C_f L/D_H)]} \sqrt{1 - \left(\frac{p_e}{p_{02}}\right)^{(\gamma-1)/\gamma}}. \qquad (15\text{–}22)$$

If the heat-transfer channels are choked, $M_2 = 1$ and

$$\frac{T}{A p_{01}} = \frac{\gamma[(\gamma + 1)/(\gamma - 1)]^{1/2}[1 - (p_e/p_{02})^{(\gamma-1)/\gamma}]^{1/2}}{1 + \gamma[1 + (C_f L/D_H)]}. \qquad (15\text{–}23)$$

For $M_2 = 1$, Eq. (15–11) becomes

$$\frac{p_{02}}{p_{01}} = \frac{[(\gamma + 1)/2]^{\gamma/(\gamma-1)}}{1 + \gamma[1 + (C_f L/D_H)]}. \qquad (15\text{–}24)$$

Using the value $p_{01}/p_e = 100$, which is a reasonable value for the overall pressure ratio, the thrust coefficient $T/A p_{01}$ may be calculated from Eqs. (15–23) and (15–24) for the particular case $M_2 = 1$. (Note that thrust coefficient as normally defined would be $T/A^* p_{02}$.) The result is shown in Fig. 15–8.

15–5 ADVANCED NUCLEAR ROCKET CONCEPTS

The great energy release accompanying fission would, in itself, seem to make possible very high propellant temperatures and exhaust velocities. It is interesting to note further that propellant dissociation does not necessarily detract from performance in the case of a nuclear rocket. This follows from the fact that dissociation does not require a reduction in stagnation temperature, as in a chemical rocket. The additional energy input to dissociate the propellant is easily available in a nuclear reactor. In fact, dissociation would increase performance (for a given fluid temperature), since it would result in a lower propellant molecular weight, and possibly some reassociation and energy release on expansion. However, as discussed in Section 15–1, such high performance is prohibited, at least in solid-core reactors, by rather conventional material limitations. In particular, the requirement that heat be transferred *from* (or through) a solid *to* the propellant limits the allowable propellant temperatures. As a consequence, efforts are being made to devise processes which bring about energy transfer from fission products to propellant by means other than heat conduction through a solid. Two processes are possible: radiation or direct molecular collision.

Radiative heat transfer would require that the reacting material be maintained at an extremely high temperature; hence the reactor would be in a gaseous state. The gaseous fission material must then be held within a cooled container whose walls are transparent to the (thermal) radiation. Propellant-cooled quartz has been considered as the wall material.

Direct molecular collision would require that the fission fragments, which carry most of the fission energy, be stopped within the propellant itself rather than within a solid material. One device would use fissionable material dispersed over the reactor surface or throughout thin fins, in such a way that some or most of the fission fragments would enter the coolant. Bussard and DeLauer [3] discuss such a reactor, with the conclusion that prohibitively large reactors would be required to satisfy the requirements of criticality and to provide propellant passages sufficiently wide for the fission fragments to be decelerated by the propellant itself.

A simpler arrangement would be to mix the nuclear fuel in gaseous form with the propellant, perhaps within a solid reactor or reflector, and expel the hot mixture of propellant, fission fragments, and unburned fuel. In this way, nearly all the fission fragments would transfer energy directly to the propellant. However, to provide criticality, very high fuel density would be required, and the resultant loss of fissionable material would be economically prohibitive. Hence some method of retaining the fissionable material within the reactor would be necessary.

One of the ideas suggested for retaining fissionable material is vortex containment [8]. As shown schematically in Fig. 15–9, a strong vortex is maintained

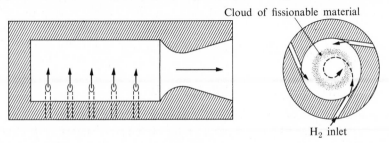

Cloud of fissionable material

H₂ inlet

FIG. 15–9.   Schematic diagram of a gaseous reactor, in which the fissionable material is contained by a vortex.

within a cylindrical chamber by tangential injection of propellant (hydrogen). The hydrogen diffuses radially inward through a rotating cloud of fissioning material, where it is heated. Then it moves axially along the vortex axis and out the nozzle. The fissionable material, which has a very much greater molecular weight, has, for the same temperature, considerably less thermal motion and is thus less likely to diffuse inward against the large radial inertial forces. The relative radial motion of light and heavy molecules is similar to that by which hydrogen leaves the earth's atmosphere, except that gravitational forces are replaced by much stronger inertial forces.

It seems safe to say that successful implementation of these and other advanced nuclear-rocket concepts must await the successful completion of considerable theoretical and experimental work. The concepts are presented here primarily to illustrate the severity of the temperature limitation imposed by the necessity of heat transfer through a solid, and to indicate that attempts are being made to avoid this limitation.

References

1. KAPLAN, I., *Nuclear Physics*. Reading, Mass.: Addison-Wesley, 1955
2. SCHREIBER, R. E., "Kiwi Tests Pave Way to Rover." *Nucleonics*, **19**, 4, April 1961, pages 77–79
3. BUSSARD, R. W., and R. D. DeLAUER, *Nuclear Rocket Propulsion*. New York: McGraw-Hill, 1958
4. GLASTONE, S., and M. C. EDLUND, *The Elements of Nuclear Reactor Theory*. Princeton, N. J.: Van Nostrand, 1952
5. SCHREIBER, R. E., "A Review of Project Rover," presented at the International Symposium on Aero-Space Propulsion, sponsored by IRE, AEC, and NASA; Las Vegas, Nevada, October 23–27, 1961
6. FINGER, H. B., J. LAZAR, and J. J. LYNCH, "A Survey of Space Nuclear Propulsion." AIAA Paper No. 64–523, presented at first annual meeting of the American Institute of Aeronautics and Astronautics, Washington, D.C., June 29–July 2, 1964.
7. STENNING, A. H., "Rapid Approximate Method for Analyzing Nuclear Rocket Performance." *ARS Journal*, **30**, 2, February 1960, pages 169–172
8. KERREBROCK, J. L., and R. V. MEGHREBLIAN, "Vortex Containment for the Gaseous-Fission Rocket." *Journal of the Aerospace Sciences*, **28**, 9, September 1961, pages 710–724

Problems

1. Compare the specific impulse of a nuclear-heated hydrogen rocket, with and without dissociation, if the chamber temperature and pressure are 6500°R and 50 atm, respectively. It may be assumed that the composition is constant in the nozzle. The ratio of specific heats is 7/5 for H_2 and 5/3 for H, and the nozzle pressure ratio is 50:1.

2. Chemical rockets may be expected to achieve vacuum specific impulses (calculated for $p_e = 0$) of the order of 400 sec, with a combustion chamber pressure of 500 psia and a stagnation temperature of 5500°R. What is the specific impulse which may be achieved in a nuclear-heated hydrogen rocket at the same peak temperature? The universal gas constant is $\bar{R} = 1545$ ft·lb$_f$/lb$_m$-mole·°R. Consider two cases: (a) no dissociation in reactor or nozzle, and (b) equilibrium dissociation in reactor and frozen flow in nozzle. Use Fig. 2–4 to estimate the specific heats of H_2 and H, and assume $\gamma = 1.3$.

3. The reactor core of a certain nuclear-heated hydrogen rocket engine is restricted to 10 ft in length. It has a cosine distribution of axial power density. If the maximum wall surface temperature is 5000°R, approximately how much will the specific impulse be affected by a change in the diameter of the heat exchanger tube from 0.500 in. to 0.100 in.? The reactor inlet temperature is 100°R.

4. The illustration shows a 0.139-in. diameter tube through which gaseous hydrogen flows while receiving heat from the walls. Assuming isentropic flow from stagnation conditions to ① and simple heating from ① to ②, calculate T_{02} when $\dot{m}$ is the maximum possible. The properties of the hydrogen may be taken to be $c_p = 3.45$ Btu/lb$_m$·°R and $\gamma = 1.4$. The total rate of heat transfer to the hydrogen is 300 Btu/sec.

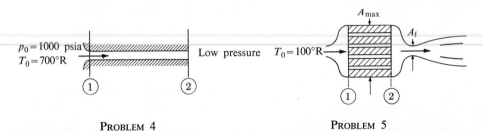

PROBLEM 4   PROBLEM 5

5. A gas ($\gamma = 1.4$, $c_p = 0.24$) passes through a number of cylindrical holes in the solid core of a nuclear reactor, where it is heated from 100°R to 4000°R (stagnation temperatures). If the total hole area is 30% of the core cross section A_{max}, what must the area ratio A_{max}/A_t be to assure a stagnation pressure loss due only to heating effects of no more than 10%? Assume that within the core the flow can be approximated by the steady one-dimensional flow of a perfect gas.

6. For some of the very high temperature nuclear rockets it may be expected that considerable increase in specific impulse can be derived from dissociation of the hydrogen propellant. Suppose that the chamber pressure is 20 atm and the ambient pressure is 1 atm. What stagnation temperature would be required to dissociate 90% of the hydrogen? Assuming frozen flow in the nozzle, determine the specific impulse. For space flight of nuclear rockets, the thrust required would be relatively low and a stagnation pressure of much less than 20 atm would be suitable. For a stagnation pressure of only 1 atm, what stagnation temperature would be required for 90% dissociation?

16

Electrical Rocket Propulsion

16–1 INTRODUCTION

The apparent suitability of electrical rockets for propelling space vehicles was discussed in Chapter 9. In general, electrical rockets should be able to develop considerably higher specific impulse than chemical or nuclear ones. It has been shown, however, that this gain in specific impulse requires massive energy-conversion equipment, so that electrical rockets generally have low thrust, perhaps only one-thousandth of the vehicle weight in the earth's gravitational field. For this reason they are mainly restricted to space missions during which gravitational forces are very nearly balanced by inertial forces. Low accelerations are quite acceptable, since the journeys are of long duration.

The propellant of an electrical rocket consists of either discrete charged particles which are accelerated by electrostatic forces, or a stream of electrically conducting fluid which is accelerated by electromagnetic or pressure forces (or both). Figure 16–1 is a classification of electrical rockets which are being considered for future spaceships. .

The electrostatic rockets are subdivided into categories depending on the nature of their propellant; charged single atoms or ions, and charged, relatively massive,

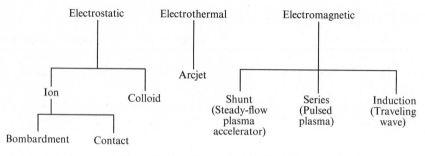

FIG. 16–1. Electrical accelerators.

multi-atom particles called colloids. Ion rockets are further subdivided according to the process by which the ions are formed. Bombardment ionization results from the collision of a high-speed electron with a single propellant atom in the gaseous phase. Contact ionization is an interaction between a single propellant atom and a suitable hot surface. Both processes are discussed in Section 16–2.

The arcjet, or plasmajet as it is sometimes called, is not really an electrical acceleration device. Electrical power is used simply to heat the propellant to a very high temperature, after which it is accelerated in a conventional nozzle. The fact that the propellant is electrically conductive is incidental to the acceleration process. Arcjets are discussed in Section 16–6.

Electromagnetic accelerators utilize the conductivity of the plasma propellant to create an electromagnetic accelerating force acting as a "body force" within the plasma. There are three principal methods of accelerating plasmas electromagnetically; these are analogous to the behavior of shunt, series, and induction motors, respectively. In the first, the current which produces the magnetic field is independent of the current which flows through the plasma. In the second, the plasma current induces its own magnetic field and the interaction of these two gives rise to the accelerating force. In the third, the plasma is not subjected to an external applied voltage, so it does not transmit current. The accelerating force is due to the interaction of an externally applied unsteady magnetic field with the circulating currents induced within the plasma by this field.

It should be emphasized that all the devices which are discussed in this chapter are in experimental stages at present. With such a variety of possible electrical rockets, it is difficult to say at this point which types will eventually prove most satisfactory. Numerous problems remain and space flight tests are only beginning. Nevertheless it seems clear that electrical phenomena will play a role in space flight propulsion. The purpose of this chapter is to demonstrate possible applications and to show how these may be understood with the aid of fundamental laws. However, we must recognize that not all the experimental devices in existence at the present time are well understood; hence this discussion will be restricted mainly to certain steady-state acceleration mechanisms.

Let us now discuss the subject of *electrostatic propulsion*.

16–2 ELECTROSTATIC PROPELLANT ACCELERATION

Figure 16–2 shows, schematically, a typical ion rocket. Neutral propellant is pumped to an ion production chamber, from which ions and electrons are withdrawn in separate streams. The ions pass through the strong electrostatic field established by the acceleration electrodes shown in Fig. 16–2. The ions accelerate to high speeds, and the thrust of the rocket is the total reaction to their accelerating forces. The ionization chamber is maintained at the positive acceleration potential, while the last electrode is "grounded" at the vehicle potential. Intermediate electrodes may be held at other potentials to aid in the extraction of ions and in the formation of a suitable exhaust beam.

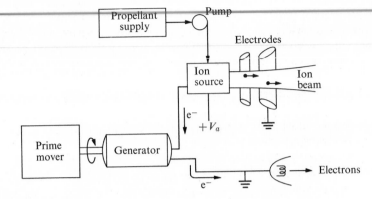

FIG. 16–2.   Ion rocket.

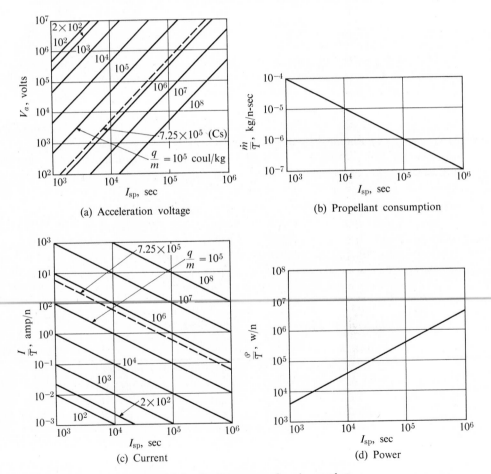

FIG. 16–3.   Performance of an ion rocket.

It is also necessary to expel the electrons in order to prevent the vehicle from acquiring a net negative charge. Otherwise ions would be attracted back to the vehicle and the thrust would vanish. To remove the electrons from the vehicle it is necessary that they first be "pumped" to ground potential by the generator. They may then be removed from the vehicle at a hot filament, although a simple filament may not be adequate. Thus the "thrust chamber" of an electrostatic rocket consists of a source of charged propellant and an electrode structure to extract, accelerate, and expel the propellant in a well-defined beam. In addition the system requires a propellant storage and supply system and an electrical power source.

The overall performance of the electrostatic rocket can be derived from basic principles reviewed in Chapter 5. If a particle of mass m, charge q, and negligible initial velocity passes through a potential difference V_a, it will acquire an exhaust kinetic energy of

$$\frac{mu_e^2}{2} = qV_a,$$

where u_e is the exhaust velocity. Therefore the specific impulse of a beam of such particles is

$$I_{\text{sp}} = \frac{1}{g_e}\sqrt{2\frac{q}{m}V_a}, \tag{16-1}$$

where g_e is the acceleration due to gravity near the surface of the earth. This relationship is shown graphically in Fig. 16-3.

Since the thrust due to the beam acceleration is $T = \dot{m}u_e$, the mass flow per unit thrust is

$$\frac{\dot{m}}{T} = \frac{1}{g_e I_{\text{sp}}}. \tag{16-2}$$

The total beam current I is given by $I = (q/m)\dot{m}$, or, from Eq. (16-2),

$$\frac{I}{T} = \frac{q}{m}\frac{1}{g_e I_{\text{sp}}}. \tag{16-3}$$

The beam power $\mathcal{P}_b$ is given by

$$\mathcal{P}_b = IV_a = \frac{\dot{m}u_e^2}{2}$$

or

$$\frac{\mathcal{P}_b}{T} = \frac{g_e I_{\text{sp}}}{2}. \tag{16-4}$$

Equations (16-2), (16-3), and (16-4) are illustrated graphically in Fig. (16-3) for various values of the charge-to-mass ratio q/m. For reference, the characteristics for cesium (7.25×10^5 coulombs per kg) are shown in Figs. 16-3a and 16-3c.

The power consumed by an electrostatic rocket is greater than the beam power defined above, largely because power must be supplied to form the ions. The energy which must be supplied to each atom in order to ionize it is called the *ionization potential*. It is of the order of several electron volts per ion (see Section 16–3). Experimentally it is found that very much more than this energy must be supplied to the ionization chamber, of the order of several hundred to one thousand electron volts per ion. Most of it is lost by radiation, since either the propellant itself or portions of the chamber must be maintained at a high temperature. The ionization chamber may also require power for the generation of magnetic fields. These power requirements are collectively called the *charging power*. In addition, smaller quantities of power may be lost in beam interception by the accelerating system, and power consumed by the electron ejection system. The ratio of beam power to total electrical power consumption is defined as the *power efficiency*, η, of the ion engine:

$$\eta = \frac{\text{beam power}}{\text{total electrical power}}.$$ (16–5)

In terms of energy per ion rather than power, and neglecting beam interception and neutralization power requirements,

$$\eta \simeq \frac{\text{exhaust kinetic energy}}{\text{exhaust kinetic energy} + \text{charging energy}}$$

or

$$\eta \simeq \frac{\frac{1}{2}mu_e^2}{\frac{1}{2}mu_e^2 + e_i + e_l},$$

in which m is the ion mass, e_i is the ionization energy and e_l is the energy loss per ion. Since in general $e_i \ll e_l$,

$$\eta \approx \frac{\frac{1}{2}mu_e^2}{\frac{1}{2}mu_e^2 + e_l} = \frac{\frac{1}{2}mg_e^2 I_{\text{sp}}^2}{\frac{1}{2}mg_e^2 I_{\text{sp}}^2 + e_l}.$$

Figure 16–4 shows the variation of power efficiency for mercury as a function of specific impulse and energy loss per ion.

It may be seen from the above equation that if the energy loss from the ionization chamber, e_l per ion, is independent of the propellant, the power efficiency at a *given specific impulse* is greater for more massive ions. For example, consider a specific impulse of 2000 seconds, a value which was shown in Chapter 10 to be desirable for an earth-moon transfer mission. Equation (16–1) shows that, with cesium, the accelerating voltage is only 260 volts, assuming the cesium to be singly ionized. Thus the beam energy is only 260 eV per ion. In this case, then, the radiation losses would be at least as large as the beam energy, and the overall thrust-chamber efficiency would be low. The acceleration voltage for mercury at the same specific impulse is 400 volts. Hence, if the charging energy per ion is roughly the same, power efficiency would improve with the use of mercury. Unless very much

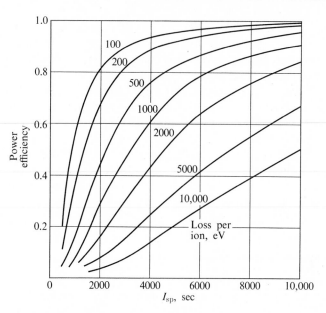

Fig. 16–4.   Theoretical variation of power efficiency with specific impulse for mercury ion propellant. (Courtesy Kaufman and Reader [32].)

heavier particles are used, however, the electrostatic rocket may well be restricted to specific impulse values above 5000 seconds.   It is for this reason that heavy molecular ion sources would be desirable.   Even more massive multiple-charged colloidal particles accelerated by potential differences as high as 10^6 volts could provide very good power efficiency even at a specific impulse as low as 1000 seconds, if the charging energy per electronic charge remained comparable to the values for ionization of single atoms.

16–3 SOURCES OF CHARGED PARTICLES

The electron structure of an atom consists of a number of shells, each containing a specified number of electrons.   The removal of one of these electrons to create a positive ion requires a quantity of energy called the *ionization potential*. (Negative ions produced by electron attachment have not yet been successfully produced for ion rockets.)   As might be expected, those atoms containing single electrons in unfilled outer shells are especially easily ionized; that is, they have relatively low ionization potentials.   The alkali metal elements—lithium, sodium, potassium, rubidium, and cesium—have particularly low ionization potentials.   Table 16–1 lists first and second ionization potentials (pertaining to the removal of the first and second electrons, respectively) for these elements, along with some others for reference.   Note the relatively high ionization potentials of the inert elements, reflecting the fact that electrons must be extracted from a stable outer shell which is

Table 16–1*

Element		Atomic number	First ionization potential, eV	Second ionization potential, eV
Alkali metals	Li	3	5.363	75.26
	Na	11	5.12	47.06
	K	19	4.318	31.66
	Rb	37	4.159	27.36
	Cs	55	3.87	23.4
Inert elements	He	2	24.46	54.14
	Ne	10	21.47	40.9
	A	18	15.68	27.76
	H	1	13.527	—
	C	6	11.217	24.27
	Hg	80	10.39	18.65

* From *Handbook of Physics and Chemistry*, 43rd edition. Cleveland, O.: Chemical Rubber Publishing Company, 1961.

completely filled. The second electron extracted from an alkali metal must also come from a full shell, which accounts for the high ratio of second to first ionization potentials of these materials. Mercury, often considered as a propellant because of its high atomic mass and relatively easy handling characteristics, normally contains two electrons in its outermost shell. Hence its first two ionization potentials are relatively close together.

Two ionization processes are of importance for ion rockets: electron-bombardment ionization, occurring as a result of direct collision between an energetic electron and a single propellant atom in the gaseous phase; and contact ionization, occurring as an interaction between a single propellant atom and a suitable solid surface. In the former, low ionization potential favors low charging power, but it is not absolutely essential. In the latter, as we shall see, it is essential that the ionization potential be quite low.

Bombardment ionization

If an electron current is conducted through a gas, electron-atom collisions will occur and, if the electrons are of the proper energy, some of these collisions will result in ionization. Although a single atom might conceivably be ionized as the result of several electron collisions, it is reasonable to expect that no appreciable ionization will occur until the bombarding electrons each possess kinetic energies equal to or greater than the ionization potential. As the electron energy is increased above this value, it is experimentally observed that the rate of formation of ions increases sharply to a maximum at an electron energy from three to five times the ionization energy, and then falls off slowly with further increases of electron energy.

As discussed in Chapter 5, the collision frequency can be described in terms of a collision cross section pertaining to electron-atom collisions. In this case we are

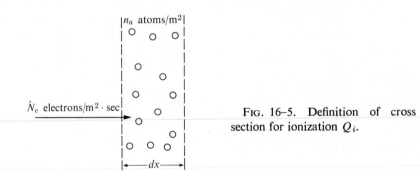

FIG. 16–5. Definition of cross section for ionization Q_i.

interested only in those collisions resulting in ionization as described by the *cross section for ionization.* If a flux of $\dot{N}_e$ electrons per second per unit area enters a region of length dx (Fig. 16–5), where the atom density is n_a (the atoms being assumed stationary), the rate of formation of ions dn_i/dt will be proportional to $\dot{N}_e$ and n_a. The proportionality constant is the cross section for ionization Q_i:

$$\frac{dn_i}{dt} = Q_i \dot{N}_e n_a \text{ ions per second per unit volume.} \qquad (16\text{–}6)$$

Just as the total collision cross section is energy-dependent, the cross section for ionization by electron impact is highly dependent on the energy of the bombarding electron. Figure 16–6 indicates the energy dependence of the cross section for various degrees of ionization of mercury. Note that each curve begins approximately at the ionization potential, and the successive ionization potentials are 10.39, 18.65, 34.3, 72, and 82 eV [1]*. From Fig. 16–6 it may be seen that at the electron energy yielding the peak cross section for single ionization some double- and even triple-charged ions would also be formed. In a given accelerator their charge-to-mass ratios would be much too high, so that they would develop exces-

FIG. 16–6. Cross section for different degrees of ionization of mercury by electron impact. (Courtesy Bleakney [2].)

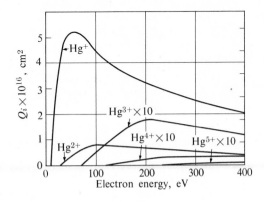

* Bleakney [2] gives ionization potentials of 10.4, 30, 71, 143, and 225 eV in agreement with his figure, but the more recent data compiled in [1] appear to be more accurate.

sive exhaust velocities and perhaps seriously alter the focusing characteristics of the exhaust beam. It is for this reason that propellants with a large difference between first and second ionization potentials are especially desirable.

Kaufman [3] has proposed an ion engine with a low current density utilizing an electron-bombardment ion source which is shown schematically in Fig. 16-7, along with typical operating potentials of the various electrodes. Propellant, in most cases mercury vapor, flows axially through the ionization chamber. Electrons, generated from a cathode at the center line, flow to the concentric anode. However, an axial magnetic field prevents direct flow from cathode to anode and causes the electrons to move in curved paths. At the gas densities utilized, the electron mean free path is much longer (of the order of meters) than the chamber radius (centimeters). However, if its path is sufficiently curved, it is held in the ionization chamber until (on the average) it suffers at least one collision. Not all collisions result in the formation of an ion, of course. Small axial variations in potential within the ionization chamber prevent the axial flow of electrons.

It is of interest to consider the approximate strength required of the magnetic field. For simplicity, consider an ionization chamber which is uniform in the axial direction and in which the magnetic field is uniform in the radial direction (the magnetic field in actual sources may intentionally be varied in both directions). In cylindrical coordinates, the forces on an electron in such a region are given by:

$$\text{Radial direction} \qquad F_r = m_e(\ddot{r} - r\dot{\theta}^2) = e\frac{dV}{dr} \pm er\dot{\theta}B, \qquad (16\text{-}7)$$

in which m_e and e are the mass and charge of an electron, and

$$\text{Tangential direction} \qquad F_\theta = m_e(2\dot{r}\dot{\theta} + r\ddot{\theta}) = \pm e\dot{r}B$$

or

$$rF_\theta = m_e\frac{d}{dt}(r^2\dot{\theta}) = \pm r(e\dot{r}B), \qquad (16\text{-}8)$$

where the plus-or-minus sign depends on the direction of the axial magnetic field. While the determination of the radial variation of potential within the ionization chamber (dV/dr in Eq. 16-7) is not a simple matter, Eq. (16-8) may be integrated, yielding the required field strength in terms of the total potential drop regardless of the intervening shape of the potential distribution. We may write (16-8) as

$$m_e\frac{d}{dt}(r^2\dot{\theta}) = \pm eB\frac{d}{dt}\left(\frac{1}{2}r^2\right),$$

which, after we multiply both sides by dt, integrates directly to

$$m_e r^2\dot{\theta} = \pm\frac{eB}{2}r^2 + \text{const.}$$

At the cathode, $r = r_c$ and $\dot{\theta} \approx 0$, hence the constant of integration is $\pm(eB/2)r_c^2$,

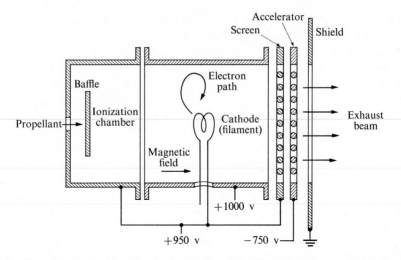

FIG. 16–7. Schematic of low-density ion engine of Kaufman, showing typical operating voltages.

and the solution may be written

$$\dot\theta = \pm \frac{eB}{2m_e}\left(1 - \frac{r_c^2}{r^2}\right).$$ (16–9)

At the anode, if the electron velocity is nearly tangential, then

$$\dot\theta = \frac{u_a}{r_a} = \frac{\sqrt{2e/m_e\,\Delta V_a}}{r_a}$$

where r_a = anode radius and ΔV_a = anode potential relative to cathode.
Substituting this expression in (16–9), we obtain

$$B = \frac{\sqrt{8(m_e/e)\,\Delta V_a}}{r_a[1 - (r_c^2/r_a^2)]}.$$ (16–10)

A small correction may be made to this expression to account for the presence of a sheath at the surface of the anode. As described in Chapter 5, the plasma within the ionization chamber will be at a slightly higher potential than its container. Thus, just away from the anode, the potential will be higher by an amount ΔV_s (of the order of 5 v [3]) and ΔV_a should be replaced by $\Delta V_a + \Delta V_s$ in Eq. (16–10).

The choice of ΔV_a depends on the ionization potentials of the propellant being used. Thus, for mercury, ΔV_a should be greater than 10.39 v. Since no Hg^{2+} ions will be formed below 18.65 v, ΔV_a can be raised to at least this value. In practice it is even higher, for two reasons. First, the cross section for single ionization continues to rise rapidly (up to energies of approximately 100 eV for mercury), while that for double ionization does not rise so fast. Second, the electron energy

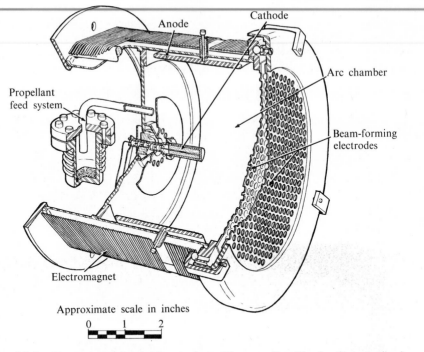

FIG. 16–8.   Low current density ion engine.   (Courtesy Ion Physics Corporation.)

is everywhere less than ΔV_a (or $\Delta V_a + \Delta V_s$), since the potential increases in the radial direction.   Actually, the presence of the conducting plasma within the chamber permits only slight potential variation throughout most of the volume; thus a large fraction of the potential rise (90% or so) occurs in a relatively thin cathode sheath.   A value of ΔV_a between 20 and 100 v would appear reasonable for mercury propellant.

Ions which reach the downstream end of the ionization chamber are accelerated through a simple grid system.   Typically this consists of a series of parallel wires or a pair of perforated plates.   For the potentials indicated in Fig. 16–7, the ions are first accelerated through 1750 v and then decelerated through 750 v.

Figure 16–8 is a cutaway view of an actual ion engine of this type, showing the windings which produce the magnetic field.   The acceleration grid in this case consists of perforated plates.   The cathode consists of an internally heated composite structure. The mercury feed system is a simple boiler arrangement. Engines of this type have produced ions with an energy consumption of the order of 450 eV per ion (based only on arc power consumption).   Propellant efficiencies (ratio of ion mass flow to total propellant flow) greater than 90% have been achieved with less than 0.5% beam interception on the accelerating electrode.   Current densities of the order of 1.5 to 2.0 ma/cm^2 (averaged over the total accelerator area) are achieved.

Contact ionization

The electron structure of a metallic atom becomes modified when the atom is part of a solid or crystalline structure. The outermost electron, the same one which is most easily removed to form an ion, often loses its attachment to any particular atom and becomes capable of moving about within the solid. The relatively free motion of such "conduction electrons" accounts for the electrical conductivity of metals. In fact, almost all those atoms containing single electrons in their outer shells (the alkali metals plus copper, silver, and gold) contribute this electron to the conduction process when they are in the solid state, so that conductors made of these metals contain very nearly one conduction electron per atom [5]. Other metals may contain a fraction of one conducting electron per atom. Thus, if an alkali metal atom contacts a metallic solid, its outermost electron is very likely to be lost to the large sea of conduction electrons within the solid. If the atom is removed from the surface (for example, by heating the surface and "boiling" it off), it might leave without its electron (or any other electron from the sea of conduction electrons) if it is easier to extract an electron from the atom than from the solid.

It takes energy to remove an electron from a solid. The energy to move a single electron from its position as a conduction electron to an infinite distance from the solid is called the *work function*, ϕ, of the surface. It can be partially accounted for by the method of images as illustrated in Fig. 16–9, which shows the field between an electron and a nearby conductor. The lines of force emanating from the electron intercept the conductor surface at right angles, since the conductor is an equipo-

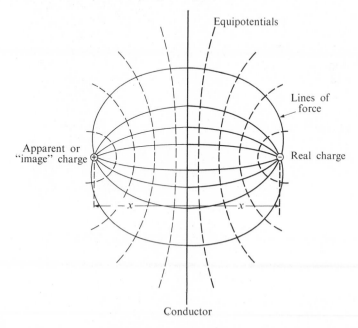

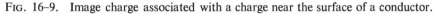

FIG. 16–9. Image charge associated with a charge near the surface of a conductor.

tential. This boundary condition makes the field between the electron and the conductor equivalent in all respects to half of the field between unlike charges. That is, the real charge behaves exactly as if there were an equal and opposite "image charge" replacing the conductor. In particular, the attractive force on the real charge is

$$F = -\frac{e^2}{4\pi\epsilon_0(2x)^2}.$$

The work done on a charge in moving it to infinity from a distance x from the surface against such a force is

$$W_{x\to\infty} = \int_x^\infty -F\,dx = \int_x^\infty \frac{e^2}{16\pi\epsilon_0}\frac{dx}{x^2} = \frac{e^2}{16\pi\epsilon_0 x}.$$

As the point x approaches the surface, it would appear that infinite work must be done to move the electron from the surface. This expression is inadequate, however, since as x becomes very small, the assumption of a solid continuum becomes invalid as regards both the definition of surface and the distribution of charges within the solid. A satisfactory solution must treat interactions between individual particles as x approaches zero. Using the methods of quantum mechanics [5], the above expression may be modified to make it valid for vanishingly small values of x. The result of this modification is

$$W_{x\to\infty} = \frac{e^2}{16\pi\epsilon_0 x + e^2/\phi}, \tag{16-11}$$

in which the empirical constant ϕ, called the work function, is the work necessary to remove an electron from the surface to infinity.

It is to supply some electrons with this energy that cathode filaments must be heated to bring about electron emission. Table 16-2 gives the work functions and melting temperatures of several *clean* metals. As ϕ is ultimately related to interparticle phenomena on the surface, it is to be expected that surface contamination

Table 16-2 [6]

Metal	Work function, eV	Melting temperature, °C
Ag	4.7	960
C	4.81	3550
Cr	4.37	1890
Ir	4.9	2454
Mo	4.3	2620
Pt	5.2	1773
Re	5.1	3167
Ta	4.1	3027
W	4.5	3370

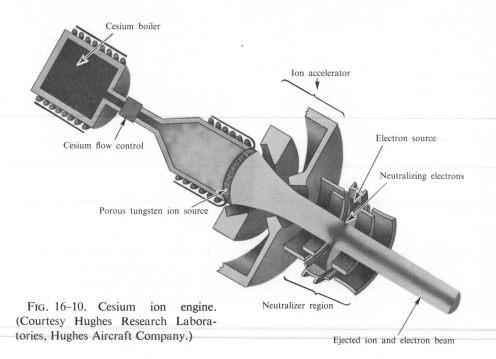

Cesium boiler

Ion accelerator

Electron source

Cesium flow control

Neutralizing electrons

Porous tungsten ion source

FIG. 16-10. Cesium ion engine. (Courtesy Hughes Research Laboratories, Hughes Aircraft Company.)

Neutralizer region

Ejected ion and electron beam

can strongly influence the work function. In fact, surface contamination by propellant atoms decreases ϕ, and this effect is of major importance in the operation of contact ionization sources.

We have already discussed the energy required to remove an electron from an atom, the ionization potential. Hence the requirements for surface ionization may be more succinctly stated as follows: Surface ionization will occur if the work function of the surface is greater than the ionization potential of the propellant. Thus combinations of the low-work-function alkali metals with high-work-function surfaces are utilized. In particular, cesium is attractive as a propellant, since it has both a lower ionization potential and a higher mass than the others (except perhaps francium, which is extremely rare).

In practice, ions (or atoms) are removed from the surface by raising the surface temperature to a value sufficient to boil off the propellant. The rather high temperatures required preclude the use of a platinum surface with its attractively high work function. Carbon, with a high work function and melting temperature, has not displayed adequate mechanical properties in the configurations tried so far. The most popular combination has been cesium and tungsten. To avoid neutral cesium atoms in the ion extraction and acceleration region, the tungsten surface is provided by a porous tungsten plug through which the neutral cesium diffuses to the ionizing face. Figure 16-10 is a schematic view of such an ion source, along with the acceleration and beam-handling electrodes. The presence of the cesium atoms on the tungsten surface tends to lower the work function.

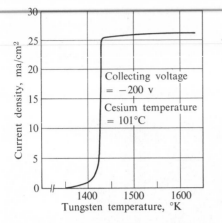

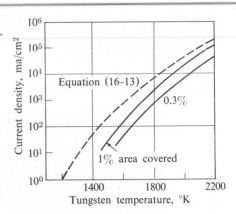

FIG. 16–11.   Ion current density drawn from tungsten filament as a function of filament temperature for constant cesium mass flow rate. (Courtesy Shelton, *et al.* [7].)

FIG. 16–12.   Equation (16–13) and theoretical dependence of ion current on tungsten temperature for constant area coverages. (Courtesy Husmann [9].)

It has been found that, to maintain a sufficiently high work function, only about 0.5% surface coverage is permissible. This rather severely limits the current density which may be drawn from such an ion source. The surface density is related to the propellant mass flow rate and surface dwell time as follows. In general, the continuity equation may be expressed as

$$\dot{m} = \rho u A = \rho A \frac{dx}{dt}.$$

If the incremental distance dx is set equal to the surface "thickness," then $\rho\, dx$ is the surface mass density σ. The time dt to travel dx then becomes the dwell time t_d that the atom spends on the surface. Thus

$$\sigma = \frac{\dot{m}}{A} t_d. \tag{16–12}$$

Surface mass density then increases with increasing mass flow rate and dwell time. The time necessary to gain sufficient thermal energy to leave the surface increases with decreasing surface temperature. The restriction on maximum allowable σ and the temperature dependence of t_d lead to a critical temperature relationship as illustrated in Fig. 16–11. Below the critical temperature, about 1435°K in this example, the cesium flow is almost entirely in the form of neutral atoms. As the temperature is raised, t_d and thus σ decrease and ϕ increases until essentially all the cesium flow is in the form of ions. Higher mass flow rates or current densities require proportionately lower t_d or higher temperature to maintain the same σ. The relationship between current density j (or mass flow rate) and the critical

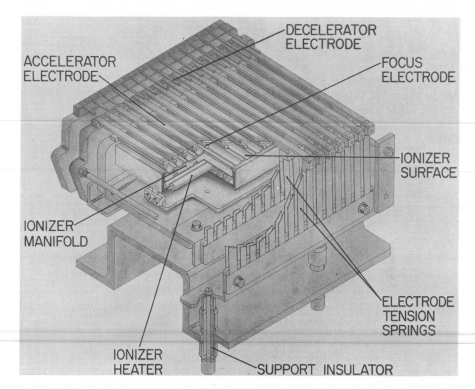

FIG. 16–13. Cutaway view of a linear-strip cesium contact-ionization ion engine. (Courtesy Hughes Research Laboratories, Hughes Aircraft Company.)

temperature T_c necessary to produce it is, for cesium on tungsten [8],

$$\log_{10} j = 8.99 - \frac{14{,}350}{T_c}, \tag{16–13}$$

where j is in amps/cm^2 and T_c is in degrees Kelvin. This relationship, along with theoretical curves for constant σ based on experimental correlations [9] of t_d with temperature, is shown in Fig. 16–12. It is felt that the experimental dwell times used in these computations might be too long, and that the magnitude of the theoretical current would therefore be too low. However, the similar curve shapes verify the concept of a maximum allowable surface coverage for ion formation.

The cesium contact ion engine has been developed sufficiently to indicate relatively high performance. Current densities of the order of 20 to 80 ma/cm^2 are supplied at reasonable power efficiency and with very low beam interception by the acceleration system [4]. Figure 16–13 is a cutaway view of such an engine in which the beam is a series of linear strips. The acceleration and beam neutralization elements are discussed in Section 16–5. Figure 16–14 shows how power efficiency depends on specific impulse for such an engine.

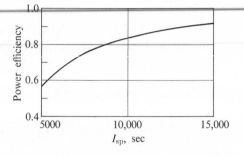

FIG. 16–14. Power efficiency of a cesium ion engine versus specific impulse for a current density of 20 ma/cm^2. Calculated from measured power requirements, including neutralizer electron emitter. (Courtesy Brewer, *et al.* [4].)

The following section, which considers the internal behavior of a simple acceleration system, indicates that ion current may be limited by electrical behavior rather than by source capabilities. It is important, therefore, that the acceleration system be capable of handling the current density of the entire propellant flow in ionized form, lest the propellant be forced to leave as neutral atoms. Neutral propellant flow, in addition to being wasteful of propellant, can seriously interfere with the acceleration of the ions.

Attempts have been made to form charged colloidal particles (so-called because of the size, roughly 1 to 0.01 μ in diameter). Either solid particles or liquid droplets would be suitable if their charge-to-mass ratio could be suitably tailored to the engine requirements. For example, if a specific impulse of 2000 sec is desired, along with an acceleration voltage of the order of 1,000,000 v, then a minimum charge-to-mass ratio of about 200 coul/kgm is necessary. Thus far, attempts to produce such propellant particles with reasonable charging power have been unsuccessful. A major drawback has been the presence of appreciable numbers of ions mixed in with the heavier particles, and also an undesirable nonuniformity of the heavy particles themselves.

16–4 THE PLANE DIODE

The electrostatic field and the behavior of charged particles in an actual accelerator may be difficult to determine, principally because of geometrical complexity. It is therefore helpful to study, in a simplified model, the acceleration of charges between two parallel infinite-plane electrodes. The electrostatic field in such a region depends on the distribution of moving charges between the electrodes, as well as on the electrode voltage and spacing, but the dependence is one-dimensional.

The plane diode is of course a greatly simplified model of the actual accelerator, since all the positive charges impinge on the negative electrode. However, this is not as great a restriction as it might seem, for we shall presently show that a beam of finite width can be accelerated through holes in finite electrodes, and that it will exhibit approximately the same behavior as the charges within the plane diode.

Consider the flow of charged particles such as ions or electrons between very large plane electrodes which have fixed spacing and potential difference. The cloud of particles in transit establishes a space charge which effectively limits the maximum current density which may be transmitted. Figure 16–15 shows typical potential distributions between two electrodes whose potential difference is V_a. If no current exists between them, the inter-electrode potential distribution will be linear. If positive particles are made available at the high potential electrode, the potential gradient at the surface will "extract" the particles and accelerate them to the opposite electrode. These charges in transit constitute a current between the electrodes and, at the same time, the cloud of charges within the gap raises the potential in that region. As the charge density increases, the extraction gradient at the charge-emitting surface decreases. If an unlimited supply of charged particles is available, the current will increase until the extraction gradient is reduced to zero, as in Fig. 16–15. This current is called the space-charge-limited current for the potential difference V_a and electrode spacing L.

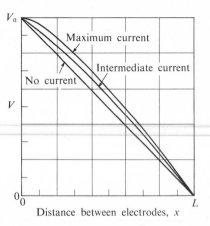

FIG. 16–15. Potential distribution between infinite parallel-plane electrodes with a current of positive particles.

The effect of the charge distribution on the potential distribution is given by Eq. (5–8), which for this problem becomes

$$\frac{d^2V}{dx^2} = \frac{-\rho_e}{\epsilon_0}, \tag{16–14}$$

where ρ_e = local electron charge density.

The current and the charge density are related by the one-dimensional form of Eq. (5–10), which is

$$j = \rho_e u, \tag{16–15}$$

where j is the current density and u is the charge velocity. The charge velocity in turn is related to the voltage by the energy equation

$$qV + \frac{mu^2}{2} = qV_a, \tag{16–16}$$

assuming that the charges have negligible velocity as they leave the positive electrode.

Equations (16–14), (16–15), and (16–16) may be used with appropriate boundary
conditions to determine the distribution of potential and charge velocity, as well
as the current density between the electrodes. The solution for maximum current
density is of particular interest. Using Eqs. (16–15) and (16–16) to eliminate the
variables u and ρ_e, we obtain from Eq. (16–14)

$$\frac{d^2V}{dx^2} = -\frac{j}{\epsilon_0\sqrt{2q/m}}(V_a - V)^{-1/2}. \qquad (16\text{–}17)$$

Making the substitutions

$$V_a - V = V^*$$

and

$$\frac{j}{\epsilon_0\sqrt{2q/m}} = \alpha,$$

we may write Eq. (16–17) as

$$\frac{d^2V^*}{dx^2} = \alpha(V^*)^{-1/2}.$$

Multiplying both sides by $2\,dV^*/dx$, we have

$$2\left(\frac{dV^*}{dx}\right)\frac{d^2V^*}{dx^2} = 2\alpha\left(\frac{dV^*}{dx}\right)(V^*)^{-1/2}.$$

After rearrangement, this becomes

$$d\left[\left(\frac{dV^*}{dx}\right)^2\right] = 2\alpha(V^*)^{-1/2}\,dV^*,$$

which may be integrated to

$$\left(\frac{dV^*}{dx}\right)^2 = 4\alpha(V^*)^{1/2} + C_1,$$

in which C_1 is a constant of integration. For maximum current,

$$\frac{dV^*}{dx} = V^* = 0, \qquad \text{at } x = 0.$$

Therefore $C_1 = 0$, and hence

$$\frac{dV^*}{dx} = 2\sqrt{\alpha}\,(V^*)^{1/4}.$$

Integrating again and using the boundary condition $V^* = 0$ at $x = 0$, we have

$$\tfrac{4}{3}(V^*)^{3/4} = 2\sqrt{\alpha}\,x.$$

From this equation we may obtain the potential distribution for space-charge-limited current:

$$V = V_a - \left(\frac{9}{4} \frac{j}{\epsilon_0 \sqrt{2q/m}}\right)^{2/3} x^{4/3}. \tag{16–18}$$

For electrodes of spacing L and potential difference V_a, the maximum current is

$$j = \frac{4}{9} \epsilon_0 \sqrt{2q/m} \, \frac{V_a^{3/2}}{L^2}. \tag{16–19}$$

If the beam could pass through the negative electrode without impingement, the maximum thrust per unit area of beam would be given by

$$\frac{T}{A} = \frac{\dot{m}u_e}{A},$$

where u_e is the particle velocity at $x = L$. The mass flow per unit area is

$$\frac{\dot{m}}{A} = \frac{\rho_e u_e}{q/m} = \frac{j}{q/m},$$

the exit velocity is

$$u_e = \sqrt{2(q/m)V_a},$$

and the maximum thrust per unit area is

$$\frac{T}{A} = \sqrt{\frac{2}{q/m}} \, jV_a^{1/2}$$

or, using Eq. (16–19),

$$\frac{T}{A} = \frac{8}{9} \epsilon_0 \left(\frac{V_a}{L}\right)^2. \tag{16–20}$$

Thus the maximum thrust per unit area depends only on the average electrostatic field intensity. Field intensity, in turn, is limited by the voltage-breakdown characteristics of the acceleration gap.

Using Eq. (16–1), we obtain Eq. (16–21):

$$\frac{T}{A} = \frac{2}{9} \frac{\epsilon_0}{L^2} \frac{(g_e I_{sp})^4}{(q/m)^2}, \tag{16–21}$$

where I_{sp} is the specific impulse and g_e is the acceleration due to gravity at the earth's surface. For a finite beam of diameter D which is essentially one-dimensional, the thrust may be written

$$T = \frac{2\pi}{9} \epsilon_0 V_a^2 R^2 \tag{16–22}$$

in which $R = D/L$ is called the *beam aspect ratio*.

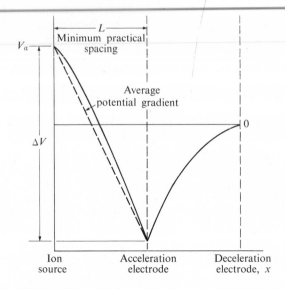

V_a

L

Minimum practical spacing

Average potential gradient

ΔV

0

Ion source

Acceleration electrode

Deceleration electrode, x

FIG. 16–16.   Plane triode accelerator potential distribution.

Equation (16–21) demonstrates the strong dependence of thrust per unit area on specific impulse for the space-charge-limited plane diode. For a given propellant the specific impulse determines the acceleration potential which, as Eq. (16–19) shows, limits the current density for any given spacing of the electrodes, provided the source is capable of providing this current density. It is advantageous, therefore, to reduce the electrode spacing as much as possible. Practical considerations limit this spacing to several millimeters or so, and at the acceleration voltages used this does not come close to the limiting average potential gradient which could be tolerated. The current density can be greatly increased for a given specific impulse (again provided the source is capable of supplying it) by using a so-called accel-decel system employing three electrodes which allow operation at maximum potential gradient, as illustrated in Fig. 16–16. In this way the current density is not limited by the specific impulse. The maximum current density, instead of being given by Eq. (16–19), is given by

$$j = \frac{4}{9}\,\epsilon_0 \sqrt{2q/m}\,\frac{(\Delta V)^{3/2}}{L^2}, \tag{16–23}$$

and the thrust per unit area by

$$\frac{T}{A} = \frac{8}{9}\,\epsilon_0 \left(\frac{\Delta V}{L}\right)^2 \left(\frac{V_a}{\Delta V}\right)^{1/2} \tag{16–24}$$

or

$$\frac{T}{A} = \frac{8}{9}\,\epsilon_0 \left(\frac{\Delta V}{L}\right)^2 \frac{g_e I_{\text{sp}}}{[2(q/m)\,\Delta V]^{1/2}}. \tag{16–25}$$

The maximum average electric field strength $\Delta V/L$ is limited by the possibility of electrical breakdown, especially in the presence of easily ionizable vapors. Experimental mercury and cesium engines have been reported operating at a field strength of 3×10^6 v/meter.

Even if additional current density is not achieved, some slight deceleration of the ions just upstream of the exhaust plane is desirable. The resultant potential "valley" serves to prevent the electrons from flowing upstream from the exhaust region. Such "backstreaming" electrons, if accelerated through the full ion acceleration voltage, could quickly overheat and destroy the ion source.

16–5   BEAM OPTICS

In general the design of the accelerator electrodes for the formation of a charged particle beam is not an easy task. Both analytical and experimental methods are used to design the electrode system or lens, as it is called. Useful analytical and analog methods have been developed by Pierce [10] for two- and three-dimensional beams.

Figure 16–17 shows a parallel two-dimensional beam surrounded by charge-free space. Pierce reasoned that the beam could be made to accelerate in the x-direction in the same way as the beam in the plane diode if the potential distribution along its boundaries duplicated that of an infinite beam. For flow limited by space charge, this would require that the potential along the edge vary according to Eq. (16–18), while (for any current) the potential gradient normal to the beam edge, $\partial V/\partial y$, must be zero. This boundary potential distribution must be established by a suitable arrangement of electrodes in the charge-free regions. Since electrodes constitute equipotentials, they should be located on equipotential lines corresponding to the desired boundary potential distribution.

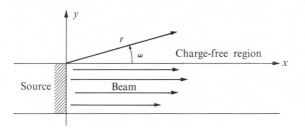

FIG. 16–17.   Two-dimensional beam.

In the charge-free region the potential must satisfy the equation of Laplace, [Eq. (5–8) with $\rho_q = 0$], which in the r, ω coordinates of Fig. 16–17 is

$$\frac{\partial^2 V}{\partial r^2} + \frac{1}{r}\frac{\partial V}{\partial r} + \frac{1}{r^2}\frac{\partial^2 V}{\partial \omega^2} = 0. \qquad (16\text{–}26)$$

From Eq. (16–18), the desired boundary potential distribution is

$$V(r, 0) = V_a - \left(\frac{9}{4}\frac{j}{\epsilon_0\sqrt{2q/m}}\right)^{2/3} r^{4/3}. \qquad (16\text{–}27)$$

The second boundary condition is

$$\frac{\partial V}{\partial y}(r, 0) = 0. \qquad (16\text{–}28)$$

A solution of Eq. (16–26) which will satisfy these conditions may be obtained as follows. Assuming that the potential may be written

$$V_a - V = f(r)g(\omega),$$

in which f and g are functions only of the indicated variable, we may transform Eq. (16–26) to

$$\frac{1}{f}(r^2 f'' + r f') = -\frac{g''}{g},$$

in which the primes signify differentiation with respect to r or ω, whichever is appropriate. The left-hand side of this equation is not a function of ω, the right-hand side is not a function of r. Therefore, since r and ω are independent variables, the equality can be meaningful only if both sides are independent of r and ω; i.e., if they are equal to some constant k^2. This being the case, two differential equations result,

$$r^2 f'' + r f' - k^2 f = 0, \qquad g'' + k^2 g = 0,$$

which are satisfied by the solutions

$$f(r) = Ar^k + Br^{-k}, \qquad g(\omega) = C\cos k\omega + D\sin k\omega.$$

Equation (16–26) is indeed satisfied by

$$V_a - V(r, \omega) = (Ar^k + Br^{-k})(C\cos k\omega + D\sin k\omega),$$

so that the original assumption of the form $f(r)g(\omega)$ is valid. The boundary conditions are satisfied by

$$B = 0, \qquad D = 0, \qquad AC = \left(\frac{9}{4}\frac{j}{\epsilon_0\sqrt{2q/m}}\right)^{2/3}, \qquad k = \frac{4}{3}.$$

Thus the potential distribution in the charge-free region may be written

$$V_a - V = \left(\frac{9}{4}\frac{j}{\epsilon_0\sqrt{2q/m}}\right)^{2/3} r^{4/3}\cos\left(\frac{4}{3}\omega\right). \qquad (16\text{–}29)$$

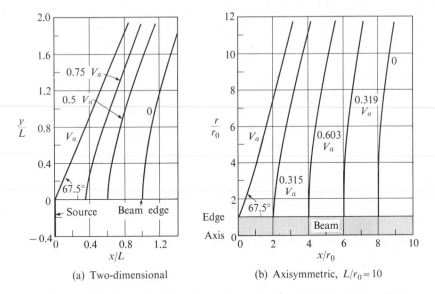

FIG. 16–18. Equipotentials for electrode geometry. (Courtesy Pierce [10].)

This solution is illustrated in Fig. 16–18(a), which shows the location of particular equipotentials. The theoretical solution is independent of the width of the beam. However, if the equipotential surfaces within the beam are to remain approximately planar, the beam width should be small compared to the acceleration length L, unless the last electrode can include a plane grid extending through the beam. Electrodes of the shapes illustrated in Fig. 16–18 might find use in two-dimensional slit sources or in annular sources where beam width is much less than the annular diameter.

Circular or axisymmetric beams are also common. For the circular parallel beam the same boundary conditions apply at the beam surface, but a closed-form solution of Laplace's equation corresponding to Eq. (16–29) has not been found. However, the solution may be obtained by the interesting and generally useful electrolytic-tank analog technique [10]. The charge-free region is represented experimentally by an electrolyte in a tank. The electrolyte carries a very weak current between two sheet-metal electrodes (equipotentials) which may be easily shaped to represent the positive and negative electrode of an accelerator. The potential distribution in the electrolyte also satisfies Laplace's equation. The potential distribution in the electrode is measured by means of a movable probe. A wedge-shaped section of electrolyte is used to represent a portion of the axisymmetric accelerator, the thin edge of the wedge representing the axis of the beam. The beam is represented by an insulating body. Since current cannot enter or leave the insulating body, the equipotentials in the electrolyte must be normal to its surface, thus automatically satisfying the second boundary condition. The first boundary condition is satisfied by a trial-and-error shaping of the two electrodes until the desired potential distribution is measured along the beam edge. Intermediate equipotentials are then easily plotted in the electrolyte. Figure 16–18(b) shows electrode shapes determined by this method for the particular case in which the accelerator length is ten times the beam radius. The method may be applied to determine electrode shape for any desired potential distribution.

Unless a grid is placed across the beam at the grounded electrode, the beam emerging from either of these devices will diverge somewhat. A grid would force the equipotentials within the beam to be plane, as assumed, in the region near the exhaust aperture. In the fieldfree region of exhaust space there will be a potential variation through the beam, the beam edge being at ground potential and the beam axis being at a higher potential due to the space charge of the beam itself. In the absence of an exhaust plane grid this distorts the potential distribution near the exhaust aperture in such a way as to cause beam divergence. The exhaust aperture is then said to act as a divergent lens. A grid is highly undesirable in a propulsion application, of course, since it would be rapidly destroyed by the high-energy beam.

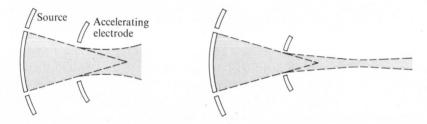

FIG. 16–19.   Conical beam geometries.   (Courtesy Pierce [10].)

Other geometries, some analytical, some empirical, are used. To offset the divergence caused by the exhaust aperture, an initially converging beam issuing from a conical geometry might be used. The characteristics of space-charge-limited current between spherical electrodes are known [10], and a conical section cut from such a current represents a finite beam which, in conjunction with a spherical ion source and suitable external electrodes, might be used as a propulsion device. Suitable electrodes are found using the electrolyte tank already discussed. Again the hole in the accelerating electrode distorts the flow from a truly conical pattern, but an initially converging beam can be produced. The space charge of an unneutralized exhaust beam would inevitably bring about divergence, as illustrated in Fig. 16–19. Note the similarity between these figures and the actual engines of Section 16–3.

These geometries require ion origination on a well-defined plane or spherical equipotential surface. This or any other geometric condition is easily satisfied by the contact ionization source which emits ions at a metallic surface. Other sources, such as the duoplasmatron, do not have well-defined emitting surfaces.

Exhaust neutralization

To avoid a charge buildup in the vehicle that would attract exhausted propellant and reduce thrust, it is necessary to expel charge of both signs at an equal rate. It may be necessary to neutralize the exhaust-space charge very near the exhaust plane, since a large positive potential "hill" arising from inadequately neutralized space charge could seriously perturb the beam. On the other hand the high charge-to-mass ratio of the electrons may cause them to be very quickly attracted toward the positive charge if they are emitted in the vicinity of the ion beam, thus eliminating any potential hill which might tend to build up.

Analytical treatment of neutralization techniques is extremely difficult, but two simple analyses of unneutralized beam behavior should serve to illustrate the problem. These analyses treat the limiting cases of very broad and very narrow beams. Experience with beam power tubes indicates that unneutralized beams of aspect ratio unity or less are easily projected without danger of beam "turnaround." With special magnetic restraining fields aspect ratios up to three can be handled, but beyond this particles in the center of the beam are brought to rest and turned around by the potential hill in the beam "exhaust" [8]. Hence beams of aspect ratio less than unity may be termed narrow, while beams of aspect ratios greater than about three may be said to be broad.

Except for edge effects, the behavior of very broad beams can be approximated by the flow between infinite-plane parallel grids at the same potential. The upstream grid represents the exhaust plane of the thrust chamber, while the downstream grid, as it moves off to an infinite distance, represents the conditions existing in the exhaust if no attempt at beam neutralization is made (but the space vehicle is kept neutral by some other means). Schematically these electrodes would appear as in Fig. 16–20, which includes the source electrode at the accelerating potential. The problem is to find the exhaust beam behavior as L_1/L_0 increases toward infinity. The solution of Fay, Samuel, and Shockley [11] is mentioned briefly in the discussion below.

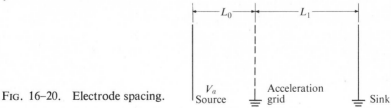

FIG. 16–20.   Electrode spacing.

For moderate sink distance ($L_1/L_0 < 2\sqrt{2}$), all current will reach the sink, while the space charge creates a symmetric potential hill between the acceleration grid and the sink, as in Fig. 16–21. If L_1/L_0 is now increased, the height of the potential hill will increase. As the spacing increases the maximum potential rises because there is more positive charge between the grids. Because of the increase in potential the velocity of the charges decreases, so that the charge density rises and this in turn leads to a further increase in the potential hill. At $L_1/L_0 = 2\sqrt{2}$,

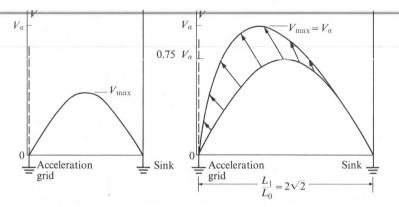

FIG. 16–21.   Potential difference between electrodes.

it is found that the maximum potential $V_{max} = 0.75\ V_a$. Any slight additional increase in L_1/L_0 leads to $V_{max} = V_a$ and a reduction in beam current. At this potential the particle kinetic energy is reduced to zero, and some particles drift forward while others are reflected. The presence of the reflected charges on the upstream side increases the charge density in that region and shifts the maximum potential V_{max} in the upstream direction, as in Fig. 16–21. The net effect of increasing L_1/L_0 beyond $2\sqrt{2}$ is thus an abrupt change from a symmetric potential distribution with full current transmission to an unsymmetric distribution with *reduced* current transmission. As L_1/L_0 is increased beyond about 3, the exhaust beam current goes rather rapidly to zero.

In the actual beam there is no sink electrode, of course (although there is one in all vacuum test chambers), and there will be an attempt at neutralization. The hypothetical grounded sink then corresponds to that plane where space charge is neutralized. Thus it would seem that neutralization must occur within a distance comparable to the acceleration spacing in a space-charge-limited broad beam.

Particle trajectory in a narrow beam is approximately described by the so-called paraxial-ray equation. This equation treats narrow beams in which the radial velocity is small compared with the axial, and in which the potential difference from beam axis to edge multiplied by the electron charge is small compared with the kinetic energy of the charged particles. As a result, one can assume that the radial potential gradient which causes beam divergence does not appreciably disturb the axial potential distribution or the velocity component. The reader is referred to Pierce [10] for the derivation and solution of this equation for the special case in which the potential is constant along the beam axis. Figure 16–22 illustrates typical unneutralized beam shapes for space-charge-limited cylindrical ion engines as a function of accelerator aspect ratio R, which is defined as the ratio of the initial diameter of the beam to the distance between accelerating electrodes. It can be seen that beams of aspect ratio unity or less are indeed only slightly divergent.

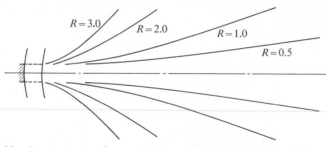

FIG. 16–22. Beam shapes for space-charge-limited flow of various aspect ratios.

Experimentally, adequate space-charge neutralization has been easily achieved in vacuum test chambers by a simple hot-filament electron emitter placed in the beam exhaust, even though an electron possessing merely thermal energy (2 or 3 eV) travels much faster than the ions in the exhaust beam. Figure 16–23 illustrates a slightly more complex neutralization scheme which is intended to exhaust electrons at near ion velocity. It represents schematically the neutralization system of the engine shown in Fig. 16–10, along with a typical potential distribution along the axis. Electrons are injected at right angles to the ion beam in a region of slightly positive potential which serves to "trap" the electrons. The electrons remain within this trap until they become intimately mixed with the ions, although it is unlikely that they will emerge with the same velocity as the ions. The presence of electrons within the electron trap tends to lower the potential there until the potential barrier at the exhaust passes electrons at the same rate they are injected. Note that the accel-decel ion system prevents backstreaming of electrons to the ionizer. This neutralization system has also performed satisfactorily in vacuum chamber tests. However, there are unresolved questions about the influence of the chamber walls and the general abundance of electrons in such test chambers. One of the major objectives of early space tests is the resolution of this problem.

We shall now discuss *electrothermal propulsion*.

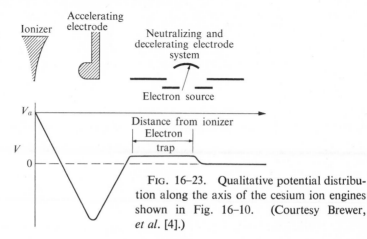

FIG. 16–23. Qualitative potential distribution along the axis of the cesium ion engines shown in Fig. 16–10. (Courtesy Brewer, *et al.* [4].)

16–6 THE ARCJET

In Chapter 10 it was shown that for certain space missions it would be desirable to have a specific impulse considerably higher than can be obtained with a chemical rocket. This requires much greater energy content per unit mass of propellant than is supplied by the process of combustion. It has been suggested that one way to obtain the necessary energy content is to use an electric-arc discharge to heat the fluid in a device called an arcjet or plasmajet. Much attention has been given to arcjets for a variety of applications [12], and their general characteristics and limitations have been identified. It is not yet clear whether they will be ultimately suitable for propulsion, but they are discussed briefly here as the simplest possible use of a conducting fluid in a rocket.

Figure 16–24 shows an arcjet schematically. Propellant is heated to a high temperature in an electric arc and then expanded in a conventional nozzle. If the propellant were to behave as a perfect gas and if the expansion were isentropic, the exhaust velocity would be given by

$$u_e = \sqrt{2 \frac{\gamma}{\gamma - 1} \frac{\overline{R}}{M} T_0 \left[1 - \left(\frac{p_e}{p_0} \right)^{(\gamma - 1)/\gamma} \right]}.$$

Actually this represents a considerable oversimplification of actual performance, but it does indicate the desirability of propellants which have high stagnation temperature and low molecular weight. Although hydrogen has the lowest molecular weight, it will be shown that helium is actually a more attractive propellant largely because it is monatomic and, therefore, cannot dissociate.

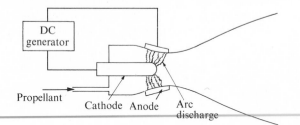

Fig. 16–24. Schematic diagram of an arcjet.

The high-current-density arc discharge is maintained by a sufficient voltage difference between cathode and anode. The behavior of high current discharges in relatively high-pressure gases is not well understood. In general, the arc consists of three zones: a central "arc column" and zones of high potential gradient immediately near the anode and cathode surfaces. The arc column occupies most of the discharge space. Although the potential gradients within it are relatively low, the arc column should have most of the total potential drop in order that a large fraction of the arc power be liberated within the main flow. The potential drops across the anode and cathode zones may be several volts, while the total potential across the arc may be of the order of hundreds of volts. The anode and cathode zones are, however, very thin, so that heat transfer into the electrodes can be severe. Near the cathode the discharge is nonuniform, tending to concentrate in "spots," especially at high current density. If the local current density increases slightly, so does the joule heating and the local conductivity. This leads to further increase in current density and to concentrated heat transfer. Current densities within such spots may be as high as 10^6 amp/cm^2 [12].

Near the anode the electrons acquire considerable energy as they fall through the potential drop. If the density is sufficiently high, this energy may be largely consumed through collisions within the gas. Otherwise it is transferred to the anode by direct collision with the surface. Likewise, ion acceleration in the cathode sheath can lead to ion energies sufficient to cause "sputtering" of the cathode, that is, removal of surface material as a result of ion impact. Thus a number of important phenomena govern the behavior of the arc in the vicinity of the electrode surfaces. These may have very strong effects on electrode life. It cannot be said that these effects are well understood, but considerable experience with carbon, copper, and tungsten electrodes has been obtained. It has been found that arc spots which would otherwise cause burnout can be tolerated if kept in motion on the electrode surface. Spot motion may be produced by imparting vortex motion to the propellant gas before it enters the arc, or by the presence of a suitable magnetic field which interacts with the arc current and induces motion of the arc transverse to the current direction.

Most of the transmission of energy to the propellant occurs in the arc column. Electrons and ions, accelerated in the electrostatic field, transfer their energy by collision to neutral gas particles. Since gas conductivity increases rapidly with temperature, attempts to produce high exhaust velocity (high T_0) are hampered by the associated decrease of arc-column resistance relative to the rest of the arc circuit. This results in a decreasing fraction of total electrical power liberated in the arc column.

The efficiency of an arcjet may be defined as the ratio of kinetic energy flux in the exhaust to electrical power supplied. As an example of experimental arcjet performance, Fig. 16–25 shows the data of Brogan [13] for arc-heated helium, indicating the expected falloff of efficiency at high specific impulse. It may be seen that efficiencies above 50% were obtained with specific impulse values up to about 1200.

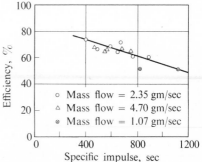

FIG. 16–25.   Efficiency of experimental arcjet.   (Courtesy Brogan [13].)

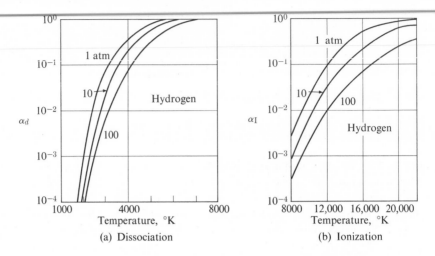

FIG. 16–26. Equilibrium dissociation and ionization of hydrogen. (Courtesy Camac [14].)

The choice of helium as propellant stems from several considerations. Although hydrogen has a lower molecular weight, it tends to dissociate at the high temperature involved in arcjet operation. Energy consumed by dissociation and ionization processes is, of course, not available for propellant acceleration unless recombination occurs during expansion. In general, little information is available for recombination rates of typical gases, but it appears that the fluid will expand with much the same composition it had in the chamber (or arc). The fraction f of the total energy supplied to the propellant which appears as thermal energy may be defined, following Camac [14] as

$$f = \frac{e_k}{e_k + \alpha_d e_d + \alpha_{\mathrm{I}} e_{\mathrm{I}} + \alpha_{\mathrm{II}} e_{\mathrm{II}}},$$

in which e_k is the internal energy added to the propellant per unit mass, and e_d, e_{I}, and e_{II} are the energies of dissociation, first ionization, and second ionization, respectively. The fractions α_d, α_{I}, and α_{II} are the respective dissociation and ionization fractions. Figure 16–26 illustrates the results of calculations of the equilibrium dissociation and ionization of hydrogen. The ionization calculation has been made with the aid of the Saha equation [Eq. (5–22)]. It may be seen that pressure has an important effect.

Using these results and similar ones for the ionization of helium, Figure 16–27 shows how the efficiency of an otherwise perfect thrust chamber is influenced by dissociation. This efficiency, which considers only losses due to dissociation and ionization, is called the "frozen-flow efficiency." It seems clear that helium is superior to hydrogen for specific impulse values up to at least about 2000 sec.

In addition to the lack of dissociation losses, helium exhibits several other favorable characteristics. It is chemically inert, of course. It has a very high ionization potential (see Table 16–1). This contributes to low ionization loss in

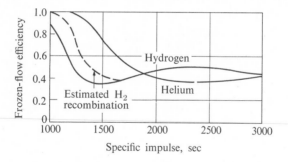

FIG. 16–27. Frozen-flow efficiency as a function of specific impulse, applicable to arcjet propulsion devices. (Courtesy Camac [14].)

the event of frozen flow in the nozzle. In addition, high arc column resistance can be maintained at relatively high gas temperatures, thus promoting good efficiency at high specific impulse. It also seems that potential drops at electrodes are low when helium is used [14].

Heat transfer to the walls is of course one of the major limitations on arcjet performance. As pointed out already, the electrode surfaces are subject to heat transfer from the zones of high joule heating immediately adjacent to them. Convective and radiative heat transfer to the rest of the nozzle may have total rates of 1 to 10 kw per cm^2. It has proved possible to water-cool copper electrode surfaces with heat-transfer rates as high as 2 kw/cm^2. At very high temperatures the energy loss by the propellant on account of radiation may be quite significant. Figure 16–28 indicates a characteristic radiation time (the time it would take to radiate all the energy of the gas at the rate corresponding to the initial temperature). It may, at high pressures and temperatures, reduce to the same order of magnitude as the time the propellant takes to pass through the nozzle, perhaps 10^{-3} or 10^{-4} sec.

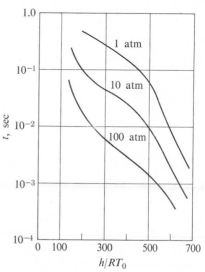

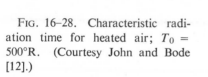

FIG. 16–28. Characteristic radiation time for heated air; $T_0 = 500°R$. (Courtesy John and Bode [12].)

In summary, the principal limitations on the performance of the arcjet are heat transfer, energy losses through dissociation and ionization, and loss of material from the electrode surfaces. The last point is particularly serious for a space-vehicle propulsion unit which should be capable of long periods of operation. More effort will be required to make the arcjet an operational rocket.

We now move on to a consideration of *electromagnetic propulsion*.

Some introductory principles of magnetogasdynamic acceleration of a plasma were discussed in Chapter 5. In particular, Eqs. (5–43) to (5–47) treated steady one-dimensional flow of a plasma. Although the general principles of plasma acceleration are now becoming fairly well understood, the ultimate utility of plasma accelerators in space propulsion is not yet clear. In the following sections we must content ourselves with a brief examination of current research and a few of the future possibilities.

16–7   STEADY CROSSED-FIELD ACCELERATORS

Figure 16–29 is a schematic view of a steady-flow magnetogasdynamic accelerator bounded on diverging walls by continuous electrodes. Plasma, perhaps from a source very similar to an arcjet, enters the acceleration region from the left. A magnetic field is provided by an electromagnet consisting of a coil and, very likely, a ferromagnetic pole piece.

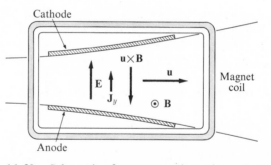

FIG. 16–29.   Schematic of magnetogasdynamic accelerator.

In the absence of appreciable Hall effects, a current j_y flows through the plasma in response to the applied and induced electrostatic fields according to

$$j_y = \sigma(E - uB),$$

where σ is the plasma conductivity and B is the magnetic field intensity. In order to achieve useful conductivities (of the order of 1 mho/cm or more) it may be necessary to "seed" the propellant gas with an alkali metal, as discussed in Chapter 5. Acceleration of the plasma occurs primarily as a result of the $\mathbf{j} \times \mathbf{B}$ or Lorentz force per unit volume, although pressure forces may also contribute to the momentum increase.

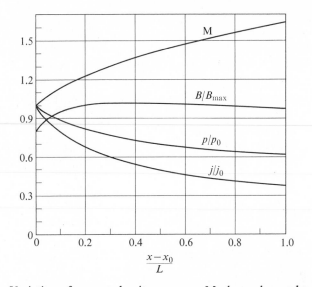

FIG. 16–30.   Variation of current density, pressure Mach number and magnetic field strength for the one-dimensional flow of a constant-temperature plasma. Constant electric field, $B_{\max} = 1$ weber/m^2, $L = 10$ cm, $\sigma = 320$ mho/m; atmospheric inlet. (Courtesy Oates [15].)

The channel-flow equations of Chapter 5, Eqs. (5–43) through (5–47), provide some insight into the plasma behavior within an idealized acceleration region. For example, a typical variation of plasma properties, for the case of constant area, constant temperature (constant σ), and constant E, as described by Eq. (5–54), is shown in Fig. 16–30.

Numerous other one-dimensional channel flow solutions which are available [15, 16, 17] may be used to show the general or bulk characteristics of crossed-field plasma accelerators. However, these solutions do not account for certain important factors. If the size of the apparatus is too small, viscous effects may be so important that they cannot be ignored and the flow cannot be considered one-dimensional. In addition, if the acceleration region is not at least several channel widths in length, it is difficult to establish fields of desired strength variation along the axis. For example, the constant-temperature solution of Fig. 16–30 requires specific distributions of E and B which would be difficult to provide in a short acceleration channel.

Demetriades, *et al.* [18, 19, 20, 21] have reported measurements on an experimental crossed-field accelerator. Figure 16–31(a) is a schematic diagram of their apparatus, which enabled direct measurement of the thrust created by the electromagnetic acceleration device. Figure 16–31(b) is a photograph of the plasma flow with the magnet pole piece removed to show the acceleration region. The plasma source and the accelerator electrodes are clearly visible in the photograph.

The thrust developed by the accelerator may be determined by application of Eq. (2–4) for a control volume enclosing the test section. In the experiment of

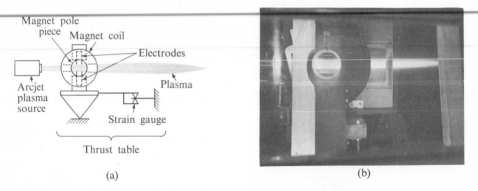

FIG. 16–31. Schematic diagram and photograph of a magnetogasdynamic accelerator which permits direct thrust measurements. (Courtesy Demetriades [19].)

Demetriades [19], pressure forces were negligible compared to the body force X on the fluid. Hence the thrust on the accelerator could be written

$$\mathcal{T} = X = \dot{m}(u_2 - u_1), \tag{16–30}$$

in which the subscripts 1 and 2 denote entrance to, and exit from, the test section. The body force X is the integral of $j \times B$ throughout the acceleration region. Assuming that current density varies only in the stream direction and that the current is perpendicular to the B-fields, we have

$$X = \int_0^L jBA \, dx = \int_0^L jBwh \, dx, \tag{16–31}$$

where w = channel width (parallel to B), h = channel height (parallel to j), and L = channel length. In general, j will vary with x in a complex manner. For example, if Hall currents are negligible,

$$j = \sigma(E - uB),$$

and at least u will vary. However, for the special case of constant B and constant h (as approximated in these experiments), it will now be shown that the current distribution is not important. Let the increment of total current between x and $x + dx$ be dI. Then

$$dI = jw \, dx,$$

and from Eq. (16–31), for constant B and h,

$$X = Bh \int_0^L dI = BhI_t, \tag{16–32}$$

in which I_t is the total current. Thus

$$\mathcal{T} = X = BhI_t. \tag{16–33}$$

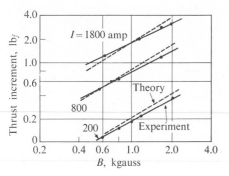

FIG. 16–32. Experimental accelerator thrust as a function of magnetic field strength, as compared with Eq. (16–33). (Courtesy Demetriades [19].)

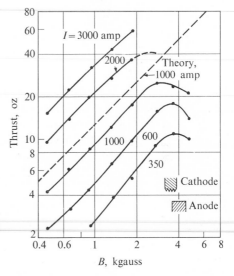

FIG. 16–33. Experimental accelerator performance at high magnetic field strength. (Courtesy Demetriades [20].)

Despite the foregoing approximations, Eq. (16–33) has been experimentally verified, as shown in Fig. 16–32. At higher magnetic field strengths the simple expression is inadequate, as indicated by Fig. 16–33. The falloff of thrust is attributed to at least two effects. First, Hall currents become appreciable, so that the current is no longer parallel to the E-field (Chapter 5) and the thrust is no longer parallel to the axis. At very high magnetic field strengths the exhaust beam was visibly deflected by the transverse force associated with the flow of Hall currents. Second, and perhaps much more important, large end effects were believed to have occurred. The current, rather than traverse the high resistance within the magnetic field, may have partially bypassed the acceleration region altogether. Thus, while Eq. (16–33) does not depend on current distribution, it does require that all of the current flow through the plasma in a region where the magnetic field is essentially constant, since current outside the magnetic field does not contribute to the thrust. In a slightly more complex model [21], the assumption of current bypassing the magnetic field adequately predicts the falloff of thrust at high B.

The efficiency of the accelerator may be defined as the ratio of the power absorbed in propellant acceleration (increase of kinetic energy flux) to the total electrical power required:

$$\eta = \frac{(\dot{m}/2)(u_2^2 - u_1^2)}{\mathcal{P}}, \tag{16–34}$$

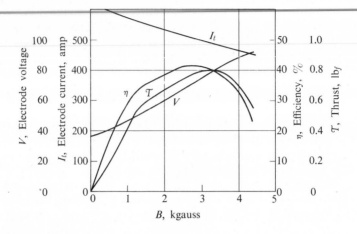

FIG. 16–34. Typical experimental performance of magnetogasdynamic accelerator with argon plasma. (Courtesy Demetriades [19].)

in which $\mathcal{P}$ is the total electrical power. Since thrust is more easily measured than exhaust velocity, we use Eq. (16–30), to express the velocity as

$$u_2 = \frac{T}{\dot{m}} + u_1.$$

The efficiency may be expressed as

$$\eta = \frac{T^2/2\dot{m} + Tu_1}{\mathcal{P}}. \qquad (16\text{–}35)$$

Experimentally only u_1 is difficult to measure, but it can be done with reasonable accuracy. Figure 16–34 indicates typical performance data for the accelerator whose thrust variation was presented in Fig. 16–32. Later accelerators operated at 54% efficiency at a thrust of 3.6 lb and specific-impulse increment of 1200 sec [20].

The extreme simplicity of these thrust and efficiency expressions can be misleading unless it is remembered that I_t is an experimentally determined quantity. Actual calculation of I_t would entail all the complexities and uncertainties of estimating plasma conductivity and current distribution.

Several qualitative observations may be based on the operation of these accelerators. In most cases the electrodes employed were fabricated of one-inch-square tungsten rods. The cathode surface was sometimes serrated, as shown schematically in Fig. 16–33. It was found that a simple field emission from a "cold" (i.e., not intentionally heated) cathode provided sufficient current density. The presence of a conducting plasma between the electrodes would tend to flatten the potential gradient in the middle while making it steeper near the electrodes. The high field strength near the cathode, then, contributed to the satisfactory field-emission characteristics. Although the electrodes were not operated with axial potential variations, it was found that appropriate staggering (anode upstream of cathode) of the electrodes could significantly reduce Hall effects [22].

Often, as in Fig. 16–31(b), the electrodes were not even in contact with the luminous core of the plasma. This greatly reduced electrode heating problems while not appreciably increasing the resistance of the gap. Apparently the plasma jet is surrounded by a relatively cool layer of reasonable conductivity [19]. This, of course, could have important effects on the reliability or lifetime of such devices.

In all cases the apparent conductivity of the plasma within the electrode gap was substantially greater than would be expected from calculations of equilibrium conductivity. Possibly, as mentioned in Chapter 5, the conducting electrons have substantially higher "temperature" than the plasma.

Other preliminary experimental tests of crossed-field accelerators have been reported [23, 24], but much research and development will be needed before the feasibility of this device for space propulsion can be assessed. The information on typical efficiency and specific impulse obtained by Demetriades encourages further work.

16–8 PULSED-PLASMA ACCELERATORS

One of the disadvantages of steady crossed-field accelerators is that they require a substantial external field and therefore a massive electromagnet. It is possible to make an accelerator for which an electromagnet is unnecessary by using the plasma current itself to generate the B-field, which gives rise to the $\mathbf{j} \times \mathbf{B}$ body force. Whereas the crossed-field accelerator is analogous to a shunt motor (which has separate current circuits for j and B), the analog of this type of accelerator is the series motor in which the magnetic field is established by the same current which interacts to establish the $\mathbf{j} \times \mathbf{B}$ force. Perhaps the simplest example of such a device is the rail-type accelerator [25, 26] shown in Fig. 16–35. A thin wire is placed across two rails (relatively long, straight electrodes). A high-voltage discharge between the rails vaporizes and ionizes the wire material. The high current, perhaps 10^4 or 10^5 amp, in the loop induces a magnetic field as shown, which interacts with the current to set up a body force, $\mathbf{j} \times \mathbf{B}$ per unit volume, acting on the plasma in a direction parallel to the rails. Velocities of up to 10^7 cm/sec have been measured experimentally by magnetic pickups. Assuming that all the mass is accelerated to this speed, the efficiency may be calculated from the mass of the wire and the discharge energy. By various experiments, efficiencies of

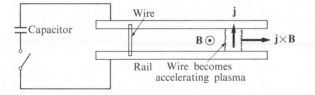

Fig. 16–35. Schematic of rail-type plasma accelerator.

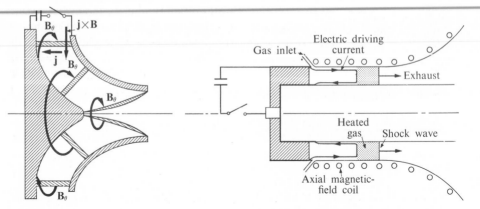

FIG. 16–36. Plasma pinch engine. (Courtesy Gloersen [26].)

FIG. 16–37. Magnetic annular shock-tube accelerator. (Courtesy Kantrowitz and Jones [28].)

up to 40% have been estimated in this way. However, it seems unlikely that the plasma is accelerated uniformly in these experiments.

Another configuration which uses the induced field of the plasma current for acceleration is the axisymmetric plasma pinch engine [26, 27] shown in Fig. 16–36. A discharge between the electrodes induces a magnetic field B_θ in the circumferential direction. The body force $\mathbf{j} \times \mathbf{B}$ then acts radially inward and accelerates the plasma into the nozzle. Successive positions of the discharge are indicated in Fig. 16–36.

Still another configuration of the pulsed-plasma accelerator is shown in Fig. 16–37. It has been referred to as the annular shock-tube accelerator [28]. In this apparatus the current discharges radially, inducing a circumferential magnetic field.

In general these devices have produced high exhaust velocities. However, a very serious problem is raised by the fact that they are essentially unsteady. Usually the heavy current during discharge comes from a capacitor. If the mass of the capacitor energy-storage reservoirs is not to become very large for a given thrust, they must be operated at high frequency, perhaps 10^3 or 10^4 cps. It has not yet been possible experimentally to achieve continuous operation with frequencies higher than 10^2 cps. Further, the accelerators are subject to severe electrode erosion because of the magnitude of the currents during the discharge period.

16–9 TRAVELING-WAVE ACCELERATORS

A third type of plasma accelerator, sometimes called the magnetic-induction plasma motor, offers potential advantages over both the foregoing accelerators. It requires neither external magnets nor electrodes, and relies on currents being induced in the plasma by a traveling magnetic wave.

Consider first a single conductor surrounding a transparent duct, as indicated in Fig. 16–38. The duct contains stationary plasma. If the current I in the conductor increases, the magnetic field strength in the plane of the loop will increase. Then, according to Faraday's induction law, an electromotive force will be induced in any loop in this plane. If the conductor current I increases rapidly enough, the induced electric field will establish a substantial plasma current, as in Fig. 16–38. The induced magnetic field and the induced plasma current then interact to cause a body force normal to both, which tends to compress the plasma toward the axis of the tube and expel it axially in both directions. This phenomenon has been demonstrated experimentally [30] for single-current loops as well as in traveling-wave accelerators.

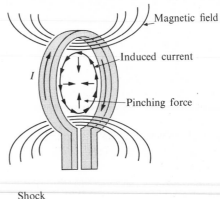

FIG. 16–38. Induced plasma current. (After Clauser [29].) (Adapted from Fig. 5, page 128 of "The Magnetic Induction Plasma Engine," a paper presented by M. U. Clauser, AGARD meeting, Pasadena, Cal., August 1960. Also published in *Advanced Propulsion Techniques*, S. S. Penner, editor; Pergamon Press, 1961.)

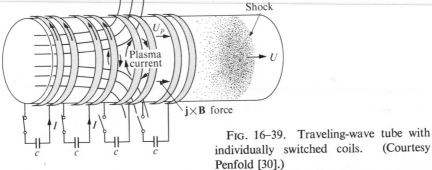

FIG. 16–39. Traveling-wave tube with individually switched coils. (Courtesy Penfold [30].)

As shown in Fig. 16–39, a traveling-wave accelerator makes use of a number of sequentially energized external conductors along the duct. As the switches are fired in turn, the magnetic field lines move axially along the tube, interacting with induced currents and imparting axial motion to the plasma. The induced currents, and therefore the force on the plasma, will be large only if the plasma conductivity is high. For infinite plasma conductivity, the plasma will move downstream at the same speed as the magnetic-field lines. For finite conductivity, there will be relative motion (slippage) between the two and, depending on the degree of slippage, the so-called magnetic piston is said to be hard or soft. Joule heating by the induced currents tends to maintain the conductivity of the plasma.

The inward radial force on the plasma in this accelerator appears to offer an advantage in keeping the high-temperature plasma away from the solid walls of the duct. The fact that no electrodes are needed is also an attractive feature.

The switching circuits indicated in Fig. 16–39 may be replaced by a single transmission line connecting a series of field coils which are energized in sequence. Alternatively, for cyclic operation, a three-phase power supply could be used, connecting adjacent coils to different phases. Several experimental devices [30] have been constructed to study this general type of accelerator, but the work is still in an early stage.

In general the electrostatic accelerators are more highly developed than the electromagnetic, and much work remains to be done to determine the future of both. For significant thrust levels, both machines require a substantial supply of electrical power. The generation of sufficient electrical power in space vehicles, employing a reliable system of low mass and long endurance, is still a very challenging problem.

References

1. *Handbook of Chemistry and Physics*, 43rd edition. Cleveland, O.: Chemical Rubber Publishing Co., 1961, pages 2586–2587

2. BLEAKNEY, W., "Probability and Critical Potentials for the Formation of Multiply Charged Ions in Hg Vapor by Electron Impact." *The Physical Review*, **35,** 2, January 1930, pages 139–148

3. KAUFMAN, H. R., "An Ion Rocket with an Electron-Bombardment Ion Source." *NASA TN* D-585, January 1961

4. BREWER, G. R., M. R. CURRIE, and R. C. KNECHTLI, "Ionic and Plasma Propulsion for Space Vehicles." *Proceedings of the I.R.E.*, **49,** 12, December 1961, pages 1789–1821

5. PUGH, E. M., and E. W. PUGH, *Principles of Electricity and Magnetism*. Reading, Mass.: Addison-Wesley, 1960, Chapter 6

6. STUHLINGER, E., and R. N. SEITZ, "Electrostatic Propulsion Systems," *Advances in Space Sciences*, Volume II. New York: Academic Press, 1960

7. SHELTON, H., R. F. WUERKER, and J. M. SELLEN, "Generation and Neutralization of Ions for Electrostatic Propulsion." American Rocket Society meeting, San Diego, Calif., June 1959; American Rocket Society Preprint 882-59

8. Staff of the Ramo-Wooldridge Research Laboratory, "Electrostatic Propulsion." *Proceedings of the I.R.E.*, **48,** 4, April 1960, pages 477–491

9. HUSMANN, O. K., "Diffusion of Cesium and Ionization on Porous Tungsten." *Electrostatic Propulsion*, edited by D. B. Langmuir, E. Stuhlinger, and J. M. Sellen, Jr. New York: Academic Press, 1961, pages 505–522

10. PIERCE, J. R., *Theory and Design of Electron Beams*, 2nd edition. New York: D. Van Nostrand, 1954

11. FAY, C. E., A. L. SAMUEL, and W. SHOCKLEY, "On the Theory of Space Charge Between Parallel Plane Electrodes." *Bell System Technical Journal*, **16,** January 1938, pages 49–79

12. JOHN, R. R., and W. L. BODE, "Recent Advances in Arc Plasma Technology." *J. Amer. Rocket Soc.*, **31,** 1, January 1961, pages 4–17

13. BROGAN, T. R., "Preliminary Evaluation of Helium as an Electric Arc-Heated Propellant." *J. Amer. Rocket Soc.*, **29,** 9, September 1959, pages 662–663

14. CAMAC, M., "Plasma Propulsion Devices," *Advances in Space Sciences*, Volume II. New York: Academic Press, 1960

15. OATES, G. C., J. K. RICHMOND, Y. AOKI, and G. GROHS, "Loss Mechanisms of a Low Temperature Plasma Accelerator." *J. Amer. Rocket Soc.*, **23,** 4, April 1962, page 541

16. DEISSLER, R. G., "A One-Dimensional Analysis of Magnetohydrodynamic Energy Conversion." *NASA TN* D-680, March 1961

17. DRAKE, J. H., "Optimum Isothermal Acceleration of a Plasma with Constant Magnetic Field." *J. Amer. Inst. Aeronautics and Astronautics*, **1,** 9, September 1963

18. DEMETRIADES, S. T., "Experimental Magnetogasdynamic Engine for Argon, Nitrogen and Air," *Engineering Aspects of Magnetogasdynamics*. New York: Columbia University Press, 1962, pages 19–44

19. DEMETRIADES, S. T., and R. W. ZIEMER, "Energy Transfer to Plasmas by Continuous Lorentz Forces." American Rocket Society Preprint 2002–61; *also* Proceedings of the Fourth Biennial Gas Dynamics Symposium, Northwestern Univ., Evanston, Ill., 1961

20. DEMETRIADES, S. T., and R. W. ZIEMER, "Direct Thrust and Efficiency Measurements of a Continuous Plasma Accelerator." *J. Amer. Rocket Soc.*, **31,** 9, September 1961, pages 1278–1280

21. DEMETRIADES, S. T., G. L. HAMILTON, R. W. ZIEMER, R. W. JARL, and P. D. LENN, "Three-Fluid Nonequilibrium Plasma Accelerators," Part I. American Rocket Society Preprint 2375–62, March 1962

22. DEMETRIADES, S. T., "Experiments with a High Specific Impulse Crossed-Field Accelerator." Presented at Third Annual Symposium on Engineering Aspects of Magnetohydrodynamics, University of Rochester, March 28–30, 1962

23. BLACKMAN, V. H., and R. J. SUNDERLAND, "Experimental Performance of a Crossed-Field Plasma Accelerator." *J. Amer. Inst. Aeronautics and Astronautics*, **1,** 9, September, 1963

24. RAGUSA, D., and J. BAKER, "Experimental Results with a Direct-Current Electromagnetic Accelerator," *Engineering Aspects of Magnetohydrodynamics*. New York: Columbia University Press, 1961

25. BOSTICK, W., "Plasma Motors." *J. Astronautical Sciences*, **5,** 2, 1958

26. GLOERSEN, P., "Pulsed-Plasma Accelerators." American Rocket Society Preprint 2129–61, October 1961

27. KIMEN, A. E., and W. MCILRAY, "The Electromagnetic Pinch Effect in Space Propulsion." Proceedings of the Third Biennial Gas Dynamics Symposium on Dynamics of Conducting Gases, Northwestern University Press, Evanston, Ill., 1960

28. KANTROWITZ, A., and G. S. JONES, "On Magnetohydrodynamic Propulsion." Presented at the 14th annual meeting of the American Rocket Society, Washington, D. C., November 16–20, 1959; American Rocket Society Preprint 1009–59

29. CLAUSER, M. U., "The Magnetic-Induction Plasma Engine," *Advanced Propulsion Techniques*. Proceedings of a technical meeting sponsored by the AGARD Combustion and Propulsion Panel, Pasadena, Cal., Aug. 24–26, 1960. New York: Pergamon Press, 1961

30. PENFOLD, A. S., "Traveling-Wave Plasma Accelerators." American Rocket Society Preprint 2130–61, October 1961

31. HYMAN, J., W. O. ECKHARDT, R. C. KNECHTLI, and C. R. BUCKLEY, "Formation of Ion Beams from Plasma Sources." *J. Amer. Inst. Aeronautics and Astronautics*, **2,** 10, October 1964

32. KAUFMAN, H. R., and P. D. READER, "Experimental Performance of Ion Rockets Employing Electron-Bombardment Ion Sources," in *Electrostatic Propulsion*, edited by David B. Langmuir, Ernst Stuhlinger, and J. M. Sellen, Jr.), Volume 5 of *Progress in Astronautics and Rocketry*. New York: Academic Press, 1961

Problems

1. An electrostatic rocket is to use heavy particles with charge-to-mass ratio of 500 coul/kg to produce a specific impulse of 3000 sec. What acceleration voltage would be necessary?

    With one-dimensional space-charge-limited current and a maximum allowable gradient of 10^5 v/cm, what is the acceleration distance and diameter of a round beam producing 0.5 lb thrust? [*Note:* 1 $lb_f = 4.45$ n.]

2. A cesium ion rocket is to be used at a specific impulse of 5000 sec. Assuming the cesium to be 100% singly ionized ($q/m = 7.25 \times 10^5$ coul/kg), what is the required acceleration voltage and the beam power ($\dot{m}u_e^2/2$) per unit thrust?

    If the cesium is 90% singly ionized and 10% doubly ionized, what is the specific impulse and beam power per unit thrust, assuming that the acceleration voltage is unchanged?

3. A cesium contact ionization rocket of 10 millipounds thrust (4.45×10^{-2} n) at a specific impulse of 5000 sec is to be built within the following two limitations: (a) minimum electrode space = 5 mm, (b) maximum average gradient between electrodes = 10^5 v/cm. Assuming the acceleration region to behave as a simple space-charge-limited plane diode, what beam area is required?

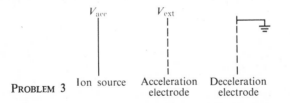

PROBLEM 3

    To reduce the beam area, it is proposed to increase the current density by first accelerating through a larger potential difference. The ions are subsequently decelerated to the proper exhaust speed (see diagram). What is the minimum beam area and the corresponding acceleration potential?

4. An arcjet heats hydrogen at 50 atm pressure from 300°K to 6000°K. At the latter temperature the reaction $\frac{1}{2}H_2 \rightarrow H$ is described by the equilibrium constant $K_p = 15.0$ (p in atmospheres), and over the temperature interval the specific heats are

    $$\bar{c}_{p_H} = 6 \text{ cal/gm-mole·°K}, \qquad \bar{c}_{p_{H_2}} = 9 \text{ cal/gm-mole·°K}.$$

    The heat of formation of H is $+ 52 \times 10^3$ cal/gm-mole at 300°K. If no recombination takes place in the nozzle, what fraction of the total energy addition to the fluid is lost due to dissociation?

5. An arcjet is used to heat and expand hydrogen from an inlet pressure and temperature of 10 atm and 100°K. How much electrical energy must be added to the fluid to raise its temperature to 3000°K, assuming equilibrium composition? It may be assumed that the mean specific heats of H and H_2 are 0.004 and 0.006 kcal/gm-mole·°K, respectively.

APPENDIXES

Optimization of Multistage Rockets

The optimization of multistage rockets with respect to the distribution of mass has been treated in a number of interesting papers [1, 2, 3, and 4]. The problem is reasonably simple if all stages have the same exhaust velocity and the same structural coefficients ϵ. In this case the increment of vehicle velocity due to the ith stage is

$$\Delta u_i = u_e \ln \frac{1 + \lambda_i}{\epsilon + \lambda_i},$$ (A–1)

and the final or nth-stage velocity is

$$\frac{u_n}{u_e} = \sum_1^n \ln \frac{1 + \lambda_i}{\epsilon + \lambda_i} = \sum_1^n F(\lambda_i).$$ (A–2)

The techniques of the calculus of variations may be employed to find the payload ratios of λ_i such that the velocity ratio u_n/u_e is maximized. For a given payload $\mathfrak{M}_{\mathfrak{L}}$ and initial mass $\mathfrak{M}_{01}$, the payload ratios λ_i are subject to the constraint of Eq. (10–33):

$$\frac{\mathfrak{M}_{\mathfrak{L}}}{\mathfrak{M}_{01}} = \prod_1^n \left(\frac{\lambda_i}{1 + \lambda_i} \right).$$

This constraint may be expressed more conveniently in logarithmic form as

$$\ln \frac{\mathfrak{M}_{\mathfrak{L}}}{\mathfrak{M}_{01}} = \sum_1^n \ln \left(\frac{\lambda_i}{1 + \lambda_i} \right) = \sum_1^n G(\lambda_i).$$ (A–3)

In order to maximize the quantity $\sum_1^n F(\lambda_i)$ subject to Eq. (A–3), it is appropriate [5] to maximize

$$L(\lambda_i) = F(\lambda_i) + \alpha G(\lambda_i),$$ (A–4)

where α is an undetermined constant, the *Lagrange multiplier*. For a maximum

$$\frac{\partial L}{\partial \lambda_i} = \frac{\partial F}{\partial \lambda_i} + \alpha \frac{\partial G}{\partial \lambda_i} = 0.$$

Using Eqs. (A–2) and (A–3), we obtain

$$+ \frac{1}{1 + \lambda_i} - \frac{1}{\epsilon + \lambda_i} + \frac{\alpha}{\lambda_i} - \frac{\alpha}{1 + \lambda_i} = 0,$$ (A–5)

which may be reduced to

$$\lambda_i = \frac{\alpha\epsilon}{1 - \alpha - \epsilon}. \tag{A-6}$$

Since all other terms in the equation are constant, λ_i must be constant. Thus, using Eq. (10–33) again, and setting $\lambda_i = \lambda$, we have

$$\frac{\mathfrak{M}_\mathfrak{L}}{\mathfrak{M}_{01}} = \prod_1^n \left(\frac{\lambda}{1 + \lambda}\right) = \left(\frac{\lambda}{1 + \lambda}\right)^n$$

or

$$\lambda = \frac{(\mathfrak{M}_\mathfrak{L}/\mathfrak{M}_{01})^{1/n}}{1 - (\mathfrak{M}_\mathfrak{L}/\mathfrak{M}_{01})^{1/n}}. \tag{A-7}$$

If, for a particular case, u_e is constant but some approximate variation of ϵ_i is assumed, perhaps according to Fig. 10–6, then a more refined calculation may be made. In this case, Eq. (A–6) becomes

$$(\epsilon_i + \alpha - 1)\lambda_i + \alpha\epsilon_i = 0$$

or

$$\lambda_i = \frac{\alpha\epsilon_i}{1 - \alpha - \epsilon_i}, \tag{A-8}$$

and Eq. (10–33) becomes

$$\frac{\mathfrak{M}_\mathfrak{L}}{\mathfrak{M}_{01}} = \prod_1^n \frac{-\alpha\epsilon_i}{\epsilon_i + \alpha - 1 - \alpha\epsilon_i}. \tag{A-9}$$

From Eq. (A–9), α is found in terms of $\mathfrak{M}_\mathfrak{L}$, $\mathfrak{M}_{01}$, and the assumed ϵ_i. Then the λ_i are found from Eq. (A–8).

For the case where both ϵ_i and u_{ei} are *known* variables, a similar procedure leads to minimum $\mathfrak{M}_{01}/\mathfrak{M}_\mathfrak{L}$ for a given final velocity in the absence of drag and gravity [2]. Using the definition of $\mathfrak{R}_i$ [Eq. (10–29)], we may write the overall mass ratio as

$$\frac{\mathfrak{M}_{01}}{\mathfrak{M}_\mathfrak{L}} = \prod_1^n \left(1 + \frac{\mathfrak{R}_i - 1}{1 - \epsilon_i\mathfrak{R}_i}\right). \tag{A-10}$$

This is to be minimized subject to the constraint

$$u_n = \sum_1^n u_{ei} \ln \mathfrak{R}_i = \sum_1^n G(u_{ei}, \mathfrak{R}_i). \tag{A-11}$$

Again, for convenience, we express the first equation in logarithmic form:

$$\ln \frac{\mathfrak{M}_{01}}{\mathfrak{M}_\mathfrak{L}} = \sum_1^n \ln \left(1 + \frac{\mathfrak{R}_i - 1}{1 - \epsilon_i\mathfrak{R}_i}\right) = \sum_1^n F(\epsilon_i, \mathfrak{R}_i). \tag{A-12}$$

If we proceed in the same way as with Eq. (A–5), the minimum occurs when $\mathfrak{R}_i$ satisfies

$$\frac{\partial}{\partial \mathfrak{R}_i} \ln \left(1 + \frac{\mathfrak{R}_i - 1}{1 - \epsilon_i \mathfrak{R}_i} \right) + \alpha \frac{\partial}{\partial \mathfrak{R}_i} (u_{ei} \ln \mathfrak{R}_i) = 0, \qquad (A–13)$$

where α is again the Lagrange multiplier, a constant. After differentiation and rearrangement, we have

$$\mathfrak{R}_i = \frac{\alpha u_{ei} + 1}{\alpha \epsilon_i u_{ei}}. \qquad (A–14)$$

A solution for α is obtained by substituting Eq. (A–14) in Eq. (A–11) to yield

$$u_n = \sum_1^n u_{ei} \ln \frac{\alpha u_{ei} + 1}{\alpha \epsilon_i u_{ei}}. \qquad (A–15)$$

After α is obtained from Eq. (A–15), the mass ratios $\mathfrak{R}_i$ can be determined from Eq. (A–14). The individual payload ratios are given by

$$\lambda_i = \frac{1 - \mathfrak{R}_i \epsilon_i}{\mathfrak{R}_i - 1}. \qquad (A–16)$$

Note that this analysis says nothing about selection of the stage exhaust velocities.

References

1. BUILDER, C. H., "General Solution for Optimization of Staging of Multistaged Boost Vehicles." *J. Amer. Rocket Society*, August 1959

2. HALL, H. H. and E. D. ZAMBELLI, "On the Optimization of the *N*-Step Rocket." *Jet Propulsion*, July 1958

3. SUBOTOWICZ, M., "The Optimization of the *N*-Step Rocket with Different Construction Parameters and Propellant Specific Impulse in Each Stage." *Jet Propulsion*, July 1958

4. WEISBORD, LEON, "Optimum Staging Techniques." *J. Amer. Rocket Society*, June 1959

5. HILDEBRAND, F. B., *Methods of Applied Mathematics for Engineers.* New York: Prentice-Hall, 1952

APPENDIX II

Tables

TABLE 1 Conversion Factors

1 meter (m) $=$ 100 cm $=$ 39.4 in. $=$ 3.28 ft

1 centimeter (cm) $=$ 10 millimeters (mm) $=$ 0.394 in.

1 kilometer (km) $=$ 1000 m $=$ 0.621 mi

1 foot (ft) $=$ 12 in. $=$ 0.305 m

1 inch (in.) $=$ 0.0833 ft $=$ 2.54 cm

1 mile (mi) $=$ 5280 ft $=$ 1.61 km

1 m/sec $=$ 3.28 ft/sec $=$ 2.24 mi/hr $=$ 3.60 km/hr

1 ft/sec $=$ 0.305 m/sec $=$ 0.682 mi/hr $=$ 1.10 km/hr
   (*Note:* it is often convenient to remember that 88 ft/sec $=$ 60 mi/hr.)

1 mi/hr $=$ 1.47 ft/sec $=$ 0.447 m/sec $=$ 1.61 km/hr

1 radian (rad) $=$ 57.30° $=$ 57°18′

1° $=$ 0.01745 rad

1 revolution/minute (rev/min) $=$ 0.1047 rad/sec

1 kilogram (kg) $=$ 1000 grams (gm) $=$ 0.0685 slug $=$ 2.21 lb_m

1 slug $=$ 14.6 kg $=$ 32.2 lb_m

1 atomic mass unit (amu) $=$ 1.66 $\times$ 10^{-27} kg $=$ 1.49 $\times$ 10^{-10} j $=$ 931 Mev

1 newton (n) $=$ 0.225 lb

1 pound (lb) $=$ 4.45 n

1 joule (j) $=$ 0.738 ft·lb $=$ 2.39 $\times$ 10^{-4} kcal $=$ 6.24 $\times$ 10^{18} ev

1 kilocalorie (kcal) $=$ 4,185 j

1 foot·pound (ft·lb) $=$ 1.36 j

1 electron volt (eV) $=$ 10^{-6} Mev $=$ 1.60 $\times$ 10^{-19} j $=$ 1.18 $\times$ 10^{-19} ft·lb
   $=$ 3.83 $\times$ 10^{-23} kcal

1 watt $=$ 1 j/sec $=$ 0.738 ft·lb/sec

TABLE 2 Physical Constants

Constant	Units		
	mks	cgs	Engineering
Normal specific volume* of ideal gas	22.414 m^3/kgm-mole	22.414 liters/gm-mole	358 ft^3/lb_m-mole
Gas constant, R	8.3146×10^3 joule/kgm-mole·°K	1.9864 cal/mole·°K 82.07 atm·cm^3/mole·°K	1545.43 ft·lb/lb_m-mole·°R
Avogadro's number, N_0	6.025×10^{26} molecules/kgm-mole	6.025×10^{23} molecules/gm-mole	2.743×10^{26} molecules/lb_m-mole
Normal atmospheric pressure	1.013246×10^5 n/m^2	1.013246×10^6 dyne/cm^2	14.697 lb/in^2
Velocity of light, c	2.9979×10^8 m/sec	2.9979×10^{10} cm/sec	9.836×10^8 ft/sec
Boltzmann's constant, k	1.3803×10^{-23} joule/°K	1.3803×10^{-16} erg/°K	7.23×10^{-27} Btu/°R
Planck's constant, h	6.6237×10^{-34} joule·sec	6.626×10^{-27} erg·sec	6.2779 Btu·sec
Electronic charge, e	1.6018×10^{-19} coul	4.8029×10^{-10} abs·esu	1.6018×10^{-19} coul
Electron mass, m	9.107×10^{-31} kgm	9.107×10^{-28} gm	2.008×10^{-30} lb_m

* For the mks and cgs systems, "normal temperature and pressure" are 0°C and 1 atm. For the engineering system, "normal temperature and pressure" are 32°F and 1 atm.

TABLE 3 Air at Low Pressures

T, °R	T, °F	c_p, $\dfrac{\text{Btu}}{\text{lb·°F}}$	c_v, $\dfrac{\text{Btu}}{\text{lb·°F}}$	$\gamma = \dfrac{c_p}{c_v}$	a, ft/sec	$\mu \times 10^7$, $\dfrac{\text{lb}_m}{\text{sec-ft}}$	k, $\dfrac{\text{Btu}}{\text{hr-ft-°F}}$	$\Pr = \dfrac{c_p\mu}{k}$
100	−359.7	0.2392	0.1707	1.402	490.5			
150	−309.7	0.2392	0.1707	1.402	600.7			
200	−259.7	0.2392	0.1707	1.402	693.6			
250	−209.7	0.2392	0.1707	1.402	775.4			
300	−159.7	0.2392	0.1707	1.402	849.4			
350	−109.7	0.2393	0.1707	1.402	917.5			
400	−59.7	0.2393	0.1707	1.402	980.9	100	0.0118	0.73
450	−9.7	0.2394	0.1708	1.401	1040.3	109	0.0130	0.72
500	40.3	0.2396	0.1710	1.401	1096.4	118	0.0143	0.71
550	90.3	0.2399	0.1713	1.400	1149.6	126	0.0156	0.70
600	140.3	0.2403	0.1718	1.399	1200.3	135	0.0168	0.70
650	190.3	0.2409	0.1723	1.398	1248.7	143	0.0180	0.69
700	240.3	0.2416	0.1730	1.396	1295.1	151	0.0191	0.68
750	290.3	0.2424	0.1739	1.394	1339.6	158	0.0202	0.68
800	340.3	0.2434	0.1748	1.392	1382.5	166	0.0213	0.68
900	440.3	0.2458	0.1772	1.387	1463.6	179	0.0237	0.67
1000	540.3	0.2486	0.1800	1.381	1539.4	192	0.026	0.66
1100	640.3	0.2516	0.1830	1.374	1610.8	205	0.028	0.66
1200	740.3	0.2547	0.1862	1.368	1678.6	218	0.030	0.66
1300	840.3	0.2579	0.1894	1.362	1743.2	230	0.032	0.66
1400	940.3	0.2611	0.1926	1.356	1805.0	242	0.035	0.65
1500	1040.3	0.2642	0.1956	1.350	1864.5	253	0.037	0.65
1600	1140.3	0.2671	0.1985	1.345	1922.0	264	0.039	0.65
1700	1240.3	0.2698	0.2013	1.340	1977.6	274	0.041	0.65
1800	1340.3	0.2725	0.2039	1.336	2032	284	0.043	0.65
1900	1440.3	0.2750	0.2064	1.332	2084	293	0.045	0.65
2000	1540.3	0.2773	0.2088	1.328	2135	302	0.046	0.65
2100	1640.3	0.2794	0.2109	1.325	2185	311	0.048	0.65
2200	1740.3	0.2813	0.2128	1.322	2234	320	0.050	0.65
2300	1840.3	0.2831	0.2146	1.319	2282	329	0.052	0.65
2400	1940.3	0.2848	0.2162	1.317	2329	338	0.053	0.65
2600	2140.3	0.2878	0.2192	1.313	2420			
2800	2340.3	0.2905	0.2219	1.309	2508			
3000	2540.3	0.2929	0.2243	1.306	2593			
3200	2740.3	0.2950	0.2264	1.303	2675			
3400	2940.3	0.2969	0.2283	1.300	2755			
3600	3140.3	0.2986	0.2300	1.298	2832			
3800	3340.3	0.3001	0.2316	1.296	2907			
4000	3540.3	0.3015	0.2329	1.294	2981			
4200	3740.3	0.3029	0.2343	1.292	3052			
4400	3940.3	0.3041	0.2355	1.291	3122			
4600	4140.3	0.3052	0.2367	1.290	3191			
4800	4340.3	0.3063	0.2377	1.288	3258			
5000	4540.3	0.3072	0.3287	1.287	3323			
5200	4740.3	0.3081	0.2396	1.286	3388			
5400	4940.3	0.3090	0.2405	1.285	3451			
5600	5140.3	0.3098	0.2413	1.284	3513			
5800	5340.3	0.3106	0.2420	1.283	3574			
6000	5540.3	0.3114	0.2428	1.282	3634			
6200	5740.3	0.3121	0.2435	1.282	3693			
6400	5940.3	0.3128	0.2442	1.281	3751			

Conversion Factor
for μ

$$1\ \frac{\text{lb}_m}{\text{sec-ft}} = 14.882 \text{ poises}$$

$$= 14.882\ \frac{\text{gm}}{\text{sec-cm}}$$

$$= 1488.2 \text{ centipoises}$$

$$= 0.031081\ \frac{\text{lb}_f\text{-sec}}{\text{ft}^2}$$

TABLE 4   Distribution of Temperature, Speed of Sound, Dynamic Viscosity, Density, and Pressure in the ARDC Model Atmosphere*

$H \times 10^{-3}$ (ft)	$\dfrac{T}{T_{SL}}$	$\dfrac{a}{a_{SL}}$	$\dfrac{\mu}{\mu_{SL}}$	$\sigma = \dfrac{\rho}{\rho_{SL}}$	$\dfrac{p}{p_{SL}}$
0	1	1	1	1	1
5	0.9656	0.9827	0.9731	8.617×10^{-1}	8.320×10^{-1}
10	0.9312	0.9650	0.9457	7.385	6.877
15	0.8969	0.9470	0.9178	6.292	5.643
20	0.8625	0.9287	0.8894	5.328	4.595
25	0.8281	0.9100	0.8605	4.481	3.711
30	0.7937	0.8909	0.8311	3.741	2.970
35	0.7594	0.8714	0.8011	3.099	2.353
36.089	0.7519	0.8671	0.7945	2.971	2.234
40	0.7519	0.8671	0.7945	2.462	1.851
45	0.7519	0.8671	0.7945	1.936	1.455
50	0.7519	0.8671	0.7945	1.522	1.145
55	0.7519	0.8671	0.7945	1.197	9.000×10^{-2}
60	0.7519	0.8671	0.7945	9.414×10^{-2}	7.078
65	0.7519	0.8671	0.7945	7.403	5.566
70	0.7519	0.8671	0.7945	5.821	4.377
75	0.7519	0.8671	0.7945	4.578	3.442
80	0.7519	0.8671	0.7945	3.600	2.707
82.021	0.7519	0.8671	0.7945	3.267	2.456
85	0.7613	0.8725	0.8028	2.798	2.130
90	0.7772	0.8816	0.8167	2.167	1.684
95	0.7931	0.8905	0.8305	1.687	1.338
100	0.8089	0.8994	0.8442	1.320	1.068

(Continued)

* $T_{SL} = 5.1869 \times 10^2$ °R; $a_{SL} = 1.1164 \times 10^3$ ft·sec^{-1}; $\mu_{SL} = 3.7373 \times 10^{-7}$ lb/sec·ft^{-2}; $\rho_{SL} = 2.3769 \times 10^{-3}$ lb/sec^2·ft^{-4}; $p_{SL} = 2.1162 \times 10^{-3}$ lb·ft^{-2}. The symbol ARDC stands for the Air Research and Development Command of the United States Air Force. The subscript SL stands for *sea level;* H = altitude, T = temperature, a = sonic speed, μ = viscosity, ρ = density, p = pressure.

TABLE 4   (Continued)

$H \times 10^{-3}$ (ft)	$\dfrac{T}{T_{\text{SL}}}$	$\dfrac{a}{a_{\text{SL}}}$	$\dfrac{\mu}{\mu_{\text{SL}}}$	$\sigma = \dfrac{\rho}{\rho_{\text{SL}}}$	$\dfrac{p}{p_{\text{SL}}}$
110	0.8470	0.9169	0.8711	8.196×10^{-3}	6.890×10^{-3}
120	0.8724	0.9340	0.8977	5.179	4.518
130	0.9041	0.9509	0.9237	3.327×10^{-3}	3.008×10^{-3}
140	0.9359	0.9674	0.9494	2.170	2.031
150	0.9676	0.9837	0.9746	1.436	1.389
154.199	0.9809	0.9904	0.9851	1.212	1.189
160	0.9809	0.9904	0.9851	9.786×10^{-4}	9.600×10^{-4}
170	0.9809	0.9904	0.9851	7.771	6.641
173.885	0.9809	0.9904	0.9851	5.868	5.756
180	0.9518	0.9756	0.9621	4.811	4.579
190	0.9042	0.9509	0.9238	3.430	3.102
200	0.8566	0.9255	0.8845	2.402	2.058
210	0.8090	0.8995	0.8442	1.648	1.333
220	0.7614	0.8726	0.8029	1.105	8.412×10^{-5}
230	0.7138	0.8449	0.7604	7.219×10^{-5}	5.153
240	0.6662	0.8162	0.7167	4.580	3.052
250	0.6186	0.7865	0.6718	2.810	1.738
259.186	0.5749	0.7582	0.6293	1.733	9.964×10^{-6}
260	0.5749	0.7582	0.6293	1.647	9.467
270	0.5749	0.7582	0.6293	8.783×10^{-6}	5.050
280	0.5749	0.7582	0.6293	4.684	2.693
290	0.5749	0.7582	0.6293	2.498	1.436
295.276	0.5749	0.7582	0.6293	1.793	1.031

APPENDIX III

List of Symbols

Page	Symbol	Definition	Page	Symbol	Definition
322	I	total impulse	103	$\mathbf{r}$	unit vector
322	I_{sp}	specific impulse	86	r	radius
113	j	current density	383	r	burning rate
108	k	Boltzmann's constant	155	r_c	combustor stagnation pressure ratio
93	k	thermal conductivity			
429	k_w	thermal conductivity of the wall material	154	r_d	diffuser stagnation pressure ratio
32	K_n	equilibrium constant (based on mole fractions)	155	r_n	nozzle stagnation pressure ratio
31	K_p	equilibrium constant (based on partial pressures)	86	R	radius
			16	R	gas constant
128	L	length	16	$\overline{R}$	universal gas constant
144	L	lift	53	Re	Reynolds number
129	L	characteristic dimension	259	s	blade spacing
54	L^*	length of duct required for friction to change M to unity	14	s	entropy per unit mass
			30	$\bar{s}$	entropy per mole
378	L^*	characteristic rocket chamber length	14	S	entropy
			455	S	suction specific speed
8	m	mass	97	St	Stanton number
113	m_e	electron mass $(9.11 \times 10^{-31} \text{ kg})$	8	t	time
			324	t_b	burning period
12	$\dot{m}_a$	air flow rate	14	T	temperature
12	$\dot{m}_f$	fuel flow rate	114	T_e	electron temperature
46	M	Mach number	23	T_f	reference temperature
250	M_{rel}	relative Mach number	429	T_L	liquid coolant temperature
21	$\overline{M}$	molecular weight	46	T_0	stagnation temperature
8	$\mathbf{n}$	a unit vector	91	T_w	wall temperature
21	n	number of moles	91	T_∞	free stream temperature
108	n	particle density	427	T_{aw}	adiabatic wall temperature
331	n	number of rocket stages	427	T_{wh}	wall surface temperature, hot side
111	n_e	electron density			
183	N	rotor speed	429	T_{wc}	wall surface temperature, cold side
108	N_0	Avogadro's number			
452	N_s	specific speed	8, 72	$u, \mathbf{u}$	velocity, velocity component in the x-direction
95	Nu	Nusselt number			
11	p	pressure	114	u_d	drift velocity
13	p_a	ambient pressure	12	u_e	exhaust velocity
13	p_e	exhaust pressure	14	u_i	velocity in the inlet plane
46	p_0	stagnation pressure	321	u_{eq}	equivalent exhaust velocity
95	Pr	Prandtl number	341	u_{esc}	escape velocity
15	q	heat transfer per unit mass	342	u_{circ}	velocity in a circular orbit
103	q	electric charge	240	U	blade velocity
109	Q	collision cross section	67	U	velocity just outside the boundary layer
451	Q	volume flow rate			
10	Q	heat interaction	88	$\overline{U}$	bulk mean velocity
26	Q_R	heat of reaction	87	U_c	center-line velocity in a duct
23	Q_f	heat of formation	89	U_∞	free-stream velocity

Page	Symbol	Definition	Page	Symbol	Definition
72	v	velocity component in the y-direction	84	ϵ	eddy viscosity
104	V	electrical potential	103	ϵ	permittivity
493	V_a	acceleration potential	103	ϵ_0	permittivity of free space
244	$w, \mathbf{w}$	relative velocity	328	ϵ	structural coefficient
13, 72	x	streamwise coordinate	339	ϵ	eccentricity of conic
			113	ϵ_i	ionization potential
11	$\mathbf{X}$	body force per unit volume	336	η	thrust chamber efficiency
72	y	coordinate normal to the wall	156	η_b	burner efficiency
239	z	coordinate in the axial direction	162	η_c	adiabatic compressor efficiency
			162	η_d	adiabatic diffuser efficiency
11	z	height above an arbitrary datum	175	η_g	gear efficiency
			458	η_m	mechanical efficiency
			162	η_n	adiabatic nozzle efficiency
			144	η_o	overall efficiency
		SCRIPT LETTERS	143	η_p	propulsion efficiency
			448	η_p	pump efficiency
144	$\mathfrak{D}$	drag force	162	η_t	adiabatic turbine efficiency
34	$\mathfrak{K}_n$	equilibrium constant based on number of moles	271	η_{pc}	polytropic compression efficiency
53	$\mathfrak{K}$	typical roughness height	144	η_{pr}	propeller efficiency
21	$\mathfrak{M}$	mass	175	η_{pt}	adiabatic power turbine efficiency
328	$\mathfrak{M}_\wp$	mass of payload	248	η_{st}	stage efficiency
322	$\mathfrak{M}_p$	mass of propellant	143	η_{th}	thermal efficiency
328	$\mathfrak{M}_s$	structural mass	302	η_{ts}	total-to-static turbine efficiency
323	$\mathfrak{M}_0$	total initial mass			
323	$\mathfrak{M}_b$	burnout mass	302	η_{tt}	total-to-total turbine efficiency
11	$\mathcal{P}$	power	22	χ	mole fraction
11, 143	$\mathcal{P}_s$	shear power, shaft power	239	θ	angle
			76	θ	momentum thickness
11	$\mathfrak{Q}$	heat transfer per unit area, per unit time	110	λ	mean free path
			328	λ	payload ratio
323	$\mathfrak{R}$	mass ratio	53	μ	viscosity
12	$\mathcal{T}$	thrust	105	μ	permeability
144	$\mathcal{T}_{pr}$	thrust due to propeller	106	μ_0	permeability of free space
9	$\mathfrak{V}$	volume	74	ν	kinematic viscosity $\nu = \mu/\rho$
			109	ν	collision frequency
			114	ν_e	electron-neutral particle collision frequency
		GREEK LETTERS	255	ξ	stagnation pressure loss coefficient
245	α	absolute flow angle	283	Π_M	Mach index
398	α	Mach angle	8	ρ	density
111	α	degree of ionization	103	ρ_q	charge density
336	α	specific mass of power plant	311	σ	stress
245	β	relative flow angle	59	σ	wave angle
18	γ	specific heat ratio	113	σ_q	surface charge density
66	δ	boundary layer thickness			
76	δ^*	displacement thickness			

Page	Symbol	Definition	Page	Symbol	Definition
127	σ	plasma conductivity	120	τ_e	time between electron collisions
113	σ_0	plasma conductivity in the absence of magnetic fields	15	v	specific volume
			15	ϕ	specific heat function
72	τ	shear stress	448	ϕ	flow coefficient
44, 75	τ_0	wall shear stress	502	ϕ	work function
			448	ψ	pressure coefficient
84	τ_t	turbulent shear stress	117	ω_e	cyclotron frequency of the electron
240	τ	torque			
453	τ	cavitation coefficient	267	Ω	rotational speed

Answers to Selected Problems

CHAPTER 2

1. (a) 1.13×10^6 ft-lb$_f$/lb$_m$ (b) 8650 fps
3. 0.3   4. 5600 lb$_f$, 35,900 lb$_f$ 6. 3.84, 1.48   7. 2520°K
8. 2230°R   10. 0.5 mole CO_2

CHAPTER 3

3. (a) 6160 fps (b) 3.06 (c) 1740°R (d) 0.0066 ft^2·sec/lb$_m$
4. (a) 0.728 (b) 0.68   5. 2.8   6. 1990°R   8. 0.0257 sec

CHAPTER 4

2. 1.5 in. (turbulent)   4. 4:1   5. 115°R   6. 690°R
8. $Q = \frac{7}{8}\delta U \left(\frac{2\,\Delta p}{\rho U^2}\right)^4$ 10. (1) 0.0053 in. (2) 0.0094 in.

CHAPTER 5

1. (a) 1 (b) $\sqrt{2}$ (c) $\dfrac{\theta_2}{\theta_1} = \dfrac{E + u_2 B}{E + u_1 B}$, $u = \sqrt{2qV_{\text{acc}}/m}$
2. (a) 3.9×10^{20} m^{-3} (b) 33.5 amp/cm^2 (c) 9400 m/sec (d) 5310 m/sec
3. $V(r) = -\dfrac{I}{4\pi\epsilon_0 u_i}\left(\dfrac{r}{R}\right)^2$ 5. $\sigma = 0.188 \times 10^{-28}$ mhos/m (with no potassium)
8. (a) $\omega_e \tau_e = 7.0$ (b) 5.3×10^{-6} cm, 1.13×10^{-6} cm (c) Yes

CHAPTER 6

3. (a) 32.0 lb$_f$-sec/lb$_m$, 39.2 lb$_f$-sec/lb$_m$ (b) 1.31
4. 0.40, 0.55   5. 0.886 lb$_m$/lb$_f$-hour, 2.10 lb$_m$/lb$_f$-hour   7. 0.57
8. (a) 942 lb$_f$ (b) -457 lb$_f$ (c) 0

CHAPTER 7

1. 0.8   2. (a) 1.33 (b) 1.11 (c) 0.44   4. 0.82
6. $\dfrac{p_{04}}{p_{02}} = 0.963$ (M$_2$ = 0.10), 0.785 (M$_2$ = 0.20); M$_4$ = 1 when M$_2$ = 0.215
8. $A_2^*/A_1^* = 1.44$ 9. 1830°R

CHAPTER 8

2. 4.7×10^5 ft-lb$_f$/sec, 9900 rpm
3. -1730, 5780, 18,900 ft-lb$_f$/lb$_m$ (an undesirable condition)
4. (a) 1.47 (b) 40° toward tangential direction
6. $F_\theta = 7.56$ lb$_f$, $F_z = 8.75$ lb$_f$ ($C'_{\mathfrak{D}} = 0$), 7.59 lb$_f$ ($C'_{\mathfrak{D}} = 0.10$)

CHAPTER 9

2. (a) 2.49 (b) 2.93   3. 2.08   5. (a) 6920 rpm (b) $\beta_1 = 43.6°, \beta_2 = 61.0°$
6. (a) 1920°R (b) 1890°R   7. (c) -12.5, 50, 73   8. 884 fps, 648 fps

CHAPTER 10

1. 1.45   2. $\mathfrak{M}_{01} = 5 \times 10^6$ lb$_m$ 3. 4100 m (diameter)
4. (a) 5.6×10^3 m/sec, 255 days   (b) 1.4×10^4 m/sec   5. 1:1   7. 0.56

CHAPTER 11

1. 3460 psi
2. (a) 192 sec   (b) 162 slugs/sec   (c) 3.5 ft   (d) 7.04 ft   (e) 1.08×10^6 lb$_f$
   (f) 2.08×10^6 lb$_f$ (g) 10^6 lb$_f$ (h) 10^6 lb$_f$
3. (1) 7180 lb$_f$ (2) 10,300 lb$_f$
4. 96.6 in^2, 480 in^2, 1.043×10^5 lb$_f$, 1.055×10^5 lb$_f$
5. 0.148 in., 2.25 sec

CHAPTER 12

1. Increase by a factor of 1.98   2. $\dfrac{dA_b}{dt} = 2n \dfrac{kA_b}{D^*}$ 3. 0.00263 sec
4. 63 psia, 7 psia (with oblique shock)   5. 3540°R, 352 sec   6. 7850°R

CHAPTER 13

1. 45 psia   3. 1.2 : 1
4. $p_e/p_a > 0.36$; therefore little likelihood of shock-induced separation

CHAPTER 14

1. 10,100 hp   2. 55.6 ft^3; $\mathfrak{M}_{\text{helium}}/\mathfrak{M}_{\text{air}} = 0.165$ 5. 26 psia
6. 1.8%   7. (a) 2690°F   (b) 0.034 in.
9. No ($T_{\text{wh}} = 2710$°R and hoop stress $\approx$ 60,000 psia)
10. 600 psia   11. 2.66×10^7 Btu/hr-ft^2

CHAPTER 15

1. 1.01 : 1   2. (a) 1070 sec   (b) 1330 sec
3. It will increase by a factor of 1.14.   4. 3900°R (M$_1$ = 0.2)   5. 4.7
6. 6000°K; 1520 sec; 4500°K

CHAPTER 16

1. 8.65×10^5 v; 8.65 cm, 6 cm
2. 180 v, 24.5 watts/newton; 5210 sec, 25.9 watts/newton
3. (a) 515 cm^2 (b) 3.13 cm^2 4. 52% ($Q = 140,000$ cal/gm-mole of H$_2$)
5. 28.2 kcal/gm-mole H$_2$

INDEX

Index

Ablation cooling, 426
Absolute velocity, 244
Adiabatic, 10
Adiabatic efficiency, 162
Adiabatic flame temperature, 27
Adiabatic wall temperature, 427, 428
Aerodynamic losses, compressors, 255
  turbines, 303
Afterburners, 218
Aircraft range, 144
Alpha particle, 471
Angular momentum, 239
  equation for, 240, 242, 282
Annular combustion chambers, 227
Apogee, 343
Apparent electrostatic field, 106
Arc column, 518
Arcjet electrode heat transfer, 521
Arcjets, 490, 518
Aspect ratio (beam), 509
Atmospheric density, 325
Atomic mass unit, 470
Average velocity, 108
Avogadro's number, 108
Axial compressors, 238
  aerodynamic losses, 261
  blade chord, 253
  blade spacing, 253
  blow-off valves, 276
  cascade aerodynamics, 253
  comparison with centrifugal
    compressors, 293
  degree of reaction, 265
  dimensional analysis, 264
  effects of compressibility on, 250
  effects of Reynolds number on, 265
  free-vortex velocity diagrams, 269
  limiting stage pressure ratio, 250
  matching to axial turbines, 308
  multistage machines, 270
  off-design performance of, 262
  overall adiabatic efficiency of, 271
  polytropic efficiency of, 271
  pressure coefficient, 249

pressure rise limitations, 249
radial variations in, 267
rotating stall, 274
rotors, 244
schematic, 242
single-stage, 244
solidity of, 253
stage characteristics of, 263
stage design points, 277
stagger angle, 253
stagnation pressure loss coefficients, 255
starting, 275
surge, 272
typical cutaway, 243
twin spools, 276
variable stators, 276
work-done factor, 270
Axial turbines, 294
  aerodynamic losses, 297
  deviation angle, 301
  dimensional analysis, 305
  efficiency of, 301
  ideal performance, 298
  impulse, 298, 456
  matching to axial compressors, 308
  reaction, 300, 456
  rocket turbopump, 441, 446
  schematic diagram, 296
  stage dynamics, 297
  stresses in, 310
  total-to-static efficiency, 302, 457
  total-to-total efficiency, 302
  typical blading, 296
  typical rotors, 295
  velocity compound, 456
Axial velocity component, 245

Backward-leaning blades, centrifugal
  compressors, 285
Barn, 474
Beam aspect ratio, 509
Beam shapes, electrostatic rockets, 517
Bending stresses, 310
Beta particle, 471

553

$$Q_{1-2} + h_1 + \frac{v_1^2}{2g} + z_1 = W_{1-2} + h_2 + \frac{v_2^2}{2g} + z_2$$

$$778 \text{ ft } lb_f = 1 B$$